Great Physicists

Great Physicists

The Life and Times of Leading Physicists
from Galileo to Hawking

William H. Cropper

OXFORD
UNIVERSITY PRESS

Oxford New York
Auckland Bangkok Buenos Aires
Cape Town Chennai Dar es Salaam Delhi Hong Kong Istanbul
Karachi Kolkata Kuala Lumpur Madrid Melbourne Mexico City Mumbai
Nairobi São Paulo Shanghai Taipei Tokyo Toronto

First published by Oxford University Press, Inc., 2001
First issued as an Oxford University Press paperback, 2004
198 Madison Avenue, New York, New York 10016
www.oup.com

Library of Congress Cataloging-in-Publication Data
Cropper, William H.
Great Physicists: the life and times of leading physicists from Galileo to Hawking /
William H. Cropper.
p. cm Includes bibliographical references and index.
ISBN 0–19–513748–5 (cloth) ISBN 0–19–517324–4 (pbk.)
1. Physicists—Biography. I. Title.
QC15 .C76 2001 530'.092'2—dc21 [B] 2001021611

9 8 7 6 5 4 3 2
Printed in the United States of America
on acid-free paper

Contents

Preface

This book tells about lives in science, specifically the lives of thirty from the pantheon of physics. Some of the names are familiar (Newton, Einstein, Curie, Heisenberg, Bohr), while others may not be (Clausius, Gibbs, Meitner, Dirac, Chandrasekhar). All were, or are, extraordinary human beings, at least as fascinating as their subjects. The short biographies in the book tell the stories of both the people and their physics.

The chapters are varied in format and length, depending on the (sometimes skimpy) biographical material available. Some chapters are equipped with short sections (entitled "Lessons") containing background information on topics in mathematics, physics, and chemistry for the uninformed reader.

Conventional wisdom holds that general readers are frightened of mathematical equations. I have not taken that advice, and have included equations in some of the chapters. Mathematical equations express the language of physics: you can't get the message without learning something about the language. That should be possible if you have a rudimentary (high school) knowledge of algebra, and, if required, you pay attention to the "Lessons" sections. The glossary and chronology may also prove helpful. For more biographical material, consult the works cited in the "Invitation to More Reading" section.

No claim is made that this is a comprehensive or scholarly study; it is intended as recreational reading for scientists and students of science (formal or informal). My modest hope is that you will read these chapters casually and for entertainment, and learn the lesson that science is, after all, a human endeavor.

William H. Cropper

Acknowledgments

It is a pleasure to acknowledge the help of Kirk Jensen, Helen Mules, and Jane Lincoln Taylor at Oxford University Press, who made an arduous task much more pleasant than it might have been. I am indebted to my daughters, Hazel and Betsy, for many things, this time for their artistry with computer software and hardware.

I am also grateful for permission to reprint excerpts from the following publications:

Subtle is the Lord: The Science and Life of Albert Einstein, by Abraham Pais, copyright © 1983 by Abraham Pais. Used by permission of Oxford University Press, Inc.; *The Quantum Physicists*, by William H. Cropper, copyright © 1970 by Oxford University Press, Inc. Used by permission of Oxford University Press, Inc.; *Ludwig Boltzmann:The Man Who Trusted Atoms*, by Carlo Cercignani, copyright © 1998 by Carlo Cercignani. Used by permission of Oxford University Press, Inc.; *Lise Meitner: A Life in Physics*, by Ruth Lewin Sime, copyright © 1996 by the Regents of the University of California. Used by permission of the University of California Press; *Marie Curie: A Life*, by Susan Quinn, copyright © 1996, by Susan Quinn. Used by permission of the Perseus Books Group; *Atoms in the Family: My Life with Enrico Fermi*, by Laura Fermi, copyright © 1954 by The University of Chicago. Used by permission of The University of Chicago Press; *Enrico Fermi, Physicist*, by Emilio Segrè, copyright © 1970 by The University of Chicago. Used by permission of The University of Chicago Press; *Strange Beauty: Murray Gell-Mann and the Revolution in Twentieth-Century Physics,* by George Johnson, copyright © 1999 by George Johnson. Used by permission of Alfred A. Knopf, a division of Random House, Inc. Also published in the United Kingdom by Jonathan Cape, and used by permission from the Random House Group, Limited; *QED and the Men Who Made It*, by Silvan S. Schweber, copyright © 1994 by Princeton University Press. Used by permission of Princeton University Press; *Surely You're Joking, Mr. Feynman* by Richard Feynman as told to Ralph Leighton, copyright © 1985 by Richard Feynman and Ralph Leighton. Used by permission of W.W. Norton Company, Inc. Also published in the United Kingdom by Century, and used by permission from the Random House Group, Limited; *What Do You Care What Other People Think?*, by Richard Feynman as told to Ralph Leighton, copyright © 1988 by Gweneth Feynman and Ralph Leighton. Used by permission of W.W. Norton Company, Inc.; *The Feynman Lectures on Physics*, by Richard Feynman, Robert Leighton, and Matthew Sands, copyright © 1988 by Michelle Feynman and Carl Feynman. Used by permission of the Perseus Books Group; *Chandra: A Biography of S. Chandrasekhar*, by Kameshwar Wali, copyright © 1991 by The University of Chicago. Used by permission of The University of Chicago Press; *Edwin Hubble: Mariner of the Nebulae*, by Gale E. Christianson, copyright © 1995 by Gale E. Christianson. Used by permission of

Farrar, Straus and Giroux, L.L.C. Published in the United Kingdom by the Institute of Physics Publishing. Used by permission of the Institute of Physics Publishing; "Rudolf Clausius and the Road to Entropy," by William H. Cropper, *American Journal of Physics* **54**, 1986, pp. 1068–1074, copyright © 1986 by the American Association of Physics Teachers. Used by permission of the American Institute of Physics; "Walther Nernst and the Last Law," by William H. Cropper, *Journal of Chemical Education* **64**, 1987, pp. 3–8, copyright © 1987 by the Division of Chemical Education, American Chemical Society. Used by permission of the *Journal of Chemical Education*; "Carnot's Function, Origins of the Thermodynamic Concept of Temperature," by William H. Cropper, *American Journal of Physics* **55**, 1987, pp. 120–129, copyright © 1987 by the American Association of Physics Teachers. Used by permission of the American Institute of Physics; "James Joule's Work in Electrochemistry and the Emergence of the First Law of Thermodynamics," by William H. Cropper, *Historical Studies in the Physical and Biological Sciences* **19**, 1986, pp. 1–16, copyright © 1988 by the Regents of the University of California. Used by permission of the University of California Press.

All of the portrait photographs placed below the chapter headings were supplied by the American Institute of Physics Emilio Segrè Visual Archives, and are used by permission of the American Institute of Physics. Further credits are: Chapter 2 (Newton), Massachusetts Institute of Technology Burndy Library; Chapter 4 (Mayer), Massachusetts Institute of Technology Burndy Library; Chapter 5 (Joule), *Physics Today* Collection; Chapter 7 (Thomson), Zeleny Collection; Chapter 8 (Clausius), *Physics Today* Collection; Chapter 10 (Nernst), Photograph by Francis Simon; Chapter 11 (Faraday), E. Scott Barr Collection; Chapter 13 (Boltzmann), *Physics Today* Collection; Chapter 14 (Einstein), National Archives and Records Administration; Chapter 16 (Bohr), Segrè Collection; Chapter 19 (Schrödinger), W.F. Meggers Collection; Chapter 20 (Curie), W. F. Meggers Collection; Chapter 21 (Rutherford), *Nature;* Chapter 22 (Meitner), Herzfeld Collection; Chapter 23 (Fermi), Fermi Film Collection; Chapter 24 (Dirac), photo by A. Börtzells Tryckeri; Chapter 25 (Feynman), WGBH-Boston; Chapter 26 (Gell-Mann), W.F. Meggers Collection; Chapter 27 (Hubble), Hale Observatories; Chapter 28 (Chandrasekhar), K.G. Somsekhar, *Physics Today* Collection; Chapter 29 (Hawking), *Physics Today* Collection.

i MECHANICS

Historical Synopsis

Physics builds from observations. No physical theory can succeed if it is not confirmed by observations, and a theory strongly supported by observations cannot be denied. For us, these are almost truisms. But early in the seventeenth century these lessons had not yet been learned. The man who first taught that observations are essential and supreme in science was Galileo Galilei.

Galileo first studied the motion of terrestrial objects, pendulums, free-falling balls, and projectiles. He summarized what he observed in the mathematical language of proportions. And he extrapolated from his experimental data to a great idealization now called the "inertia principle," which tells us, among other things, that an object projected along an infinite, frictionless plane will continue forever at a constant velocity. His observations were the beginnings of the science of motion we now call "mechanics."

Galileo also observed the day and night sky with the newly invented telescope. He saw the phases of Venus, mountains on the Moon, sunspots, and the moons of Jupiter. These celestial observations dictated a celestial mechanics that placed the Sun at the center of the universe. Church doctrine had it otherwise: Earth was at the center. The conflict between Galileo's telescope and Church dogma brought disaster to Galileo, but in the end the telescope prevailed, and the dramatic story of the confrontation taught Galileo's most important lesson.

Galileo died in 1642. In that same year, his greatest successor, Isaac Newton, was born. Newton built from Galileo's foundations a system of mechanics based on the concepts of mass, momentum, and force, and on three laws of motion. Newton also invented a mathematical language (the "fluxion" method, closely related to our present-day calculus) to express his mechanics, but in an odd historical twist, rarely applied that language himself.

Newton's mechanics had—and still has—cosmic importance. It applies to the motion of terrestrial objects, and beyond that to planets, stars, and galaxies. The grand unifying concept is Newton's theory of universal gravitation, based on the concept that all objects, small, large, and astronomical (with some exotic exceptions), attract one another with a force that follows a simple inverse-square law.

Galileo and Newton were the founders of modern physics. They gave us the rules of the game and the durable conviction that the physical world is comprehensible.

1

How the Heavens Go

Galileo Galilei

The Tale of the Tower

Legend has it that a young, ambitious, and at that moment frustrated mathematics professor climbed to the top of the bell tower in Pisa one day, perhaps in 1591, with a bag of ebony and lead balls. He had advertised to the university community at Pisa that he intended to disprove by experiment a doctrine originated by Aristotle almost two thousand years earlier: that objects fall at a rate proportional to their weight; a ten-pound ball would fall ten times faster than a one-pound ball. With a flourish the young professor signaled to the crowd of amused students and disapproving philosophy professors below, selected balls of the same material but with much different weights, and dropped them. Without air resistance (that is, in a vacuum), two balls of different weights (and made of any material) would have reached the ground at the same time. That did not happen in Pisa on that day in 1591, but Aristotle's ancient principle was clearly violated anyway, and that, the young professor told his audience, was the lesson. The students cheered, and the philosophy professors were skeptical.

The hero of this tale was Galileo Galilei. He did not actually conduct that "experiment" from the Tower of Pisa, but had he done so it would have been entirely in character. Throughout his life, Galileo had little regard for authority, and one of his perennial targets was Aristotle, the ultimate authority for university philosophy faculties at the time. Galileo's personal style was confrontational, witty, ironic, and often sarcastic. His intellectual style, as the Tower story instructs, was to build his theories with an ultimate appeal to observations.

The philosophers of Pisa were not impressed with either Galileo or his methods, and would not have been any more sympathetic even if they had witnessed the Tower experiment. To no one's surprise, Galileo's contract at the University of Pisa was not renewed.

Padua

But Galileo knew how to get what he wanted. He had obtained the Pisa post with the help of the Marquis Guidobaldo del Monte, an influential nobleman and competent mathematician. Galileo now aimed for the recently vacated chair of mathematics at the University of Padua, and his chief backer in Padua was Gianvincenzio Pinelli, a powerful influence in the cultural and intellectual life of Padua. Galileo followed Pinelli's advice, charmed the examiners, and won the approval of the Venetian senate (Padua was located in the Republic of Venice, about twenty miles west of the city of Venice). His inaugural lecture was a sensation.

Padua offered a far more congenial atmosphere for Galileo's talents and lifestyle than the intellectual backwater he had found in Pisa. In the nearby city of Venice, he found recreation and more—aristocratic friends. Galileo's favorite debating partner among these was Gianfrancesco Sagredo, a wealthy nobleman with an eccentric manner Galileo could appreciate. With his wit and flair for polemics, Galileo was soon at home in the city's salons. He took a mistress, Marina Gamba, described by one of Galileo's biographers, James Reston, Jr., as "hot-tempered, strapping, lusty and probably illiterate." Galileo and Marina had three children: two daughters, Virginia and Livia, and a son, Vincenzo. In later life, when tragedy loomed, Galileo found great comfort in the company of his elder daughter, Virginia.

During his eighteen years in Padua (1592–1610), Galileo made some of his most important discoveries in mechanics and astronomy. From careful observations, he formulated the "times-squared" law, which states that the vertical distance covered by an object in free fall or along an inclined plane is proportional to the square of the time of the fall. (In modern notation, the equation for free fall is expressed $s = \frac{gt^2}{2}$, with s and t the vertical distance and time of the fall, and g the acceleration of gravity.) He defined the laws of projected motion with a controlled version of the Tower experiment in which a ball rolled down an inclined plane on a table, then left the table horizontally or obliquely and dropped to the floor. Galileo found that he could make calculations that agreed approximately with his experiments by resolving projected motion into two components, one horizontal and the other vertical. The horizontal component was determined by the speed of the ball when it left the table, and was "conserved"—that is, it did not subsequently change. The vertical component, due to the ball's weight, followed the times-squared rule.

For many years, Galileo had been fascinated by the simplicity and regularity of pendulum motion. He was most impressed by the constancy of the pendulum's "period," that is, the time the pendulum takes to complete its back-and-forth cycle. If the pendulum's swing is less than about 30°, its period is, to a good approximation, dependent only on its length. (Another Galileo legend pictures him as a nineteen-year-old boy in church, paying little attention to the service, and timing with his pulse the swings of an oil lamp suspended on a wire from a high ceiling.) In Padua, Galileo confirmed the constant-period rule with experiments, and then uncovered some of the pendulum's more subtle secrets.

In 1609, word came to Venice that spectacle makers in Holland had invented an optical device—soon to be called a telescope—that brought distant objects much closer. Galileo immediately saw a shining opportunity. If he could build a prototype and demonstrate it to the Venetian authorities before Dutch entrepreneurs arrived on the scene, unprecedented rewards would follow. He knew enough about optics to guess that the Dutch design was a combination of a convex and a concave lens, and he and his instrument maker had the exceptional skill needed to grind the lenses. In twenty-four hours, according to Galileo's own account, he had a telescope of better quality than any produced by the Dutch artisans. Galileo could have demanded, and no doubt received, a large sum for his invention. But fame and influence meant more to him than money. In an elaborate ceremony, he gave an eight-power telescope to Niccolò Contarini, the doge of Venice. Reston, in *Galileo,* paints this picture of the presentation of the telescope: "a celebration of Venetian genius, complete with brocaded advance men, distinguished heralds and secret operatives. Suddenly, the tube represented the flowering of Paduan learning." Galileo was granted a large bonus, his salary was doubled, and he was reappointed to his faculty position for life.

Then Galileo turned his telescope to the sky, and made some momentous, and as it turned out fateful, discoveries. During the next several years, he observed the mountainous surface of the Moon, four of the moons of Jupiter, the phases of Venus, the rings of Saturn (not quite resolved by his telescope), and sunspots. In 1610, he published his observations in *The Starry Messenger*, which was an immediate sensation, not only in Italy but throughout Europe.

But Galileo wanted more. He now contrived to return to Tuscany and Florence, where he had spent most of his early life. The grand duke of Tuscany was the young Cosimo de Medici, recently one of Galileo's pupils. To further his cause, Galileo dedicated *The Starry Messenger* to the grand duke and named the four moons of Jupiter the Medicean satellites. The flattery had its intended effect. Galileo soon accepted an astonishing offer from Florence: a salary equivalent to that of the highest-paid court official, no lecturing duties—in fact, no duties of any kind—and the title of chief mathematician and philosopher for the grand duke of Tuscany. In Venice and Padua, Galileo left behind envy and bitterness.

Florence and Rome

Again the gregarious and witty Galileo found intellectual companions among the nobility. Most valued now was his friendship with the young, talented, and skeptical Filippo Salviati. Galileo and his students were regular visitors at Salviati's beautiful villa fifteen miles from Florence. But even in this idyll Galileo was restless. He had one more world to conquer: Rome—that is, the Church. In 1611, Galileo proposed to the grand duke's secretary of state an official visit to Rome in which he would demonstrate his telescopes and impress the Vatican with the importance of his astronomical discoveries.

This campaign had its perils. Among Galileo's discoveries was disturbing evidence against the Church's doctrine that Earth was the center of the universe. The Greek astronomer and mathematician Ptolemy had advocated this cosmology in the second century, and it had long been Church dogma. Galileo could see in his observations evidence that the motion of Jupiter's moons centered on Jupiter, and, more troubling, in the phases of Venus that the motion of that planet centered on the Sun. In the sixteenth century, the Polish astronomer Nicolaus Copernicus had proposed a cosmology that placed the Sun at the center of the universe. By 1611, when he journeyed to Rome, Galileo had become largely converted to Copernicanism. Holy Scripture also regarded the Moon and the Sun as quintessentially perfect bodies; Galileo's telescope had revealed mountains and valleys on the Moon and spots on the Sun.

But in 1611 the conflict between telescope and Church was temporarily submerged, and Galileo's stay was largely a success. He met with the autocratic Pope Paul V and received his blessing and support. At that time and later, the intellectual power behind the papal throne was Cardinal Robert Bellarmine. It was his task to evaluate Galileo's claims and promulgate an official position. He, in turn, requested an opinion from the astronomers and mathematicians at the Jesuit Collegio Romano, who reported doubts that the telescope really revealed mountains on the Moon, but more importantly, trusted the telescope's evidence for the phases of Venus and the motion of Jupiter's moons.

Galileo found a new aristocratic benefactor in Rome. He was Prince Frederico Cesi, the founder and leader of the "Academy of Lynxes," a secret society whose members were "philosophers who are eager for real knowledge, and who will give themselves to the study of nature, and especially to mathematics." The members were young, radical, and, true to the lynx metaphor, sharp-eyed and ruthless in their treatment of enemies. Galileo was guest of honor at an extravagant banquet put on by Cesi, and shortly thereafter was elected as one of the Lynxes.

Galileo gained many influential friends in Rome and Florence—and, inevitably, a few dedicated enemies. Chief among those in Florence was Ludovico della Colombe, who became the self-appointed leader of Galileo's critics. *Colombe* means "dove" in Italian. Galileo expressed his contempt by calling Colombe and company the "Pigeon League."

Late in 1611, Colombe, whose credentials were unimpressive, went on the attack and challenged Galileo to an intellectual duel: a public debate on the theory of floating bodies, especially ice. A formal challenge was delivered to Galileo by a Pisan professor, and Galileo cheerfully responded, "Ever ready to learn from anyone, I should take it as a favor to converse with this friend of yours and reason about the subject." The site of the debate was the Pitti Palace. In the audience were two cardinals, Grand Duke Cosimo, and Grand Duchess Christine, Cosimo's mother. One of the cardinals was Maffeo Barberini, who would later become Pope Urban VIII and play a major role in the final act of the Galileo drama.

In the debate, Galileo took the view that ice and other solid bodies float because they are lighter than the liquid in which they are immersed. Colombe held to the Aristotelian position that a thin, flat piece of ice floats in liquid water because of its peculiar shape. As usual, Galileo built his argument with demonstrations. He won the audience, including Cardinal Barberini, when he showed that pieces of ebony, even in very thin shapes, always sank in water, while a block of ice remained on the surface.

The Gathering Storm

The day after his victory in the debate, Galileo became seriously ill, and he retreated to Salviati's villa to recuperate. When he had the strength, Galileo summarized in a treatise his views on floating bodies, and, with Salviati, returned to the study of sunspots. They mapped the motion of large spots as the spots traveled across the sun's surface near the equator from west to east.

Then, in the spring of 1612, word came that Galileo and Salviati had a competitor. He called himself Apelles. (He was later identified as Father Christopher Scheiner, a Jesuit professor of mathematics in Bavaria.) To Galileo's dismay, Apelles claimed that his observations of sunspots were the first, and explained the spots as images of stars passing in front of the sun. Not only was the interloper encroaching on Galileo's priority claim, but he was also broadcasting a false interpretation of the spots. Galileo always had an inclination to paranoia, and it now had the upper hand. He sent a series of bold letters to Apelles through an intermediary, and agreed with Cesi that the letters should be published in Rome by the Academy of Lynxes. In these letters Galileo asserted for the first time his adherence to the Copernican cosmology. As evidence he recalled his observations of the planets: "I tell you that [Saturn] also, no less than the horned Venus agrees admirably with the great Copernican system. Favorable winds are now blowing on that system. Little reason remains to fear crosswinds and shadows on so bright a guide."

Galileo soon had another occasion to proclaim his belief in Copernicanism. One of his disciples, Benedetto Castelli, occupied Galileo's former post, the chair of mathematics at Pisa. In a letter to Galileo, Castelli wrote that recently he had had a disturbing interview with the pious Grand Duchess Christine. "Her Ladyship began to argue against me by means of the Holy Scripture," Castelli wrote. Her particular concern was a passage from the Book of Joshua that tells of God commanding the Sun to stand still so Joshua's retreating enemies could not escape into the night. Did this not support the doctrine that the Sun moved around Earth and deny the Copernican claim that Earth moved and the Sun was stationary?

Galileo sensed danger. The grand duchess was powerful, and he feared that he was losing her support. For the first time he openly brought his Copernican views to bear on theological issues. First he wrote a letter to Castelli. It was sometimes a mistake, he wrote, to take the words of the Bible literally. The Bible had to be interpreted in such a way that there was no contradiction with direct observations: "The task of wise interpreters is to find true meanings of scriptural passages that will agree with the evidence of sensory experience." He argued that God could have helped Joshua just as easily under the Copernican cosmology as under the Ptolemaic.

The letter to Castelli, which was circulated and eventually published, brought no critical response for more than a year. In the meantime, Galileo took more drastic measures. He expanded the letter, emphasizing the primacy of observations over doctrine when the two were in conflict, and addressed it directly to Grand Duchess Christine. "The primary purpose of the Holy Writ is to worship God and save souls," he wrote. But "in disputes about natural phenomena, one must not begin with the authority of scriptural passages, but with sensory experience and necessary demonstrations." He recalled that Cardinal Cesare Baronius had once said, "The Bible tells us how to go to Heaven, not how the heavens go."

The first attack on Galileo from the pulpit came from a young Dominican priest named Tommaso Caccini, who delivered a furious sermon centering on the miracle of Joshua, and the futility of understanding such grand events without faith in established doctrine. This was a turning point in the Galileo story. As Reston puts it: "Italy's most famous scientist, philosopher to the Grand Duke of Tuscany, intimate of powerful cardinals in Rome, stood accused publicly of heresy from an important pulpit, by a vigilante of the faith." Caccini and Father Niccolò

Lorini, another Dominican priest, now took the Galileo matter to the Roman Inquisition, presenting as evidence for heresy the letter to Castelli.

Galileo could not ignore these events. He would have to travel to Rome and face the inquisitors, probably influenced by Cardinal Bellarmine, who had, four years earlier, reported favorably on Galileo's astronomical observations. But once again Galileo was incapacitated for months by illness. Finally, in late 1615 he set out for Rome.

As preparation for the inquisitors, a Vatican commission had examined the Copernican doctrine and found that its propositions, such as placing the Sun at the center of the universe, were "foolish and absurd and formally heretical." On February 25, 1616, the Inquisition met and received instructions from Pope Paul to direct Galileo not to teach or defend or discuss Copernican doctrine. Disobedience would bring imprisonment.

In the morning of the next day, Bellarmine and an inquisitor presented this injunction to Galileo orally. Galileo accepted the decision without protest and waited for the formal edict from the Vatican. That edict, when it came a few weeks later, was strangely at odds with the judgment delivered earlier by Bellarmine. It did not mention Galileo or his publications at all, but instead issued a general restriction on Copernicanism: "It has come to the knowledge of the Sacred Congregation that the false Pythagorean doctrine, namely, concerning the movement of the Earth and immobility of the Sun, taught by Nicolaus Copernicus, and altogether contrary to the Holy Scripture, is already spread about and received by many persons. Therefore, lest any opinion of this kind insinuate itself to the detriment of Catholic truth, the Congregation has decreed that the works of Nicolaus Copernicus be suspended until they are corrected."

Galileo, always an optimist, was encouraged by this turn of events. Despite Bellarmine's strict injunction, Galileo had escaped personal censure, and when the "corrections" to Copernicus were spelled out they were minor. Galileo remained in Rome for three months, and found occasions to be as outspoken as ever. Finally, the Tuscan secretary of state advised him not to "tease the sleeping dog further," adding that there were "rumors we do not like."

Comets, a Manifesto, and a Dialogue

In Florence again, Galileo was ill and depressed during much of 1617 and 1618. He did not have the strength to comment when three comets appeared in the night sky during the last four months of 1618. He was stirred to action, however, when Father Horatio Grassi, a mathematics professor at the Collegio Romano and a gifted scholar, published a book in which he argued that the comets provided fresh evidence against the Copernican cosmology. At first Galileo was too weak to respond himself, so he assigned the task to one of his disciples, Mario Guiducci, a lawyer and graduate of the Collegio Romano. A pamphlet, *Discourse on Comets*, was published under Guiducci's name, although the arguments were clearly those of Galileo.

This brought a worthy response from Grassi, and in 1621 and 1622 Galileo was sufficiently provoked and healthy to publish his eloquent manifesto, *The Assayer*. Here Galileo proclaimed, "Philosophy is written in this grand book the universe, which stands continually open to our gaze. But the book cannot be understood unless one first learns to comprehend the language and to read the alphabet in which it is composed. It is written in the language of mathematics, and its characters are triangles, circles and other geometric figures, without which it is humanly impossible to understand a single word of it; without these, one wanders about in a dark labyrinth."

The Assayer received Vatican approval, and Cardinal Barberini, who had supported Galileo in his debate with della Colombe, wrote in a friendly and reassuring letter, "We are ready to serve you always." As it turned out, Barberini's good wishes could hardly have been more opportune. In 1623, he was elected pope and took the name Urban VIII.

After recovering from a winter of poor health, Galileo again traveled to Rome in the spring of 1624. He now went bearing microscopes. The original microscope design, like that of the telescope, had come from Holland, but Galileo had greatly improved the instrument for scientific uses. Particularly astonishing to the Roman cognoscenti were magnified images of insects.

Shortly after his arrival in Rome, Galileo had an audience with the recently elected Urban VIII. Expecting the former Cardinal Barberini again to promise support, Galileo found to his dismay a different persona. The new pope was autocratic, given to nepotism, long-winded, and obsessed with military campaigns. Nevertheless, Galileo left Rome convinced that he still had a clear path. In a letter to Cesi he wrote, "On the question of Copernicus His Holiness said that the Holy Church had not condemned, nor would condemn his opinions as heretical, but only rash. So long as it is not demonstrated as true, it need not be feared."

Galileo's strategy now was to present his arguments hypothetically, without claiming absolute truth. His literary device was the dialogue. He created three characters who would debate the merits of the Copernican and Aristotelian systems, but ostensibly the debate would have no resolution. Two of the characters were named in affectionate memory of his Florentine and Venetian friends, Gianfrancesco Sagredo and Filippo Salviati, who had both died. In the dialogue Salviati speaks for Galileo, and Sagredo as an intelligent layman. The third character is an Aristotelian, and in Galileo's hands earns his name, Simplicio.

The dialogue, with the full title *Dialogue Concerning the Two Chief World Systems*, occupied Galileo intermittently for five years, between 1624 and 1629. Finally, in 1629, it was ready for publication and Galileo traveled to Rome to expedite approval by the Church. He met with Urban and came away convinced that there were no serious obstacles.

Then came some alarming developments. First, Cesi died. Galileo had hoped to have his *Dialogue* published by Cesi's Academy of Lynxes, and had counted on Cesi as his surrogate in Rome. Now with the death of Cesi, Galileo did not know where to turn. Even more alarming was an urgent letter from Castelli advising him to publish the *Dialogue* as soon as possible in Florence. Galileo agreed, partly because at the time Rome and Florence were isolated by an epidemic of bubonic plague. In the midst of the plague, Galileo found a printer in Florence, and the printing was accomplished. But approval by the Church was not granted for two years, and when the *Dialogue* was finally published it contained a preface and conclusion written by the Roman Inquisitor. At first, the book found a sympathetic audience. Readers were impressed by Galileo's accomplished use of the dialogue form, and they found the dramatis personae, even the satirical Simplicio, entertaining.

In August 1632, Galileo's publisher received an order from the Inquisition to cease printing and selling the book. Behind this sudden move was the wrath of

Urban, who was not amused by the clever arguments of Salviati and Sagredo, and the feeble responses of Simplicio. He even detected in the words of Simplicio some of his own views. Urban appointed a committee headed by his nephew, Cardinal Francesco Barberini, to review the book. In September, the committee reported to Urban and the matter was handed over to the Inquisition.

Trial

After many delays—Galileo was once again seriously ill, and the plague had returned—Galileo arrived in Rome in February 1633 to defend himself before the Inquisition. The trial began on April 12. The inquisitors focused their attention on the injunction Bellarmine had issued to Galileo in 1616. Francesco Niccolini, the Tuscan ambassador to Rome, explained it this way to his office in Florence: "The main difficulty consists in this: these gentlemen [the inquisitors] maintain that in 1616 he [Galileo] was commanded neither to discuss the question of the earth's motion nor to converse about it. He says, to the contrary, that these were not the terms of the injunction, which were that that doctrine was not to be held or defended. He considers that he has the means of justifying himself since it does not appear at all from his book that he holds or defends the doctrine . . . or that he regards it as a settled question." Galileo offered in evidence a letter from Bellarmine, which bolstered his claim that the inquisitors' strict interpretation of the injunction was not valid.

Historians have argued about the weight of evidence on both sides, and on a strictly legal basis, concluded that Galileo had the stronger case. (Among other things, the 1616 injunction had never been signed or witnessed.) But for the inquisitors, acquittal was not an option. They offered what appeared to be a reasonable settlement: Galileo would admit wrongdoing, submit a defense, and receive a light sentence. Galileo agreed and complied. But when the sentence came on June 22 it was far harsher than anything he had expected: his book was to be placed on the Index of Prohibited Books, and he was condemned to life imprisonment.

Last Act

Galileo's friends always vastly outnumbered his enemies. Now that he had been defeated by his enemies, his friends came forward to repair the damage. Ambassador Niccolini managed to have the sentence commuted to custody under the Archbishop Ascanio Piccolomini of Siena. Galileo's "prison" was the archbishop's palace in Siena, frequented by poets, scientists, and musicians, all of whom arrived to honor Galileo. Gradually his mind returned to the problems of science, to topics that were safe from theological entanglements. He planned a dialogue on "two new sciences," which would summarize his work on natural motion (one science) and also address problems related to the strengths of materials (the other science). His three interlocutors would again be named Salviati, Sagredo, and Simplicio, but now they would represent three ages of the author: Salviati, the wise Galileo in old age; Sagredo, the Galileo of the middle years in Padua; and Simplicio, a youthful Galileo.

But Galileo could not remain in Siena. Letters from his daughter Virginia, now Sister Maria Celeste in the convent of St. Matthew in the town of Arcetri, near Florence, stirred deep memories. Earlier he had taken a villa in Arcetri to be near

Virginia and his other daughter, Livia, also a sister at the convent. He now appealed to the pope for permission to return to Arcetri. Eventually the request was granted, but only after word had come that Maria Celeste was seriously ill, and more important, after the pope's agents had reported that the heretic's comfortable "punishment" in Siena did not fit the crime. The pope's edict directed that Galileo return to his villa and remain guarded there under house arrest.

Galileo took up residence in Arcetri in late 1633, and for several months attended Virginia in her illness. She did not recover, and in the spring of 1634, she died. For Galileo this was almost the final blow. But once again work was his restorative. For three years he concentrated on his *Discourses on Two New Sciences.* That work, his final masterpiece, was completed in 1637, and in 1638 it was published (in Holland, after the manuscript was smuggled out of Italy). By this time Galileo had gone blind. Only grudgingly did Urban permit Galileo to travel the short distance to Florence for medical treatment.

But after all he had endured, Galileo never lost his faith. "Galileo's own conscience was clear, both as Catholic and as scientist," Stillman Drake, a contemporary science historian, writes. "On one occasion he wrote, almost in despair, that he felt like burning all his work in science; but he never so much as thought of turning his back on his faith. The Church turned its back on Galileo, and has suffered not a little for having done so; Galileo blamed only some wrong-headed individuals in the Church for that."

Methods

Galileo's mathematical equipment was primitive. Most of the mathematical methods we take for granted today either had not been discovered or had not come into reliable use in Galileo's time. He did not employ algebraic symbols or equations, or, except for tangents, the concepts of trigonometry. His numbers were always expressed as positive integers, never as decimals. Calculus, discovered later by Newton and Gottfried Leibniz, was not available. To make calculations he relied on ratios and proportionalities, as defined in Euclid's *Elements.* His reasoning was mostly geometric, also learned from Euclid.

Galileo's mathematical style is evident in his many theorems on uniform and accelerated motion; here a few are presented and then "modernized" through translation into the language of algebra. The first theorem concerns uniform motion:

> If a moving particle, carried uniformly at constant speed, traverses two distances, the time intervals required are to each other in the ratio of these distances.

For us (but not for Galileo) this theorem is based on the algebraic equation $s = vt$, in which s represents distance, v speed, and t time. This is a familiar calculation. For example, if you travel for three hours ($t = 3$ hours) at sixty miles per hour ($v = 60$ miles per hour), the distance you have covered is 180 miles ($s = 3 \times 60 = 180$ miles). In Galileo's theorem, we calculate two distances, call them s_1 and s_2, for two times, t_1 and t_2, at the same speed, v. The two calculations are

$$s_1 = vt_1 \text{ and } s_2 = vt_2$$

Dividing the two sides of these equations into each other, we get the ratio of Galileo's theorem,

$$\frac{t_1}{t_2} = \frac{s_1}{s_2}.$$

Here is a more complicated theorem, which does not require that the two speeds be equal:

> If two particles are moved at a uniform rate, but with unequal speeds, through unequal distances, then the ratio of time intervals occupied will be the product of the ratio of the distances by the inverse ratio of the speeds.

In this theorem, there are two different speeds, v_1 and v_2, involved, and the two equations are

$$s_1 = v_1 t_1 \text{ and } s_2 = v_2 t_2.$$

Dividing both sides of the equations into each other again, we have

$$\frac{s_1}{s_2} = \frac{v_1}{v_2}\frac{t_1}{t_2}.$$

To finish the proof of the theorem, we multiply both sides of this equation by $\frac{v_2}{v_1}$ and obtain

$$\frac{t_1}{t_2} = \frac{s_1}{s_2}\frac{v_2}{v_1}.$$

On the right side now is a product of the direct ratio of the distances $\frac{s_1}{s_2}$ and the inverse ratio of the speeds $\frac{v_2}{v_1}$, as required by the theorem.

These theorems assume that any speed v is constant; that is, the motion is not accelerated. One of Galileo's most important contributions was his treatment of uniformly accelerated motion, both in free fall and down inclined planes. "Uniformly" here means that the speed changes by equal amounts in equal time intervals. If the uniform acceleration is represented by a, the change in the speed v in time t is calculated with the equation $v = at$. For example, if you accelerate your car at the uniform rate $a = 5$ miles per hour per second for $t = 10$ seconds, your final speed will be $v = 5 \times 10 = 50$ miles per hour. A second equation, $s = \frac{at^2}{2}$, calculates s, the distance covered in time t under the uniform acceleration a. This equation is not so familiar as the others mentioned. It is most easily justified with the methods of calculus, as will be demonstrated in the next chapter.

The motion of a ball of any weight dropping in free fall is accelerated in the vertical direction, that is, perpendicular to Earth's surface, at a rate that is conventionally represented by the symbol g, and is nearly the same anywhere on Earth. For the case of free fall, with $a = g$, the last two equations mentioned are $v = gt$, for the speed attained in free fall in the time t, and $s = \frac{gt^2}{2}$ for the corresponding distance covered.

Galileo did not use the equation $s = \frac{gt^2}{2}$, but he did discover through experimental observations the times-squared (t^2) part of it. His conclusion is expressed in the theorem,

> The spaces described by a body falling from rest with a uniformly accelerated motion are to each other as the squares of the time intervals employed in traversing these distances.

Our modernized proof of the theorem begins by writing the free-fall equation twice,

$$s_1 = \frac{gt_1^2}{2} \text{ and } s_2 = \frac{gt_2^2}{2},$$

and combining these two equations to obtain

$$\frac{s_1}{s_2} = \frac{t_1^2}{t_2^2}.$$

In addition to his separate studies of uniform and accelerated motion, Galileo also treated a composite of the two in projectile motion. He proved that the trajectory followed by a projectile is parabolic. Using a complicated geometric method, he developed a formula for calculating the dimensions of the parabola followed by a projectile (for example, a cannonball) launched upward at any angle of elevation. The formula is cumbersome compared to the trigonometric method we use today for such calculations, but no less accurate. Galileo demonstrated the use of his method by calculating with remarkable precision a detailed table of parabola dimensions for angles of elevation from 1° to 89°.

In contrast to his mathematical methods, derived mainly from Euclid, Galileo's experimental methods seem to us more modern. He devised a system of units that parallels our own and that served him well in his experiments on pendulum motion. His measure of distance, which he called a *punto*, was equivalent to 0.094 centimeter. This was the distance between the finest divisions on a brass rule. For measurements of time he collected and weighed water flowing from a container at a constant rate of about three fluid ounces per second. He recorded weights of water in grains (1 ounce = 480 grains), and defined his time unit, called a *tempo*, to be the time for 16 grains of water to flow, which was equivalent to 1/92 second. These units were small enough so Galileo's measurements of distance and time always resulted in large numbers. That was a necessity because decimal numbers were not part of his mathematical equipment; the only way he could add significant digits in his calculations was to make the numbers larger.

Legacy

Galileo took the metaphysics out of physics, and so begins the story that will unfold in the remaining chapters of this book. As Stephen Hawking writes, "Galileo, perhaps more than any single person, was responsible for the birth of modern science. . . . Galileo was one of the first to argue that man could hope to understand how the world works, and, moreover, that he could do this by observing the real world." No practicing physicist, or any other scientist for that matter, can do his or her work without following this Galilean advice.

I have already mentioned many of Galileo's specific achievements. His work in mechanics is worth sketching again, however, because it paved the way for his greatest successor. (Galileo died in January 1642. On Christmas Day of that same year, Isaac Newton was born.) Galileo's mechanics is largely concerned with bodies moving at constant velocity or under constant acceleration, usually that of gravity. In our view, the theorems that define his mechanics are based on the equations $v = gt$ and $s = \frac{gt^2}{2}$, but Galileo did not write these, or any other, algebraic equations; for his numerical calculations he invoked ratios and proportionality. He saw that projectile motion was a resultant of a vertical component governed by the acceleration of gravity and a constant horizontal component given to the projectile when it was launched. This was an early recognition that physical quantities with direction, now called "vectors," could be resolved into rectangular components.

I have mentioned, but not emphasized, another building block of Galileo's mechanics, what is now called the "inertia principle." In one version, Galileo put it this way: "Imagine any particle projected along a horizontal plane without friction; then we know . . . that this particle will move along this plane with a motion which is uniform and perpetual, provided the plane has no limits." This statement reflects Galileo's genius for abstracting a fundamental idealization from real behavior. If you give a real ball a push on a real horizontal plane, it will not continue its motion perpetually, because neither the ball nor the plane is perfectly smooth, and sooner or later the ball will stop because of frictional effects. Galileo neglected all the complexities of friction and obtained a useful postulate for his mechanics. He then applied the postulate in his treatment of projectile motion. When a projectile is launched, its horizontal component of motion is constant in the absence of air resistance, and remains that way, while the vertical component is influenced by gravity.

Galileo's mechanics did not include definitions of the concepts of force or energy, both of which became important in the mechanics of his successors. He had no way to measure these quantities, so he included them only in a qualitative way. Galileo's science of motion contains most of the ingredients of what we now call "kinematics." It shows us how motion occurs without defining the forces that control the motion. With the forces included, as in Newton's mechanics, kinematics becomes "dynamics."

All of these specific Galilean contributions to the science of mechanics were essential to Newton and his successors. But transcending all his other contributions was Galileo's unrelenting insistence that the success or failure of a scientific theory depends on observations and measurements. Stillman Drake leaves us with this trenchant synopsis of Galileo's scientific contributions: "When Galileo was born, two thousand years of physics had not resulted in even rough measurements of actual motions. It is a striking fact that the history of each science shows continuity back to its first use of measurement, before which it exhibits no ancestry but metaphysics. That explains why Galileo's science was stoutly opposed by nearly every philosopher of his time, he having made it as nearly free from metaphysics as he could. That was achieved by measurements, made as precisely as possible with means available to Galileo or that he managed to devise."

2

A Man Obsessed

Isaac Newton

Continual Thought

In his later years, Isaac Newton was asked how he had arrived at his theory of universal gravitation. "By thinking on it continually," was his matter-of-fact response. "Continual thinking" for Newton was almost beyond mortal capacity. He could abandon himself to his studies with a passion and ecstasy that others experience in love affairs. The object of his study could become an obsession, possessing him nonstop, and leaving him without food or sleep, beyond fatigue, and on the edge of breakdown.

The world Newton inhabited in his ecstasy was vast. Richard Westfall, Newton's principal biographer in this century, describes this "world of thought": "Seen from afar, Newton's intellectual life appears unimaginably rich. He embraced nothing less than the whole of natural philosophy [science], which he explored from several vantage points, ranging all the way from mathematical physics to alchemy. Within natural philosophy, he gave new direction to optics, mechanics, and celestial dynamics, and he invented the mathematical tool [calculus] that has enabled modern science further to explore the paths he first blazed. He sought as well to plumb the mind of God and His eternal plan for the world and humankind as it was presented in the biblical prophecies."

But, after all, Newton was human. His passion for an investigation would fade, and without synthesizing and publishing the work, he would move on to another grand theme. "What he thought on, he thought on continually, which is to say exclusively, or nearly exclusively," Westfall continues, but "[his] career was episodic." To build a coherent whole, Newton sometimes revisited a topic several times over a period of decades.

Woolsthorpe

Newton was born on Christmas Day, 1642, at Woolsthorpe Manor, near the Lincolnshire village of Colsterworth, sixty miles northwest of Cambridge and one hundred miles from London. Newton's father, also named Isaac, died three months before his son's birth. The fatherless boy lived with his mother, Hannah, for three years. In 1646, Hannah married Barnabas Smith, the elderly rector of North Witham, and moved to the nearby rectory, leaving young Isaac behind at Woolsthorpe to live with his maternal grandparents, James and Mary Ayscough. Smith was prosperous by seventeenth-century standards, and he compensated the Ayscoughs by paying for extensive repairs at Woolsthorpe.

Newton appears to have had little affection for his stepfather, his grandparents, his half-sisters and half-brother, or even his mother. In a self-imposed confession of sins, made after he left Woolsthorpe for Cambridge, he mentions "Peevishness with my mother," "with my sister," "Punching my sister," "Striking many," "Threatning my father and mother Smith to burne them and the house over them," "wishing death and hoping it to some."

In 1653, Barnabas Smith died, Hannah returned to Woolsthorpe with the three Smith children, and two years later Isaac entered grammar school in Grantham, about seven miles from Woolsthorpe. In Grantham, Newton's genius began to emerge, but not at first in the classroom. In modern schools, scientific talent is often first glimpsed as an outstanding aptitude in mathematics. Newton did not have that opportunity; the standard English grammar school curriculum of the time offered practically no mathematics. Instead, he displayed astonishing mechanical ingenuity. William Stukely, Newton's first biographer, tells us that he quickly grasped the construction of a windmill and built a working model, equipped with an alternate power source, a mouse on a treadmill. He constructed a cart that he could drive by turning a crank. He made lanterns from "crimpled paper" and attached them to the tails of kites. According to Stukely, this stunt "Wonderfully affrighted all the neighboring inhabitants for some time, and caus'd not a little discourse on market days, among the country people, when over their mugs of ale."

Another important extracurricular interest was the shop of the local apothecary, remembered only as "Mr. Clark." Newton boarded with the Clark family, and the shop became familiar territory. The wonder of the bottles of chemicals on the shelves and the accompanying medicinal formulations would help direct him to later interests in chemistry, and beyond that to alchemy.

With the completion of the ordinary grammar school course of studies, Newton reached a crossroads. Hannah felt that he should follow in his father's footsteps and manage the Woolsthorpe estate. For that he needed no further education, she insisted, and called him home. Newton's intellectual promise had been noticed, however. Hannah's brother, William Ayscough, who had attended Cambridge, and the Grantham schoolmaster, John Stokes, both spoke persuasively on Newton's behalf, and Hannah relented. After nine months at home with her restless son, Hannah no doubt recognized his ineptitude for farm management. It probably helped also that Stokes was willing to waive further payment of the forty-shilling fee usually charged for nonresidents of Grantham. Having passed this crisis, Newton returned in 1660 to Grantham and prepared for Cambridge.

Cambridge

Newton entered Trinity College, Cambridge, in June 1661, as a "subsizar," meaning that he received free board and tuition in exchange for menial service. In the Cambridge social hierarchy, sizars and subsizars were on the lowest level. Evidently Hannah Smith could have afforded better for her son, but for some reason (possibly parsimony) chose not to make the expenditure.

With his lowly status as a subsizar, and an already well developed tendency to introversion, Newton avoided his fellow students, his tutor, and most of the Cambridge curriculum (centered largely on Aristotle). Probably with few regrets, he went his own way. He began to chart his intellectual course in a "Philosophical Notebook," which contained a section with the Latin title *Quaestiones quaedam philosophicam* (Certain Philosophical Questions) in which he listed and discussed the many topics that appealed to his unbounded curiosity. Some of the entries were trivial, but others, notably those under the headings "Motion" and "Colors," were lengthy and the genesis of later major studies.

After about a year at Cambridge, Newton entered, almost for the first time, the field of mathematics, as usual following his own course of study. He soon traveled far enough into the world of seventeenth-century mathematical analysis to initiate his own explorations. These early studies would soon lead him to a geometrical demonstration of the fundamental theorem of calculus.

Beginning in the summer of 1665, life in Cambridge and in many other parts of England was shattered by the arrival of a ghastly visitor, the bubonic plague. For about two years the colleges were closed. Newton returned to Woolsthorpe, and took with him the many insights in mathematics and natural philosophy that had been rapidly unfolding in his mind.

Newton must have been the only person in England to recall the plague years 1665–66 with any degree of fondness. About fifty years later he wrote that "in those days I was in the prime of my age for invention & minded Mathematicks & Philosophy more then than at any time since." During these "miracle years," as they were later called, he began to think about the method of fluxions (his version of calculus), the theory of colors, and gravitation. Several times in his later years Newton told visitors that the idea of universal gravitation came to him when he saw an apple fall in the garden at Woolsthorpe; if gravity brought the apple down, he thought, why couldn't it reach higher, as high as the Moon?

These ideas were still fragmentary, but profound nevertheless. Later they would be built into the foundations of Newton's most important work. "The miracle," says Westfall, "lay in the incredible program of study undertaken in private and prosecuted alone by a young man who thereby assimilated the achievement of a century and placed himself at the forefront of European mathematics and science."

Genius of this magnitude demands, but does not always receive, recognition. Newton was providentially lucky. After graduation with a bachelor's degree, the only way he could remain at Cambridge and continue his studies was to be elected a fellow of Trinity College. Prospects were dim. Trinity had not elected fellows for three years, only nine places were to be filled, and there were many candidates. Newton was not helped by his previous subsizar status and unorthodox program of studies. But against all odds, he was included among the elected. Evidently he had a patron, probably Humphrey Babington, who was related to Clark, the apothecary in Grantham, and a senior fellow of Trinity.

The next year after election as a "minor" fellow, Newton was awarded the Master of Arts degree and elected a "major" fellow. Then in 1668, at age twenty-seven and still insignificant in the college, university, and scientific hierarchy, he was appointed Lucasian Professor of Mathematics. His patron for this surprising promotion was Isaac Barrow, who was retiring from the Lucasian chair and expecting a more influential appointment outside the university. Barrow had seen enough of Newton's work to recognize his brilliance.

Newton's Trinity fellowship had a requirement that brought him to another serious crisis. To keep his fellowship he regularly had to affirm his belief in the articles of the Anglican Church, and ultimately be ordained a clergyman. Newton met the requirement several times, but by 1675, when he could no longer escape the ordination rule, his theological views had taken a turn toward heterodoxy, even heresy. In the 1670s Newton immersed himself in theological studies that eventually led him to reject the doctrine of the Trinity. This was heresy, and if admitted, meant the ruination of his career. Although Newton kept his heretical views secret, ordination was no longer a possibility, and for a time, his Trinity fellowship and future at Cambridge appeared doomed.

But providence intervened, once again in the form of Isaac Barrow. Since leaving Cambridge, Barrow had served as royal chaplain. He had the connections at Court to arrange a royal dispensation exempting the Lucasian Professor from the ordination requirement, and another chapter in Newton's life had a happy ending.

Critics

Newton could not stand criticism, and he had many critics. The most prominent and influential of these were Robert Hooke in England, and Christiaan Huygens and Gottfried Leibniz on the Continent.

Hooke has never been popular with Newton partisans. One of his contemporaries described him as "the most ill-natured, conceited man in the world, hated and despised by most of the Royal Society, pretending to have all other inventions when once discovered by their authors." There is a grain of truth in this concerning Hooke's character, but he deserves better. In science he made contributions to optics, mechanics, and even geology. His skill as an inventor was renowned, and he was a surveyor and an architect. In personality, Hooke and Newton were polar opposites. Hooke was a gregarious extrovert, while Newton, at least during his most creative years, was a secretive introvert. Hooke did not hesitate to rush into print any ideas that seemed plausible. Newton shaped his concepts by thinking about them for years, or even decades. Neither man could bear to acknowledge any influence from the other. When their interests overlapped, bitter confrontations were inevitable.

Among seventeenth-century physicists, Huygens was most nearly Newton's equal. He made important contributions in mathematics. He invented the pendulum clock and developed the use of springs as clock regulators. He studied telescopes and microscopes and introduced improvements in their design. His studies in mechanics touched on statics, hydrostatics, elastic collisions, projectile motion, pendulum theory, gravity theory, and an implicit force concept, including the concept of centrifugal force. He pictured light as a train of wave fronts transmitted through a medium consisting of elastic particles. In matters relating to physics, this intellectual menu is strikingly similar to that of Newton. Yet Huygens's influence beyond his own century was slight, while Newton's was enormous. One of Huygens's limitations was that he worked alone and had few disciples. Also, like Newton, he often hesitated to publish, and when the work finally saw print others had covered the same ground. Most important, however, was his philosophical bias. He followed René Descartes in the belief that natural phenomena must have mechanistic explanations. He rejected Newton's theory of universal gravitation, calling it "absurd," because it was no more than mathematics and proposed no mechanisms.

Leibniz, the second of Newton's principal critics on the Continent, is remembered more as a mathematician than as a physicist. Like that of Huygens, his physics was limited by a mechanistic philosophy. In mathematics he made two major contributions, an independent (after Newton's) invention of calculus, and an early development of the principles of symbolic logic. One manifestation of Leibniz's calculus can be seen today in countless mathematics and physics textbooks: his notation. The basic operations of calculus are differentiation and integration, accomplished with derivatives and integrals. The Leibniz symbols for derivatives (e.g., $\frac{dy}{dx}$) and integrals (e.g., $\int y dx$) have been in constant use for more than three hundred years. Unlike many of his scientific colleagues, Leibniz never held an academic post. He was everything but an academic, a lawyer, statesman, diplomat, and professional genealogist, with assignments such as arranging peace negotiations, tracing royal pedigrees, and mapping legal reforms. Leibniz and Newton later engaged in a sordid clash over who invented calculus first.

Calculus Lessons

The natural world is in continuous, never-ending flux. The aim of calculus is to describe this continuous change mathematically. As modern physicists see it, the methods of calculus solve two related problems. Given an equation that expresses a continuous change, what is the equation for the rate of the change? And, conversely, given the equation for the rate of change, what is the equation for the change? Newton approached calculus this way, but often with geometrical arguments that are frustratingly difficult for those with little geometry. I will avoid Newton's complicated constructions and present here for future reference a few rudimentary calculus lessons more in the modern style.

Suppose you want to describe the motion of a ball falling freely from the Tower in Pisa. Here the continuous change of interest is the trajectory of the ball, expressed in the equation

$$s = \frac{gt^2}{2} \tag{1}$$

in which t represents time, s the ball's distance from the top of the tower, and g a constant we will interpret later as the gravitational acceleration. One of the problems of calculus is to begin with equation (1) and calculate the ball's rate of fall at every instant.

This calculation is easily expressed in Leibniz symbols. Imagine that the ball is located a distance s from the top of the tower at time t, and that an instant later, at time $t + dt$, it is located at $s + ds$; the two intervals dt and ds, called "differentials" in the terminology of calculus, are comparatively very small. We have equation (1) for time t at the beginning of the instant. Now write the equation for time $t + dt$ at the end of the instant, with the ball at $s + ds$,

$$\begin{aligned} s + ds &= \frac{g(t + dt)^2}{2} \\ &= \frac{g}{2}[t^2 + 2tdt + (dt)^2] \\ &= \frac{gt^2}{2} + gtdt + \frac{g}{2}(dt)^2. \end{aligned} \tag{2}$$

Notice the term s on the left side of the last equation and the term $\frac{gt^2}{2}$ the right. According to equation (1), these terms are equal, so they can be canceled from the last equation, leaving

$$ds = gtdt + \frac{g}{2}(dt)^2. \tag{3}$$

In the realm where calculus operates, the time interval dt is very small, and $(dt)^2$ is much smaller than that. (Squares of small numbers are much smaller numbers; for example, compare 0.001 with $(0.001)^2 = 0.000001$.) Thus the term containing $(dt)^2$ in equation (3) is much smaller than the term containing dt, in fact, so small it can be neglected, and equation (3) finally reduces to

$$ds = gtdt. \tag{4}$$

Dividing by the dt factor on both sides of this equation, we have finally

$$\frac{ds}{dt} = gt. \tag{5}$$

(As any mathematician will volunteer, this is far from a rigorous account of the workings of calculus.)

This result has a simple physical meaning. It calculates the instantaneous speed of the ball at time t. Recall that speed is always calculated by dividing a distance interval by a time interval. (If, for example, the ball falls 10 meters at constant speed for 2 seconds, its speed is $\frac{10}{2} = 5$ meters per second.) In equation (5), the instantaneous distance and time intervals ds and dt are divided to calculate the instantaneous speed $\frac{ds}{dt}$.

The ratio $\frac{ds}{dt}$ in equation (5) is called a "derivative," and the equation, like any other containing a derivative, is called a "differential equation." In mathematical physics, differential equations are ubiquitous. Most of the theories mentioned in this book rely on fundamental differential equations. One of the rules of theoretical physics is that (with a few exceptions) its laws are most concisely stated in the common language of differential equations.

The example has taken us from equation (1) for a continuous change to equation (5) for the rate of the change at any instant. Calculus also supplies the means to reverse this argument and derive equation (1) from equation (5). The first step is to return to equation (4) and note that the equation calculates only one differential step, ds, in the trajectory of the ball. To derive equation (1) we must add all of these steps to obtain the full trajectory. This summation is an "integration" operation and in the Leibniz notation it is represented by the elongated-S symbol $\int$. For integration of equation (4) we write

$$\int ds = \int gtdt. \tag{6}$$

We know that this must be equivalent to equation (1), so we infer that the rules for evaluating the two "integrals" in equation (6) are

$$\int ds = s, \tag{7}$$

and

$$\int gtdt = \frac{gt^2}{2}. \tag{8}$$

Integrals and integration are just as fundamental in theoretical physics as differential equations. Theoreticians usually compose their theories by first writing differential equations, but those equations are likely to be inadequate for the essential further task of comparing the predictions of the theory with experimental and other observations. For that, integrated equations are often a necessity. The great misfortune is that some otherwise innocent-looking differential equations are extremely difficult to integrate. In some important cases (including one Newton struggled with for many years, the integration of the equations of motion for the combined system comprising Earth, the Moon, and the Sun), the equations cannot be handled at all without approximations.

A glance at a calculus textbook will reveal the differentiation rule used to arrive at equation (5), the integration rules (7) and (8), and dozens of others. As its name implies, calculus is a scheme for calculating, in particular for calculations involving derivatives and differential equations. The scheme is organized around the differentiation and integration rules.

Calculus provides a perfect mathematical context for the concepts of mechanics. In the example, the derivative $\frac{ds}{dt}$ calculates a speed. Any speed v is calculated the same way,

$$v = \frac{ds}{dt}. \tag{9}$$

If the speed changes with time—if there is an acceleration—that can be expressed as the rate of change in v, as the derivative $\frac{dv}{dt}$. So the acceleration differential equation is

$$a = \frac{dv}{dt}, \tag{10}$$

in which a represents acceleration. The freely falling ball accelerates, that is, its speed increases with time, as equation (5) combined with equation (9), which is written

$$v = gt, \tag{11}$$

shows. The constant factor g is the acceleration of free fall, that is, the gravitational acceleration.

This discussion has used the Leibniz notation throughout. Newton's calculus notation was similar but less convenient. He emphasized rates of change with time, called them "fluxions," and represented them with an overhead dot notation. For example, in Newton's notation, equation (5) becomes

$$\dot{s} = gt,$$

in which $\dot{s}$, Newton's symbol for $\frac{ds}{dt}$, is the distance fluxion, and equation (10) is

$$a = \dot{v},$$

with $\dot{v}$ representing $\frac{dv}{dt}$, the speed fluxion.

Optics

The work that first brought Newton to the attention of the scientific community was not a theoretical or even a mathematical effort; it was a prodigious technical achievement. In 1668, shortly before his appointment as Lucasian Professor, Newton designed and constructed a "reflecting" telescope. In previous telescopes, beginning with the Dutch invention and Galileo's improvement, light was refracted and focused by lenses. Newton's telescope *reflected* and focused light with a concave mirror. Refracting telescopes had limited resolution and to achieve high magnification had to be inconveniently long. (Some refracting telescopes at the time were a hundred feet long, and a thousand-footer was planned.) Newton's design was a considerable improvement on both counts.

Newton's telescope project was even more impressive than that of Galileo. With no assistance (Galileo employed a talented instrument maker), Newton cast and ground the mirror, using a copper alloy he had prepared, polished the mirror, and built the tube, the mount, and the fittings. The finished product was just six inches in length and had a magnification of forty, equivalent to a refracting telescope six feet long.

Newton was not the first to describe a reflecting telescope. James Gregory, professor of mathematics at St. Andrews University in Scotland, had earlier published a design similar to Newton's, but could not find craftsmen skilled enough to construct it.

No less than Galileo's, Newton's telescope was vastly admired. In 1671, Barrow demonstrated it to the London gathering of prominent natural philosophers known as the Royal Society. The secretary of the society, Henry Oldenburg, wrote to Newton that his telescope had been "examined here by some of the most eminent in optical science and practice, and applauded by them." Newton was promptly elected a fellow of the Royal Society.

Before the reflecting telescope, Newton had made other major contributions in the field of optics. In the mid-1660s he had conceived a theory that held that ordinary white light was a mixture of pure colors ranging from red, through orange, yellow, green, and blue, to violet, the rainbow of colors displayed by a prism when it receives a beam of white light. In Newton's view, the prism separated the pure components by refracting each to a different extent. This was a contradiction of the prevailing theory, advocated by Hooke, among others, that light in the purest form is white, and colors are modifications of the pristine white light.

Newton demonstrated the premises of his theory in an experiment employing two prisms. The first prism separated sunlight into the usual red-through-violet components, and all of these colors but one were blocked in the beam received by the second prism. The crucial observation was that the second prism caused no further modification of the light. "The purely red rays refracted by the second prism made no other colours but red," Newton observed in 1666, "& the purely blue no other colours but blue ones." Red and blue, and other colors produced by the prism, were the pure colors, not the white.

Soon after his sensational success with the reflecting telescope in 1671, Newton sent a paper to Oldenburg expounding this theory. The paper was read at a meeting of the Royal Society, to an enthusiastically favorable response. Newton was then still unknown as a scientist, so Oldenburg innocently took the additional step of asking Robert Hooke, whose manifold interests included optics, to comment on Newton's theory. Hooke gave the innovative and complicated paper about three hours of his time, and told Oldenburg that Newton's arguments were not convincing.

This response touched off the first of Newton's polemical battles with his critics. His first reply was restrained; it prompted Hooke to give the paper in question more scrutiny, and to focus on Newton's hypothesis that light is particle-like. (Hooke had found an inconsistency here; Newton claimed that he did not rely on hypotheses.) Newton was silent for awhile, and Hooke, never silent, claimed that he had built a reflecting telescope before Newton. Next, Huygens and a Jesuit priest, Gaston Pardies, entered the controversy. Apparently in support of Newton, Huygens wrote, "The theory of Mr. Newton concerning light and colors appears highly ingenious to me." In a communication to the *Philosophical Transactions* of the Royal Society, Pardies questioned Newton's prism experiment, and Newton's reply, which also appeared in the *Transactions*, was condescending. Hooke complained to Oldenburg that Newton was demeaning the debate, and Oldenburg wrote a cautionary letter to Newton. By this time, Newton was aroused enough to refute all of Hooke's objections in a lengthy letter to the Royal Society, later published in the *Transactions*. This did not quite close the dispute; in a final episode, Huygens reentered the debate with criticisms similar to those offered by Hooke.

In too many ways, this stalemate between Newton and his critics was petty, but it turned finally on an important point. Newton's argument relied crucially on experimental evidence; Hooke and Huygens would not grant the weight of that evidence. This was just the lesson Galileo had hoped to teach earlier in the century. Now it was Newton's turn.

Alchemy and Heresy

In his nineteenth-century biography of Newton, David Brewster surprised his readers with an astonishing discovery. He revealed for the first time that Newton's papers included a vast collection of books, manuscripts, laboratory notebooks, recipes, and copied material on alchemy. How could "a mind of such power . . . stoop to be even the copyist of the most contemptible alchemical poetry," Brewster asked. Beyond that he had little more to say about Newton the alchemist.

By the time Brewster wrote his biography, alchemy was a dead and unlamented endeavor, and the modern discipline of chemistry was moving forward at a rapid pace. In Newton's century the rift between alchemy and chemistry was just beginning to open, and in the previous century alchemy *was* chemistry.

Alchemists, like today's chemists, studied conversions of substances into other substances, and prescribed the rules and recipes that governed the changes. The ultimate conversion for the alchemists was the transmutation of metals, including the infamous transmutation of lead into gold. The theory of transmutation had many variations and refinements, but a fundamental part of the doctrine was the belief that metals are compounded of mercury and sulfur—not ordinary mercury and sulfur but principles extracted from them, a "spirit of sulfur" and a "philosophic mercury." The alchemist's goal was to extract these principles from impure natural mercury and sulfur; once in hand, the pure forms could be combined to achieve the desired transmutations. In the seventeenth century, this program was still plausible enough to attract practitioners, and the practitioners patrons, including kings.

The alchemical literature was formidable. There were hundreds of books (Newton had 138 of them in his library), and they were full of the bizarre terminology and cryptic instructions alchemists devised to protect their work from competitors. But Newton was convinced that with thorough and discriminating study, coupled with experimentation, he could mine a vein of reliable observations beneath all the pretense and subterfuge. So, in about 1669, he plunged into the world of alchemy, immediately enjoying the challenges of systematizing the chaotic alchemical literature and mastering the laboratory skills demanded by the alchemist's fussy recipes.

Newton's passion for alchemy lasted for almost thiry years. He accumulated more than a million words of manuscript material. An assistant, Humphrey Newton (no relation), reported that in the laboratory the alchemical experiments gave Newton "a great deal of satisfaction & Delight. . . . The Fire [in the laboratory furnaces] scarcely going out either Night or Day. . . . His Pains, his Dilligence at those sett times, made me think, he aim'd at something beyond y[e] Reach of humane Art & Industry."

What did Newton learn during his years in company with the alchemists? His transmutation experiments did not succeed, but he did come to appreciate a fundamental lesson still taught by modern chemistry and physical chemistry: that the particles of chemical substances are affected by the forces of attraction and repulsion. He saw in some chemical phenomena a "principle of sociability" and in others "an endeavor to recede." This was, as Westfall writes, "arguably the most advanced product of seventeenth-century chemistry." It presaged the modern theory of "chemical affinities," which will be addressed in chapter 10.

For Newton, the attraction forces he saw in his crucibles were of a piece with the gravitational force. There is no evidence that he equated the two kinds of forces, but some commentators have speculated that his concept of universal gravitation was inspired, not by a Lincolnshire apple, but by the much more complicated lessons of alchemy.

During the 1670s, Newton had another subject for continual study and thought; he was concerned with biblical texts instead of scientific texts. He became convinced that the early Scriptures expressed the Unitarian belief that although Christ was to be worshipped, he was subordinate to God. Newton cited historical evidence that this text was corrupted in the fourth century by the introduction of the doctrine of the Trinity. Any form of anti-Trinitarianism was considered heresy in the seventeenth century. To save his fellowship at Cambridge, Newton kept his unorthodox beliefs secret, and, as noted, he was rescued by a special dispensation when he could no longer avoid the ordination requirement of the fellowship.

Halley's Question

In the fall of 1684, Edmond Halley, an accomplished astronomer, traveled to Cambridge with a question for Newton. Halley had concluded that the gravitational force between the Sun and the planets followed an inverse-square law—that is, the connection between this "centripetal force" (as Newton later called it) and the distance r between the centers of the planet and the Sun is

$$\text{centripetal force} \propto \frac{1}{r^2}.$$

(Read "proportional to" for the symbol $\propto$.) The force decreases by $\frac{1}{2}^2 = \frac{1}{4}$ if r doubles, by $\frac{1}{3}^2 = \frac{1}{9}$ if r triples, and so forth. Halley's visit and his question were later described by a Newton disciple, Abraham DeMoivre:

> In 1684 Dr Halley came to visit [Newton] at Cambridge, after they had some time together, the D^r asked him what he thought the curve would be that would be described by the Planets supposing the force of attraction towards the Sun to be reciprocal to the square of their distance from it. Sr Isaac replied immediately that it would be an [ellipse], the Doctor struck with joy & amazement asked him how he knew it, why saith he I have calculated it, whereupon Dr Halley asked him for his calculation without farther delay, Sr Isaac looked among his papers but could not find it, but he promised him to renew it, & then send it to him.

A few months later Halley received the promised paper, a short, but remarkable, treatise, with the title *De motu corporum in gyrum* (On the Motion of Bodies in Orbit). It not only answered Halley's question, but also sketched a new system of celestial mechanics, a theoretical basis for Kepler's three laws of planetary motion.

Kepler's Laws

Johannes Kepler belonged to Galileo's generation, although the two never met. In 1600, Kepler became an assistant to the great Danish astronomer Tycho Brahe,

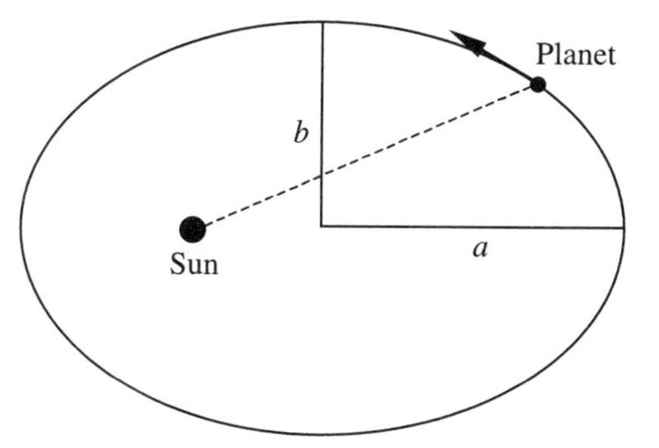

Figure 2.1. An elliptical planetary orbit. The orbit shown is exaggerated. Most planetary orbits are nearly circular.

and on Tycho's death, inherited both his job and his vast store of astronomical observations. From Tycho's data Kepler distilled three great empirical laws:

> 1. *The Law of Orbits: The planets move in elliptical orbits, with the Sun situated at one focus.*

Figure 2.1 displays the geometry of a planetary ellipse. Note the dimensions a and b of the semimajor and semiminor axes, and the Sun located at one focus.

> 2. *The Law of Equal Areas: A line joining any planet to the Sun sweeps out equal areas in equal times.*

Figure 2.2 illustrates this law, showing the radial lines joining a planet with the Sun, and areas swept out by the lines in equal times with the planet traveling different parts of its elliptical orbit. The two areas are equal, and the planet travels faster when it is closer to the Sun.

> 3. *The Law of Periods: The square of the period of any planet about the Sun is proportional to the cube of the length of the semimajor axis.*

A planet's period is the time it requires to travel its entire orbit—365 days for Earth. Stated as a proportionality, with P representing the period and a the length of the semimajor axis, this law asserts that

$$P^2 \propto a^3.$$

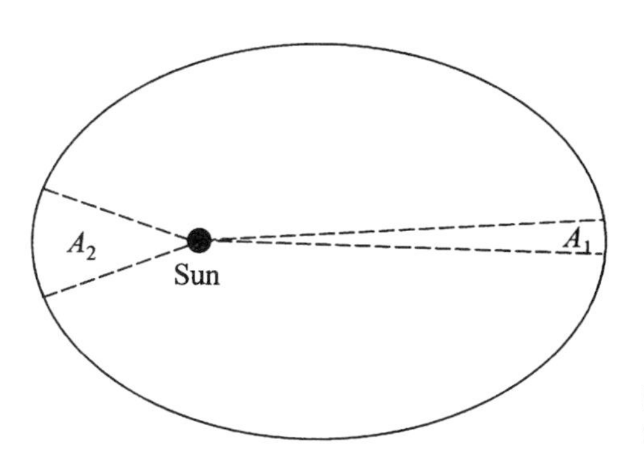

Figure 2.2. Kepler's law of equal areas. The area A_1 equals the area A_2.

Halley's Reward

"I keep [a] subject constantly before me," Newton once remarked, "and wait 'till the first dawnings open slowly, by little and little, into a full and clear light." Kepler's laws had been on Newton's mind since his student days. In "first dawnings" he had found connections between the inverse-square force law and Kepler's first and third laws, and now in *De motu* he was glimpsing in "a full and clear light" the entire theoretical edifice that supported Kepler's laws and other astronomical observations. Once more, Newton's work was "the passionate study of a man obsessed." His principal theme was the mathematical theory of universal gravitation.

First, he revised and expanded *De motu*, still focusing on celestial mechanics, and then aimed for a grander goal, a general dynamics, including terrestrial as well as celestial phenomena. This went well beyond *De motu*, even in title. For the final work, Newton chose the Latin title *Philosophiae naturalis principia mathematica* (Mathematical Principles of Natural Philosophy), usually shortened to the *Principia*.

When it finally emerged, the *Principia* comprised an introduction and three books. The introduction contains definitions and Newton's candidates for the fundamental laws of motion. From these foundations, book 1 constructs extensive and sophisticated mathematical equipment, and applies it to objects moving without resistance—for example, in a vacuum. Book 2 treats motion in resisting mediums—for example, in a liquid. And book 3 presents Newton's cosmology, his "system of the world."

In a sense, Halley deserves as much credit for bringing the *Principia* into the world as Newton does. His initial Cambridge visit reminded Newton of unfinished business in celestial mechanics and prompted the writing of *De motu*. When Halley saw *De motu* in November 1684, he recognized it for what it was, the beginning of a revolution in the science of mechanics. Without wasting any time, he returned to Cambridge with more encouragement. None was needed. Newton was now in full pursuit of the new dynamics. "From August 1684 until the spring of 1686," Westfall writes, "[Newton's] life [was] a virtual blank except for the *Principia*."

By April 1686, books 1 and 2 were completed, and Halley began a campaign for their publication by the Royal Society. Somehow (possibly with Halley exceeding his limited authority as clerk of the society), the members were persuaded at a general meeting and a resolution was passed, ordering "that Mr. Newton's *Philosophiae naturalis principia mathematica* be printed forthwith." Halley was placed in charge of the publication.

Halley now had the *Principia* on the road to publication, but it was to be a bumpy ride. First, Hooke made trouble. He believed that he had discovered the inverse-square law of gravitation and wanted recognition from Newton. The acknowledgment, if any, would appear in book 3, now nearing completion. Newton refused to recognize Hooke's priority, and threatened to suppress book 3. Halley had not yet seen book 3, but he sensed that without it the *Principia* would be a body without a head. "S[r] I must now again beg you, not to let your resentment run so high, as to deprive us of your third book," he wrote to Newton. The beheading was averted, and Halley's diplomatic appeals may have been the decisive factor.

In addition to his editorial duties, Halley was also called upon to subsidize the publication of the *Principia*. The Royal Society was close to bankruptcy and unable even to pay Halley his clerk's salary of fifty pounds. In his youth, Halley had been wealthy, but by the 1680s he was supporting a family and his means were reduced. The *Principia* was a gamble, and it carried some heavy financial risks.

But finally, on July 5, 1687, Halley could write to Newton and announce that "I have at length brought your Book to an end." The first edition sold out quickly. Halley at least recovered his costs, and more important, he received the acknowledgment from Newton that he deserved: "In the publication of this work the most acute and universally learned Mr Edmund Halley not only assisted me in correcting the errors of the press and preparing the geometrical figures, but it was through his solicitations that it came to be published."

The *Principia*

What Halley coaxed from Newton is one of the greatest masterpieces in scientific literature. It is also one of the most inaccessible books ever written. Arguments in the *Principia* are presented formally as propositions with (sometimes sketchy) demonstrations. Some propositions are theorems and others are developed as illustrative calculations called "problems." The reader must meet the challenge of each proposition in sequence to grasp the full argument.

Modern readers of the *Principia* are also burdened by Newton's singular mathematical style. Propositions are stated and demonstrated in the language of geometry, usually with reference to a figure. (In about five hundred pages, the *Principia* has 340 figures, some of them extremely complicated.) To us this seems an anachronism. By the 1680s, when the *Principia* was under way, Newton had already developed his fluxional method of calculus. Why did he not use calculus to express his dynamics, as we do today?

Partly it was an aesthetic choice. Newton preferred the geometry of the "ancients," particularly Euclid and Appolonius, to the recently introduced algebra of Descartes, which played an essential role in fluxional equations. He found the geometrical method "much more elegant than that of Descartes . . . [who] attains the result by means of an algebraic calculus which, if one transcribed it in words (in accordance with the practice of the Ancients in their writings) is revealed to be boring and complicated to the point of provoking nausea, and not be understood."

There was another problem. Newton could not use the fluxion language he had invented twenty years earlier for the practical reason that he had never published the work (and would not publish it for still another twenty years). As the science historian François De Gandt explains, "[The] innovative character [of the *Principia*] was sure to excite controversy. To combine with this innovative character another novelty, this time mathematical, and to make unpublished procedures in mathematics the foundation for astonishing physical assertions, was to risk gaining nothing."

So Newton wrote the *Principia* in the ancient geometrical style, modified when necessary to represent continuous change. But he did not reach his audience. Only a few of Newton's contemporaries read the *Principia* with comprehension, and following generations chose to translate it into a more transparent, if less elegant, combination of algebra and the Newton-Leibniz calculus. The fate of the *Principia*, like that of some of the other masterpieces of scientific literature

(Clausius on thermodynamics, Maxwell on the electromagnetic field, Boltzmann on gas theory, Gibbs on thermodynamics, and Einstein on general relativity), was to be more admired than read.

The fearsome challenge of the *Principia* lies in its detailed arguments. In outline, free of the complicated geometry and the maddening figures, the work is much more accessible. It begins with definitions of two of the most basic concepts of mechanics:

> *Definition 1: The quantity of matter is the measure of the same arising from its density and bulk conjointly.*

> *Definition 2: The quantity of motion is the measure of the same, arising from the velocity and quantity of matter conjointly.*

By "quantity of matter" Newton means what we call "mass," "quantity of motion" in our terms is "momentum," "bulk" can be measured as a volume, and "density" is the mass per unit volume (lead is more dense than water, and water more dense than air). Translated into algebraic language, the two definitions read

$$m = \rho V, \tag{12}$$

and

$$p = mv, \tag{13}$$

in which mass is represented by m, density by ρ, volume by V, momentum by p, and velocity by v.

Following the definitions are Newton's axioms, his famous three laws of motion. The first is Galileo's law of inertia:

> *Law 1: Every body continues in its state of rest, or of uniform motion in a right [straight] line, unless it is compelled to change that state by forces impressed upon it.*

The second law of motion has more to say about the force concept:

> *Law 2: The change of motion is proportional to the motive force impressed; and is made in the direction of the right line in which the force is impressed.*

By "change of motion" Newton means the instantaneous rate of change in the momentum, equivalent to the time derivative $\frac{dp}{dt}$. In the modern convention, force is *defined* as this derivative, and the equation for calculating a force f is simply

$$f = \frac{dp}{dt}, \tag{14}$$

or, with the momentum p evaluated by equation (13),

$$f = \frac{d(mv)}{dt}. \tag{15}$$

The first two laws convey simple physical messages. Imagine that your car is coasting on a flat road with the engine turned off. If the car meets no resistance (for example, in the form of frictional effects), Newton's first law tells us that the car will continue coasting with its original momentum and direction forever. With the engine turned on, and your foot on the accelerator, the car is driven by the engine's force, and Newton's second law asserts that the momentum increases at a rate ($= \frac{dp}{dt}$) equal to the force. In other words: increase the force by depressing the accelerator and the car's momentum increases.

Newton's third law asserts a necessary constraint on forces operating mutually between two bodies:

> *Law 3: To every action there is always opposed an equal reaction: or, the mutual actions of two bodies upon each other are always equal, and directed to contrary parts.*

Newton's homely example reminds us, "If you press on a stone with your finger, the finger is also pressed by the stone." If this were not the case, the stone would be soft and not stonelike.

Building from this simple, comprehensible beginning, Newton takes us on a grand tour of terrestrial and celestial dynamics. In book 1 he assumes an inverse-square centripetal force and derives Kepler's three laws. Along the way (in proposition 41), a broad concept that we now recognize as conservation of mechanical energy emerges, although Newton does not use the term "energy," and does not emphasize the conservation theme.

Book 1 describes the motion of bodies (for example, planets) moving without resistance. In book 2, Newton approaches the more complicated problem of motion in a resisting medium. This book was something of an afterthought, originally intended as part of book 1. It is more specialized than the other two books, and less important in Newton's grand scheme.

Book 3 brings the *Principia* to its climax. Here Newton builds his "system of the world," based on the three laws of motion, the mathematical methods developed earlier, mostly in book 1, and empirical raw material available in astronomical observations of the planets and their moons.

The first three propositions put the planets and their moons in elliptical orbits controlled by inverse-square centripetal forces, with the planets orbiting the Sun, and the moons their respective planets. These propositions define the centripetal forces mathematically but have nothing to say about their physical nature.

Proposition 4 takes that crucial step. It asserts "that the Moon gravitates towards the earth, and is always drawn from rectilinear [straight] motion, and held back in its orbit, by the force of gravity." By the "force of gravity" Newton means the force that causes a rock (or apple) to fall on Earth. The proposition tells us that the Moon is a rock and that it, too, responds to the force of gravity.

Newton's demonstration of proposition 4 is a marvel of simplicity. First, from the observed dimensions of the Moon's orbit he concludes that to stay in its orbit the Moon falls toward Earth 15.009 "Paris feet" (= 16.000 of our feet) every second. Then, drawing on accurate pendulum data observed by Huygens, he calculates that the number of feet the Moon (or anything else) would fall in one second on the surface of Earth is 15.10 Paris feet. The two results are close enough to each other to demonstrate the proposition.

Proposition 5 simply assumes that what is true for Earth and the Moon is true for Jupiter and Saturn and their moons, and for the Sun and its planets.

Finally, in the next two propositions Newton enunciates his universal law of gravitation. I will omit some subtleties and details here and go straight to the algebraic equation that is equivalent to Newton's inverse-square calculation of the gravitational attraction force F between two objects whose masses are m_1 and m_2,

$$F = G\frac{m_1 m_2}{r^2}, \tag{16}$$

where r is the distance separating the centers of the two objects, and G, called the "gravitational constant," is a universal constant. With a few exceptions, involving such bizarre objects as neutron stars and black holes, this equation applies to any two objects in the universe: planets, moons, comets, stars, and galaxies. The gravitational constant G is always given the same value; it is the hallmark of gravity theory. Later in our story, it will be joined by a few other universal constants, each with its own unique place in a major theory.

In the remaining propositions of book 3, Newton turns to more-detailed problems. He calculates the shape of Earth (the diameter at the equator is slightly larger than that at the poles), develops a theory of the tides, and shows how to use pendulum data to demonstrate variations in weight at different points on Earth. He also attempts to calculate the complexities of the Moon's orbit, but is not completely successful because his dynamics has an inescapable limitation: it easily treats the mutual interaction (gravitational or otherwise) of two bodies, but offers no exact solution to the problem of three or more bodies. The Moon's orbit is largely, but not entirely, determined by the Earth-Moon gravitational attraction. The full calculation is a "three-body" problem, including the slight effect of the Sun. In book 3, Newton develops an approximate method of calculation in which the Earth-Moon problem is first solved exactly and is then modified by including the "perturbing" effect of the Sun. The strategy is one of successive approximations. The calculations dictated by this "perturbation theory" are tedious, and Newton failed to carry them far enough to obtain good accuracy. He complained that the prospect of carrying the calculations to higher accuracy "made his head ache."

Publication of the *Principia* brought more attention to Newton than to his book. There were only a few reviews, mostly anonymous and superficial. As De Gandt writes, "Philosophers and humanists of this era and later generations had the feeling that great marvels were contained in these pages; they were told that Newton revealed truth, and they believed it. . . . But the *Principia* still remained a sealed book."

The *Opticks*

Newton as a young man skirmished with Hooke and others on the theory of colors and other aspects of optics. These polemics finally drove him into a silence of almost thirty years on the subject of optics, with the excuse that he did not want to be "engaged in Disputes about these Matters." What persuaded him to break the silence and publish more of his earlier work on optics, as well as some remarkable speculations, may have been the death of his chief adversary, Hooke, in 1703. In any case, Newton published his other masterpiece, the *Opticks*, in 1704.

The *Opticks* and the *Principia* are contrasting companion pieces. The two books have different personalities, and may indeed reflect Newton's changing persona. The *Principia* was written in the academic seclusion of Cambridge, and the *Opticks* in the social and political environment Newton entered after moving to London. The Opticks is a more accessible book than the *Principia*. It is written in English, rather than in Latin, and does not burden the reader with difficult mathematical arguments. Not surprisingly, Newton's successors frequently mentioned the *Opticks*, but rarely the *Principia*.

In the *Opticks*, Newton presents both the experimental foundations, and an attempt to lay the theoretical foundations, of the science of optics. He describes experiments that demonstrate the main physical properties of light rays: their reflection, "degree of refrangibility" (the extent to which they are refracted), "inflexion" (diffraction), and interference.

The term "interference" was not in Newton's vocabulary, but he describes interference effects in what are now called "Newton's rings." In the demonstration experiment, two slightly convex prisms are pressed together, with a thin layer of air between them; a striking pattern of colored concentric rings appears, surrounding points where the prisms touch.

Diffraction effects are demonstrated by admitting into a room a narrow beam of sunlight through a pinhole and observing that shadows cast by this light source on a screen have "Parallel Fringes or Bands of colour'd Light" at their edges.

To explain this catalogue of optical effects, Newton presents in the *Opticks* a theory based on the concept that light rays are the trajectories of small particles. As he puts it in one of the "queries" that conclude the *Opticks*: "Are not the Rays of Light very small Bodies emitted from shining Substances? For such Bodies will pass through Mediums in right Lines without bending into the Shadow, which is the Nature of the Rays of Light."

In another query, Newton speculates that particles of light are affected by optical forces of some kind: "Do not Bodies act upon Light at a distance, and by their action bend its Rays; and is not this action strongest at the least distance?"

With particles and forces as the basic ingredients, Newton constructs in the *Opticks* an optical mechanics, which he had already sketched at the end of book 1 of the *Principia*. He explains reflection and refraction by assuming that optical forces are different in different media, and diffraction by assuming that light rays passing near an object are more strongly affected by the forces than those more remote.

To explain the rings, Newton introduces his theory of "fits," based on the idea that light rays alternate between "Fits of easy Reflexion, and . . . Fits of easy Transmission." In this way, he gives the rays periodicity, that is, wavelike character. However, he does not abandon the particle point of view, and thus arrives at a complicated duality.

We now understand Newton's rings as an interference phenomenon, arising when two trains of waves meet each other. This theory was proposed by Thomas Young, one of the first to see the advantages of a simple wave theory of light, almost a century after the *Opticks* was published. By the 1830s, Young in England and Augustin Fresnel in France had demonstrated that all of the physical properties of light known at the time could be explained easily by a wave theory.

Newton's particle theory of light did not survive this blow. For seventy-five years the particles were forgotten, until 1905, when, to everyone's astonishment, Albert Einstein brought them back. (But we are getting about two centuries beyond Newton's story. I will postpone until later [chapter 19] an extended excursion into the strange world of light waves and particles.)

The queries that close the *Opticks* show us where Newton finally stood on two great physical concepts. In queries 17 through 24, he leaves us with a picture of the universal medium called the "ether," which transmits optical and gravitational forces, carries light rays, and transports heat. Query 18 asks, "Is not this medium exceedingly more rare and subtile than the Air, and exceedingly more elastick and active? And doth not it readily pervade all Bodies? And is it not (by its elastick force) expanded through all the Heavens?" The ether concept in one form or another appealed to theoreticians through the eighteenth and nineteenth centuries. It met its demise in 1905, that fateful year when Einstein not only resurrected particles of light but also showed that the ether concept was simply unnecessary.

In query 31, Newton closes the *Opticks* with speculations on atomism, which he sees (and so do we) as one of the grandest of the unifying concepts in physics. He places atoms in the realm of another grand concept, that of forces: "Have not the small particles of Bodies certain Powers, Virtues or Forces, by which they act at a distance, not only upon the Rays of Light for reflecting, refracting, and inflecting them [as particles], but also upon one another for producing a great Part of the Phaenomena of Nature?"

He extracts, from his intimate knowledge of chemistry, evidence for attraction and repulsion forces among particles of all kinds of chemical substances, metals, salts, acids, solvents, oils, and vapors. He argues that the particles are kinetic and indestructible: "All these things being considered, it seems probable to me, that God in the Beginning form'd Matter in solid, massy, hard, impenetrable, moveable Particles, of such Sizes and Figures, and in such Proportion to Space, as most conduced to the End for which he form'd them; even so very hard, as never to wear or break in pieces; no ordinary Power being able to divide what God himself made one in the First Creation."

London

There were two great divides in Newton's adult life: in the middle 1660s from the rural surroundings of Lincolnshire to the academic world of Cambridge, and thirty years later, when he was fifty-four, from the seclusion of Cambridge to the social and political existence of a well-placed civil servant in London. The move to London was probably inspired by a feeling that his rapidly growing fame deserved a more material reward than anything offered by the Lucasian Professorship. We can also surmise that he was guided by an awareness that his formidable talent for creative work in science was fading.

In March 1696, Newton left Cambridge, took up residence in London, and started a new career as warden of the Mint. The post was offered by Charles Montague, a former student and intimate friend who had recently become chancellor of the exchequer. Montague described the warden's office to Newton as a sinecure, noting that "it has not too much bus'nesse to require more attendance than you may spare." But that was not what Newton had in mind; it was not in his character to perform any task, large or small, superficially.

Newton did what he always did when confronted with a complicated problem: he studied it. He bought books on economics, commerce, and finance, asked searching questions, and wrote volumes of notes. It was fortunate for England that he did. The master of the Mint, under whom the warden served, was Thomas Neale, a speculator with more interest in improving his own fortune than in coping with a monumental assignment then facing the Mint. The English currency, and with it the Treasury, were in crisis. Two kinds of coins were in circulation, those produced by hammering a metal blank against a die, and those made by special machinery that gave each coin a milled edge. The hammered coins were easily counterfeited and clipped, and thus worth less than milled coins of the same denomination. Naturally, the hammered coins were used and the milled coins hoarded.

An escape from this threatening problem, general recoinage, had already been mandated before Newton's arrival at the Mint. He quickly took up the challenge of the recoinage, although it was not one of his direct responsibilities as warden. As Westfall comments, "[Newton] was a born administrator, and the Mint felt the benefit of his presence." By the end of 1696, less than a year after Newton went to the Mint, the crisis was under control. Montague did not hesitate to say later that, without Newton, the recoinage would have been impossible. In 1699 Neale died, and Newton, who was by then master in fact if not in name, succeeded him.

Newton's personality held many puzzles. One of the deepest was his attitude toward women. Apparently he never had a cordial relationship with his mother. Aside from a woman with whom he had a youthful infatuation and to whom he may have made a proposal of marriage, there was one other woman in Newton's life. She was Catherine Barton, the daughter of Newton's half-sister Hannah Smith. Her father, the Reverend Robert Barton, died in 1693, and sometime in the late 1690s she went to live with Newton in London. She was charming and beautiful and had many admirers, including Newton's patron, Charles Montague. She became Montague's mistress, no doubt with Newton's approval. The affair endured; when he died, Montague left her a generous income. She was also a friend of Jonathan Swift's, and he mentioned her frequently in his collection of letters, called *Journal to Stella*. Voltaire gossiped: "I thought . . . that Newton made his fortune by his merit. . . . No such thing. Isaac Newton had a very charming niece . . . who made a conquest of Minister Halifax [Montague]. Fluxions and gravitation would have been of no use without a pretty niece." After Montague's death, Barton married John Conduitt, a wealthy man who had made his fortune in service to the British army. The marriage placed him conveniently (and he was aptly named) for another career: he became an early Newton biographer.

Newton the administrator was a vital influence in the rescue of two institutions from the brink of disaster. In 1703, long after the recoinage crisis at the Mint, he was elected to the presidency of the Royal Society. Like the Mint when Newton arrived, the society was desperately in need of energetic leadership. Since the early 1690s its presidents had been aristocrats who were little more than figureheads. Newton quickly changed that image. He introduced the practice of demonstrations at the meetings in the major fields of science (mathematics, mechanics, astronomy and optics, biology, botany, and chemistry), found the society a new home, and installed Halley as secretary, followed by other disciples. He restored the authority of the society, but he also used that authority to get his way in two infamous disputes.

On April 16, 1705, Queen Anne knighted Newton at Trinity College, Cambridge. The ceremony appears to have been politically inspired by Montague (Newton was then standing for Parliament), rather than being a recognition of Newton's scientific achievements. Political or not, the honor was the climactic point for Newton during his London years.

More Disputes

Newton was contentious, and his most persistent opponent was the equally contentious Robert Hooke. The Newton story is not complete without two more accounts of Newton in rancorous dispute. The first of these was a battle over astronomical data. John Flamsteed, the first Astronomer Royal, had a series of observations of the Moon, which Newton believed he needed to verify and refine his lunar perturbation theory. Flamsteed reluctantly supplied the requested observations, but Newton found the data inaccurate, and Flamsteed took offense at his critical remarks.

About ten years later, Newton was still not satisfied with his lunar theory and still in need of Flamsteed's Moon data. He was now president of the Royal Society, and with his usual impatience, took advantage of his position and attempted to force Flamsteed to publish a catalogue of the astronomical data. Flamsteed resisted. Newton obtained the backing of Prince George, Queen Anne's husband, and Flamsteed grudgingly went ahead with the catalogue.

The scope of the project was not defined. Flamsteed wanted to include with his own catalogue those of previous astronomers from Ptolemy to Hevelius, but Newton wanted just the data needed for his own calculations. Flamsteed stalled for several years, Prince George died, and as president of the Royal Society, Newton assumed dictatorial control over the Astronomer Royal's observations. Some of the data were published as *Historia coelestis* (History of the Heavens) in 1712, with Halley as the editor. Neither the publication nor its editor was acceptable to Flamsteed.

Newton had won a battle but not the war. Flamsteed's political fortunes rose, and Newton's declined, with the deaths of Queen Anne in 1714 and Montague in 1715. Flamsteed acquired the remaining copies of *Historia coelestis*, separated Halley's contributions, and "made a sacrifice of them to Heavenly Truth" (meaning that he burned them). He then returned to the project he had planned before Newton's interference, and had nearly finished it when he died in 1719. The task was completed by two former assistants and published as *Historia coelestis britannica* in 1725. As for Newton, he never did get all the data he wanted, and was finally defeated by the sheer difficulty of precise lunar calculations.

Another man who crossed Newton's path and found himself in an epic dispute was Gottfried Leibniz. This time the controversy concerned one of the most precious of a scientist's intellectual possessions: priority. Newton and Leibniz both claimed to be the inventors of calculus.

There would have been no dispute if Newton had published a treatise composed in 1666 on his fluxion method. He did not publish that, or indeed any other mathematical work, for another forty years. After 1676, however, Leibniz was at least partially aware of Newton's work in mathematics. In that year, Newton wrote two letters to Leibniz, outlining his recent research in algebra and on fluxions. Leibniz developed the basic concepts of his calculus in 1675, and published a sketchy account restricted to differentiation in 1684 without mentioning Newton. For Newton, that publication and that omission were, as Westfall puts it, Leibniz's "original sin, which not even divine grace could justify."

During the 1680s and 1690s, Leibniz developed his calculus further to include integration, Newton composed (but did not publish) his *De quadratura* (*quadrature* was an early term for integration), and John Wallis published a brief account of fluxions in volume 2 of his *Algebra*. In 1699, a former Newton protégé, Nicholas Fatio de Duillier, published a technical treatise, *Lineae brevissimi* (Line of Quickest Descent), in which he claimed that Newton was the first inventor, and Leibniz the second inventor, of calculus. A year later, in a review of Fatio's *Lineae*, Leibniz countered that his 1684 book was evidence of priority.

The dispute was now ignited. It was fueled by another Newton disciple, John Keill, who, in effect, accused Leibniz of plagiarism. Leibniz complained to the secretary of the Royal Society, Hans Sloane, about Keill's "impertinent accusations." This gave Newton the opportunity as president of the society to appoint a committee to review the Keill and Leibniz claims. Not surprisingly, the committee found in Newton's favor, and the dispute escalated. Several attempts to bring Newton and Leibniz together did not succeed. Leibniz died in 1716; that cooled the debate, but did not extinguish it. Newtonians and Leibnizians confronted each other for at least five more years.

Nearer the Gods

Biographers and other commentators have never given us a consensus view of Newton's character. His contemporaries either saw him as all but divine or all but monstrous, and opinions depended a lot on whether the author was friend or foe. By the nineteenth century, hagiography had set in, and Newton as paragon emerged. In our time, the monster model seems to be returning.

On one assessment there should be no doubt: Newton was the greatest creative genius physics has ever seen. None of the other candidates for the superlative (Einstein, Maxwell, Boltzmann, Gibbs, and Feynman) has matched Newton's combined achievements as theoretician, experimentalist, *and* mathematician.

Newton was no exception to the rule that creative geniuses lead self-centered, eccentric lives. He was secretive, introverted, lacking a sense of humor, and prudish. He could not tolerate criticism, and could be mean and devious in the treatment of his critics. Throughout his life he was neurotic, and at least once succumbed to breakdown.

But he was no monster. He could be generous to colleagues, both junior and senior, and to destitute relatives. In disputes, he usually gave no worse than he received. He never married, but he was not a misogynist, as his fondness for Catherine Barton attests. He was reclusive in Cambridge, where he had little admiration for his fellow academics, but entertained well in the more stimulating intellectual environment of London.

If you were to become a time traveler and meet Newton on a trip back to the seventeenth century, you might find him something like the performer who first exasperates everyone in sight and then goes on stage and sings like an angel. The singing is extravagantly admired and the obnoxious behavior forgiven. Halley, who was as familiar as anyone with Newton's behavior, wrote in an ode to Newton prefacing the *Principia* that "nearer the gods no mortal can approach." Albert Einstein, no doubt equal in stature to Newton as a theoretician (and no paragon), left this appreciation of Newton in a foreword to an edition of the *Opticks*:

> Fortunate Newton, happy childhood of science! He who has time and tranquility can by reading this book live again the wonderful events which the great Newton experienced in his young days. Nature to him was an open book, whose letters he could read without effort. The conceptions which he used to reduce the material of experience to order seemed to flow spontaneously from experience itself, from the beautiful experiments which he ranged in order like playthings and describes with an affectionate wealth of details. In one person he combined the experimenter, the theorist, the mechanic and, not least, the artist in exposition. He stands before us strong, certain, and alone: his joy in creation and his minute precision are evident in every word and in every figure.

ii

Thermodynamics

Historical Synopsis

Our history now turns from mechanics, the science of motion, to thermodynamics, the science of heat. The theory of heat did not emerge as a quantitative science until late in the eighteenth century, when heat was seen as a weightless fluid called "caloric." The fluid analogy was suggested by the apparent "flow" of heat from a high temperature to a low temperature. Eighteenth-century engineers knew that with cleverly designed machinery, this heat flow could be used in a "heat engine" to produce useful work output.

The basic premise of the caloric theory was that heat was "conserved," meaning that it was indestructible and uncreatable; that assumption served well the pioneers in heat theory, including Sadi Carnot, whose heat engine studies begin our story of thermodynamics. But the doctrine of heat conservation was attacked in the 1840s by Robert Mayer, James Joule, Hermann Helmholtz, and others. Their criticism doomed the caloric theory, but offered little guidance for construction of a new theory.

The task of building the rudiments of the new heat science, eventually called thermodynamics, fell to William Thomson and Rudolf Clausius in the 1850s. One of the basic ingredients of their theory was the concept that any system has an intrinsic property Thomson called "energy," which he believed was somehow connected with the random motion of the system's molecules. He could not refine this molecular interpretation because in the mid–nineteenth century the structure and behavior—and even the existence—of molecules were controversial. But he could see that the energy of a system—not the heat—was conserved, and he expressed this conclusion in a simple differential equation.

In modern thermodynamics, energy has an equal partner called "entropy." Clausius introduced the entropy concept, and supplied the name, but he was ambivalent about recognizing its fundamental importance. He showed in a second simple differential equation how entropy is connected with heat and temperature, and stated formally the law now known as the second law of thermodynamics: that in an isolated system, entropy increases to a maximum value. But he hesitated to go further. The dubious status of the molecular hypothesis was again a concern.

Thermodynamics had its Newton: Willard Gibbs. Where Clausius hesitated, Gibbs did not. Gibbs recognized the energy-entropy partnership, and added to it a concept of great utility in the study of chemical change, the "chemical potential." Without much guidance from experimental results—few were available—Gibbs applied his scheme to a long list of disparate phenomena. Gibbs's masterpiece was a lengthy, but compactly written, treatise on thermodynamics, published in the 1870s.

Gibbs's treatise opened theoretical vistas far beyond the theory of heat sought by Clausius and Thomson. Once Gibbs's manifold messages were understood (or rediscovered), the new territory was explored. One of the explorers was Walther Nernst, who was in search of a theory of chemical affinity, the force that drives chemical reactions. He found his theory by taking a detour into the realm of low-temperature physics and chemistry.

3

A Tale of Two Revolutions

Sadi Carnot

Reflections

The story of thermodynamics begins in 1824 in Paris. France had been rocked to its foundations by thirty-five years of war, revolution, and dictatorship. A king had been executed, constitutions had been written, Napoleon had come and gone twice, and the monarchy had been restored twice. Napoleon had successfully marched his armies through the countries of Europe and then disastrously into Russia. France had been invaded and occupied and had paid a large war indemnity.

In 1824, a technical memoir was published by a young military engineer who had been born into this world of social, military, and political turmoil. The engineer's name was Sadi Carnot, and his book had the title *Reflections on the Motive Power of Fire*. By "motive power" he meant work, or the rate of doing work, and "fire" was his term for heat. His goal was to solve a problem that had hardly even been imagined by his predecessors. He hoped to discover the general operating principles of steam engines and other heat engine devices that supply work output from heat input. He did not quite realize his purpose, and his work was largely ignored at the time it was published, but after Carnot's work was rediscovered more than twenty years later it became the main inspiration for subsequent work in thermodynamics.

Lazare Carnot

Although he always worked on the fringes of the scientific world of his time, Sadi Carnot did not otherwise live in obscurity. His father, Lazare, was one of the most powerful men in France during the late eighteenth and early nineteenth centuries. Sadi was born in 1796 in the Paris Luxembourg Palace when Lazare was a member of the five-man executive Directory. Lazare Carnot served in high-level positions for only about four years, but his political accomplishments and longevity were extraordinary for those turbulent times. Before joining the government of the Directory, he was an influential member of the all-powerful Committee of Public Safety led by Maximilien de Robespierre. In that capacity, Lazare was responsible for the revolutionary war efforts. His brilliant handling of logistics and strategy salvaged what might otherwise have been a military disaster; in French history textbooks he is known as "the great Carnot" and "the organizer of victory." He was the only member of the Committee of Public Safety to survive the fall of Robespierre in 1794 and to join the Directory. A leftist coup in 1797 forced him into exile, but he returned as Napoleon's war minister. (He had given Napoleon command of the Italian army in 1797.) Napoleon's dictatorial ways soon became evident, however, and Lazare, unshakable in his republican beliefs, resigned after a few months. But he returned once more in 1814, near the end of the Napoleonic regime, first as the governor of Antwerp and then as Napoleon's last minister of the interior.

Lazare Carnot's status in history may be unique. Not only was he renowned for his practice of politics and warfare; he also made important discoveries in science and engineering. A memoir published in 1783 was, according to Lazare's biographer, Charles Gillispie, the first attempt to deal in a theoretical way with the subject of engineering mechanics. Lazare's goal in this and in later work in engineering science was to abstract general operating principles from the mechanical workings of complicated machinery. His aim, writes Gillispie, "was to specify in a completely general way the optimal conditions for the operation of machines of every sort." Instead of probing the many detailed elements of machinery design, as was customary at the time, he searched for theoretical methods whose principles had no need for the details.

Lazare Carnot's main conclusion, which Gillispie calls the "principle of continuity of power," asserts that accelerations and shocks in the moving parts of machinery are to be avoided because they lead to losses of the "moment of activity" or work output. The ideal machine is one in which power is transmitted continuously, in very small steps. Applied to water machines (for instance, waterwheels), Lazare's theorem prescribes that for maximum efficiency there must be no turbulent or percussive impact between the water and the machine, and the water leaving the machine should not have appreciable velocity.

Lazare's several memoirs are not recognized today as major contributions to engineering science, but in an important sense his work survives. His approach gave his son Sadi a clear indication of where to begin his own attack on the theory of heat engines. Lazare's views on the design of water engines seem to have been particularly influential. Waterwheels and other kinds of hydraulic machinery are driven by falling water, and the greater the fall, the greater the machine's work output per unit of water input. Sadi Carnot's thinking was guided by an analogy between falling water in water engines and falling heat in heat engines: he reasoned that a heat engine could not operate unless its design included a high-temperature body and a low-temperature body between which heat dropped while it drove the working parts of the machine.

Heat Engines, Then and Now

The heat engines of interest to Sadi Carnot were steam engines applied to such tasks as driving machinery, ships, and conveyors. The steam engine invented by a Cornishman, Arthur Woolf, was particularly admired in France in the 1810s and 1820s. Operation of the Woolf engine is diagrammed in figure 3.1. Heat Q_2

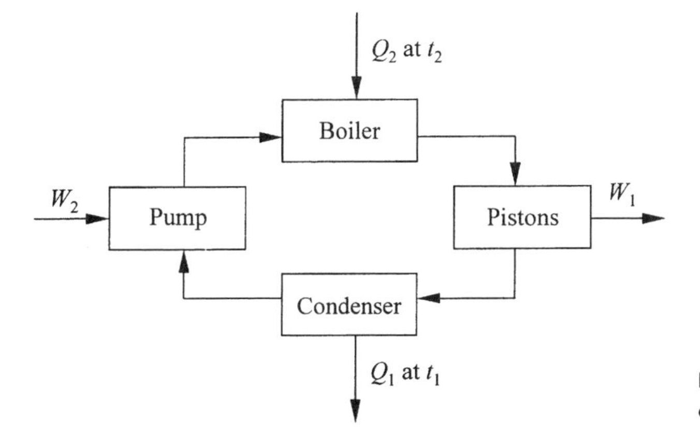

Figure 3.1. Diagram of the Woolf steam engine.

was supplied at a high temperature t_2 by burning a fuel, and this heat generated steam at a high pressure in a boiler. The steam drove two pistons and they provided the work output W_1. (In this chapter and elsewhere in this part of the book, keep in mind that the symbol t represents temperature and not time, as in chapters 1 and 2.) The steam leaves the pistons at a decreased pressure and temperature. Heat Q_1 was then extracted in a condenser where the steam was further cooled to a still lower temperature t_1 and condensed to liquid water. Finally, the liquid water passed through a pump, which restored the high pressure by expending work W_2, and low-temperature, pressurized water was returned to the boiler. This is a cycle of operations, and its net effect is the dropping of heat from the high temperature t_2 to the low temperature t_1, with work output W_1 from the pistons and a much smaller work input W_2 to the pump.

The Woolf steam engine and its variations have evolved into a vast modern technology. Most contemporary power plants operate similarly. The scale is much larger in the modern plants, the operating steam pressures and temperatures are higher, and the working device is a turbine rather than pistons. But the concept of heat falling between a high and a low temperature with net work output again applies.

Carnot's Cycle

Sadi Carnot had the same ambitions as his father. He hoped to abstract, from the detailed complexities of real machinery, general principles that dictated the best possible performance. Lazare's analysis had centered on ideal mechanical operation; Sadi aimed for the mechanical ideal, and also for ideal thermal operation.

He could see, first of all, that when heat was dropped from a high temperature to a low temperature in a heat engine it could accomplish something. His conceptual model was based on an analogy between heat engines and water engines. He concluded that for maximum efficiency a steam engine had to be designed so it operated with no direct fall of heat from hot to cold, just as the ideal water engine could not have part of the water stream spilling over and falling directly rather than driving the waterwheel. This meant that in the perfect heat engine, hot and cold parts in contact could differ only slightly in temperature. One can say, to elaborate somewhat, that the thermal driving forces (that is, temperature differences) in Carnot's ideal heat engine have to be made very small. This design had more than an accidental resemblance to Lazare Carnot's principle of continuity in the transmission of mechanical power.

To make it more specific, Carnot imagined that his ideal heat engine used a gaseous working substance put through cyclic changes—something like the steam in the pistons of the Woolf steam engine. Carnot's cycles consisted of four stages:

1. An isothermal (constant-temperature) expansion in which the gas absorbed heat from a heat "reservoir" kept at a high temperature t_2.
2. An adiabatic (insulated) expansion that lowered the temperature of the gas from t_2 to t_1.
3. An isothermal compression in which the gas discarded heat to a reservoir kept at the low temperature t_1.
4. An adiabatic compression that brought the gas back to the original high temperature t_2.

Stages 1 and 3 accomplish the heat fall by absorbing heat at a high temperature and discarding it at a low temperature. More work is done *by* the gas in the expansion of stage 1 than *on* the gas in the compression of stage 3; and amounts of work done on and by the gas in stages 2 and 4 nearly cancel each other. Thus, for each turn of the cycle, heat is dropped from a high temperature to a low temperature, and there is net work output.

Carnot's Principle

To summarize, Carnot constructed his ideal heat engine, as Lazare had made *his* ideal machinery, so that all its parts and stages functioned continuously in very small steps under very small thermal and mechanical driving forces. This and the necessity for operating in cycles between two fixed temperatures were, Carnot realized, the main features required for all ideal heat engine operation. The special features of the four-stage gas cycle were convenient but unnecessary; other ways could be found to drop the heat between the two heat reservoirs and produce work output.

Carnot's point of view insists that the forces driving an ideal heat engine be so small they can be reversed with no additional external effect and the engine made to operate in the opposite direction. Run forward, in its normal mode of operation as a heat *engine*, the ideal machine *drops* heat, let's say between the temperatures t_2 and t_1, and provides work *output*. Run backward, with all its driving forces reversed, the ideal machine requires work *input* and it *raises* heat from t_1 to t_2. This is a heat *pump,* analogous to a mechanical device capable of pumping water from a low level to a high level. Carnot reached the fundamental conclusion that any ideal heat engine, operated as it had to be by very small driving forces, was literally "reversible." All of its stages could be turned around and, with no significant effect in the surroundings, the heat engine made into a heat pump, or vice versa.

This reversibility aspect of ideal heat engine operation led Carnot to his main result, a proof that *any* ideal heat engine operating between heat reservoirs maintained at t_2 and t_1, had to supply the same work output W for a given heat input Q_2. If two ideal heat engines had different work outputs W and W' with W' larger than W, say, the engine with higher work output W' could be used to drive the

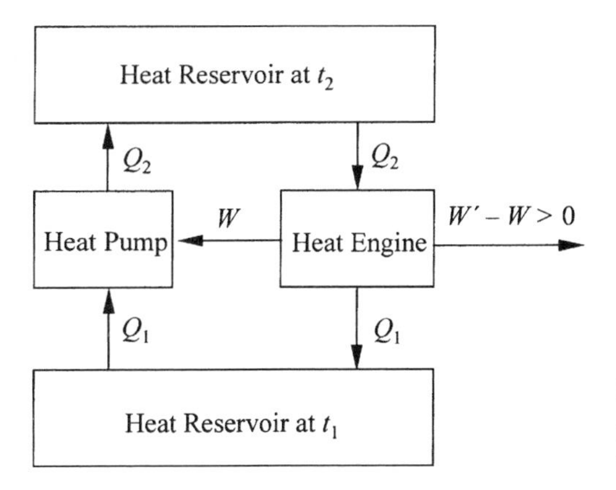

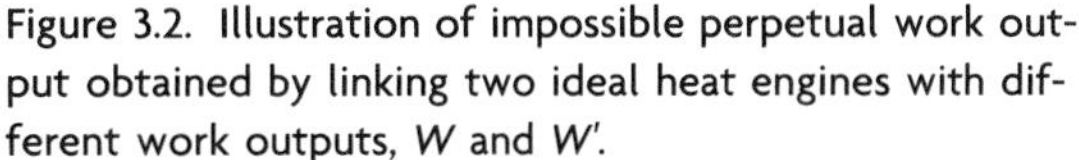
Figure 3.2. Illustration of impossible perpetual work output obtained by linking two ideal heat engines with different work outputs, W and W'.

engine with lower work output W in reverse to pump the heat Q_2 back to its original thermal level in the upper heat reservoir, and with net work output $W' - W$ (fig. 3.2).

If this composite device had been possible, it would have served as a perpetual-motion machine because it supplied work output with no need to replenish the heat supply in the upper heat reservoir; every unit of heat dropped through the heat engine was restored to the upper reservoir by the heat pump. In other words, this composite heat engine could have worked endlessly without having to burn fuel. Lazare Carnot had relied heavily on the axiom that perpetual motion of any kind was physically impossible, and this was another one of the father's lessons learned by the son. Sadi Carnot also categorically rejected the possibility of perpetual motion and therefore concluded that the two ideal heat engines in the composite machine had to have the same work output, that is, $W = W'$.

Put more formally, Carnot's conclusion was that all ideal heat engines operating in cycles between the two temperatures t_1 and t_2 with the heat input Q_2 have the same work output W. Design details make no difference. The working material can be steam, air, or even a liquid or solid; the working part of the cycle can be a gas expansion, as in Carnot's cycle, or it can be something else. The work output W of the ideal heat engine is *precisely* determined by just three things, the heat input Q_2 and the temperatures t_1 and t_2 of the two reservoirs between which the heat engine operates. This statement expresses "Carnot's principle." It was an indispensable source of inspiration for all of Carnot's successors.

Carnot's Function

To continue with his analysis, Carnot had to deduce what he could concerning the physical and mathematical nature of ideal engine operation. Here he seems to have exploited further his idea that heat engines do work by dropping heat from a higher to a lower temperature. It seemed that the ability of heat to do work in a heat engine depended on its thermal level expressed by the temperature t, just as the ability of water to do work in a water engine depends on its gravitational level.

Carnot emphasized a function $F(t)$ that expressed the ideal heat engine's operating efficiency at the temperature t. He made three remarkable calculations of numerical values for his function $F(t)$. These calculations were based on three different heat engine designs that used air, boiling water, and boiling alcohol as the working materials. Carnot's theory required that ideal heat engine behavior be entirely independent of the nature of the working material and other special design features: values obtained for $F(t)$ in the three cases had to be dependent *only* on the temperature t. Although the primitive data available to Carnot for the calculation limited the accuracy, his results for $F(t)$ seemed to satisfy this requirement. No doubt this success helped convince Carnot that his heat engine theory was fundamentally correct.

To complete his theory, Carnot had to find not just numbers but a mathematical expression for his function $F(t)$. In this effort, he was unsuccessful; he could see only that $F(t)$ decreased with increasing temperature. Many of Carnot's successors also became fascinated with this problem. Although in the end Carnot's function was found to be nothing more complicated than the reciprocal of the temperature expressed on an absolute scale, it took no fewer than eight thermodynamicists, spanning two generations, to establish this conclusion unequivocally; five of them (Carnot, Clausius, Joule, Helmholtz, and Thomson) were major figures in nineteenth-century physics.

Publication and Neglect

Sadi Carnot's work was presented as a privately published memoir in 1824, one year after Lazare Carnot's death, and it met a strange fate. The memoir was published by a leading scientific publisher, favorably reviewed, mentioned in an important journal—and then for more than twenty years all but forgotten. With one fortunate exception, none of France's esteemed company of engineers and physicists paid any further attention to Carnot's memoir.

One can only speculate concerning the reasons for this neglect. Perhaps Carnot's immediate audience did not appreciate his scientific writing style. Like his father, whose scientific work was also ignored at first, Carnot wrote in a semipopular style. He rarely used mathematical equations, and these were usually relegated to footnotes; most of his arguments were stated verbally. Evidently Carnot, like his father, was writing for engineers, but his book was still too theoretical for the steam-engine engineers who should have read it. Others of the scientific establishment, looking for the analytical mathematical language commonly used at the time in treatises on mechanics, probably could not take seriously this unknown youth who insisted on using verbal science to formulate his arguments. It didn't help either that Carnot was personally reserved and wary of publicity of any kind. One of his rules of conduct was, "Say little about what you know and nothing at all about what you don't know." In the end, like Newton with the *Principia*, Carnot missed his audience.

In time, Carnot probably would have seen his work recognized, if not in France, perhaps elsewhere where theoretical research on heat and heat engines was more active. But Carnot never had the opportunity to wait for the scientific world to catch up. In 1831, he contracted scarlet fever, which developed into "brain fever." He partially recovered and went to the country for convalescence. But later, in 1832, while studying the effects of a cholera epidemic, he became a cholera victim himself. The disease killed him in hours; he was thirty-six years old. Most of his papers and other effects were destroyed at the time of his death, the customary precaution following a cholera casualty.

After Carnot

The man who rescued Carnot's work from what certainly would otherwise have been oblivion was Émile Clapeyron, a former classmate of Carnot's at the École Polytechnique. It was Clapeyron who, in a paper published in the *Journal de l'École Polytechnique* in 1834, put Carnot's message in the acceptable language of mathematical analysis. Most important, Clapeyron translated into differential equations Carnot's several verbal accounts of how to calculate his efficiency function $F(t)$.

Clapeyron's paper was translated into German and English, and for ten years or so it was the only link between Carnot and his followers. Carnot's theory, in the mathematical translation provided by Clapeyron, was to become the point of departure in the 1840s and early 1850s for two second-generation thermodynamicists, a young German student at the University of Halle, Rudolf Clausius, and a recent graduate of Cambridge University, William Thomson (who became Lord Kelvin). Thomson spent several months in 1845 in the Paris laboratory of Victor Regnault. He scoured the Paris bookshops for a copy of Carnot's memoir with no success. No one remembered either the book or its author.

In different ways, Clausius and Thomson were to extend Carnot's work into the science of heat that Thomson eventually called thermodynamics. One of Clapeyron's differential equations became a fixture in Thomson's approach to thermodynamics; Thomson found a way to use the equation to define an absolute temperature scale. Later, he introduced the concept of energy, and with it resolved a basic flaw in Carnot's theory: its apparent reliance on the caloric theory. Among Clausius's contributions was an elaboration of Carnot's heat engine analysis, which recognized that heat is not only dropped in the heat engine from a high temperature to a low temperature but is also partially converted to work. This was a departure from Carnot's water engine analogy, and in later research it led to the concept of entropy.

Recognition

So, in the end, Sadi Carnot's theory was resurrected, understood, and used. And it finally became clear that Carnot, no less than his father Lazare, should be celebrated as a great revolutionary. Born into a political revolution, Carnot started a scientific revolution. His theory was radically new and completely original. None of Carnot's predecessors had exploited, or even hinted at, the idea that heat fall was the universal driving force of heat engines.

If Carnot's contemporaries lacked the vision to appreciate his work, his numerous successors have, at least for posterity, repaired the damage of neglect. Science historians now regard Carnot as one of the most inventive of scientists. In his history of thermodynamics, *From Watt to Clausius*, Donald Cardwell assesses for us Sadi Carnot's astonishing success in achieving Lazare Carnot's grand goal, the abstraction of general physical principles from the complexities of machinery: "Perhaps one of the truest indicators of Carnot's greatness is the unerring skill with which he abstracted, from the highly complicated mechanical contrivance that was the steam engine . . . the essentials, and the essentials alone, of his argument. Nothing unnecessary is included, and nothing essential is missed out. It is, in fact, very difficult to think of a more efficient piece of abstraction in the history of science since Galileo taught . . . the basis of the procedure."

Scant records of Carnot's life and personality remain. In the two published portraits, we see a sensitive, intelligent face, with large eyes regarding us with a steady, slightly melancholy gaze. Most of the biographical material on Carnot comes from a brief article written by Sadi's brother Hippolyte. (Lazare Carnot was partial to exotic names for his sons.) Hippolyte's anecdotes tell of Carnot's independence and courage, even in childhood. As a youngster, he sometimes accompanied his father on visits to Napoleon's residence; while Lazare and Bonaparte conducted business, Sadi was put in the care of Madame Bonaparte. On one occasion, she and other ladies were amusing themselves in a rowboat on a pond when Bonaparte appeared and splashed water on the rowers by throwing stones near the boat. Sadi, about four years old at the time, watched for a while, then indignantly confronted Bonaparte, called him "beast of a First Consul," and demanded that he desist. Bonaparte stared in astonishment at his tiny attacker, and then roared with laughter.

The child who challenged Napoleon later entered the École Polytechnique at about the same time the French military fortunes began to collapse. Two years later Napoleon was in full retreat, and France was invaded. Hippolyte relates that Sadi could not remain idle. He petitioned Napoleon for permission to form a brigade to fight in defense of Paris. The students fought bravely at Vincennes, but Paris fell to the Allied armies, and Napoleon was forced to abdicate.

Hippolyte records one more instance of his brother's courage. Sadi was walking in Paris one day when a mounted drunken soldier galloped down the street, "brandishing his saber and striking down passers-by." Sadi ran forward, dodged the sword and the horse, grabbed the soldier, and "laid him in the gutter." Sadi then "continued on his way to escape from the cheers of the crowd, amazed at this daring deed."

Sadi Carnot lived in a time of unsurpassed scientific activity, most of it centered in Paris. The list of renowned physicists, mathematicians, chemists, and engineers who worked in Paris during Carnot's lifetime includes Pierre-Simon Laplace, André-Marie Ampère, Augustin Fresnel, Siméon-Denis Poisson, Adrien-Marie Legendre, Pierre Dulong, Alexis Petit, Evariste Galois, and Gaspard de Coriolis. Many of these names appeared on the roll of the faculty and students at the École Polytechnique, where Carnot received his scientific training. Except as a student, Carnot was never part of this distinguished company. Like some other incomparable geniuses in the history of science (notably, Gibbs, Joule, and Mayer in our story), Carnot did his important work as a scientific outsider. But there is no doubt that Carnot's name belongs on anyone's list of great French physicists. He may have been the greatest of them all.

4 On the Dark Side
Robert Mayer

Something Is Conserved

To the modern student, the term *energy* has a meaning that is almost self-evident. This meaning was far from clear, however, to scientists of the early nineteenth century. The many effects that would finally be unified by the concept of energy were still seen mostly as diverse phenomena. It was suspected that mechanical, thermal, chemical, electrical, and magnetic effects had something in common, but the connections were incomplete and confused.

What was most obvious by the 1820s and 1830s was that strikingly diverse effects were interconvertible. Alessandro Volta's electric cell, invented in 1800, produced electrical effects from chemical effects. In 1820, Hans Christian Oersted observed magnetic effects produced by electrical effects. Magnetism produces motion (mechanical effects), and for many years it had been known that motion can produce electrical effects through friction. This sequence is a chain of "conversions":

> Chemical effect → electrical effect → magnetic effect → mechanical effect → electrical effect.

In 1822, Thomas Seebeck demonstrated that a bimetallic junction produces an electrical effect when heated, and twelve years later Jean Peltier reported the reverse conversion: cooling produced by an electrical effect. Heat engines perform as conversion devices, converting a thermal effect (heat) into a mechanical effect (work).

Most of the major theories of science have been discovered by one scientist, or at most by a few. The search for broad theoretical unities tends to be difficult, solitary work, and important scientific discoveries are usually subtle enough that special kinds of genius are needed to recognize and develop them. But, as Thomas Kuhn points out, there is at least one prominent exception to this rule. The theoretical studies inspired by the discoveries of conversion processes, which finally gave us the energy concept, were far from a singular effort. Kuhn lists twelve scientists who contributed importantly during the early stages of this "simultaneous discovery."

The idea that occurred to all twelve—not quite simultaneously, but independently—was that *conversion* was somehow linked with *conservation*. When one effect was converted to another, some measure of the first effect was quantitatively replaced by the same kind of measure of the second. This measure, applicable to all the various interconvertible effects, was *conserved*: throughout a conversion process its total amount, whether it assessed one effect, the other effect, or both, was precisely constant.

The twelve simultaneous discoverers were not the first to make important use of a conservation principle. In one form or another, conservation principles had been popular, almost intuitive it seems, with scientists for many years. Theorists had counted among their most impressive achievements discoveries of quantities that were both indestructible and uncreatable. Adherents of the caloric theory of heat had postulated conservation of heat. In the late eighteenth century, Antoine-Laurent Lavoisier and others had established that mass is conserved in chemical reactions; when a chemical reaction proceeds in a closed container, there is no change in total mass.

So it was natural for theorists who studied conversion processes to attempt to build their theories from a conservation law. But, as always in the formulation of a conservation principle, a difficult question had to be asked at the outset: what is the quantity conserved? As it turned out, a workable answer to this question was practically impossible without some knowledge of the conservation law itself, because the most obvious property of the conserved quantity, ultimately identified as energy, was that it was conserved. No direct measurement like that of mass could be made for verification of the conservation property. This was a search for something that could not be fully defined until it was actually found.

Voyage of Discovery

One of the first to penetrate this conceptual tangle was Robert Mayer, a German physician and physicist who spent most of his life in Heilbronn, Germany. Mayer was a contemporary of James Joule (chapter 5), and like Joule, he was an amateur in the scientific fields that most absorbed his interest. His university training was in medicine, and what is known of his student record at the University of Tübingen shows little sign of intellectual genius. He was good at billiards and cards, devoted to his fraternity, and inclined to be rebellious and unpopular with the university authorities; eventually he was suspended for a year. With hindsight, we can see in Mayer's reaction to the suspension—a six-day hunger strike—evidence for his stubbornness and sensitivity to criticism, and even some forewarning of his later mental problems.

Mayer's youthful behavior was not that of an unmitigated rebel, however; when the Tübingen authorities permitted, he returned, finished his dissertation, and passed the doctoral examination. But he was still too restless to plan his future according to conventional (and family) expectations. Instead of settling into a routine medical practice, he decided to travel by taking a position as ship's surgeon on a Dutch vessel sailing for the East Indies. He found little inspiration on this trip, either in the company of his fellow officers or in the quality and quantity of the ship's food. But to Mayer the voyage was worth any amount of hunger and boredom.

Mayer tells us, in an exotic tale of scientific imagination, of an event in Java that set him on the intellectual path he followed for the rest of his life. On several occasions in 1840, when he let blood from sailors in an East Java port, Mayer noticed that venous blood had a surprisingly bright red color. He surmised that this unusual redness of blood in the tropics indicated a slower rate of metabolic oxidation. He became convinced that oxidation of food materials produced heat internally and maintained a constant body temperature. In a warm climate, he reasoned, the oxidation rate was reduced.

For those of us who are inclined toward the romantic view that theoreticians make their most inspired advances in intuitive leaps, this story and the sequel are fascinating. Mayer's assumed connection between blood color and metabolic oxidation rate was certainly oversimplified and partly wrong, but this germ of a theory brought an intellectual excitement and stimulation Mayer had never before experienced. It did not take him long to see his discovery as much more than a new medical fact: metabolic oxidation was a physiological conversion process in which heat was produced from food materials, a chemical effect producing a thermal effect. Mayer was convinced that the chemical effect and the thermal effect were somehow related; to use the terminology he adopted to express his theory, the chemical reaction was a "force" that changed its form but not its magnitude in the metabolic process. And most important in Mayer's view, this interpretation of metabolic oxidation was just one instance of a general principle.

Conservation of Force (Energy)

In 1841 Mayer, now back in Heilbronn, began a paper that summarized his point of view in the broadest terms. He wrote that "all bodies are subject to change . . . [which] cannot happen without a cause . . . [that] we call force," that "we can derive all phenomena from a basic force," and that "forces, like matter, are invariable." His intention, he said, was to write physics as a science concerned with "the nature of the existence of force." The program of this physics paralleled that of chemistry. Chemists dealt with the properties of matter, and relied on the principle that mass is conserved. Physicists should similarly study forces and adopt a principle of conservation of force. Both chemistry and physics were based on the principle that the "quantity of [their] entities is invariable and only the quality of these entities is variable."

Mayer's use of the term *force* requires some explanation. It was common for nineteenth-century physicists to give the force concept a dual meaning. They used it at times in the Newtonian sense, to denote a push or pull, but just as often the usage implied that *force* was synonymous with the modern term *energy*. The modern definition of the word "energy"—the capacity to do work—was not introduced until the 1850s, by William Thomson. In the above quotations, and throughout most of Mayer's writings, it is appropriate to assume the second usage, and to read "energy" for "force." With that simple but significant change, Mayer's thesis becomes an assertion of the principle of the conservation of energy.

Rejection

Mayer submitted his 1841 paper to Johann Poggendorff's *Annalen der Physik und Chemie.* It was not accepted for publication, or even returned with an acknowledgment. But, according to one of Mayer's biographers, R. Bruce Lindsay, the careless treatment was a blessing in disguise. Mayer's detailed arguments in the paper were "based on a profound misunderstanding of mechanics." Although the rejection was a blow to Mayer's pride, "it was a good thing for [his] subsequent reputation that [the paper] did not see the light of day."

If Mayer had great pride, he had even more perseverance. With help from his friend Carl Baur (later a professor of mathematics in Stuttgart), he improved the paper, expanded it in several ways, and at last saw it published in Justus von Liebig's *Annalen der Chemie und Pharmacie* in 1842. Mayer's most important addition to the paper was a calculation of the mechanical effect, work done in the expansion of a gas, produced by a thermal effect, the heating of the gas. This was an evaluation of the "mechanical equivalent of heat," a concern independently occupying Joule at about the same time. Whether or not Mayer made the *first* such calculation became the subject of a celebrated controversy. One thing that weakened Mayer's priority claim was that he omitted all details but the result in his calculation in the 1842 paper. Not until 1845, in a more extended paper, did he make his method clear. By 1845, Joule was reporting impressive experimental measurements of the mechanical equivalent of heat.

In the 1842 paper, Mayer based his ultimately famous calculation on the experimental fact that it takes more heat to raise the temperature of a gas held at constant pressure than at constant volume. Mayer could see in the difference between the constant-pressure and constant-volume results a measure of the heat converted to an equivalent amount of work done by the gas when it expands against constant pressure. He could also calculate that work, and the work-to-heat ratio, was a numerical evaluation of the mechanical equivalent of heat. His calculation showed that 1 kilocalorie of heat converted to work could lift 1 kilogram 366 meters. In other words, the mechanical equivalent of heat found by Mayer was 366 kilogram-meters per kilocalorie.

This was the quantity Joule had measured, or was about to measure, in a monumental series of experiments started in 1843. Joule's best result (labeled as it was later with a J) was

$$J = 425 \text{ kilogram-meters per kilocalorie.}$$

Mayer's calculation was incorrect principally because of errors in heat measurements. More-accurate measurements by Victor Regnault in the 1850s brought Mayer's calculation much closer to Joule's result,

$$J = 426 \text{ kilogram-meters per kilocalorie.}$$

In addition to clarifying his determination of the mechanical equivalent of heat, Mayer's 1845 paper also broadened his speculations concerning the conservation of energy, or force, as Mayer's terminology had it. Two quotations will show how committed Mayer had become to the conservation concept: "What chemistry performs with respect to matter, physics has to perform in the case of force. The only mission of physics is to become acquainted with force in its various forms and to investigate the conditions governing its change. The creation or destruction of force, if [either has] any meaning, lies outside the domain of human thought and action." And: "In truth there exists only a *single* force. In never-ending exchange this circles through all dead as well as living nature. In the latter as well as the former nothing happens without form variation of force!"

Mayer submitted his 1845 paper to Liebig's *Annalen;* it was rejected by an assistant editor, apparently after a cursory reading. The assistant's advice was to try Poggendorff's *Annalen*, but Mayer did not care to follow that publication route again. In the end, he published the paper privately, and hoped to gain recognition by distributing it widely. But beyond a few brief journal listings, the paper, Mayer's magnum opus, went unnoticed.

Over the Edge and Back

Although by this time Mayer was losing ground in his battle against discouragement, perseverance still prevailed. In 1846, he wrote another paper (this one, on celestial mechanics, anticipated work done much later by William Thomson), and again had to accept private publication.

Professional problems were now compounded by family and health problems. During the years 1846 to 1848, three of Mayer's children died, and his marriage began to deteriorate. Finally, in 1850, he suffered a nearly fatal breakdown. An attack of insomnia drove him to a suicide attempt; the attempt was unsuccessful, but from the depths of his despair Mayer might have seen this as still another failure.

In an effort to improve his condition, Mayer voluntarily entered a sanatorium. Treatment there made the situation worse, and finally he was committed to an asylum, where his handling was at best careless and at times brutal. The diagnosis of his mental and physical condition became so bleak that the medical authorities could offer no hope, and he was released from the institution in 1853.

It may have been Mayer's greatest achievement that he survived, and even partially recovered from, this appalling experience. After his release, he returned to Heilbronn, resumed his medical practice in a limited way, and for about ten years deliberately avoided all scientific activity. In slow stages, and with occasional relapses, his health began to return. That Mayer could, by an act of will it seems, restore himself to comparatively normal health, demonstrated, if nothing else did, that his mental condition was far from hopelessly unbalanced. To abandon entirely for ten years an effort that had become an obsession was plainly an act of sanity.

The period of Mayer's enforced retirement, the 1850s, was a time of great activity in the development of thermodynamics. Energy was established as a concept, and the energy conservation principle was accepted by most theorists. This work was done mostly by James Joule in England, by Rudolf Clausius in Germany, and by William Thomson and Macquorn Rankine in Scotland, with little appreciation of Mayer's efforts. Not only was Mayer's theory ignored during this time, but in 1858 Mayer himself was reported by Liebig to have died in an asylum. Protests from Mayer did not prevent the appearance of his official death notice in Poggendorff's *Handwörterbuch*.

Strange Success

The final episode in this life full of ironies will seem like the ultimate irony. Recognition of Mayer's achievements finally came, but hardly in a way deserved by a man who had endured indifference, rejection, breakdown, cruel medical treatment, and reports of his own death. In the early 1860s Mayer, now peacefully tending his vineyards in Heilbronn, suddenly became the center of a famous scientific controversy.

It all started when John Tyndall, a popular lecturer, professor, and colleague of Michael Faraday at the Royal Institution in London, prepared himself for a series of lectures on heat. He wrote to Hermann Helmholtz and Rudolf Clausius in Germany for information. Included in Clausius's response was the comment that Mayer's writings were not important. Clausius promised to send copies of Mayer's papers nevertheless, and before mailing the papers he read them, apparently for the first time with care. Clausius wrote a second letter with an entirely different assessment: "I must retract the statements in my last letter that you would not find much of importance in Mayer's writings; I am astonished at the multitude of beautiful and correct thoughts which they contain." Clausius was now convinced that Mayer had been one of the first to understand the energy concept and its conservation doctrine. Helmholtz also sent favorable comments on Mayer, pointing especially to the early evaluation of the mechanical equivalent of heat.

Tyndall was a man who loved controversy and hated injustice. Because his ideas concerning the latter were frequently not shared by others who were equally adept in the practice of public controversy, he was often engaged in arguments that were lively, but not always friendly. When Tyndall decided to be Mayer's champion, he embarked on what may have been the greatest of all his controversies. As usual, he chose as his forum the popular lectures at the Royal Institution. He had hastily decided to broaden his topic from heat to the general subject of energy, which was by then, in the 1860s, mostly understood; the title of his lecture was "On Force." (Faraday and his colleagues at the Royal Institution still preferred to use the term "force" when they meant "energy.")

Tyndall began by listing many examples of energy conversion and conservation, and then summarized Mayer's role with the pronouncement, "All that I have brought before you has been taken from the labors of a German physician, named Mayer." Mayer should, he said, be recognized as one of the first thermodynamicists, "a man of genius arriving at the most important results some time in advance of those whose lives were entirely devoted to Natural Philosophy." Tyndall left no doubt that he felt Mayer had priority claims over Joule: "Mr. Joule published his first paper 'On the Mechanical Value of Heat' in 1843, but in 1842 Mayer had actually calculated the mechanical equivalent of heat." In the gentlemanly world of nineteenth-century scientific discourse, this was an invitation to verbal combat. It brought quick responses from Joule and Thomson, and also from Thomson's close friend Peter Guthrie Tait, professor of natural philosophy at the University of Edinburgh, and Tyndall's match in the art of polemical debate.

Joule was the first to reply, in a letter published in the *Philosophical Magazine*. He could not, he said, accept the view that the "dynamical theory of heat" (that is, the theory of heat that, among other things, was based on the heat-work connection) was established by Mayer, or any of the other authors who speculated on the meaning of the conversion processes. Reliable conclusions "require experiments," he wrote, "and I therefore fearlessly assert my right to the position which has been generally accorded to me by my fellow physicists as having been the first to give decisive proof of the correctness of this theory."

Tyndall responded to Joule in another letter to the *Philosophical Magazine*, protesting that he did not wish to slight Joule's achievements: "I trust you will find nothing [in my remarks] which indicates a desire on my part to question your claim to the honour of being the experimental demonstrator of the equivalence of heat and work." Tyndall was willing to let Mayer speak for himself; at Tyndall's suggestion, Mayer's papers on the energy theme were translated and published in the *Philosophical Magazine*.

But this did not settle the matter. An article with both Thomson and Tait listed as authors (although the style appears to be that of Tait) next appeared in a popular magazine called *Good Words*, then edited by Charles Dickens. In it, Mayer's 1842 paper was summarized as mainly a recounting of previous work with a few suggestions for new experiments; "a method for finding the mechanical equivalent of heat [was] propounded." This was, the authors declared, a minor achievement, and they could find no reason to surrender British claims:

> On the strength of this publication an attempt has been made to claim for Mayer the credit of being the first to establish in all its generality the principle of the Conservation of Energy. It is true that *la science n'a pas de patrie* and it is highly creditable to British philosophers that they have so liberally acted according to this maxim. But it is not to be imagined that on this account there should be no scientific patriotism, or that, in our desire to do justice to a foreigner, we should depreciate or suppress the claims of our countrymen.

Tyndall replied, again in the *Philosophical Magazine*, pointedly directing his remarks to Thomson alone, and questioning the wisdom of discussing weighty matters of scientific priority in the pages of a popular magazine. He now relaxed his original position and saw Joule and Mayer more in a shared role:

> Mayer's labors have in some measure the stamp of profound intuition, which rose, however, to the energy of undoubting conviction in the author's mind. Joule's labours, on the contrary, are in an experimental demonstration. True to the speculative instinct of his country, Mayer drew large and weighty conclusions from slender premises, while the Englishman aimed, above all things, at the firm establishment of facts. And he did establish them. The future historian of science will not, I think, place these men in antagonism.

Tait was next heard from. He wrote to one of the editors of the *Philosophical Magazine*, first offering the observation that if *Good Words* was not a suitable medium for the debate of scientific matters, neither were certain popular lecture series at the Royal Institution. He went on: "Prof. Tyndall is most unfortunate in the possession of a mental bias which often prevents him . . . from recognizing the fact that claims of individuals whom he supposes to have been wronged have, before his intervention, been fully ventilated, discussed, and settled by the general award of scientific men. Does Prof. Tyndall know that Mayer's paper has *no claim to novelty or correctness at all*, saving this, that by a lucky analogy he got an approximation to a true result from an *utterly false analogy*?"

Even if the polemics had been avoided, any attempt to resolve Joule's and Mayer's conflicting claims would have been inconclusive. If the aim of the debate was to identify once and for all the discoverer of the energy concept, neither Joule nor Mayer should have won the contest. The story of the energy concept does not end, nor does it even begin, with Mayer's speculations and Joule's experimental facts. Several of Kuhn's simultaneous discoverers were earlier, although more tentative, than Joule and Mayer. In the late 1840s, after both men had made their most important contributions, the energy concept was still only about half understood; the modern distinction between the terms *force* and *energy* had not even been made clear. Helmholtz, Clausius, and Thomson still had fundamentally important contributions to make.

Those who spend their time fighting priority wars should forget their individual claims and learn to appreciate a more important aspect of the sociology of science: that the scientific community, with all its diversity cutting across race, class, and nationality, can, as often as it does, arrive at a consensus acceptable to all. The final judgment in the Joule-Mayer controversy teaches this lesson. In 1870, almost a decade after the last Tyndall or Tait outburst, the Royal Society awarded its prestigious Copley medal to Joule—and a year later to Mayer.

5

A Holy Undertaking

James Joule 1818–1889

The Scientist as Amateur

James Joule's story may seem a little hard to believe. He lived near Manchester, England—in the scientific hinterland during much of Joule's career—where his family operated a brewery, making ale and porter. He did some of his most important work in the early morning and evening, before and after a day at the brewery. He had no university education, and hardly any formal training at all in science. As a scientist he was, in every way, an amateur. Like Mayer, who was also an amateur as a physicist, Joule was ignored at first by the scientific establishment. Yet, despite his amateur status, isolation, and neglect, he managed to probe more deeply than anyone else at the time (the early and middle 1840s) the tantalizing mysteries of conversion processes. And (unlike Mayer) he did not suffer prolonged neglect. The story of Joule's rapid progress, from dilettante to a position of eminence in British science, can hardly be imagined in today's world of research factories and prolonged scientific apprenticeships.

Equivalences

The theme that dominated Joule's research from beginning to end, and served as his guiding theoretical inspiration, was the belief that quantitative equivalences could be found among thermal, chemical, electrical, and mechanical effects. He was convinced that the extent of any one of these effects could be assessed with the units of any one of the other effects. He studied such quantitative connections in no less than eight different ways: in investigations of chemical effects converted to thermal, electrical, and mechanical effects; of electrical effects converted to thermal, chemical, and mechanical effects; and of mechanical effects converted to thermal and electrical effects.

At first, Joule did not fully appreciate the importance of mechanical effects in this scheme of equivalences. His earliest work centered on chemical, electrical, and thermal effects. In 1840, when he was twenty-two, he started a series of five investigations that was prompted by his interest in electrochemistry. (Joule was an electrochemist before he was a physicist.) First, he demonstrated accurately that the heating produced by an electrical current in a wire is proportional to the square of the current I and to the electrical resistance R—the "I^2R-heating law." His experimental proof required temperature measurements in a "calorimeter" (a well-insulated, well-stirred vessel containing water or some other liquid), electrical current measurements with an instrument of his own design, and the invention of a system of absolute electrical units.

Joule then invested considerable effort in various studies of the role played by his heating law in the chemical processes produced in electric cells. He worked with "voltaic cells," which supply an electrical *output* (the modern flashlight battery is an example), and "electrolysis cells," which consume an electrical *input* (for example, a cell that decomposes water into hydrogen gas and oxygen gas). In these experiments, Joule operated an electrolysis cell with a battery of voltaic cells. He eventually arrived at the idea that the electrical currents generated by the chemical reaction in the voltaic cell carried the reaction's "calorific effect" or "chemical heat" away from the primary reaction site either to an external resistance where it could be converted to "free heat," according to the I^2R-heating law, or to an electrolysis cell where it could be invested, all or partly, as "latent heat" in the electrolysis reaction.

To determine the total chemical heat delivered to the electrolysis cell from the voltaic cells, call it Q_e, Joule found the resistance R_e of a wire that could replace the electrolysis cell without causing other electrical changes, measured the current I in the wire, and calculated Q_e with the heating law as I^2R_e. He also measured the temperature rise in the electrolysis cell doubling as a calorimeter, and from it calculated the free heat Q_t generated in the cell. He always found that Q_e substantially exceeded Q_t; in extreme cases, there was no heating in the cell and Q_t was equal to zero. The difference $Q_e - Q_t$ represented what Joule wanted to calculate: chemical heat converted to the latent heat of the electrolysis reaction. Representing the electrolysis reaction's latent heat with Q_r, Joule's calculation was

$$Q_r = Q_e - Q_t.$$

This is the statement Joule used in 1846 to determine several latent heats of electrolysis reactions with impressive accuracy. It is a complicated and exact application of the first law of thermodynamics, which Joule seems to have understood in terms of inputs and outputs to the electrolysis cell. That is evident in the last equation rearranged to

$$Q_t = Q_e - Q_r,$$

with Q_e an input to the cell, Q_r an output because it is lost to the reaction, and Q_t the difference between the input and output (see fig. 5.1). This was a balancing or bookkeeping kind of calculation, and it implied a conservation assumption: the balanced entity could not be created or destroyed *within* the cell. Joule did not have a name for the conserved entity. It would be identified six years later by Rudolf Clausius and William Thomson, and called "energy" by Thomson. Although he had not arrived at the energy concept, Joule clearly did have, well

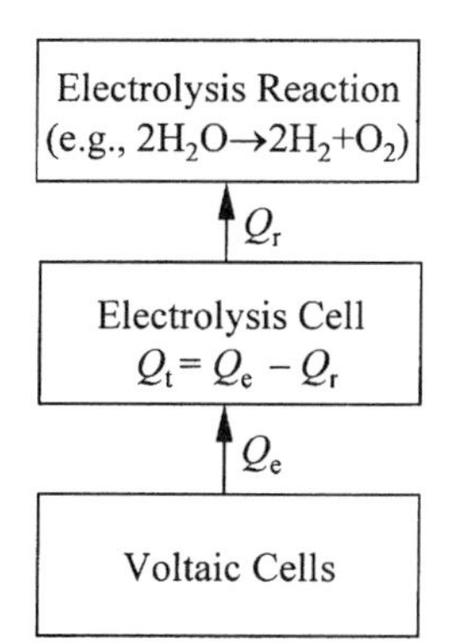

Figure 5.1. Input to and output from an electrolysis cell, according to Joule. The measured free heat Q_t in the cell depends only on the input Q_e from the voltaic cell and the output Q_r to the electrolysis reaction. It is equal to the input Q_e minus the output Q_r, that is, $Q_t = Q_e - Q_r$.

ahead of his contemporaries, a working knowledge of the first law of thermodynamics.

Joule's electrochemistry papers aroused little interest when they were first published, neither rejection nor acceptance, just silence. One reason for the indifference must have been the extraordinary nature of Joule's approach. The input-output calculation was difficult enough to comprehend at the time, but in addition to that, Joule used his measured heats of electrolysis reactions to calculate heats of combustion reactions (that is, reactions with oxygen gas). For example, he obtained an accurate heat for the hydrogen combustion reaction,

$$2\,H_2 + O_2 \rightarrow 2\,H_2O,$$

which is just the reverse of the water electrolysis reaction,

$$2\,H_2O \rightarrow 2\,H_2 + O_2,$$

and therefore, Joule assumed, its heat had the same magnitude as that of the electrolysis reaction.

This was an exotic way to study a combustion reaction. Joule's first biographer, Osborne Reynolds, remarks that "the views they [the electrochemistry papers and others of Joule's early papers] contained were so much in advance of anything accepted at the time that no one had sufficient confidence in his own opinion or was sufficiently sure of apprehending the full significance of the discoveries on which these views were based, to venture an expression of acceptance or rejection." We can imagine a contemporary reader puzzling over the papers and finally deciding that the author was either a genius or a crank.

But for Joule—apparently unconcerned about the accessibility or inaccessibility of his papers for readers—the complicated method was natural. His primary interest at the time was the accurate determination of equivalences among thermal, electrical, and chemical effects. He could imagine no better way to tackle this problem than to use electrical and calorimetric measurements to calculate the thermal effect of a chemical effect.

Mechanical Equivalents

Joule made the crucial addition of mechanical effects to his system of equivalences by following a time-honored route to scientific discovery: he made a fortunate mistake. In the fourth of his electrochemistry papers he reported electric potential data (voltages, in modern units) measured on voltaic cells whose electrode reactions produced oxidation of zinc and other metals. He believed, mistakenly, that these reaction potentials could be used in much the same way as reaction heats: that for a given reaction the potential had the same value no matter how the reaction was carried out. This interpretation is not sanctioned by modern thermodynamics unless cell potentials are measured carefully (reversibly). Joule and his contemporaries were unaware of this limitation, however, and the mistake led Joule to calculate electrical and thermal equivalents for the process in which dissolved oxygen is given "its elastic condition," the reaction

$$O_2 \text{ (solution)} \rightarrow O_2 \text{ (gas)}.$$

Joule's result was an order of magnitude too large. But mistaken as it was quantitatively, the calculation advanced Joule's conceptual understanding immensely, because he believed he had obtained electrical and thermal equivalents for a *mechanical* effect, the evolution of oxygen gas from solution. In Joule's fertile imagination, this was suggestive. In the fourth electrochemistry paper, he remarked that he had already thought of ways to measure mechanical equivalents. He hoped to confirm the conclusion that "the mechanical and heating powers of a current are proportional to each other."

In this serendipitous way, Joule began the determinations of the mechanical equivalent of heat for which he is best known today. The first experiments in this grand series were performed in 1843, when Joule was twenty-four. In these initial experiments, he induced an electrical current in a coil of wire by rotating it mechanically in a strong magnetic field. The coil was contained in a glass tube filled with water and surrounded by insulation, so any heating in the coil could be measured by inserting a thermometer in the tube before and after rotating it in the magnetic field. The induced current in the coil was measured by connecting the coil to an external circuit containing a galvanometer. Although its origin was entirely different, the induced current behaved the same way as the voltaic current Joule had studied earlier: in both cases the current caused heating that followed the I^2R-law.

In the final experiments of this design, the wheel of the induction device was driven by falling weights for which the mechanical effect, measured as a mechanical work calculation, could be made directly in foot-pounds (abbreviated ft-lb): one unit was equivalent to the work required to raise one pound one foot. Heat was measured by a unit that fit the temperature measurements: one unit raised the temperature of one pound of water 1° Fahrenheit (F). We will use the term later attached to this unit, "British thermal unit," or Btu.

In one experiment, Joule dropped weights amounting to 4 lb 12 oz (= 4.75 lb) 517 feet (the weights were raised and dropped many times), causing a temperature rise of 2.46° F. He converted the weight of the glass tube, wire coil, and water in which the temperature rise occurred all into a thermally equivalent weight of water, 1.114 lb. Thus the heating effect was 2.46° F in 1.114 lb of water. If this same amount of heat had been generated in 1 lb of water, the heating effect would have been $\frac{(2.46)(1.114)}{1} = 2.74°$ F. Joule concluded that in this case (517)(4.75) ft-lb was equivalent to 2.74 Btu. He usually determined the mechanical work equivalent to 1 Btu. That number, which Thomson later labeled J to honor Joule, was

$$J = \frac{(4.75)(517)}{(2.74)} = 896 \text{ ft-lb per Btu}$$

for this experiment. This was one determination of the mechanical equivalent of heat. Joule did thirteen experiments of this kind and obtained results ranging from $J = 587$ to 1040 ft-lb per Btu, for which he reported an average value of 838 ft-lb per Btu. The modern "correct" value, it should be noted, is $J = 778$ ft-lb per Btu.

If the $\pm 27\%$ precision achieved by Joule in these experiments does not seem impressive, one can sympathize with Joule's critics, who could not believe his claims concerning the mechanical equivalent of heat. But the measurements Joule was attempting set new standards for experimental difficulty. According to Reynolds, the 1843 paper reported experiments that were more demanding than any previously attempted by a physicist.

In any case, Joule was soon able to do much better. In 1845, he reported another, much different determination of the mechanical equivalent of heat, which agreed surprisingly well with his earlier measurement. In this second series of experiments, he measured temperature changes, and calculated the heat produced, when air was compressed. From the known physical behavior of gases he could calculate the corresponding mechanical effect as work done on the air during the compression.

In one experiment involving compression of air, Joule calculated the work at 11230 ft-lb and a heating effect of 13.628 Btu from a measured temperature rise of 0.344°F. The corresponding mechanical equivalent of heat was

$$J = \frac{11230}{13.628} = 824 \text{ ft-lb per Btu.}$$

Another experiment done the same way, in which Joule measured the temperature change 0.128°F, gave the result $J = 796$ ft-lb per Btu. Joule's average for the two experiments was 810 ft-lb per Btu. This was in impressive, if somewhat fortuitous, agreement with the result $J = 838$ ft-lb per Btu reported in 1843.

Joule also allowed compressed air to expand and do work against atmospheric pressure. Temperature measurements were again made, this time with a temperature decrease being measured. In one of these expansion experiments, Joule measured the temperature change −0.1738°F and reduced this to 4.085 Btu. The corresponding work calculation gave 3357 ft-lb, so

$$J = \frac{3357}{4.085} = 822 \text{ ft-lb per Btu.}$$

Joule did two more experiments of this kind and measured the temperature changes −0.081°F and −0.0855°F, giving $J = 814$ and $J = 760$ ft-lb per Btu.

When Joule's colleagues looked at these results, the first thing they noticed was the accuracy claimed for measurements of very small temperature changes. In Joule's time, accurate measurement of one-degree temperature changes was difficult enough. Joule reported temperature changes of tenths of a degree with three or four significant digits, and based his conclusions on such tiny changes. As William Thomson remarked, "Joule had nothing but hundredths of a degree to prove his case by." Yet, most of Joule's claims were justified. He made temperature measurements with mercury thermometers of unprecedented sensitivity and accuracy. He told the story of the thermometers in an autobiographical note: "It was needful in these experiments to use thermometers of greater exactness and delicacy than any that could be purchased at that time. I therefore determined to get some calibrated on purpose after the manner they had been by Regnault. In this I was ably seconded by Mr. Dancer [J. B. Dancer, a well-known Manchester instrument maker], at whose workshop I attended every morning for some time until we completed the first accurate thermometers which were ever made in England."

Joule demonstrated the heat-mechanical-work equivalence with a third gas expansion experiment that incorporated one of his most ingenious experimental designs. In this experiment, two constant-volume copper vessels, one evacuated and the other pressurized with air, were connected with a valve. The connected vessels were placed in a calorimeter, the valve opened, and the usual temperature measurements made. In this case, Joule could detect *no* net temperature change. Air expanding from the pressurized vessel was cooled slightly, and air flowing into the evacuated vessel was slightly heated, but no net temperature change was observed.

This was what Joule expected. Because the combined system consisting of the two connected vessels was closed and had a fixed volume, all of the work was done internally, in tandem between the two vessels. Work done *by* the gas in one vessel was balanced by work done *on* the gas in the other; no net work was done. Heat equivalent to zero work was also zero, so Joule's concept of heat-mechanical-work equivalence demanded that the experiment produce no net thermal effect, as he observed.

The next stage in Joule's relentless pursuit of an accurate value for the mechanical equivalent of heat, which he had begun in 1847, was several series of experiments in which he measured heat generated by various frictional processes. The frictional effects were produced in a water-, mercury-, or oil-filled calorimeter by stirring with a paddle-wheel device, the latter being driven by falling weights, as in the 1843 experiments. The work done by the weights was converted directly by the paddle-wheel stirrer into heat, which could be measured on a thermometer in the calorimeter.

Of all Joule's inventions, this experimental design, which has become the best-known monument to his genius, made the simplest and most direct demonstration of the heat-mechanical-work equivalence. This was the Joule technique reduced to its essentials. No complicated induction apparatus was needed, no calculational approximations, just falling weights and one of Joule's amazingly accurate thermometers.

With the paddle-wheel device and water as the calorimeter liquid, Joule obtained $J = 773.64$ ft-lb per Btu from a temperature rise of 0.563°F. Using mercury in the calorimeter, he obtained $J = 773.762$ and 776.303 ft-lb per Btu. In two further series of experiments, Joule arranged his apparatus so the falling weights caused two cast-iron rings to rub against each other in a mercury-filled calorimeter; the results $J = 776.997$ and 774.880 ft-lb per Btu were obtained.

Joule described his paddle-wheel experiments in 1847 at an Oxford meeting of the British Association for the Advancement of Science. Because his previous papers had aroused little interest, he was asked to make his presentation as brief as possible. "This I endeavored to do," Joule recalled later, "and a discussion not being invited the communication would have passed without comment if a young man had not risen in the section, and by his intelligent observations created a lively interest in the new theory."

The silence was finally broken. The young man was William Thomson, recently installed as professor of natural philosophy at Glasgow University. Thomson had reservations about Joule's work, but he also recognized that it could not be ignored. "Joule is, I am sure, wrong in many of his ideas," Thomson wrote to his father, "but he seems to have discovered some facts of extreme importance, as for instance, that heat is developed by the friction of fluids." Thomson recalled in 1882 that "Joule's paper at the Oxford meeting made a great sensation. Faraday was there, and was much struck by it, but did not enter fully into the new views. . . . It was not long after when Stokes told me he was inclined to be a Joulite." George Stokes was another rising young physicist and mathematician, in 1847 a fellow at Pembroke College, Cambridge, and in two years to be appointed Lucasian Professor of Mathematics, the chair once occupied by Newton.

During the three years following the Oxford meeting, Joule rose from obscurity to a prominent position in the British scientific establishment. Recognition came first from Europe: a major French journal, *Comptes Rendu*, published a short account of the paddle-wheel experiments in 1847, and in 1848 Joule was elected a corresponding member of the Royal Academy of Sciences at Turin. Only two other British scientists, Faraday and William Herschel, had been honored by the Turin Academy. In 1850, when he was thirty-one, Joule received the badge of British scientific acceptance: election as a fellow of the Royal Society.

After these eventful years, Joule's main research effort was a lengthy collaboration with Thomson, focusing on the behavior of expanding gases. This was one of the first collaborative efforts in history in which the talents of a theorist and those of an experimentalist were successfully and happily united.

Living Force and Heat

Joule believed that water at the bottom of a waterfall should be slightly warmer than water at the top, and he made attempts to detect such effects (even on his honeymoon in Switzerland, according to an apocryphal, or at any rate embellished, story told by Thomson). For Joule this was an example of the conservation principle that "heat, living force, and attraction through space . . . are mutually convertible into one another. In these conversions nothing is ever lost." This statement is almost an expression of the conservation of mechanical and thermal energy, but it requires some translation and elaboration.

Newtonian mechanics implies that mechanical energy has a "potential" and a "kinetic" aspect, which are linked in a fundamental way. "Potential energy" is evident in a weight held above the ground. The weight has energy because work was required to raise it, and the work can be completely recovered by letting the weight fall very slowly and drive machinery that has no frictional losses. As one might expect, the weight's potential energy is proportional to its mass and to its height above the ground: if it starts at a height of 100 feet it can do twice as much work as it can if it starts at 50 feet.

If one lets the weight fall freely, so that it is no longer tied to machinery, it does no work, but it accelerates and acquires "kinetic energy" from its increasing speed. Kinetic energy, like potential energy, can be converted to work with the right kind of machinery, and it is also proportional to the mass of the weight. Its relationship to speed, however, as dictated by Newton's second law of motion, is to the *square* of the speed.

In free fall, the weight has a mechanical energy equal to the sum of the kinetic and potential energies,

$$\text{mechanical energy} = \text{kinetic energy} + \text{potential energy}. \tag{1}$$

As it approaches the ground the freely falling weight loses potential energy, and at the same time, as it accelerates, it gains kinetic energy. Newton's second law informs us that the two changes are exactly compensating, and that the total mechanical energy is conserved, if we define

$$\text{kinetic energy} = \frac{mv^2}{2} \tag{2}$$

$$\text{potential energy} = mgz. \tag{3}$$

In equations (2) and (3), m is the mass of the weight, v its speed, z its distance above the ground, and g the constant identified above as the gravitational acceleration. If we represent the total mechanical energy as E, equation (1) becomes

$$E = \frac{mv^2}{2} + mgz, \tag{4}$$

and the conservation law justified by Newton's second law guarantees that E is always constant. This is a conversion process, of potential energy to kinetic energy, as illustrated in figure 5.2. In the figure, before the weight starts falling it has 10 units of potential energy and no kinetic energy. When it has fallen halfway to the ground, it has 5 units of both potential and kinetic energy, and in the instant before it hits the ground it has no potential energy and 10 units of kinetic energy. At all times its total mechanical energy is 10 units.

Joule's term "living force" (or *vis viva* in Latin) denotes mv^2, almost the same

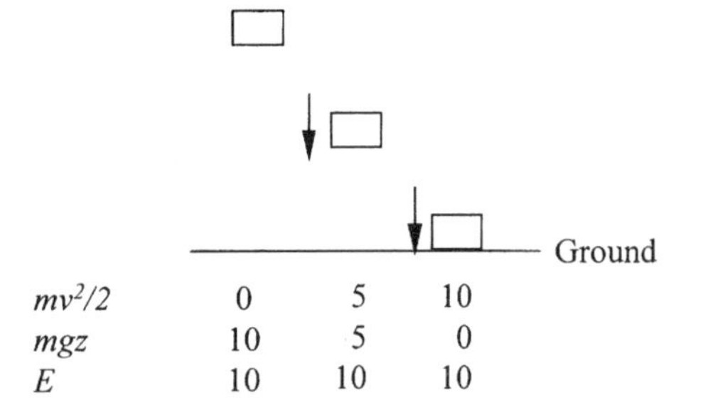

Figure 5.2. Illustration of the conversion of potential energy to kinetic energy by a freely falling weight, and the conservation of total mechanical energy.

thing as the kinetic energy $\frac{mv^2}{2}$, and his phrase "attraction through space" means the same thing as potential energy. So Joule's assertions that living force and attraction through space are interconvertible and that nothing is lost in the conversion are comparable to the Newtonian conservation of mechanical energy. Water at the top of the falls has potential energy only, and just before it lands in a pool at the bottom of the falls, it has kinetic energy only. An instant later the water is sitting quietly in the pool, and according to Joule's principle, with the third conserved quantity, heat, included, the water is warmer because its mechanical energy has been converted to heat. Joule never succeeded in confirming this waterfall effect. The largest waterfall is not expected to produce a temperature change of more than a tenth of a degree. Not even Joule could detect that on the side of a mountain.

Joule's mechanical view of heat led him to believe further that in the conversion of the motion of an object to heat, the motion is not really lost because heat is itself the result of motion. He saw heat as the internal, random motion of the constituent particles of matter. This general idea had a long history, going back at least to Robert Boyle and Daniel Bernoulli in the seventeenth century.

Joule pictured the particles of matter as atoms surrounded by rapidly rotating "atmospheres of electricity." The centrifugal force of the atmospheres caused a gas to expand when its pressure was decreased or its temperature increased. Mechanical energy converted to heat became rotational motion of the atomic atmospheres. These speculations of Joule's mark the beginning of the development of what would later be called the "molecular (or kinetic) theory of gases." Following Joule, definitive work in this field was done by Clausius, Maxwell, and Boltzmann.

A Joule Sketch

Osborne Reynolds, who met Joule in 1869, gives us this impression of his manner and appearance in middle age: "That Joule, who was 51 years of age, was rather under medium height; that he was somewhat stout and rounded in figure; that his dress, though neat, was commonplace in the extreme, and that his attitude and movements were possessed of no natural grace, while his manner was somewhat nervous, and he possessed no great facility of speech, altogether conveyed an impression of simplicity, and utter absence of all affectation which had characterized his life."

Joule married Amelia Grimes in 1847, when he was twenty-nine and she thirty-three; they had two children, a son and a daughter. Amelia died in 1854, and "the shock took a long time to wear off," writes Joule's most recent biographer, Donald Cardwell. "His friends and contemporaries agreed that this never very assertive man became more withdrawn." About fourteen years later, Joule fell in love again, this time with his cousin Frances Tappenden, known as "Fanny." In a letter to Thomson he writes "an affection has sprung up between me and my cousin you saw when last here. There are hindrances in the way so that nothing may come of it." The "hindrances" prevented marriage, and eventually Fanny married another man.

Joule's political leanings were conservative. He had a passionate, sometimes irrational, dislike of reform-minded Liberal politicians such as William Gladstone and John Bright. In a letter to John Tyndall, he wrote, "The fact is that Mr. Gladstone was fashioning a neat machine of 'representation' with the object of keeping himself in power. . . . Posterity will judge him as the worst 'statesman' that England ever had and the verdict with regard to that Parliament will be ditto, ditto."

Joule had a personality that was "finely poised," as another biographer, J. G. Crowther, puts it. On the one hand he was conducting experiments with unlimited care and patience, and on the other hand fulminating against Liberal politicians. He feared that too much mental effort would threaten his health. In 1860, a new professorship of physics was created at Owens College in Manchester, and Joule could have had it, but he decided not to apply, as he explained in a letter to Thomson: "I have not the courage to apply for the Owens professorship. The fact is that I do not feel it would do for me to overtask my brain. A few years ago, I felt a very small mental effort too much for me, and in consequence spared myself from thought as much as possible. I have felt a gradual improvement, but I do not think it would be well for me to build too much on it. I shall do a great deal more in the long run by taking things easily."

Joule's life was hectic and burdensome at this time, and he may have felt that he was near breakdown. Amelia died in 1854, the brewery was sold in the same year, and the experiments with Thomson were in progress. During the next six years, he moved his household and laboratory twice. After the second move, he was upset by an acrimonious dispute with a neighbor who objected to the noise and smoke made by a three-horsepower steam engine Joule included in his apparatus. The neighbor was "a Mr Bowker, an Alderman of Manchester and chairman of the nuisances committee, a very important man in his own estimation like most people who have risen from the dregs of society."

During this same period, Joule narrowly escaped serious injury in a train wreck, and after that he had an almost uncontrollable fear of railway travel. At the same time, he loved to travel by sea, even when it was dangerous. In a letter to Fanny, he described a ten-mile trip to Tory Island, in the Atlantic off the coast of Ireland, where his brother owned property: "Waves of 4 to 600 feet from crest to crest and 20 feet high. Dr Brady who was with us and had yachted in the ocean for 25 years said he was never in a more dangerous sea. However the magnificence of it took away the disagreeable sense of danger which might have prevailed."

In some measures of scientific ability, Joule was unimpressive. As a theorist, he was competent but not outstanding. He was not an eloquent speaker, and he was not particularly important in the scientific establishment of his time. But Joule had three things in extraordinary measure—experimental skill, independence, and inspiration.

He was the first to understand that unambiguous equivalence principles could be obtained only with the most inspired attention to experimental accuracy. He accomplished his aim by carefully selecting the measurements that would make his case. Crowther marvels at the directness and simplicity of Joule's experimental strategies: "He did not separate a quantity of truth from a large number of groping unsuccessful experiments. Nearly all of his experiments seem to have been perfectly conceived and executed, and the first draft of them could be sent almost without revision to the journals for publication."

For most of his life, Joule had an ample independent income. That made it possible for him to pursue a scientific career privately, and to build the kind of intellectual independence he needed. Crowther tells us about this facet of Joule's background:

> As a rich young man he needed no conventional training to qualify him for a career, or introduce him to powerful future friends. His early researches were pursued partly in the spirit of a young gentleman's entertainment, which happened to be science instead of fighting or politics or gambling. It is difficult to believe that any student who had received a lengthy academic training could have described researches in Joule's tone of intellectual equality. The gifted student who has studied under a great teacher would almost certainly adopt a less independent tone in his first papers, because he would have the attitude of a pupil to his senior, besides a deference due to appreciation of his senior's achievements. A student without deference after distinguished tuition is almost always mediocre.

Joule was not entirely without distinguished tuition. Beginning in 1834, and continuing for three years, Joule and his brother Benjamin studied with John Dalton, then sixty-eight and, as always, earning money teaching children the rudiments of science and mathematics. The Joules' studies with Dalton were not particularly successful pedagogically. Dalton took them through arithmetic and geometry (Euclid) and then proceeded to higher mathematics, with little attention to physics and chemistry. Dalton's syllabus did not suit Joule, but he benefited in more-informal ways. Joule wrote later in his autobiographical note, "Dalton possessed a rare power of engaging the affection of his pupils for scientific truth; and it was from his instruction that I first formed a desire to increase my knowledge by original researches." In his writings, if not in his tutoring, Dalton emphasized the ultimate importance of accurate measurements in building the foundations of physical science, a lesson that Joule learned and used above all others. The example of Dalton, internationally famous for his theories of chemical action, yet self-taught, and living and practicing in Manchester, must have convinced Joule that he, too, had prospects.

Joule's independence and confidence in his background and talents, natural or learned from Dalton, were tested many times in later years, but never shaken. His first determination, in 1843, of the mechanical equivalent of heat was ignored, and subsequent determinations were given little attention until Thomson and Stokes took notice at the British Association meeting in 1847.

When Joule submitted a summary of his friction experiments for publication, he closed the paper with three conclusions that asserted the heat-mechanical-work equivalence in the friction experiments, quoted his measured value of J, and stated that "the friction consisted in the conversion of mechanical power to heat." The referee who reported on the paper (believed to have been Faraday) requested that the third conclusion be suppressed.

Joule's first electrochemistry paper was rejected for publication by the Royal Society, except as an abstract. Arthur Schuster reported that, when he asked Joule what his reaction was when this important paper was rejected, Joule's reply was characteristic: "I was not surprised. I could imagine those gentlemen sitting around a table in London and saying to each other: 'What good can come out of a town [Manchester] where they dine in the middle of the day?' "

But with all his talents, material advantages, and intellectual independence,

Joule could never have accomplished what he did if he had not been guided in his scientific work by inspiration of an unusual kind. For Joule "the study of nature and her laws" was "essentially a holy undertaking." He could summon the monumental patience required to assess minute errors in a prolonged series of measurements, and at the same time transcend the details and see his work as a quest "for acquaintance with natural laws . . . no less than an acquaintance with the mind of God therein expressed." Great theorists have sometimes had thoughts of this kind—one might get the same meaning from Albert Einstein's remark that "the eternal mystery of the world is its comprehensibility"—but experimentalists, whose lives are taken up with the apparently mundane tasks of reading instruments and designing apparatuses, have rarely felt that they were communicating with the "mind of God."

It would be difficult to find a scientific legacy as simple as Joule's, and at the same time as profoundly important in the history of science. One can summarize Joule's major achievement with the single statement

$$J = 778 \text{ ft-lb per Btu},$$

and add that this result was obtained with extraordinary accuracy and precision. This is Joule's monument in the scientific literature, now quoted as 4.1840 kilogram-meters per calorie, used routinely and unappreciatively by modern students to make the quantitative passage from one energy unit to another.

In the 1840s, Joule's measurements were far more fascinating, or disturbing, depending on the point of view. The energy concept had not yet been developed (and would not be for another five or ten years), and Joule's number had not found its niche as the hallmark of energy conversion and conservation. Yet Joule's research made it clear that something *was* converted and conserved, and provided vital clues about what the something was.

6

Unities and a Unifier

Hermann Helmholtz 1821–1894

Unifiers and Diversifiers

Science is largely a bipartisan endeavor. Most scientists have no difficulty identifying with one of two camps, which can be called, with about as much accuracy as names attached to political parties, theorists and experimentalists. An astute observer of scientists and their ways, Freeman Dyson, has offered a roughly equivalent, but more inspired, division of scientific allegiances and attitudes. In Dyson's view, science has been made throughout its history in almost equal measure by "unifiers" and "diversifiers." The unifiers, mostly theorists, search for the principles that reveal the unifying structure of science. Diversifiers, likely to be experimentalists, work to discover the unsorted facts of science. Efforts of the scientific unifiers and diversifiers are vitally complementary. From the great bodies of facts accumulated by the diversifiers come the unifier's theories; the theories guide the diversifiers to new observations, sometimes with disastrous results for the unifiers.

The thermodynamicists celebrated here were among the greatest scientific unifiers of the nineteenth and early twentieth centuries. Three of their stories have been told above: of Sadi Carnot and his search for unities in the bewildering complexities of machinery; of Robert Mayer and his grand speculations about the energy concept; of James Joule's precise determination of equivalences among thermal, electrical, chemical, and mechanical effects. Continuing now with the chronology, we focus on the further development of the energy concept. The thermodynamicist who takes the stage is Hermann Helmholtz, the most confirmed of unifiers.

Medicine and Physics

Helmholtz, like Mayer, was educated for a medical career. He would have preferred to study physics and mathematics, but the only hope for scientific training, given his father's meager salary as a gymnasium teacher, was a government scholarship in medicine. With the scholarship, Helmholtz studied at the Friedrich-Wilhelm Institute in Berlin and wrote his doctoral dissertation under Johannes Müller. At that time, Müller and his circle of gifted students were laying the groundwork for a physical and chemical approach to the study of physiology, which was the beginning of the disciplines known today as biophysics and biochemistry. Müller's goal was to rid medical science of all the metaphysical excesses it had accumulated, and retain only those principles with sound empirical foundations. Helmholtz joined forces with three of Müller's students, Emil du Bois-Reymond, Ernst Brücke, and Carl Ludwig; the four, known later as the "1847 group," pledged their talents and careers to the task of reshaping physiology into a physicochemical science.

Die Erhaltung der Kraft

If medicine was not Helmholtz's first choice, it nevertheless served him (and he served medicine) well, even when circumstances were trying. His medical scholarship stipulated eight years of service as an army surgeon. He took up this service without much enthusiasm. Life as surgeon to the regiment at Potsdam offered little of the intellectual excitement he had found in Berlin. But to an extraordinary degree, Helmholtz had the ability to supply his own intellectual stimulation. Although severely limited in resources, and unable to sleep after five o'clock in the morning when the bugler sounded reveille at his door, he quickly started a full research program concerned with such topics as the role of metabolism in muscle activity, the conduction of heat in muscle, and the rate of transmission of the nervous impulse.

During this time, while he was mostly in scientific isolation, Helmholtz wrote the paper on energy conservation that brings him to our attention as one of the major thermodynamicists. (Once again, as in the stories of Carnot, Mayer, and Joule, history was being made by a scientific outsider.) Helmholtz's paper had the title *Über die Erhaltung der Kraft* (On the Conservation of Force), and it was presented to the Berlin Physical Society, recently organized by du Bois-Reymond, and other students of Müller's, and Gustav Magnus, in July 1847.

As the title indicates, Helmholtz's 1847 paper was concerned with the concept of "force"—in German, "Kraft"—which he defined as "the capacity [of matter] to produce effects." He was concerned, as Mayer before him had been, with a composite of the modern energy concept (not clearly defined in the thermodynamic context until the 1850s) and the Newtonian force concept. Some of Helmholtz's uses of the word "Kraft" can be translated as "energy" with no confusion. Others cannot be interpreted this way, especially when directional properties are assumed, and in those instances "Kraft" means "force," with the Newtonian connotation.

Helmholtz later wrote that the original inspiration for his 1847 paper was his reaction as a student to the concept of "vital force," current at the time among physiologists, including Müller. The central idea, which Helmholtz found he could not accept, was that life processes were controlled not only by physical and chemical events, but also by an "indwelling life source, or vital force, which controls the activities of [chemical and physical] forces. After death the free action of [the] chemical and physical forces produces decomposition, but during life their action is continually being regulated by the life soul." To Helmholtz this was metaphysics. It seemed to him that the vital force was a kind of biological perpetual motion. He knew that physical and chemical processes did not permit perpetual motion, and he felt that the same prohibition must be extended to all life processes.

Helmholtz also discussed in his paper what he had learned about mechanics from seventeenth- and eighteenth-century authors, particularly Daniel Bernoulli and Jean d'Alembert. It is evident from this part of the paper that a priori beliefs are involved, but the most fundamental of these assumptions are not explicitly stated. The science historian Yehuda Elkana fills in for us what was omitted: "Helmholtz was very much committed—a priori—to two fundamental beliefs: (a) that all phenomena in physics are reducible to mechanical processes (no one who reads Helmholtz can doubt this), and (b) that there be some basic entity in Nature which is being conserved ([although] this does not appear in so many words in Helmholtz's work)." To bring physiology into his view, a third belief was needed, that "all organic processes are reducible to physics." These general ideas were remarkably like those Mayer had put forward, but in 1847 Helmholtz had not read Mayer's papers.

Helmholtz's central problem, as he saw it, was to identify the conserved entity. Like Mayer, but independently of him, Helmholtz selected the quantity "Kraft" for the central role in his conservation principle. Mayer had not been able to avoid the confused dual meaning of "Kraft" adopted by most of his contemporaries. Helmholtz, on the other hand, was one of the first to recognize the ambiguity. With his knowledge of mechanics, he could see that when "Kraft" was cast in the role of a conserved quantity, the term could no longer be used in the sense of Newtonian force. The theory of mechanics made it clear that Newtonian forces were not in any general way conserved quantities.

This reasoning brought Helmholtz closer to a workable identification of the elusive conserved quantity, but he (and two other eminent thermodynamicists, Clausius and Thomson) still had some difficult conceptual ground to cover. He could follow the lead of mechanics, note that mechanical energy had the conservation property, and assume that the conserved quantity he needed for his principle had some of the attributes (at least the units) of mechanical energy. Helmholtz seems to have reasoned this way, but there is no evidence that he got any closer than this to a full understanding of the energy concept. In any case, his message, as far as it went, was important and eventually accepted. "After [the 1847 paper]," writes Elkana, "the concept of energy underwent the fixing stage; the German 'Kraft' came to mean simply 'energy' (in the conservation context) and later gave place slowly to the expression 'Energie.' The Newtonian 'Kraft' with its dimensions of mass times acceleration became simply our 'force.' "

I have focused on the central issue taken up by Helmholtz in his 1847 paper. The paper was actually a long one, with many illustrations of the conservation principle in the physics of heat, mechanics, electricity, magnetism, and (briefly, in a single paragraph) physiology.

Pros and Cons

Helmholtz's youthful effort in his paper (he was twenty-six in 1847), read to the youthful members of the Berlin Physical Society, was received with enthusiasm. Elsewhere in the scientific world the reception was less favorable. Helmholtz submitted the paper for publication to Poggendorff's *Annalen*, and, like Mayer five years earlier, received a rejection. Once again an author with important things to say about the energy concept had to resort to private publication. With du Bois-Reymond vouching for the paper's significance, the publisher G. A. Reimer agreed to bring it out later in 1847.

Helmholtz commented several times in later years on the peculiar way his memoir was received by the authorities. "When I began the memoir," he wrote in 1881, "I thought of it only as a piece of critical work, certainly not as an original discovery. . . . I was afterwards somewhat surprised over the opposition which I met with among the experts . . . among the members of the Berlin academy only C. G. J. Jacobi, the mathematician, accepted it. Fame and material reward were not to be gained at that time with the new principle; quite the opposite." What surprised him most, he wrote in 1891 in an autobiographical sketch, was the reaction of the physicists. He had expected indifference ("We all know that. What is the young doctor thinking about who considers himself called upon to explain it all so fully?"). What he got was a sharp attack on his conclusions: "They [the physicists] were inclined to deny the correctness of the law . . . to treat my essay as a fantastic piece of speculation."

Later, after the critical fog had lifted, priority questions intruded. Mayer's papers were recalled, and obvious similarities between Helmholtz and Mayer were pointed out. Possibly because resources in Potsdam were limited, Helmholtz had not read Mayer's papers in 1847. Later, on a number of occasions, he made it clear that he recognized Mayer's, and also Joule's, priority.

The modern assessment of Helmholtz's 1847 paper seems to be that it was, in some ways, limited. It certainly did cover familiar ground (as Helmholtz had intended), but it did not succeed in building mathematical and physical foundations for the energy conservation principle. Nevertheless, there is no doubt that the paper had an extraordinary influence. James Clerk Maxwell, prominent among British physicists in the 1860s and 1870s, viewed Helmholtz's general program as a conscience for future developments in physical science. In an appreciation of Helmholtz, written in 1877, Maxwell wrote: "To appreciate the full scientific value of Helmholtz's little essay . . . we should have to ask those to whom we owe the greatest discoveries in thermodynamics and other branches of modern physics, how many times they have read it over, and how often during their researches they felt the weighty statements of Helmholtz acting on their minds like an irresistible driving-power."

What Maxwell and other physicists were paying attention to was passages such as this: "The task [of theoretical science] will be completed when the reduction of phenomena to simple forces has been completed and when, at the same time, it can be proved that the reduction is the only one which the phenomena will allow. This will then be established as the conceptual form necessary for understanding nature, and we shall be able to ascribe objective truth to it." To a large extent, this is still the program of theoretical physics.

Physiology

After 1847, Helmholtz was only intermittently concerned with matters relating to thermodynamics. His work now centered on medical science, specifically the physical foundations of physiology. He wanted to build an edifice of biophysics on the groundwork laid by Müller, his Berlin professor, and by his colleagues du Bois-Reymond, Ludwig, and Brücke, of the 1847 school. Helmholtz's rise in the scientific and academic worlds was spectacular. For six years, he was professor of physiology at Königsberg, and then for three years professor of physiology and anatomy at Bonn. From Bonn he went to Heidelberg, one of the leading scientific centers in Europe. During his thirteen years as professor of physiology at Heidelberg, he did his most finished work in biophysics. His principal concerns were theories of vision and hearing, and the general problem of perception. Between 1856 and 1867, he published a comprehensive work on vision, the three-volume *Treatise on Physiological Optics*, and in 1863, his famous *Sensations of Tone*, an equally vast memoir on hearing and music.

Helmholtz's work on perception was greatly admired during his lifetime, but more remarkable, for the efforts of a scientist working in a research field hardly out of its infancy, is the respect for Helmholtz still found among those who try to understand perception. Edward Boring, author of a modern text on sensation and perception, dedicated his book to Helmholtz and then explained: "If it be objected that books should not be dedicated to the dead, the answer is that Helmholtz is not dead. The organism can predecease its intellect, and conversely. My dedication asserts Helmholtz's immortality—the kind of immortality that remains the unachievable aspiration of so many of us."

Physics

By 1871, the year he reached the age of fifty, Helmholtz had accomplished more than any other physiologist in the world, and he had become one of the most famous scientists in Germany. He had worked extremely hard, often to the detriment of his mental and physical health. He might have decided to relax his furious pace and become an academic ornament, as others with his accomplishments and honors would have done. Instead, he embarked on a new career, and an intellectual migration that was, and is, unique in the annals of science. In 1871, he went to Berlin as professor of physics at the University of Berlin.

The conversion of the physiologist to the physicist was not a miraculous rebirth, however. Physics had been Helmholtz's first scientific love, but circumstances had dictated a career in medicine and physiology. Always a pragmatist, he had explored the frontier between physics and physiology, earned a fine reputation, and more than anyone else, established the new science of biophysics. But his fascination with mathematical physics, and his ambition, had not faded. With the death of Gustav Magnus, the Berlin professorship was open. Helmholtz and Gustav Kirchhoff, professor of physics at Heidelberg, were the only candidates; Kirchhoff preferred to remain in Heidelberg. "And thus," wrote du Bois-Reymond, "occurred the unparalleled event that a doctor and professor of physiology was appointed to the most important physical post in Germany, and Helmholtz, who called himself a born physicist, at length obtained a position suited to his specific talents and inclinations, since he had, as he wrote to me, become indifferent to physiology, and was really only interested in mathematical physics."

So in Berlin Helmholtz was a physicist. He focused his attention largely on the topic of electrodynamics, a field he felt had become a "pathless wilderness" of contending theories. He attacked the work of Wilhelm Weber, whose influence then dominated the theory of electrodynamics in Germany. Before most of his colleagues on the Continent, Helmholtz appreciated the studies of Faraday and Maxwell in Britain on electromagnetic theory. Heinrich Hertz, a student of Helmholtz's and later his assistant, performed experiments that proved the existence of electromagnetc waves and confirmed Maxwell's theory. Also included among Helmholtz's remarkable group of students and assistants were Ludwig Boltzmann, Wilhelm Wien, and Albert Michelson. Boltzmann was later to lay the foundations for the statistical interpretation of thermodynamics (see chapter 13). Wien's later work on heat radiation gave Max Planck, professor of theoretical physics at Berlin and a Helmholtz protégé, one of the clues he needed to write a revolutionary paper on quantum theory. Michelson's later experiments on the velocity of light provided a basis for Einstein's theory of relativity. Helmholtz, the "last great classical physicist," had gathered in Berlin some of the theorists and experimentalists who would discover a new physics.

A Dim Portrait

This has been a portrait of Helmholtz the scientist and famous intellect. What was he like as a human being? In spite of his extraordinary prominence, that question is difficult to answer. The authorized biography, by Leo Königsberger, is faithful to the facts of Helmholtz's life and work, but too admiring to be reliably whole in its account of his personal traits. Helmholtz's writings are not much help either, even though many of his essays were intended for lay audiences. His style is too severely objective to give more than an occasional glimpse of the feeling and inspiration he brought to his work. We are left with fragments of the human Helmholtz, and, like archaeologists, we must try to piece them together.

We know that Helmholtz had a marvelous scientific talent, and an immense capacity for hard work. Sessions of intense mental effort were likely to leave him exhausted and sometimes disabled with a migraine attack, but he always recovered, and throughout his life had the working habits of a workaholic.

He was blessed with two happy marriages. The death of his first wife, Olga, after she spent many years as a semiinvalid, left him incapacitated for months with headaches, fever, and fainting fits. As always, though, work was his tonic, and in less than two years he had married again. His second wife, Anna, was young and charming, "one of the beauties of Heidelberg," Helmholtz wrote to Thomson. She was a wife, wrote Königsberger, "who responded to all [of Helmholtz's] needs . . . a person of great force of character, talented, with wide views and high aspirations, clever in society, and brought up in a circle in which intelligence and character were equally well developed." Anna's handling of the household and her husband's rapidly expanding social commitments contributed substantially to the Helmholtz success story in Heidelberg and Berlin.

To achieve what he did, Helmholtz must have been intensely ambitious. Yet he seems to have traveled the road to success without pretension and with no question about his integrity, scientific or otherwise. Max Planck, a man whose opinion can be trusted on the subjects of integrity and intellectual leadership without pretension, wrote about his friendship with Helmholtz in the 1890s in Berlin:

> I learned to know Helmholtz . . . as a human being, and to respect him as a scientist. For with his entire personality, integrity of convictions and modesty of character, he was the very incarnation of the dignity and probity of science. These traits of character were supplemented by a true human kindness, which touched my heart deeply. When during a conversation he would look at me with those calm, searching, penetrating, and yet so benign eyes, I would be

> overwhelmed by a feeling of boundless filial trust and devotion, and I would feel that I could confide in him, without reservation, everything I had on my mind.

Others, who saw Helmholtz from more of a distance, had different impressions. Englebert Broda comments that Boltzmann "had the greatest respect for Helmholtz the universal scientist, [but] Helmholtz the man . . . left him cold." Among his students and lesser colleagues, Helmholtz was called the "Reich Chancellor of German Physics."

There can hardly be any doubt that Helmholtz had a passionate interest in scientific investigation and an encyclopedic grasp of the facts and principles of science. Yet something contrary in his character made it difficult for him to communicate his feelings and knowledge to a class of students. We are again indebted to Planck's frankness for this picture of Helmholtz in the lecture hall (in Berlin): "It was obvious that Helmholtz never prepared his lectures properly. He spoke haltingly, and would interrupt his discourse to look for the necessary data in his small notebook; moreover, he repeatedly made mistakes in his calculations at the blackboard, and we had the unmistakable impression that the class bored him at least as much as it did us. Eventually, his classes became more and more deserted, and finally they were attended by only three students; I was one of the three."

Helmholtz viewed scientific study in a special, personal way. The conventional generalities required by students in a course of lectures may not have been for him the substance of science. At any rate, Helmholtz was not the first famous scientist to fail to articulate in the classroom the fascination of science, and (as those who have served university scientific apprenticeships can attest) not the last.

The intellectual driving force of Helmholtz's life was his never-ending search for fundamental unifying principles. He was one of the first to appreciate that most impressive of all the unifying principles of physics, the conservation of energy. In 1882, he initiated one of the first studies in the interdisciplinary field that was soon to be called physical chemistry. His work on perception revealed the unity of physics and physiology. Beyond that, his theories of vision and hearing probed the aesthetic meaning of color and music, and built a bridge between art and science. He expressed, as few had before or have since, a unity of the subjective and the objective, of the aesthetic and the intellectual.

He had hoped to find a great principle from which all of physics could be derived, a unity of unities. He devoted many years to this effort; he thought that the "least-action principle," discovered by the Irish mathematician and physicist William Rowan Hamilton, would serve his grand purpose, but Helmholtz died before the work could be completed. At about the same time, Thomson was failing in an attempt to make his dynamical theory all-encompassing. In the twentieth century, Albert Einstein was unsuccessful in a lengthy attempt to formulate a unified theory of electromagnetism and gravity. In the 1960s, the particle physicists Sheldon Glashow, Abdus Salam, and Steven Weinberg developed a unified theory of electromagnetism and the nuclear weak force. The search goes on for still-broader theories, uniting atomic, nuclear, and particle physics with the physics of gravity. We can hope that these quests for a "theory of everything" will eventually succeed. But we may have to recognize that there are limits. Scientists may never see the day when the unifiers are satisfied and the diversifiers are not busy.

7

The Scientist as Virtuoso

William Thomson

A Problem Solver

William Thomson was many things—physicist, mathematician, engineer, inventor, teacher, political activist, and famous personality—but before all else he was a problem solver. He thrived on scientific and technological problems of all kinds. Whatever the problem, abstract or applied, Thomson usually had an original insight and a valuable solution. As a scientist and technologist, he was a virtuoso.

Even Helmholtz, another famous problem solver, was amazed by Thomson's virtuosic performances. After meeting Thomson for the first time, Helmholtz wrote to his wife, "He far exceeds all the great men of science with whom I have made personal acquaintance, in intelligence and lucidity, and mobility of thought, so that I felt quite wooden beside him sometimes." Helmholtz later wrote to his father, "He is certainly one of the first mathematical physicists of his day, with powers of rapid invention such as I have seen in no other man."

Thomson and Helmholtz became good friends, and in later years Thomson made their discussions on subjects of mutual interest into an extended competition, which we can assume Thomson usually won. On one occasion, when Helmholtz was visiting on board Thomson's sailing yacht in Scotland, the subject for marathon discussion was the theory of waves, which, as Helmholtz wrote (again in a letter to his wife), "he loved to treat as a kind of race between us." When Thomson had to go ashore for a few hours, he told his guest, "Now mind, Helmholtz, you're not to work at waves while I'm away."

Much of Thomson's problem-solving talent was based on his extraordinary mathematical aptitude. He must have been a mathematical prodigy. While in his teens, he matriculated at the University of Glasgow (where his father was a professor of mathematics) and won prizes in natural philosophy and astronomy. When he was sixteen he read Joseph Fourier's *Analytical Theory of Heat*, and correctly defended Fourier's mathematical methods against the criticism of Philip Kelland, professor of mathematics at the University of Edinburgh. This work was published in the *Cambridge Mathematical Journal* in 1841, the year Thomson entered Cambridge as an undergraduate. By the time he graduated, Thomson had published twelve research papers, all on topics in pure and applied mathematics. Most of the papers were written under the pseudonym "P.Q.R.," since it was considered unsuitable for an undergraduate to spend his time writing original papers.

Another element of Thomson's talent that certainly contributed to his success was his huge, single-minded capacity for hard work. He wrote 661 papers and held patents on 69 inventions. Every year between 1841 and 1908 he published at least two papers, and sometimes as many as twenty-five. He carried proofs and research notebooks wherever he traveled and worked on them whenever the spirit moved him, which evidently was often. Helmholtz wrote (in another of his lively letters to his wife) of life on board the Thomson yacht when the host had "calculations" on his mind:

> W. Thomson presumed so far on the freedom of his surroundings that he carried his mathematical note-books about with him, and as soon as anything occurred to him, in the midst of company, he would begin to calculate, which was treated with a certain awe by the party. How would it be if I accustomed the Berliners to the same proceedings? But the greatest *naïvete* of all was when on Friday he had invited all the party to the yacht, and then as soon as the ship was on her way, and every one was settled on deck as securely as might be in view of the rolling, he vanished into the cabin to make calculations there, while the company were left to entertain each other so long as they were in the vein; naturally they were not exactly very lively.

Thomson may not have been a considerate host, but he was able to work with great effectiveness within the scientific, industrial, and academic establishments of his time. He became a professor of natural philosophy at the University of Glasgow when he was twenty-one. One of his first scientific accomplishments was the founding of the first British physical laboratory. His researches quickly became famous, not only in Britain but also in Europe. At the age of twenty-seven, he was elected to fellowship in the Royal Society. By the time he was thirty-one, he had published 96 papers, and his most important achievements in physics and mathematics were behind him.

In 1855, he embarked on a new career, one for which his talents were, if anything, more spectacularly suited than for scientific research; he became a director of the Atlantic Telegraph Company, formed to accomplish the Herculean task of laying and operating a telegraph cable spanning two thousand miles across the Atlantic Ocean from Ireland to Newfoundland. The cable became one of the world's technological marvels, but without Thomson's advice on instrument design, and on cable theory and manufacture, it might well have been a spectacular failure.

After the Atlantic cable saga, which went on for ten years before its final success, Thomson's fame spread far beyond academic and scientific circles. He was the most famous British scientist, as Helmholtz was later to become the most famous German scientist. Income from the cable company and from his inventions made him wealthy, and he managed his investments wisely. In 1866, the year the cable project was completed, Thomson was knighted. In 1892, partly for political reasons—he was active in the Liberal Unionist Party, which opposed home rule for Ireland—he was elevated to the peerage, as Baron Kelvin of Largs. (Largs, a small town on the Firth of Clyde, was the location of Thomson's estate, Netherall; the River Kelvin flows past the University of Glasgow.)

As one of his biographers, Silvanus Thompson, tells us, Thomson was "a man lost in his work." But he was a devoted husband and family member. He was always close to his father, his sister Elizabeth, and his brother James, an engineering professor who shared his interest in thermodynamics. He was married twice. His first wife, Margaret Crum, was an invalid throughout the marriage, in need of frequent attention, which Thomson gave generously. Her death in 1870 was a severe blow. A few years later he married Frances Blandy, always called "Fanny," the daughter of a wealthy Madeira landowner. The second marriage was as blessed as the first was tragic. Fanny was gregarious and gifted; she became an efficient manager of the Thomson household and found a rich social life in Glasgow as the second Lady Thomson and then as Lady Kelvin.

The Carnot-Joule Problem

The aspect of Thomson's many-faceted career that concerns us here is his work on the principles of thermodynamics. This chapter in Thomson's life began in 1846. He had just graduated from Cambridge and had gone to Paris for a stay of about six months to meet French mathematicians and experimentalists. As always, he needed little more than his talent to open important doors. He met J. B. Biot and A. L. Cauchy, had long conversations with Joseph Liouville and C. F. Sturm, and during the summer months worked in the laboratory of Victor Regnault. But the two Frenchmen who impressed him most were no longer living.

In Paris, Thomson began to think seriously about the work of Sadi Carnot. Clapeyron's paper on Carnot's method first caught his attention, and he searched Paris in vain for a copy of Carnot's original memoir. As we saw in chapter 3, Carnot's theory concerned heat engine devices such as steam engines that work in cycles and produce work output from heat input. Carnot had concluded that heat engines were driven by the "falling" of heat from high temperatures to low temperatures, in much the same way waterwheels are driven by water falling from high to low gravitational levels. Carnot had also deduced that the ideal heat engine—one that provided maximum work output per unit of heat input—had to be operated throughout by very small driving forces. Such an ideal device could be reversed with no net change in either the heat engine or its surroundings.

Before becoming acquainted with Carnot via Clapeyron in Paris in 1845, Thomson had been strongly influenced by another great French theoretician who was no longer living, Joseph Fourier. Even before entering Cambridge, Thomson had read Fourier's masterpiece on heat theory. Thomson particularly admired Fourier's agnostic theoretical method, based on mathematical models that were useful but at the same time noncommittal on the difficult question of the nature of heat.

The prevailing theory in Carnot's time held that heat was an indestructible, uncreatable, fluid material called "caloric." Carnot adopted the caloric theory and pictured caloric falling, waterlike, from high to low temperatures, driving heat engine machinery as it dropped. By the 1840s, the caloric theory had a small but growing number of opponents, among them James Joule, who insisted that heat was associated not with caloric but somehow with the motion of the constituent molecules of matter. According to this point of view—which Thomson would later call the "dynamical theory of heat"—the mechanical effect of a heat engine was produced not by falling caloric but directly from molecular motion.

Fourier's theory did not take sides in this controversy, but it managed nevertheless to describe accurately a wide variety of thermal phenomena. Thomson was particularly impressed by Fourier's treatment of the free "conduction" of heat from a high temperature to a low temperature without producing any mechanical effect. This case was the opposite extreme from Carnot's ideal heat engine device. Although in both cases heat passed from hot to cold, Carnot pictured maximum work output produced by the falling heat, while Fourier pictured no work output at all. To Thomson the difference between Carnot and Fourier was striking. He was sure that something of theoretical and practical importance was lost when a Carnot system, with its best possible performance, was converted into a Fourier system, with its worst possible performance.

The Carnot and Fourier influences were both crucial in the development of Thomson's views on the theory of heat. Both Frenchmen had important things to say about thermal processes, and Thomson could find no inconsistencies in their conclusions. In 1847, Thomson was suddenly confronted with a third influence. At the 1847 Oxford meeting of the British Association for the Advancement of Science, Thomson met James Joule and learned of some theoretical views and experimental results that Thomson might have preferred to ignore, because they were at odds with his interpretation of Carnot.

At the Oxford meeting, Joule reported the results obtained in his famous paddle-wheel experiments. By the time Thomson heard him in 1847, Joule was able to prove convincingly that the mechanical equivalent of heat was accurately constant in his various experiments. Joule interpreted his experiments by assuming that heat and work were directly and precisely interconvertible. Work done by the paddle wheel, and other working contrivances in his experimental designs, was not lost: it was simply converted to an equivalent amount of heat. Joule was also convinced that the opposite conversion, heat to work, was possible. In his view, this conversion was accomplished by any heat engine device. The net heat input to the heat engine was not lost; it was converted to an equivalent amount of work.

It was Joule's second claim, the conversion of heat to work in a heat engine, that disturbed Thomson. In 1847, Thomson no longer had faith in the caloric doctrine that heat was a fluid, but he saw no reason to discard another axiom of the caloric theory, that heat was conserved. For Thomson and his predecessors, including Carnot, this meant that a system in a certain state had a fixed amount of heat. If the state was determined by a certain volume V and temperature t, the heat Q contained in the system was dependent *only* on V and t. Mathematically speaking, heat was a *state function*, which could be written $Q(V, t)$, showing the strict dependence on the two state-determining variables V and t. For Thomson in 1847, this principle was an essential part of Carnot's theory, and "to deny it would be to overturn the whole theory of heat, in which it is the fundamental principle."

Useful heat engines always operate in cycles. In one full cycle, the system begins in a certain state and returns to that state. Thus, according to the heat conservation axiom, a heat engine contained the same amount of heat at the end of its cycle as at the beginning, so there could be no net loss of heat, converted to work or otherwise, in one cycle of operation. Figure 7.1 illustrates this restric-

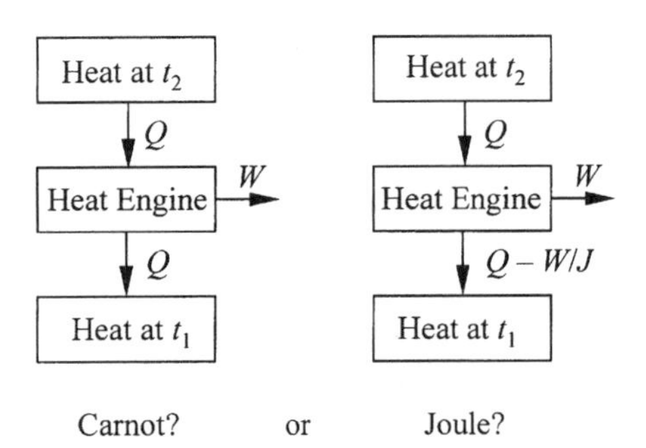

Figure 7.1. Heat engine operation between a high temperature t_2 and a low temperature t_1, as viewed by Thomson in the conflicting theories of Carnot and Joule. Q represents heat, W work, J Joule's mechanical equivalent of heat, and $\frac{W}{J}$ the heat equivalent to W. In the Carnot scheme, no heat is lost. In Joule's picture, an amount of heat $\frac{W}{J}$ is lost.

tion, and to display Thomson's dilemma, also shows heat engine operation according to Joule's claim.

It was even more difficult to reconcile Joule's theory with what apparently happened in the free-heat-conduction processes of the kind Fourier had analyzed. Heat conducted freely could always be put through a heat engine instead and made to produce work. What happened to this unused work when conduction processes were allowed to occur? In Joule's interpretation, nothing was lost in heat engine operation. But Thomson was sure that in a nonworking, purely conducting, system (or in any device allowing free heat conduction to some degree), *something* was lost. In one of his first papers on the theory of heat, published in 1849, Thomson expressed his quandary: "When 'thermal agency' is thus spent in conducting heat through a solid, what becomes of the mechanical effect which it might produce? Nothing can be lost in the operations of nature—no energy can be destroyed. What effect then is produced in place of the mechanical effect which is lost? A perfect theory of heat imperatively demands an answer to this question; yet no answer can be given in the present state of science." This was Thomson's first use of the term "energy," and a first step toward its modern meaning. At this point in the development of his ideas, Thomson could give the term only a mechanical interpretation. He was not yet willing to include heat in his energy concept.

The Thermometry Problem

At the same time he was struggling with these problems, Thomson was investigating another aspect of the Carnot legacy, the temperature-dependent function that Carnot labeled F. Thomson represented the function with μ and called it "Carnot's function." He suggested that the two fundamental properties of the function—that it was dependent only on temperature, and that in all determinations it had the same mathematical form—be used to define a new absolute temperature scale.

Previously, absolute temperatures had been expressed on a scale based on an idealization of gas behavior. If the temperature is held constant, the volume V of an ideal gas decreases as the pressure increases,

$$V \propto \frac{1}{P} \text{ (constant temperature).}$$

If the pressure is held constant, the ideal gas volume increases as the temperature increases,

$$V \propto T \text{ (constant pressure)},$$

with T representing temperature measured on an absolute scale that begins at zero and does not allow negative values. Combining the two proportionalities into one, we have in general

$$V \propto \frac{T}{P}$$

or

$$\frac{PV}{T} = \text{constant}. \quad (1)$$

The constant in this equation, since it *is* a constant, can be determined by measuring P and V at *any* temperature T. Customarily, the temperature of an ice-water mixture (0°C) is chosen. If P_0, V_0 and T_0 are measured at that temperature, equation (1) evaluates the constant as

$$\text{constant} = \frac{P_0 V_0}{T_0}$$

so

$$\frac{PV}{T} = \frac{P_0 V_0}{T_0}. \quad (2)$$

How is the absolute temperature T related to the ordinary temperature t measured, say, on the Celsius scale? Assume that the two scales differ by a constant a, that

$$T = t + a, \quad (3)$$

and substitute this in equation (2) to obtain

$$PV = \frac{P_0 V_0}{T_0}(t + a). \quad (4)$$

The expansion of a gas with increasing temperature, expressed mathematically by the derivative $\frac{dV}{dt}$, is measurable. This derivative divided by the volume V itself defines the "expansion coefficient" α, also measurable,

$$\alpha = \frac{1}{V}\frac{dV}{dt}.$$

According to this, and equation (4) applied with $P = P_0$,

$$\alpha = \frac{1}{t + a}. \qquad (5)$$

Thus a measured value of the expansion coefficient α at a known temperature evaluates the constant a in equation (3) and completes the definition of absolute temperature. Around the turn of the nineteenth century, Joseph Gay-Lussac and John Dalton independently measured α for several gases and found a value of about 267 for the constant a expressed on the Celsius scale; the corresponding modern value is 273. At zero absolute temperature $T = 0$, and according to equation (3), the Celsius temperature is $t = -a = -273$°C.

Thomson was not satisfied with this treatment of the absolute-temperature scale. He objected that it was not a satisfactory basis for a *general* theory of temperature. Real gases were never actually ideal, he argued, and that meant special elaborations of the gas law, a different one for each gas, had to be determined for accurate temperature measurements: there was no universal gas law for real gases. Carnot's function, on the other hand, had just the universality real gas laws lacked; it was always the same no matter what material was used for its determination.

Thomson proposed that Carnot's function be used as a basis for a new temperature scale. He stated this concept as a principle of absolute thermometry in 1848. His basic idea, as he put it later, was that "Carnot's function (derivable from the properties of any substance whatever, but the same for all bodies at the same temperature), or any arbitrary function of Carnot's function, may be defined as temperature and is therefore the foundation of an absolute system of thermometry." Thomson made two suggestions concerning the appropriate function, one in 1848 later abandoned, and another in 1854.

Thomson did not find it easy to make up his mind on this thermometry problem. His final decision was not made until other aspects of his theory of heat had been settled. The main obstacle to progress was still another aspect of the Carnot-Joule dilemma. Thomson found ways to derive equations from Carnot's theory that could be used to calculate Carnot's function μ, and in 1849 he prepared an extensive table of μ values. At first, this calculation had Thomson's full confidence, based as it was on the authority of Carnot's theory, but there was one loose end that he could not ignore. Joule had suggested, in a letter to Thomson in 1848, that Carnot's function was proportional to the reciprocal of the temperature according to

$$\mu = \frac{J}{T} \qquad (6)$$

in which the temperature T is determined on the ideal-gas absolute scale, and J is Joule's mechanical equivalent of heat. At about the same time, Helmholtz reached the same conclusion, but his work was not yet known in Britain.

When Thomson made comparisons between his calculations and those based on Joule's equation (6), he could get no better than approximate agreement. Again he was confronted by a problem brought on by Joule's challenge to Carnot's theory. Joule was inclined to think, correctly, that there were errors in the data used by Thomson in calculating his table of μ values.

Macquorn Rankine

Until late in the nineteenth century, most thermodynamicists developed their subject in a phenomenological vein: they concerned themselves strictly with descriptions of macroscopic events. Their thermodynamic laws were based on reasoning that did not at any point rely on the theoretical modeling of the microscopic—that is, molecular, patterns of nature that might "explain" the laws. With one noteworthy exception, all the early thermodynamicists resisted the temptation to invent speculative molecular models before the phenomenological foundations of their theories were secure.

The exceptional thermodynamicist was W. J. Macquorn Rankine, after 1855 a professor of civil engineering at the University of Glasgow, and a colleague of Thomson's. Like Clausius and Thomson, Rankine had a good grasp of the phenomenology of thermodynamics, but he preferred to derive his version of it from a complicated hypothetical model of molecular behavior. His contemporaries and successors found this approach hard to understand, and even to believe. One can, for example, read polite doubt in Willard Gibbs's assessment of Rankine's attack on the problems of thermodynamics, "in his own way, with one of those marvelous creations of the imagination of which it is so difficult to estimate the precise value."

Rankine pictured the molecules of a gas in close contact with one another. Each molecule consisted of a nucleus of high density and a spherical surrounding "elastic atmosphere" of comparatively low density. The atmospheres were held in place by attraction forces to the nuclei, and their constituent elements had several kinds of motion. Prominent in Rankine's thermodynamic calculations was the rotational motion developed by a large number of tiny, tornado-like vortices that formed around the molecule's radial directions. Rankine showed that a centrifugal force originated in these vortices, which gave individual molecules their elasticity and systems of molecules their pressure.

Rankine's contribution to thermodynamics "was ephemeral," as the science historian Keith Hutchison remarks. "It is in fact doubtful if any of Rankine's contemporaries other than Thomson had the patience to study the *details* of Rankine's work attentively." But for the attentive audience of one, if for no one else, Rankine's vortex theory was a revelation. "Even though Thomson did not accept Rankine's *specific* mechanical hypothesis of the nature of heat," write Thomson's most recent biographers, Crosbie Smith and M. Norton Wise, "he was soon prepared to accept a *general* dynamical theory of heat, namely that heat was vis viva [or kinetic energy] of some kind." Among the attractions of a dynamical theory of heat—Rankine's or any other—was that it made reasonable Joule's claim, the conversion of heat to work.

"[Rankine's] appearance was striking and prepossessing in the extreme, and his courtesy resembled almost that of a gentleman of the old school," writes Peter Guthrie Tait, another Scottish physicist. His creative output was enormous, including, in addition to many papers on thermodynamics, papers on elasticity, compressibility, energy transformations, and the oscillatory theory of light. He also published a series of engineering textbooks, four large engineering treatises, and several popular manuals. He was the Helmholtz of nineteenth-century engineering science.

A "Scot of Scots," Rankine could trace his ancestry from Robert the Bruce. He joined the company of great Scottish scientists and engineers, including Joseph Black and James Watt in the eighteenth century, and Thomson and Maxwell among his contemporaries. Like Carnot, he was trained as an engineer, and adopted the methods of physics to advance engineering science.

Rankine was, with Clausius and Thomson, one of the founders of the classical version of thermodynamics, yet his influence is all but invisible in the modern literature of thermodynamics. This failure was partly because of the impenetrable complexity of his vortex theory. But even without the vortices, his formulation of thermodynamics was obscure, and on some key points, in error. That was not good enough for his theory to survive in the competition with Clausius and Thomson.

The Carnot-Joule Problem Solved

Until about 1850, Thomson saw his theoretical problem as a Joule-or-Carnot choice; for several years the weight of Carnot's impressive successes seemed to tip the balance toward Carnot. But Thomson's theoretician's conscience kept reminding him that Joule's message could not be ignored. Sometime in 1850 or 1851, Thomson began to realize to his relief that in a dynamical theory of heat, Joule's principle of heat and work interconvertibility could be saved without discarding what was essential in Carnot's theory. He discovered that Carnot's important results were compatible with Joule's theory.

This meant proceeding without Carnot's axiom of heat conservation, but Thomson found that the conservation axiom could be excised from Carnot's theory with less damage than he had supposed. Most important, the fundamental mathematical equations he had derived from Carnot's theory—one of which he had used to calculate values of Carnot's function μ—could be derived just as well without the assumption as with it. Having taken this crucial step, Thomson could quickly, in 1851, put together and publish most of his long paper, *On the Dynamical Theory of Heat*, based on the principles of both Joule and Carnot.

As the centerpiece of his theory, Thomson introduced for the first time the idea that energy is an *intrinsic* property of any system of interest. As such, it depends on the system's volume and temperature. Increasing the temperature causes the system's energy to increase in the sense that its molecules have increased kinetic energy. Increasing the volume might cause an energy increase if the expansion were done against attraction forces among the molecules. The mathematical message is that energy is a state function. For states determined by the volume V and temperature t, Thomson's theory replaced the earlier *heat* state function $Q(V, t)$ with the new *energy* state function $e(V, t)$.

Thomson assumed that a system's energy can change *only* by means of interactions between the system and its surroundings: nature provides no internal mechanism for creating or destroying energy within the boundaries of a system. In this sense, energy is conserved. If a system is "closed," meaning that no material flows in or out, interactions with the surroundings are of just two kinds, heating and working. Heating is any thermal interaction and working any nonthermal (usually mechanical) interaction. These statements are easily compressed into an equation: if dQ and dW are small heat and work inputs to a system, the corresponding small change in the system's energy is

$$de = JdQ + dW. \qquad (7)$$

The J factor multiplying dQ is necessary to convert the heat units required for dQ to mechanical units, so it can be added to dW, also expressed in mechanical units.

Thomson's crucial contribution was to move away from his predecessor's exclusive emphasis on heat and work—this was the tradition originated by Carnot and carried on by Joule and Clausius—and to recognize that the conserved quantity, energy, is an intrinsic property of a system that changes under the influence of heating and working. This is not to say that heat and work are different forms of energy; the concept is more subtle than that. Heating and working are two different ways a system can interact with its surroundings and have its energy change.

Energy is energy, regardless of the heating or working route it takes to enter or leave a system. Maxwell made this point in a letter to Tait, criticizing Clausius and Rankine, who pictured the energy possessed by a system in more detail than Maxwell thought permissible: "With respect to our knowledge of the condition of energy within a body, both Rankine and Clausius pretend to know something about it. We certainly know how much goes in and comes out and we know whether at entrance or exit it is in the form of heat or work, but what disguise it assumes in the privacy of bodies . . . is known only to R., C. and Co."

Clausius also recognized the existence of a state function $U(V,t)$, which is equivalent to Thomson's $e(V, t)$. Clausius's work, published in 1850, had priority over Thomson's *Dynamical Theory of Heat* by about one year. But Clausius was less complete in his physical interpretation of the energy concept. In 1850, he only half understood the physical meaning of his state function $U(V,t)$.

At first, Thomson used the term "mechanical energy" for the energy of his theory. To emphasize energy as an entity possessed by a system, he introduced in 1856 the term "intrinsic energy." Later, Helmholtz used the term "internal energy" for Thomson's kind of energy.

The Fourier Problem

Thomson's *Dynamical Theory of Heat* was his magnum opus on thermodynamics. It was a complete and satisfying resolution of the Joule-Carnot conceptual conflict that had been so disturbing two years earlier. At that time, Thomson had also been worried about conflicts between the theories of Joule and Fourier. Joule had argued that nothing was really lost in heat engine operation. Any heat consumed by a heat engine—that is, not included as part of the heat output—was not lost: it was converted to an equivalent amount of work. Thomson could now accept this analysis of a heat engine performing in Carnot's ideal, reversible mode of operation. Nothing *was* lost in that case; the heat engine's efficiency and work output had maximum values, so nothing more could be obtained.

At the other extreme, however, were systems of the kind analyzed by Fourier, which conducted all their heat input to heat output and converted none of it to work. Thomson was convinced that there *were* important losses in this case; the same heat input could have been supplied to a reversible heat engine and converted to work to the maximum extent. What happened to all this work in the Fourier system? A similar question could be asked about any heat engine whose work output fell short of the maximum value. In any such case, work was lost that could have been used in a reversible mode of operation.

In 1852, Thomson published a short paper that answered these questions. His central idea was that, although energy can never be destroyed in a system, it can be wasted or "dissipated" when it might have been used as work output in a reversible operation. The extent of energy dissipation can be assessed for a system by comparing its actual work output with the calculated reversible value. The science historian Crosbie Smith, who has studied the development of Thomson's thermodynamics, describes the unusual character of Thomson's energy dissipation principle with its dependence on "arrangement" and "man's creativity." He includes quotes from Thomson's draft of his *Dynamical Theory of Heat*:

> Where conduction occurs, Thomson believes that the work which *might* have been done as a result of a temperature difference is "lost to man irrevocably" and is not available to man even if it is not lost to the material world. Such transformations therefore remove from man's control sources of power "which if the opportunity to turning them to his own account had been made use of might have been rendered available." Here the use of work or mechanical effect depends on man's creativity—on his efficient deployment of machines to transform concentrations of energy [e.g., high-temperature heat] into mechanical effect—and it is therefore a problem of arrangement, not of creation ex nihilo.

A simple example here will help clarify Thomson's meaning. A weight held above the ground can do useful work if it drops very slowly and at the same time drives machinery. If the machinery is ideal, that work can be supplied as input to another ideal machine that lifts the weight back to its original position. Thus the slow falling of the weight coupled to ideal machinery is exactly reversible—that is, the weight and its surroundings can be restored to their initial condition, and there is no dissipation of energy in the sense Thomson described.

Now suppose the weight drops to the ground in free fall, with no machinery. As the weight falls, its potential energy is converted to kinetic energy, and the kinetic energy to heat when the weight hits the ground (as in Joule's waterfall effect). Here we have an "irreversible" process. With no machinery and no work output, we cannot restore the weight to its original position above the ground without some uncompensated demands on the surroundings, and weights certainly do not rise spontaneously. This is an extreme case of irreversibility and energy dissipation: *all* of the weight's initial potential energy has been reduced to heat and rendered permanently unavailable for useful purposes.

Falling heat imitates falling weights. It, too, has potential energy (proportional to the absolute temperature), which can be completely used in a reversible heat engine operation, with no dissipation, or completely dissipated in the irreversible Fourier process of free conduction, or something in between in a real heat engine. We have a technological choice: we can design a heat engine efficiently or inefficiently, so it is wasteful or not wasteful.

The Thermometry Problem Solved

With the publication of his paper on the energy dissipation principle, Thomson could feel that he had finally brought together in harmony the concepts of Joule, Carnot, and Fourier. But the fundamentals of his thermodynamics were still not quite complete. He had not yet made a decision about the nagging thermometry problem that had been bothering him for almost five years. The specific problem was how to relate the temperature-dependent Carnot's function μ to absolute temperature.

I lack the space here to give a complete account of Thomson's work on this stubborn and frustrating problem. Thomson had hoped to be able to use equations he had derived from Carnot's theory to calculate values of Carnot's function μ. Eventually he had to admit defeat in this effort when he found that some assumptions used in the calculation were not valid. Thomson enlisted Joule's help in another, more elaborate attempt to calculate μ values. The principal aim of the Joule-Thomson work was to study real (nonideal) gas behavior, and in this it succeeded. But Thomson also tried to use Joule's data to calculate μ values, and once again he failed to muster the calculational wherewithal to complete the task.

Finally, in 1854, Thomson decided to take a different tack in his pursuit of the still-elusive Carnot function. He returned to his 1848 thermometry principle, which asserted that Carnot's function, or any function of Carnot's function, could be used as a basis for defining an absolute-temperature scale. No doubt influenced by the Joule evaluation of Carnot's function in equation (6), he defined a new absolute-temperature scale that had this same form. Representing temperatures on this scale T, his assumption was

$$T = \frac{J}{\mu}. \tag{8}$$

He also assumed that the degree on the new scale is equivalent to the degree on the Celsius scale. Even if Carnot's function μ could not be calculated accurately with the data then available, Thomson was sure that it would eventually be calculated, and that his thermometry principle was secure. The principle permitted *any* assumed mathematical relation between the absolute temperature and μ. Thomson could see that equation (6), one of the simplest possible choices, and in agreement with the ideal-gas absolute-temperature scale, was acceptable and the best choice. Thomson was rewarded for his labors on the absolute-temperature scale: the modern unit of absolute temperature is called the "kelvin" (lowercase), abbreviated "K" (uppercase).

Hazards of Virtuosity

As it comes down to us in the consensus version found in modern textbooks, the edifice of thermodynamics is based on three fundamental concepts, energy, entropy, and absolute temperature; and on three great physical laws, the first an energy law, and the second and third entropy laws. Only part of this picture is visible in Thomson's published work. He was certainly aware of the importance of the energy and absolute-temperature concepts; those parts of the story he understood better than any of his competitors. But he failed to recognize the powerful significance of entropy theory.

Actually, Thomson did touch on a calculation in 1854 that was based on the concept Clausius later explored further and eventually called entropy. As was often the case in his work, however, Thomson was inspired mainly by a special problem, in this case, thermoelectricity, or the production of electrical effects from thermal effects. He made statements of fundamental significance, and showed that he appreciated the rudiments of entropy theory; but he applied his analysis only to the special problem, and never reached the important new theoretical ground Clausius would soon explore.

Clausius did not immediately believe in the entropy concept either. It took him about ten years to have the confidence to supply a name and a symbol for his new function. At the time Thomson glimpsed the idea of entropy, he apparently did not have the patience or inspiration for such a prolonged—and possibly risky—effort.

One of Thomson's biographers, J. G. Crowther, remarks that more than once Thomson failed to "divine" the deepest significance of his discoveries. "He did not possess the highest power of scientific divination," Crowther writes. "Unlike the greatest scientists he was unable to divine what lay beyond the immediate facts. In the highest regions of scientific research he was indisciplined. That was perhaps due to his natural and habitual lack of contact with the collective stream of scientific thought. That indiscipline penetrated down into his working habits. He used to write papers in pencil, often on odd pieces of paper, and send them in this condition to the printers."

Another Thomson biographer, Joseph Larmor, gives us a picture of the scientific virtuoso, so full of brilliant solutions to technical problems of every kind he hardly had the time to write them all down, and never found the time to organize into a unified whole his greatest accomplishments. Most of his papers were "mere fragments," Larmor writes, "which overflowed from his mind . . . into the nearest channel of publication. . . . In the first half of his life, fundamental results arrived in such volume as often to leave behind all chance of effective development. In the midst of such accumulation he became a bad expositor; it is only by tracing his activity up and down through its fragmentary published records, and thus obtaining a consecutive view of his occupation, that a just idea of the vistas continually opening upon him may be reached."

Difficult as it certainly is to follow the threads of Thomson's thought "up and down through its fragmentary published records," his work certainly had vision. As Smith has emphasized, the scope of Thomson's work was as broad as that of any of his fellow physicists. At a time when other thermodynamicists were concentrating on reversible processes, Thomson was concerned with the thermodynamics of irreversible processes in flow and thermoelectric systems. Some of his methods of analysis did not come into general use until much later. Thomson overlooked the importance of the entropy concept, but he was well aware of the need for a second law of thermodynamics. His principle of energy dissipation is a consequence of the modern statement of the second law.

In his discursive way, Thomson touched on every one of the major problems of thermodynamics. But except for his temperature scale and interpretation of the energy concept, his work is not found in today's textbook version of thermodynamics. Although he ranks with Clausius and Gibbs among thermodynamicists, his scientific legacy is more limited than theirs.

The comparison with Clausius is striking. These two, of about the same age, and both in possession of the Carnot legacy, had the same thermodynamic concerns. Yet it was the Clausius thermodynamic scheme, based on the two concepts of energy and entropy and their laws, that impressed Gibbs, the principal third-generation thermodynamicist. Clausius could also be obscure, but he left no doubt about the conceptual foundations of his theories, and he gave Gibbs the requisite clues to put together the scheme we see today in thermodynamics texts.

Thomson Himself

For Thomson, however, we have a different kind of monument: we know what this man of virtuosic talent was like as a human being. Unlike Clausius, who for reasons apparently related to his contentious personality and lack of fame outside the scientific world has never attracted a skilled biographer, Thomson has been, and still is, a popular subject for biographical commentary. The first Thomson biography was *The Life of William Thomson*, written by a namesake (spelled with a *p*), Silvanus P. Thompson. Thompson is occasionally too admiring to be accurate, and one may not share his fascination with Cambridge lore, but it would be difficult to find a more enjoyable way to enter Thomson's world than to spend a few days with Silvanus Thompson's two volumes.

Even if Silvanus Thompson was overly impressed with his subject's virtues, he had the good sense to quote at length others, Thomson's friends, students, and relatives, who saw him more completely. We can hardly do better than to close this profile with comments by two young people who were impressed, amused, and a little saddened by their contact with Thomson.

We hear first from Thomson's grandniece, Margaret Gladstone, who was a favorite of Thomson's, and as a young girl often visited Netherall, the Thomson estate. (Two remarkable further aspects of Margaret Gladstone's life: she was the daughter of J. H. Gladstone, who succeeded Faraday at the Royal Institution, and she became the wife of Ramsay Macdonald, one of the founders of the British Labor Party and prime minister in the 1920s.) Her charming description of "Uncle William" and "Aunt Fanny" is naïve but at the same time perceptive:

> Aunt Fanny likes company very much: and as for Uncle William it doesn't seem to make much difference to him what happens; he works away at mathematics just the same, and in the intervals holds animated conversations with whomever is near. They were both very good to me; and the time I liked best was one day when there were no visitors at all, and we were quite by ourselves for about thirty hours.
>
> The mathematics went on vigorously in the "green book." That "green book" is a great institution. There is a series of "green books"—really notebooks made especially for Uncle William—which he uses up at the rate of 5 or 6 a year, and which are his inseparable companions. They generally go upstairs, downstairs, out of doors, and indoors, wherever he goes; and he writes in his "green book" under any circumstances. Looking through them is quite amusing; one entry will be on the train, another in the garden, a third in bed before he gets up; and so they go on, at all hours of the day and night. He always puts the place and the exact minute of beginning an entry.

In 1896, an immense celebration attended by more than two thousand guests was held in Thomson's honor in recognition of his long service at the University of Glasgow. The huge gathering hardly got what it expected in Thomson's responding remarks. At that moment, he had to tell them, his deepest feeling was a sense of failure: "One word characterizes the most strenuous efforts for the advancement of science I have made perseveringly during fifty-five years; that word is *Failure*. I know no more of electric and magnetic force, or of the relation between ether, electricity and ponderable matter, or of chemical affinity, than I knew and tried to teach to my students of natural philosophy fifty years ago in my first session as professor."

Margaret Gladstone was there and recorded some sober thoughts:

> In the evening the word "Failure" in which he characterized the results of his best efforts seemed to ring through the hall with half-sad, half-yearning emphasis. Some of the people tried to laugh incredulously, but he was too much in earnest for that. Yet at the same time he was not pessimistic, for it was evident what keen joy he had in his work, and still has, and how warmly he feels the help and affection of his fellow-workers.
>
> As for the students, I am afraid they laughed, with good cause, when he spoke of the ideal lecture as a conference, because I always hear that he goes up in the heights when he is lecturing to them, and pours forth speculations with great enthusiasm far above their heads.
>
> In thinking over Uncle William's speeches, the tone in which he gave them, and in his quiet, serious, deferential look when praise was heaped upon him, dwell in my memory. There was something pathetic about it all—a sort of wonder that people should be so kind, and a wish that he had done more to deserve it all.

Thomson rarely found the time to prepare his lectures, and as Margaret Gladstone informs us, he could not resist the temptation to tell uncomprehending student audiences about his latest discoveries. Helmholtz, who was not successful in the lecture hall either, wondered how Thomson ever made contact with his students: "He thinks so rapidly . . . that one has to get at the necessary information . . . by a long string of questions, which he shies at. How his students understand him, without keeping him as strictly to the subject as I ventured to do, is a puzzle to me."

Yet there was an affectionate bond between Thomson and his "corps" of students, able to forgive his digressions in the lecture hall and appreciate his greatness as a scientist and as an unpretentious human being. Here is a recollection by Andrew Gray, who was one of the "merry students" who attended Thomson's lectures in the 1870s, and was eventually Thomson's successor. It is an account of the last lecture of the course:

> The closing lecture of the ordinary course was usually on light, and the subject was generally the last to be taken up—for as the days lengthened in spring it was possible sometimes to obtain sunlight for the experiments—and was often relegated to the last day or two of the session. So after an hour's lecture Thomson would say, "As this is the last day of the session I will go on a little longer after those who have to leave have gone to their classes." Then he would resume after ten o'clock, and go on to eleven, when another opportunity would be given for students to leave, and the lecture would be resumed. Messengers would be sent from his house where he was wanted on business of other sorts to find what had become of him, and the answer brought would be, hour after hour, "He is still lecturing." At last he would conclude about one o'clock, and gently thank the small and devoted band who had remained to the end for their kind and prolonged attention.

8

The Road to Entropy

Rudolf Clausius 1822–1888.

Scientific Siblings

The history of thermodynamics is a story of people and concepts. The cast of characters is large. At least ten scientists played major roles in creating thermodynamics, and their work spanned more than a century. The list of concepts, on the other hand, is surprisingly small; there are just three leading concepts in thermodynamics: energy, entropy, and absolute temperature.

The three concepts were invented and first put to use during a forty-year period beginning in 1824, when Sadi Carnot published his memoir on the theory of heat engines. Carnot was the pioneer, and the conceptual tools he had available to refine his arguments were primitive. But he managed, nonetheless, to invent highly original concepts and methods that were indispensable to his successors.

Carnot died in 1832, and his scientific work almost died with him. His memoir was first ignored and then resurrected, initially by his colleague Émile Clapeyron and later by two second-generation thermodynamicists, Rudolf Clausius and William Thomson. These two men were born almost at the same time as Carnot's revolutionary memoir: they were, so to speak, Carnot's scientific progeny. Just as the generation that had ignored Carnot was passing, Clausius and Thomson came of age, ventured into the world of scientific ideas, and took full advantage of Carnot's powerful, but neglected, message. Now it is Clausius's turn, but first I must digress on some mathematical matters.

Formulas and Conventions

To describe a system in the style of thermodynamics, one must first define the system's state with suitable state-determining variables such as the volume V and temperature t (t now stands for Celsius temperature). Small changes in V and t, brought on as the system is put through some process, are represented by dV and dt. These symbols can denote either increases or decreases, and that means dV and dt are implicitly either positive or negative. In an expansion, for example, the volume of the system increases, so the change dV is positive; in compression, the volume decreases and dV is negative. Similarly, positive dt describes a temperature increase, and negative dt a temperature decrease.

Heating and working are the fundamental processes of thermodynamics. As both Clausius and Thomson understood, they involve interactions between a system and its surroundings. For example, adding a small amount of heat dQ to a system from the surroundings is a small step in a heating process. Heat added to a system is counted as positive, and dQ is implicitly positive. The reverse process removes the heat dQ from the system, and dQ is negative. These conventions are illustrated in figure 8.1.

A working process might be the compression of a gas in a piston-cylinder device, as in a car engine. A small step in the compression process is represented by the small amount of work dW done on the system (the gas), and it is counted positive. In the reverse process, expansion, the system does work on its surroundings; this is work output and dW is negative. See figure 8.2.

If we slowly add a small amount of heat dQ to a system, the response is likely to be a small temperature increase dt, accompanied by a small expansion expressed by the volume increase dV. The heat and its two effects are related by an equation that was an indispensable mathematical tool for Clausius,

$$dQ = MdV + Cdt. \tag{1}$$

The coefficient C in this equation is called a "heat capacity." We can isolate it by assuming that the volume is held constant, so there is no change in volume, $dV = 0$, and from equation (1),

$$dQ = Cdt \text{ (constant } V\text{)}. \tag{2}$$

Suppose we add $dQ = 0.1$ heat units and measure the temperature change $dt = 0.001°C$. Then the heat capacity calculated with equation (2) is

$$C = \frac{dQ}{dt} = \frac{0.1}{0.001} = 100 \text{ heat units per °C},$$

demonstrating that the heat capacity is the number of heat units required to raise the temperature of the system one degree.

If we compress a gaseous system and change its volume by dV, the small amount of work done dW is proportional to the volume change,

$$dW \propto -dV. \tag{3}$$

(Read "proportional to" for the symbol $\propto$.) The minus sign preceding dV is dictated by the sign conventions we have adopted for dW and dV. The compression

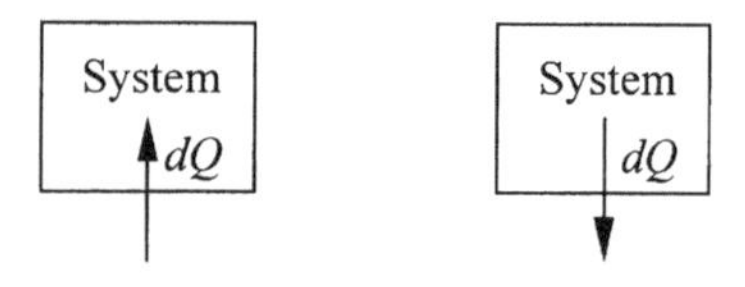

Figure 8.1. Illustration of the sign convention for dQ.

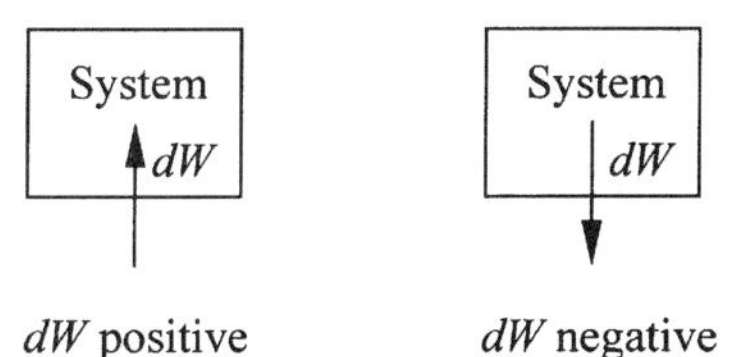

Figure 8.2. Illustration of the sign convention for dW.

provides work input, so dW is positive, but dV is negative because the compression decreases the volume. The mismatch of signs is repaired by replacing dV with $-dV$, which is positive. The same recipe applies to an expansion, with dV positive and $-dV$ negative, matched by a negative dW for work output.

The work done in compression is also proportional to a pressure factor, as one might expect, because it certainly requires less work to compress a gas at low pressure than at high pressure. If the compression is done slowly, that pressure factor is simply the pressure P of the gas. With that factor included, the proportionality (3) becomes the equation

$$dW = -PdV. \tag{4}$$

This equation is also valid for expansion of a gas, and even for expansion or compression of a liquid or solid.

Heat Transmitted and Converted

Clausius published a memoir in 1850 that reconciled Carnot's work with the discoveries of the intervening twenty-five years and formulated the first law of thermodynamics almost in its modern form. Clausius began his 1850 paper with a reference to the paper by Émile Clapeyron written two years after Carnot's death in the mathematical language understood then (and now) by theoreticians. For reasons he never had occasion to explain, Carnot had written his memoir in a mostly nonmathematical style that obscured his more subtle points.

Both Carnot and Clapeyron had been misled by the well-entrenched caloric theory of heat, which insisted that heat was indestructible, and could not therefore be converted to work in a heat engine or any other device. For them, the heat engine dropped the heat from a higher to a lower temperature without changing its amount. The time had come for Clausius, as about a year later it came for Thomson, to free the Carnot-Clapeyron work from the misconceptions of the caloric theory. Clausius did so by first making the fundamental assumption in his 1850 paper that part of the heat input to any heat engine is *converted* to work. The rest of the heat input is simply *transmitted* from a higher to a lower temperature, as in the Carnot-Clapeyron model, and it becomes the heat engine's output. In other words, heat can be affected by two kinds of transformations, *transmission* and *conversion*. Summarizing in an equation for one turn of a heat engine's cycle,

$$\text{heat input} = \text{heat converted} + \text{heat transmitted}, \tag{5}$$

or

$$\text{heat input} - \text{heat transmitted} = \text{heat converted}. \tag{6}$$

Clausius invoked a lengthy argument that put the last statement in the form of a complicated differential equation containing the two coefficients C and M.

The First Law

If Clausius had gone no further in his analysis, his 1850 paper would not have an important place in this history. The differential equation he had derived was mathematically valid, and its physical validity could be checked, but otherwise it had little significance beyond the immediate circumstances for which it was derived. Clausius was aware of these deficiencies, and his next effort was to reshape his argument into something more meaningful.

With some inspired mathematical manipulations, Clausius derived a second equation (equation [7] below) that proved a much more significant theoretical tool than his original equation. It can be found in any modern thermodynamics text as the standard mathematical version of the first law of thermodynamics. That two equations so closely related mathematically can differ so much in physical importance—one equation little more than a historical curiosity, the other now known to any physicist, engineer, or chemist—is vivid testimony that for the theoretical scientist, mathematics is a language whose message can be eloquent or dull, depending on how it is written and interpreted.

Clausius had only to integrate his original differential equation to reveal its physical message. He invoked a function of V and t, simply as a by-product of the integration, that was reminiscent of the false heat state function $Q(V,t)$, except that this function really *was* a state function. The new function, which Clausius labeled $U(V,t)$, was the first of a collection of valuable state functions that now dominate the practice of thermodynamics.

The quantity U was a proper state function, but what did it mean physically? Clausius answered by again making use of equation (1). With a few more mathematical strokes, he derived the equation

$$dQ = dU + \frac{1}{J}PdV, \tag{7}$$

where P represents pressure, and the factor $\frac{1}{J}$ converts the mechanical units attached to the PdV term into the thermal units required for dQ.

Clausius had arrived here at the equation that modern students of thermodynamics have no difficulty recognizing as a mathematical statement of the first law of thermodynamics. In modern usage, no distinction is made between thermal and mechanical units, so the factor J is unnecessary, U is recognized as internal energy, and the equation is written so it evaluates changes in U,

$$dU = dQ - PdV. \tag{8}$$

But in 1850, the energy concept was still unclear, and could not be part of Clausius's interpretation. Instead, he viewed equation (7) primarily as a contribution to the theory of heat. He understood dQ to measure the amount of heat added during a small step in a heating process. Once the heat entered the system, it could be "free" or "sensible" heat—its effect could be measured on a thermometer—or it could be converted to work. He recognized two kinds of work, that performed *internally* (against forces among molecules, in the modern interpretation) and that done *externally,* against an applied pressure in the surroundings. The term $\frac{1}{J} PdV$ in equation (7) evaluates the latter, so Clausius concluded that dU calculates two things: changes in the sensible heat (always an increase if heat is added) and the amount of internal work done, if any.

Clausius succinctly summarized his position in an appendix added to the 1850 paper in 1864, when he collected his papers in a book: "The function U, here introduced, is of great importance in the theory of heat; it will frequently come under discussion in the following memoirs. As stated, it involves two of the three quantities of heat, which enter into consideration when a body changes its condition; these are the augmentation of the so-called *sensible* or *actually present* heat, and heat *expended in interior work.*"

At about the same time Clausius was developing this interpretation of his state function $U(V,t)$, Thomson was inventing a theory based on an identical function, which he labeled $e(V,t)$. Thomson had a name for his function—"mechanical energy"—and he understood it to be a measure of the mechanical effect (molecular kinetic and potential energy) stored in a system after it has exchanged heat and work with its surroundings. Thomson later called his function "intrinsic energy," and still later Helmholtz supplied the name that has stuck, "internal energy."

It is an impressive measure of the subtlety of the energy concept—and of Thomson's insight—that Clausius was not willing to accept Thomson's energy theory for fifteen years. Not until 1865 did he adopt Thomson's interpretation and begin calling his U function "energy." He did not use Thomson's or Helmholtz's terms.

In spite of his uncertainty about the physical meaning of the U function, Clausius had in his 1850 paper come close to a complete formulation of the first law of thermodynamics. Even the mathematical notation he used is that found in modern textbooks. Clifford Truesdell summarizes Clausius's achievements in the 1850 work: "There is no doubt that Clausius with his [1850] paper created classical thermodynamics. . . . Clausius exhibits here the quality of a great discoverer; to retain from his predecessors major and minor . . . what is sound while frankly discarding the rest, to unite previously disparate theories and by one simple if drastic change to construct a complete theory that is new yet firmly based upon previous successes."

The "one simple if drastic change" made by Clausius was to assume that, in heat engines and elsewhere, heat could not only be dropped or transmitted from a higher to a lower temperature (as Carnot had assumed), but that it could also be converted into work. Others, particularly Joule, had recognized the possibility of heat-to-work conversions—much of Joule's research was based on observations of the inverse conversion, work to heat—but Clausius was the first to build the concept of such conversions into a general theory of heat.

Heat Transformations

Clausius had much more to add to his theoretical edifice based on the "simple if drastic change." In 1854, he published a second paper on heat theory, which went well beyond the realm of the first law of thermodynamics and the concept of energy, and well into the new realm of the second law of thermodynamics and the concept of entropy. His initial assumption was again that heat could undergo two kinds of transformations. I will elaborate Clausius's terminology for the two transformations and call the dropping of heat from a high to a low temperature an instance of a "transmission transformation," and the conversion of heat to work an example of a "conversion transformation." Clausius was impressed that both kinds of transformations have two possible directions, one "natural" and the other "unnatural" (again, this is not Clausius's terminology). In the natural direction, the transformation can proceed by itself, spontaneously and unaided, while the unnatural direction is not possible at all unless forced.

The natural direction for the conversion transformation can be seen in Joule's observations of heat production from work. Clausius saw the unnatural direction for the conversion transformation as the production of work from heat, a conversion that never takes place by itself, but always must be forced somehow in heat engine operation. The natural direction for the transmission transformation is the free conduction of heat from a high temperature to a low temperature. The unnatural direction is the opposite transport from a low temperature to a high temperature, which is impossible as a spontaneous process; such heat transport must be forced in a "heat pump," like those used in air conditioners.

Clausius took this reasoning one significant step further. He saw that in heat engines the two kinds of heat transformations occur at the same time. In each cycle of heat engine operation, the transmission transformation takes place in its natural direction (heat dropped from a high to a low temperature), while the conversion transformation proceeds in its unnatural direction (heat converted to work). It is as if the transmission transformation were *driving* the conversion transformation in its unnatural direction.

Moreover, Clausius concluded, the two transformations are so nearly balanced that in reversible operations either can dominate the other. They are in some sense equivalent. Clausius set out to construct a quantitative "heat transformation theory" that could follow this lead. His goal was to assess "equivalence values" for both transformations in reversible, cyclic processes. He hoped that the equivalence values could then be used to express in a new natural law the condition of balance, or "compensation," as he called it.

Although he could hardly have been aware of it at the time, Clausius had, in this simple theoretical expectation, started a line of reasoning as promising as any in the history of science. It would not be easy for him to appreciate fully the importance of what he was doing, but he now had all the theoretical clues he needed to reach the concept of entropy and its great principle, the second law of thermodynamics.

Clausius began his heat transformation theory with the axiom that heat is not transmitted spontaneously from a low temperature to a high temperature. (If you touch an icicle, heat passes from your warm hand to the cold icicle, and the icicle feels cold; icicles never feel warm.) In his 1854 paper, he stated the assumption: "Heat can never pass from a colder to a warmer body without some other change connected therewith occurring at the same time." Later he simplified his axiom to: "Heat cannot of itself pass from a colder to a warmer body."

The arguments Clausius used to develop his theory from this simple beginning are too lengthy to address here. Note that his equivalence values and condition of compensation revealed a fundamental pattern of heats and temperatures involved in any reversible, cyclic process. If t_i is the temperature at which one step in such a process takes place, and if Q_i is the heat input or output in that step, Clausius's corresponding equivalence value for the step is $f(t_i)Q_i$, where $f(t_i)$ is some *universal* function of the temperature t_i. Summation of such terms for all the steps of a process, which we write with the notation $\sum f(t_i)Q_i$ (the symbol Σ denotes a summation), then evaluates the net equivalence value for the complete process. In Clausius's condition of compensation for reversible operation, the terms in the summation exactly cancel each other, and the result is

$$\sum f(t_i)Q_i = 0 \text{ (reversible, cyclic operation).}$$

For a process consisting of many small steps, each one involving a small heat transfer dQ at the temperature t, Clausius's compensation criterion is expressed as a summation over many small steps—that is, as an integral

$$\int f(t)dQ = 0 \text{ (reversible, cyclic process).} \tag{9}$$

For Clausius, this was a crucial result: it told him that he had found a new state function. To follow Clausius's reasoning here, we represent the new function temporarily with the generic symbol F (not the same as the F used earlier for Carnot's function), and define a small change dF with

$$dF = f(t)dQ \text{ (reversible process),} \tag{10}$$

so equation (9) becomes

$$\int dF = 0 \text{ (reversible, cyclic process).} \tag{11}$$

Clausius could now turn to a mathematical theorem that guarantees from this condition that F is a state function. Paralleling Clausius's other state function $U(V,t)$, it could be identified as the function $F(V,t)$.

At this point, Clausius had the underlying mathematical ingredients of his theory, but the physical interpretation of the mathematics was anything but clear. The physical meaning of the function U was still obscure, and the new function was even more of a mystery. As a skilled theorist, Clausius was aware of the dangers of attaching too much physical meaning to quantities that might be found later to be figments of the mathematical argument. He did not offer a name for the new state function in 1854, nor did he give it a symbol.

However, Clausius felt he could trust his conclusion that his compensation condition (11) *did* define a new state function, and from that *mathematical* fact he could determine the universal function $f(t)$. A further mathematical argument led him to the conclusion that

$$f(t) = \frac{1}{t + a}, \tag{12}$$

in which $t + a$ defines absolute temperature on the ideal gas scale. Using T again to denote absolute temperature, Clausius's conclusion was that

$$f(t) = \frac{1}{T}$$

and this substituted in equation (10) completes the definition of Clausius's still nameless new thermodynamic state function,

$$dF = \frac{dQ}{T} \text{ (reversible process).} \tag{13}$$

The Second Law

When he arrived at the mathematical equivalent of equation (13), Clausius must have been aware that he had made a promising beginning toward a broader theory. But the theory was still severely limited: for one thing, equation (13) applied only to reversible processes. The condition of reversibility had originally been invented by Carnot to define an ideal mode of heat engine operation, ideal in the sense that it gives maximum efficiency. Reversibility was essential in Clausius's argument leading to equation (13) because it enabled him to assert that the two kinds of heat transformations compensate each other.

Clausius had done great things with Carnot's theoretical style. One can imagine that if Carnot had lived longer—he would have been fifty-four in 1850—and if he had recognized that heat can be transformed by conversion as well as by transmission, he would have reasoned much as Clausius did in 1850 and 1854. In the two papers, Clausius had done what Carnot demanded; and then in the 1854 paper, and later in 1865, he ventured beyond Carnot, into the realistic realm of irreversible processes, which were not of the ideal, reversible kind. Clausius's conclusion, as it is expressed by modern authors, is that for irreversible processes equation (13) is not valid, and instead it is replaced by an *inequality*,

$$dF > \frac{dQ}{T} \text{ (irreversible process).} \tag{14}$$

(Read "greater than" for the symbol $>$, and "less than" for $<$.)

Clausius had now brought forth two state functions, the function U and the $\frac{dQ}{T}$-related function we are temporarily labeling F. And he had generalized his theory so it was released from its earlier restrictions to reversible and cyclic processes. The paper in which he completed the generalization was published in 1865. By the time he wrote that paper, the last of his nine memoirs on thermodynamics, he was willing to accept the term "energy" for U, and he wrote equation (7) assuming no distinction between heat and mechanical units, so $J = 1$,

$$dQ = dU + PdV. \tag{15}$$

Or, with $dW = -PdV$ according to equation (4),

$$dQ = dU - dW. \tag{16}$$

At long last (as it seems to us, with the benefit of hindsight), Clausius had enough confidence in his second state function to give it a name and a symbol. For no specified reason, he chose the letter S and wrote equation (13)

$$dS = \frac{dQ}{T} \text{ (reversible process)}, \tag{17}$$

and the inequality (14)

$$dS > \frac{dQ}{T} \text{ (irreversible process)}. \tag{18}$$

(Clausius seems to have preferred letters from the last half of the alphabet; he used all the letters from M to Z, except for O, X, and Y, in his equations.) Because the function S calculated heat transformation equivalence values, he derived his word for it from the Greek word "trope," meaning "transformation." The word he proposed was "entropy," with an "en-" prefix and a "-y" suffix to make the word a fitting partner for "energy."

All this is familiar to the present-day student of thermodynamics, who continues the argument by deriving $dQ = TdS$ from equation (17), substituting for dQ in equation (15) and rearranging to obtain

$$dU = TdS - PdV. \tag{19}$$

We recognize this today as the master differential equation for the thermodynamic description of any system that is not changing chemically. Dozens of more specific equations can be derived from it.

Although Clausius was certainly aware of equation (19) and its mathematical power, he did not use it. He still had a curious ambivalence concerning his two state functions U and S. In a lengthy mathematical argument, he excised U and S from his equations (15) and (17), and in their place put functions of the heat Q and work W.

It appears that Clausius hesitated because he hoped to give the energy U and entropy S molecular interpretations, but had not completed that program. The fundamental ingredients of this molecular picture were the kinetic and potential energy possessed by molecules, and in the determination of entropy, a macroscopic property he called "disgregation," which measured "the degree in which the molecules of the [system] are dispersed." For example, the disgregation for a gas (with the molecules widely separated) was larger than for a liquid or solid (with the molecules much closer to each other).

In the 1860s, molecular science was in its infancy, and these molecular interpretations could be no better than speculations. Clausius was well aware of this, and did not want to jeopardize the rest of his theory by building from molecular hypotheses. Rankine had done that and lost most of his audience. Nevertheless, Clausius did not want to discard the energy and entropy concepts completely. He found a safe middle ground where energy and entropy were "summarizing concepts," as the science historian Martin Klein puts it, and the working equations of the theory were based strictly on the completely nonspeculative concepts of heat and work. Clausius never finished his molecular interpretations, but his speculations, as far as they went, were sound. Even his disgregation theory was confirmed in the later work of Maxwell, Boltzmann, and Gibbs.

Clausius's last words on thermodynamics, the last two lines of his 1865 paper, made readers aware of the grand importance of the two summarizing concepts, energy and entropy. He saw no reason why these concepts and their principles should be restricted to the earthbound problems of physics and engineering: they should have meaning for the entire universe of macroscopic phenomena. Stretching his scientific imagination to the limit, he pictured the universe with no thermal, mechanical, or other connections, so $dQ = 0$ and $dW = 0$, and then applied his statements (16) and (18) of the first and second laws of thermodynamics to this isolated system. According to equation (16), $dU = 0$ if $dQ = 0$ and $dW = 0$, so the energy of an isolated universe does not change: it is constant. With $dQ = 0$, the inequality (18) tells us that $dS > 0$, that is, all entropy changes are positive and therefore increasing. Presumably, no system, not even the universe, can change forever. When all change ceases, the increasing entropy reaches a maximum value. Clausius asked his readers to accept as "fundamental laws of the universe" his final verbal statements of the two laws of thermodynamics:

The energy of the universe is constant.

The entropy of the universe tends to a maximum.

Clausius vs. Tait et al.

Theorists need to do their work in two stages. First, they have to be sure that they themselves understand what their theories say. Then they have to make others understand. Clausius succeeded in the first stage of development of his thermodynamic theory. Rarely, if ever, did he make mistakes in the interpretations and applications that supported his theory. But for reasons that were partly his own fault, he had extraordinary difficulty when it came time to educate the rest of the scientific world about the concepts of his theory.

Clausius's critics most frequently misunderstood his quantity $\frac{dQ}{T}$, especially its sign. Here the confusion is understandable, because Clausius himself was inconsistent in the sign he gave dQ from one paper to another. He usually considered heat input as positive, but occasionally used the opposite convention. Failure to get the dQ sign right was just one of the mistakes that misled Clausius's most persistent and outspoken critic, P. G. Tait (who had jousted verbally with Tyndall in the Joule-Mayer controversy). Tait's contributions to thermodynamics were limited, but he was active in putting forward Thomson's ideas. Tait wrote a book called *Sketch of Thermodynamics*, which was a collection of conceptual bits and pieces borrowed from Thomson, Clausius, and Rankine, some of them misunderstood.

The most outstanding of Tait's misconceptions was his insistence that entropy was a measure of "available energy." It is difficult to see how he arrived at this interpretation, because entropy does not even have energy units. Perhaps the mistake originated in Clausius's association of the transformation concept with entropy. To the British, transformation meant conversion of heat to work. Tait, who never read Clausius with care, may have simply substituted this understanding of the transformation concept for Clausius's entropy definition.

Shortly after Tait's book appeared, James Clerk Maxwell published a textbook with the title *Theory of Heat*, which repeated Tait's mistaken interpretation of entropy as available energy. With some prodding from Clausius in a letter to the *Philosophical Magazine*, Maxwell recognized his error, and demonstrated, in a second edition of his book, that there actually was a connection among entropy and absolute temperature and *un*available energy.

In his example, Maxwell pictured a system whose initial absolute temperature was T interacting both mechanically and thermally with its surroundings maintained at a constant lower temperature T_0. He visualized a two-stage cyclic process in which the system exchanged an amount of heat Q with the surroundings, decreased its energy, entropy, and temperature from U, S, and T to U_0, S_0, and T_0 of the surroundings, and at the same time performed the amount of work W on the surroundings.

Maxwell's conclusion was that the total energy change $U - U_0$ in his process could never be entirely converted to work output. The maximum work obtainable, in reversible operation, was $(U - U_0) - T(S - S_0)$. Maxwell called the entropy-related quantity $T(S - S_0)$ "unavailable energy": it could not be converted to work in any case. If Maxwell's process was irreversible, the work output was diminished still more, to something less than $(U - U_0) - T(S - S_0)$. This further loss, equal to what Thomson called "dissipated energy," was avoidable with better design of the work-producing machinery.

Clausius succeeded in straightening out Maxwell's misconceptions, but he was not so fortunate with Tait and, later, Thomson. Tait had attempted in his book to carry out an analysis similar to Maxwell's just outlined. In his derivation, he managed not only to ignore the distinction between unavailable and dissipated energy, but, in one famous passage, to contradict both the first and second laws of thermodynamics. These blunders brought sharp criticism from Clausius in letters to the *Philosophical Magazine*. Finally, in retreat, Tait drew Thomson into the controversy; but Thomson's remarks were no better informed on Clausius's version of the second law than those of Tait.

A Lost Portrait

Scientists are not always objective, but the controversies—or the contestants—die eventually, and then a workable consensus is reached. When this happens (and it is a rule of science history that it always does) what is left is a textbook or "standard" version of the subject. A few names may remain, attached to theories, equations, or units, but the human story, that of the people, their claims, and their quarrels, fades. There are advantages to this practice. It would not be easy for students to appreciate the formal structure of science if they had to cope with historical misunderstandings at every turn. No doubt *some* of the historical developments, when they are misguided enough—Tait's efforts may qualify here—are dispensable. But the other side of the human story, which tells of creativity gone right, not wrong, should be remembered.

These comments are prompted by thoughts of Clausius and his place—or lack of it—in the general impression of science history. Clausius's work on the first and second laws of thermodynamics had an enormous influence on the consensus view of thermodynamics established in the late nineteenth and early twentieth centuries. Clausius's equations, some of them written almost exactly as he expressed them a century or so earlier, are on display in all modern thermodynamics textbooks, and in an astonishing variety of other texts where the methods of thermodynamics are applied. Yet Clausius himself, even his name, has all but disappeared. In a typical modern thermodynamics text we find his name associated with a single, comparatively minor, equation. His name should at least be mentioned in connection with the first-law equation (8),

$$dU = dQ - PdV,$$

and the entropy equation (17)

$$dS = \frac{dQ}{T} \text{ (reversible process).}$$

But far worse than that kind of neglect, which can, after all, be repaired, is the vanishing of Clausius as a human being. Perhaps more than any other major nineteenth-century scientist, Clausius has been neglected in biographical studies. We know that he was born in Köslin, the youngest in a family of eighteen children. His father was the principal of a small private school, where Clausius received his early education. He continued his studies at the Stettin Gymnasium, and then at the University of Berlin. He received his doctorate at the University of Halle in 1847, and did his first teaching at the Royal Artillery and Engineering School in Berlin, soon after publishing his first paper on thermodynamics. In 1855, he moved to the Polytechnicum in Zürich, where he remained for fourteen years and did some of his most important work. In 1869, he returned to Germany, first to the University of Würzburg for two years, and finally to the University of Bonn, where he remained for the rest of his life. He served as a noncombatant in the Franco-Prussian War and was severely wounded in the knee. He was married and had six children. His wife died tragically in childbirth. Late in his life, when he was in his sixties, he married again. That brief sketch reports most of what can be gathered about Clausius's personal life from the available biographical material.

The only aspect of Clausius's personality that can be inferred from comments of his contemporaries is his contentiousness. We read in letters of "old Clausius" or "that grouch Clausius." He was a lifelong rival of Helmholtz. Max Planck relates that he tried to correspond with Clausius on matters relating to the second law, but Clausius did not answer his letters. In Clausius's portraits, we see a strong, unforgiving face. It is not difficult to picture this man exchanging polemic salvos with Tait.

What we do have from Clausius is his collected papers. We can read Clausius and fully appreciate his place in the beautifully clear line of development of thermodynamics through the middle fifty years of the nineteenth century—from Carnot to Clausius and finally to Clausius's greatest successor, Willard Gibbs. Clausius's role was pivotal. He knew how to interpret and rebuild Carnot's message, and then to express his own conclusions so they could be used by another genius, Gibbs. Clausius's papers on entropy were also a major influence on Planck, who used the entropy concept as a bridge into the realm of quantum theory. The grandest theories make their own contributions, and then inspire the creation of other great theories. Clausius's achievement was of this rare kind.

This is an impressive story, but as a story it is disappointing, simply because we still do not know the main character. Most of us would consider it a great misfortune if we knew no more about Cézanne, Flaubert, and Wagner, say, than what they put on canvas or paper or in a musical score. Clausius, their contemporary and equal as a creative genius, has been taken from us as a human being in this way. We should mourn the loss.

9

The Greatest Simplicity
Willard Gibbs

1839–1903

A Natural Theorist

He held few positions of academic or scientific eminence. During his thirty-two years of teaching, no more than a hundred students in total attended his courses. For the first ten years of his tenure at Yale University, he received no salary. He rarely attended professional meetings or traveled. Except for an obligatory European trip to the scientific outside world, and annual excursions to the New England and Adirondack mountains, his life was confined to New Haven, Connecticut, and hardly spanned more than the two blocks separating his home on High Street and his office in the Sloane Laboratory.

Willard Gibbs made his life in other ways. His world was theoretical physics. He saw more and traveled further in that world than most of his contemporaries, including Clausius and Boltzmann. Just as others are natural writers or natural musicians, Gibbs was a natural theorist. His judgment was perfectly attuned to the theoretical matters he studied. He had no need—indeed, in nineteenth-century America, hardly any opportunity—for close contact with informed colleagues. He knew, and did not have to be told, when he was right simply by exercising his own intuitive response and general knowledge. Few theoretical scientists have had the talent and the assurance to do their work in such isolated fashion. Only Einstein—who wrote some of his most important papers before he had even laid eyes on another theoretical physicist—may have outdone Gibbs in this respect.

Gibbs and Clausius

Gibbs's first published work was on thermodynamics. Throughout his thermodynamic studies he was strongly influenced by Clausius, and he left no doubt concerning that debt. Gibbs's first two papers were based on Clausius's equations for heat,

$$dQ = dU + PdV, \tag{1}$$

and entropy,

$$dS = \frac{dQ}{T} \text{ (reversible process).} \tag{2}$$

Gibbs simply eliminated dQ from the two equations by solving for dQ from the second, $dQ = TdS$, substituting this in the first, and solving for dU,

$$dU = TdS - PdV. \tag{3}$$

(It can be proved that this equation does not require the reversibility restriction, but that point is not important because the equation is nearly always applied to reversible processes.) Although Gibbs was the first to appreciate the fundamental importance of equation (3), Clausius certainly *thought* about the equation, so it seems fair to call it the "Clausius equation." (Gibbs has his own more comprehensive equation.) Clausius appears to have made no comment on Gibbs's work. Had he done so, an expression of his debt to Gibbs might have been appropriate. For it was mainly Gibbs who cleared away the doubt and confusion and focused attention on Clausius's implied equation (3).

Gibbs made his case for the Clausius equation in two papers published in 1873. His style in the 1873 papers makes difficult reading for a modern student because he relies on a geometrical kind of reasoning that is no longer in fashion. But for Gibbs, and some of his contemporaries, notably Maxwell, geometrical constructions were closer to the physical truth than the analytical arguments used by Clausius, Thomson, and others. The analytical approach had brought many advances, but its lengthy, abstract arguments had also contributed a certain amount of confusion.

The entropy concept was a good example of what analytical thought could and could not accomplish in physics. Clausius had defined the entropy concept in the mathematical sense, and had not missed or misunderstood any of its formal features. Even so, he could not demonstrate to his contemporaries, or even to himself, the prime importance of entropy in thermodynamics. Others could hardly get the formalities straight. The famous Tait-Clausius entropy controversy, even when it reached the stage of open warfare, concerned matters that were, from the physical viewpoint, no more than rudimentary.

From equation (3), Gibbs could read the mathematical message that changes dU in the internal energy U are determined by changes dS and dV in the entropy S and volume V, or in other words, that internal energy is a function $U(S,V)$ of entropy and volume. He expressed this dependence of U on S and V in three-dimensional energy surfaces. Part of such a surface is sketched in figure 9.1. One point on the surface is emphasized, and arrows tangential to the surface are drawn to show how the surface is shaped at that point. The arrow on the left is constructed for a fixed value of the entropy and parallel to the V direction. In this case of constant entropy, $dS = 0$, and equation (3) becomes

$$dU = -PdV \text{ (constant } S\text{)},$$

or

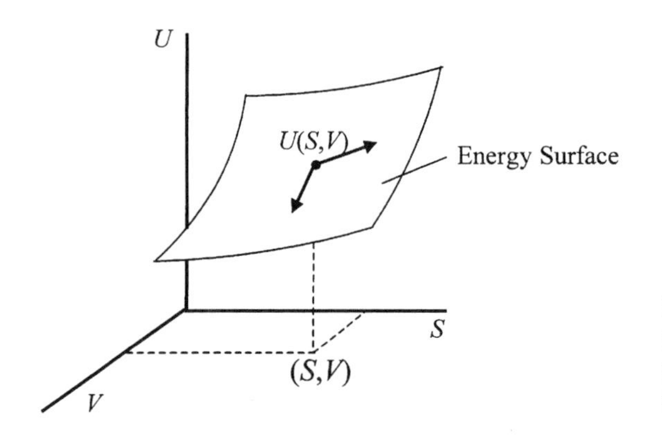

Figure 9.1. An energy surface containing points located by the entropy S, the volume V, and the internal energy regarded in the Gibbs manner, as a function $U(S,V)$ of S and V.

$$\frac{dU}{dV} = -P \text{ (constant } S). \tag{4}$$

The derivative $\frac{dU}{dV}$ in the last equation is a measure of the steepness or "slope" of the energy surface where the arrow is constructed (see fig. 9.2). According to equation (4), the same derivative is equal to $-P$, the negative of the pressure. Thus, anywhere on the energy surface the slope parallel to the V direction for some constant value of the entropy S calculates the pressure. These slopes are usually downhill, that is, negative, because pressures are usually positive, and slopes calculated as $\frac{dU}{dV} = -P$ are negative.

By a similar argument, the arrow on the right in figure 9.1 measures the slope of the energy surface parallel to the S direction for a fixed value of V. In this case, Clausius's equation (3) reduces to

$$dU = TdS \text{ (constant } V),$$

or

$$\frac{dU}{dS} = T \text{ (constant } V). \tag{5}$$

Here the slope is calculated with the derivative $\frac{dU}{dS}$ (fig. 9.3), and that derivative also equals the absolute temperature T, according to equation (5). The physical message here is that slopes of the energy surface measured parallel to the S direction calculate absolute temperatures, and those slopes are always uphill—that is, positive—because absolute temperatures are always positive.

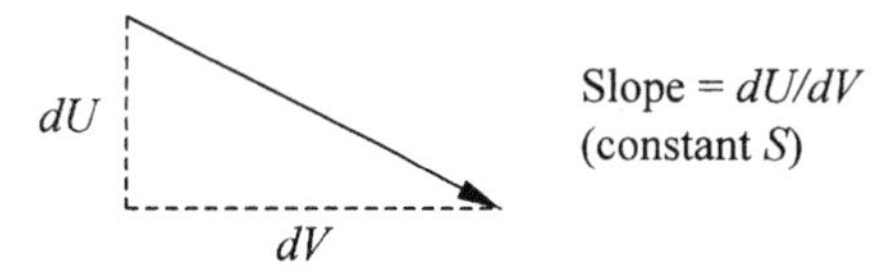

Figure 9.2. Side view of the left arrow extracted and enlarged from fig. 9.1. Like the ratio of rise to run for a staircase, the ratio of dU (rise) to dV (run), that is, the derivative $\frac{dU}{dV}$, calculates the slope of the arrow, and of the energy surface, at the point where the arrow is drawn.

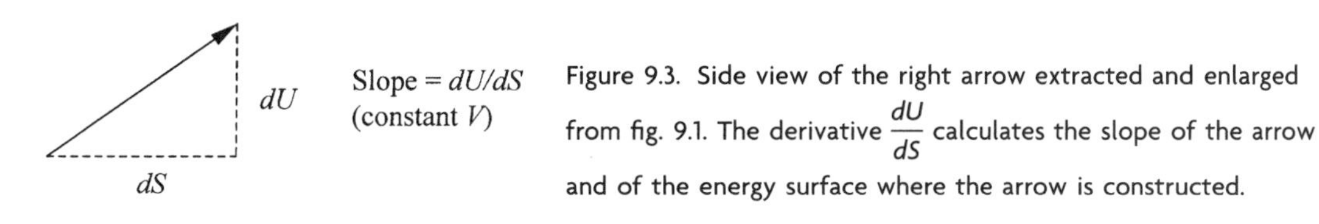

Figure 9.3. Side view of the right arrow extracted and enlarged from fig. 9.1. The derivative $\frac{dU}{dS}$ calculates the slope of the arrow and of the energy surface where the arrow is constructed.

In his first two 1873 papers, Gibbs elaborated this geometrical model in virtuosic detail. He imagined a plane containing the two tangential vectors, and pictured the plane rolling over the energy surface; at each point of contact between the plane and the surface the complete thermodynamic story is determined: a volume, an entropy, an internal energy, and from equations (4) and (5), the pressure and temperature. He showed how to project the surface into two dimensions (entropy and volume) and draw contours of constant pressure and temperature (like the constant altitude contours on a topographical map). He demonstrated that for certain conditions of pressure and temperature the rolling tangent plane has not just one but two, or even three, simultaneous points of contact. These multiple points of contact represent the coexistence of different phases (for example, solid, liquid, and vapor).

The *Principia* of Thermodynamics

These were the simple but broad conclusions reached by Gibbs in his first two papers on thermodynamics. Thus far, Gibbs had strengthened what had already been done formally, if tentatively, by Clausius. In his next work, published in several installments between 1875 and 1878, Gibbs again advertised that Clausius was his inspiration. He started with Clausius's couplet of laws: "The energy of the universe is constant. The entropy of the universe tends to a maximum." He took as his foundation the entropy rule and a simple adaptation of the Clausius equation (3). Here, however, he went far beyond the hints provided by Clausius.

Gibbs's 1875–78 "paper"—it is really a book covering about three hundred pages of compressed prose and exactly seven hundred numbered mathematical equations—has been called, without exaggeration, "the *Principia* of thermodynamics." Like Newton's masterpiece, Gibbs's *Equilibrium of Heterogeneous Substances* has practically unlimited scope. It builds from the most elementary beginnings to fundamental differential equations, and then from the fundamental equations to applications far and wide. Gibbs spells out the fundamental thermodynamic theory of gases, mixtures, surfaces, solids, phase changes (for example, boiling and freezing), chemical reactions, electrochemical cells, sedimentation, and osmosis. Each of these topics is now recognized, largely by physical chemists, as a major area of research. In the 1870s, with the discipline of physical chemistry not yet born, Gibbs's topics were unfamiliar and disparate. Gibbs's *Equilibrium* brought them together under the great umbrella of a unified theory.

But it was decades before Gibbs's book found more then a few interested readers. In another resemblance to Newton's *Principia*, Gibbs's *Equilibrium* had—and still has—a limited audience. One reason for the neglect was Gibbs's isolation, and another his decision to publish in an obscure journal, *Transactions of the Connecticut Academy of Arts and Sciences*. More important, however, was (and still is) Gibbs's writing style. Reading Gibbs is something like reading Pierre Simon Laplace (a famous mathematician and Newton's successor in the field of celestial mechanics), as E. T. Bell describes it. Laplace hated clutter in his mathematical writing, so to condense his arguments, "he frequently omits but the conclusion, with the optimistic remark, 'Il est aisé à voir' (It is easy to see). He himself would often be unable to restore the reasoning by which he had 'seen' these easy things without hours—sometimes days—of hard labor. Even gifted readers soon acquired the habit of groaning whenever the famous phrase appeared, knowing that as likely as not they were in for a week's blind work."

Gibbs did not frequently use the "famous phrase," and one doubts that he ever had trouble recalling his proofs, but he certainly left out a lot. The intrepid reader who takes on Gibbs's *Equilibrium* can expect many months of "blind work." The science historian Martin Klein quotes a letter from Lord Rayleigh, an accomplished theoretical physicist himself, suggesting to Gibbs that his *Equilibrium* was "too condensed and too difficult for most, I might say all, readers." Gibbs's response, no doubt sincere, was that the book was instead "too *long*" because he had no "sense of the value of time, of my own or others, when I wrote it."

Gibbs's writing can be faulted for its difficulty, but at the same time appreciated for its generality and unadorned directness. Gibbs expressed his ideal when he was awarded the Rumford Medal by the American Academy of Arts and Sciences: "One of the principal objects of theoretical research is to find the point of view from which the subject appears in its greatest simplicity." He always aimed for a "simpler view," which often meant perfecting the mathematical language. He said to a student, Charles Hastings, "If I have had any success in mathematical physics, it is, I think, because I have been able to dodge mathematical difficulties."

The Entropy Maximum

Clausius's entropy rule, Gibbs's principal inspiration in addition to the Clausius equation (3), asserts that any changes in an isolated system (completely disconnected from its surroundings) lead to entropy increases. These changes can be driven by any kind of nonuniformity, mechanical, thermal, chemical, or electrical. If, for example, a system has a cold part and a hot part, heat transfer from hot to cold takes place, if it can; the overall entropy increases, and continues to do so until the system is thermally uniform with a single temperature between the original high and low temperatures. The system is then in thermal equilibrium, all change ceases, and the entropy has a maximum value. A similar drive to uniformity, accompanied by an entropy increase to a maximum value at equilibrium, is found in isolated systems with nonuniformities in pressure, chemical composition, and electrical potential. Nature abhors nonuniformities, and flattens them if it can.

For a taste of Gibbs's method, here is a simple example that shows how he analyzed some of these entropy changes. Picture a gaseous system with two compartments separated by a sliding, thermally conducting partition (fig. 9.4). A rigid, insulating wall surrounds the system and keeps it isolated from its surroundings. In one compartment, the pressure and temperature are P_1 and T_1, and in the other P_2 and T_2. P_1 is greater than P_2, and T_1 greater than T_2, so the sliding partition is pushed from left to right by the pressure difference, and heat is also transported in that direction.

To find the equilibrium conditions in this situation, Gibbs noted that because the system is isolated by rigid, insulating walls its energy and volume are constant. He applied the Clausius equation (3) to both compartments, and ultimately

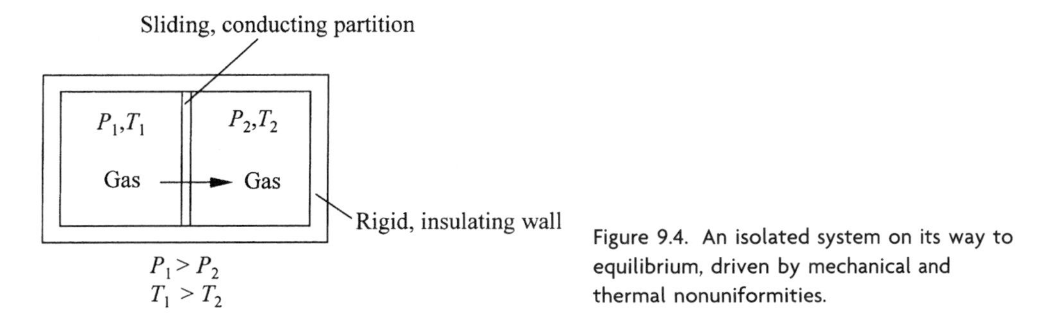

Figure 9.4. An isolated system on its way to equilibrium, driven by mechanical and thermal nonuniformities.

found, not surprisingly, that in equilibrium the pressure and temperature are equalized,

$$P_1 = P_2 \text{ and } T_1 = T_2 \text{ (equilibrium).}$$

For a second example, we elaborate the system so the partition is not only movable and conducting, but also permeable: the gas in the system can diffuse through it. The system is now considerably more complicated. We will soon look at Gibbs's general solution to the problem, but first a digression on chemical matters is in order.

Chemistry Lessons

Chemical reactions are written in a familiar language. For example,

$$2\ H_2 + O_2 \rightarrow 2\ H_2O$$

denotes the reaction of hydrogen (H_2) with oxygen (O_2) to form water (H_2O), a well-known reaction widely used in rocket engines and fuel cells. The substance formed in the reaction, H_2O, is the "product" of the reaction, and the substances consumed, H_2 and O_2, are "reactants." In modern usage, this chemical statement can be interpreted on any scale, from the microscopic to the macroscopic. At the finest microscopic level it describes two molecules of hydrogen reacting with one molecule of oxygen to form two molecules of water. These same proportions apply to any number of reactions, even a number large enough to make the H_2, O_2, and H_2O amounts macroscopic in size. For *any* number N,

$$2N \text{ molecules } H_2 + N \text{ molecules } O_2 \rightarrow 2N \text{ molecules } H_2O.$$

To do their quantitative work, chemists need a standard value of N. An arbitrary, but convenient, choice is the number of molecules in about 2 grams of H_2 (actually, 2.016 grams). Called "Avogadro's number" (for Amedeo Avogadro, who proposed in 1811—an early date in the history of molecular physics—that equal volumes of gases at the same pressure and temperature contain the same numbers of molecules), it is represented by N_A, and has the value

$$N_A = 6.022 \times 10^{23},$$

an *extremely* large number (about equal to the number of cups of water in the Pacific Ocean). This many molecules of H_2 is one "gram-molecule," or one "mole," of hydrogen. A mole of O_2 molecules, also containing N_A molecules, weighs about 32 grams, and one mole of H_2O molecules about 18 grams.

Summarizing all of this for the water reaction, we have

	$2H_2$	+	O_2	$\rightarrow$	$2\ H_2O$
	$2N_A$ molecules		N_A molecules		$2N_A$ molecules
or	2 moles		1 mole		2 moles
or	4 grams		32 grams		36 grams.

Note that in this chemical reaction—and in most others—there is no gain or loss of atoms: at the molecular level, six atoms enter into the reaction (four Hs in $2H_2$ plus two Os in O_2) and six atoms leave the reaction (four Hs and two Os in $2H_2O$). In consequence, there is no gain or loss of mass in the reaction: 36 grams of H_2 and O_2 form 36 grams of H_2O.

Potentials

Clausius's equation (3) tells us that the internal energy U changes when the volume V and entropy S change. But this is not the whole energy story. All chemical substances, or "chemical components," as Gibbs called them, have a characteristic internal energy, and if any component is added to a system, let's say through a pipe from the surrounding area, the total internal energy U changes in proportion to the amount of the component added.

Suppose a uniform system containing only one chemical component (for example, water) is isolated from its surroundings except for the pipe, and a small amount of the component measured as dn mole is added. The internal energy of the system changes in proportion to dn,

$$dU \propto dn \text{ (system isolated except for pipe).}$$

Gibbs wrote this as an equation with a proportionality factor μ included,

$$dU = \mu dn \text{ (system isolated except for pipe).} \tag{6}$$

The μ factor is a state function that Gibbs called a "potential." Maxwell gave it a better name: "chemical potential." It is to chemical changes what pressure and temperature are to mechanical and thermal changes. If a system has chemical nonuniformities for a component, that component will migrate from regions of high chemical potential to low until, in equilibrium, all the chemical nonuniformities are smoothed out.

Gibbs elaborated and generalized equation (6) by assuming that if the isolation is further broken, and the system with a pipe is allowed to communicate with its surroundings in heating and working processes, only two additional terms are required, those already familiar in the Clausius equation (3),

$$dU = Tds - PdV + \mu dn. \tag{7}$$

This is a simple version of what we will call the "Gibbs equation."

We can return now to the example of the isolated two-compartment system with a sliding, conducting, permeable partition (fig. 9.5). Gibbs analyzed this case by again recognizing that the system's total energy and volume are constant. He also assumed that the total amount of the gas is constant because the isolating walls prevent any gain or loss to the surroundings; the gas is constrained to pass between the two compartments. Two statements of equation (7), one for each compartment, then dictate that, as before, the sliding partition moves from left to right under the pressure difference, and heat is transported in the same direction under the temperature difference. At the same time, gas is transported through the permeable partition under the chemical potential difference. Finally, at equilibrium,

$$P_1 = P_2,\ T_1 = T_2,\ and\ \mu_1 = \mu_2\ \text{(equilibrium)}.$$

Here we can see the parallel roles of pressure, temperature, and chemical potential in defining mechanical, thermal, and chemical equilibrium.

For a differential equation, Gibbs's equation (7) is uncharacteristically user-friendly. Unlike most other major differential equations in physics, it is solved (integrated) with the utmost simplicity. The special mathematical structure of the equation allows one to replace dU with U, dS with S, dV with v, and dn with n, to put the equation in integrated form,

$$U = TS - PV + n\mu. \quad (8)$$

Equations (7) and (8) are still restricted to a system containing only one chemical component. Another pleasant feature of the Gibbs equation is that it can be adapted to any number of chemical components with a few more simple modifications. If there are two components in the system, call them A and B, equations (7) and (8) have two added chemical potential terms, one for each component,

$$dU = TdS - PdV + \mu_A dn_A + \mu_B \mathrm{dn}_A \quad (9)$$

and

$$U = TS - PV + n_A\mu_A + n_B\,\mu_B. \quad (10)$$

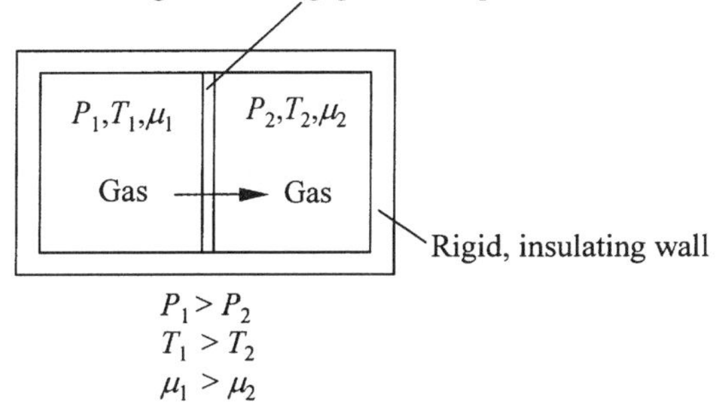

Figure 9.5. An isolated system on its way to equilibrium, driven by mechanical, thermal, and chemical nonuniformities.

Avoiding Molecules

Although he might have preferred to do so, Gibbs did not use molar quantities (the *n*s and *dn*s in equations [7]–[10]), as we have, nor did he write chemical reactions with a molecular interpretation implied. Instead, he used mass units (for example, grams) to measure quantities of chemical components. His way of writing the water reaction, which seems quaint to us, was

$$1 \text{ gram hydrogen} + 8 \text{ grams oxygen} = 9 \text{ grams water},$$

and he defined chemical potentials with respect to mass m rather than moles n, so his version of equation (8) was

$$U = TS - PV + m\mu.$$

In the 1870s, no direct experimental evidence suggested the existence of molecules, and many (but not all) physicists preferred to write their physics without molecular hypotheses. For the most part, Gibbs followed this preference in his *Equilibrium*. (As noted, so did Clausius in the 1860s.)

But when there was a fundamental point to be made, Gibbs did not hesitate to invoke molecules. He made a detailed equilibrium calculation for a chemical reaction in which two NO_2 molecules combine to form a single N_2O_4 molecule. And in the midst of a discussion of entropy changes for mixing processes, he made a prophetic remark that initiated a major discipline he would later call "statistical mechanics."

He had in mind the spontaneous mixing of two pure gases, say A and B, to form a uniform mixture,

$$\text{pure A} + \text{pure B} \rightarrow \text{A and B mixed},$$

always resulting in an entropy increase. (This is another example of an entropy increase accompanying the natural tendency for nonuniformities to evolve into uniformity.) Gibbs pictured such mixing on a molecular scale, with the random motion of A and B molecules causing them to diffuse into each other, and sooner or later, to become uniformly mixed. He also imagined the entropy-*decreasing*, *unmixing* process,

$$\text{A and B mixed} \rightarrow \text{pure A} + \text{pure B},$$

in which A molecules move in one direction, B molecules in another, and the mixture spontaneously sorts itself into phases of pure A and pure B. This is never observed, however, because once A and B molecules have mixed, their astronomical numbers and their random motion make it highly unlikely that they will ever part company.

Even so, Gibbs realized, unmixing and its associated entropy decrease are not *quite* absolute impossibilities, just fantastically improbable. "In other words," he wrote, "the impossibility of an uncompensated decrease of entropy seems to be reduced to improbability." Put more abstractly, his conclusion was that the entropy of a thermodynamic state is connected with the probability for that state; the mixed state is enormously more probable than the unmixed state.

Gibbs did not follow this reasoning further in his *Equilibrium*, but at about the same time Boltzmann was independently developing the probabilistic interpretation of entropy in quantitative terms. And much later, in 1902, Gibbs made the entropy-probability connection a centerpiece of his molecular interpretation of thermodynamics.

Gibbs Energy

When chemical potentials are added for all the chemical components in a system, a special kind of energy, now called "Gibbs energy," results. Suppose there are two components, A and B, in a system, and their molar amounts are n_A and n_B; the chemical potential sum in question is $n_A\mu_A + n_B\mu_B$, which we evaluate with equation (10) rearranged to

$$U + PV - TS = n_A\mu_A + n_B\mu_B. \tag{11}$$

The quantity on the left side of this equation defines the state function now called Gibbs energy, and represented with the symbol G (Gibbs called it the ζ function),

$$G = U + PV - TS, \tag{12}$$

which simplifies equation (11),

$$G = n_A\mu_A + n_B\mu_B. \tag{13}$$

We can see from this equation that $n_A\mu_A$ is the Gibbs energy contributed by the n_A moles of A in the system, and therefore that μ_A is the Gibbs energy for one mole of A. Similarly, $n_B\mu_B$ and μ_B are Gibbs energies for n_B and one mole of B.

One reason for defining the Gibbs energy is a simple matter of economy: it compresses into a single state function all the other state functions of importance in thermodynamics (U, S, and V) as well as the principal state-*determining* variables (P and T). It satisfies physicists' primitive instinct to make their mathematics as compact as possible. But the Gibbs energy, and its precursor chemical potentials, do much more than that.

The Second Law Transformed

As Clausius saw it, the second law of thermodynamics is a principle that shows how to calculate entropies. For a reversible process, the calculation is

$$dS = \frac{dQ}{T},$$

and for an irreversible process,

$$dS > \frac{dQ}{T}.$$

Here we combine these two statements in an equality-inequality,

$$dS \geq \frac{dQ}{T},$$

(read "greater than or equal to" for the symbol $\geq$), substitute for dQ from equation (1),

$$dS \geq \frac{dU + PdV}{T},$$

and rearrange this to

$$dU + PdV - TdS \leq 0. \tag{14}$$

If the pressure P and temperature T are constants,

$$PdV = d(PV) \text{ and } TdS = d(TS).$$

For the product ax, $adx = d(ax)$ if a is a constant. Thus for constants P and T the equality-inequality (14) becomes

$$dU + d(PV) - d(TS) \leq 0,$$

or

$$d(U + PV - TS) \leq 0,$$

or, with definition (12) recognized,

$$dG \leq 0.$$

The equality part of this statement applies to a reversible process or equilibrium,

$$dG = 0 \text{ (constant } P \text{ and } T\text{; reversible process or equilibrium)}, \tag{15}$$

and the inequality to an irreversible process,

$$dG < 0 \text{ (constant } P \text{ and } T\text{; irreversible process)}. \tag{16}$$

Although entropy (in an isolated system) increases to a maximum at equilibrium, the Gibbs energy (in a system at constant pressure and temperature) changes in the opposite direction; it decreases to a minimum.

Chemical Thermodynamics

We have pictured chemical components entering and leaving a system through pipes. (Membranes would be more elegant.) Components can also appear and disappear via chemical reactions. For instance, H_2 and O_2 are removed and H_2O is added by the water-forming reaction mentioned before,

$$2\ H_2 + O_2 \rightarrow 2\ H_2O.$$

Gibbs's equation (9), and its extensions for more than two components, apply to this and any other reacting system. Suppose a small amount, $2dx$ mole, of H_2O is produced (to keep the signs straight, we will make dx positive), so molar changes in the reactants H_2 and O_2, which are removed in the reaction, are the negative amounts $-2dx$ and $-dx$, that is,

$$\begin{aligned} dn_{H_2O} &= 2dx \\ dn_{H_2} &= -2dx \\ dn_{O_2} &= -dx, \end{aligned}$$

and Gibbs's equation for the three components H_2O, H_2 and O_2 is

$$\begin{aligned} dU &= TdS - PdV + 2\mu_{H_2O}dx - 2\mu_{H_2}dx - \mu_{O_2}dx \\ &= TdS - PdV + (2\mu_{H_2O} - 2\mu_{H_2} - \mu_{O_2})dx, \end{aligned}$$

or

$$dU + PdV - TdS = (2\mu_{H_2O} - 2\mu_{H_2} - \mu_{O_2})dx.$$

As before, if the pressure P and temperature T are constant, the left side of this equation becomes $d(U + PV - TS) = dG$, so

$$dG = (2\mu_{H_2O} - 2\mu_{H_2} - \mu_{O_2})dx \text{ (constant } P \text{ and } T).$$

The second law tells us that $dG \leq 0$ for constant pressure and temperature. Thus, according to the last equation,

$$2\mu_{H_2O} - 2\mu_{H_2} - \mu_{O_2} \leq 0 \text{ (constant } P \text{ and } T). \qquad (17)$$

(Remember that we have made dx positive.)

Here we see chemical potentials combined to characterize an entire chemical reaction by calculating the reaction's "Gibbs energy change," the difference between the Gibbs energy for the reaction's product ($2\mu_{H_2O}$) and that for the two reactants ($2\mu_{H_2} + \mu_{O_2}$), in modern usage represented

$$\Delta_r G = 2\mu_{H_2O} - \mu_{O_2} - 2\mu_{H_2}.$$

(The symbol Δ denotes finite changes; it is a finite counterpart of d, which stands for small or infinitesimal changes.) The same recipe, chemical potentials for products minus those for reactants, calculates the Gibbs energy change $\Delta_r G$ for any reaction. The general conclusion illustrated by the equality-inequality (17) is

$$\Delta_r G \leq 0 \text{ (constant } P \text{ and } T).$$

The equality describes reversible operation or equilibrium,

$$\Delta_r G = 0 \text{ (reversible operation or equilibrium; constant } P \text{ and } T), \quad (18)$$

and the inequality irreversible operation,

$$\Delta_r G < 0 \text{ (irreversible operation; constant } P \text{ and } T). \quad (19)$$

The physical picture here is easy to remember. Any chemical reaction moves downhill ($\Delta_r G < 0$ means downhill) on a Gibbs energy surface if it can, driven by the chemical potential difference between the products and the reactants. Chemical change continues until reactant and product chemical potentials are balanced, the Gibbs energy change equals zero, and chemical equilibrium is reached.

A chemical reaction descending spontaneously in Gibbs energy is something like a falling weight, and also like heat falling from a high to a low temperature in a heat engine. Like the falling weight and heat, the descending chemical reaction can be a useful source of work. That work is often used by building the reaction into an electrochemical cell, which supplies electrical work output. Flashlight batteries and fuel cells are examples.

Remember that the amount of work gotten from a falling weight or from falling heat in a heat engine depends on how well the machinery is designed; you get the most efficient performance if the device operates reversibly. The Gibbs energy concept is designed so that the Gibbs energy change $\Delta_r G$ calculates the best possible electrical work obtainable from an electrochemical cell based on the reaction running reversibly. For the water reaction, we can calculate from tabulated Gibbs energy data that a reversible hydrogen-oxygen electrochemical cell generates 1.23 volts of electrical output (if H_2O is produced as a liquid rather than as a gas). Practical electrochemical cells are always to some extent irreversible. A fuel cell using the water reaction is likely to have an output of about 0.8 volt.

The Gibbs energy change for a chemical reaction calculates the maximum energy that is available or "free" for the performance of work. For this reason, Gibbs energy is also frequently called "free energy."

Gibbs and Maxwell

Gibbs had a long mailing list for reprints of his papers, including, it seems, every established scientist in the world who could possibly have had an interest in his work. Most of these mailings, even those to Clausius, went unnoticed at first. They did, however, quickly capture the attention of Maxwell, who was more generous and alert than his colleagues, and particularly appreciated Gibbs's extensive use of geometrical reasoning. Gibbs' two 1873 papers on the geometrical interpretation of Clausius's equation prompted Maxwell to make a plaster model displaying a full energy surface for water. He located on this water "statue" areas where the liquid, vapor, and solid phases coexist, areas where two phases can coexist, and a triangle representing coexistence of all three phases. On the surface of the model, he carved contours of constant pressure and temperature, as dictated by Clausius's equation. Maxwell sent a copy of his water statue to Gibbs, who was flattered and pleased, but with typical modesty told students who asked about it that the model came from a "friend in England."

One of Gibbs's biographers, J. G. Crowther, remarks that Maxwell became, in effect, Gibbs's "intellectual publicity agent." But not for long; Maxwell died prematurely in 1879, only a year after Gibbs published the final installment of his *Equilibrium.* If Maxwell had lived, Crowther continues, "the greatness of Gibbs' discoveries might have been understood ten years sooner, and physical chemistry and chemical industry today [the 1930s] might have been twenty years in advance of its present development."

At about this same time, in 1879, Gibbs gave a series of lectures in Baltimore at a new and aspiring institution, the Johns Hopkins University. Before he returned to New Haven, Gibbs was offered a position in the Johns Hopkins physics department by the university president, D. C. Gilman (formerly the librarian at Yale). This was an attractive offer. Gilman and his department heads were recruiting a first-rate research faculty in physics and mathematics. Gibbs was still unpaid at Yale, nine years after his appointment as professor of mathematical physics; Gilman offered a respectable salary.

Gibbs planned to accept the offer. He hoped to keep his transactions with Gilman secret, but the news leaked to some of his colleagues, who promptly carried it to Yale president James Dana. A letter to Gibbs from Dana pleaded with him to "stand by us," and expressed the hope that "something will speedily be done by way of endowment to show that your services are really valued." The appeal was frank: "Johns Hopkins can get on vastly better without you than we can. *We can not.*"

Gibbs was surprised, touched, and finally persuaded. He sent his regrets to Gilman: "Within the last few days a very unexpected opposition to my departure has been manifested among my colleagues—an opposition so strong as to render it impossible for me to entertain longer the proposition which you made. . . . I remember your saying that . . . you thought it would be hard for me to break the ties that connect me with this place. Well—I have found it harder than I expected." Yale was, after all, where he belonged.

Beyond Thermodynamics

During the 1880s and 1890s, Gibbs (now receiving an annual salary of two thousand dollars) was thinking about another great theoretical problem: the molecular interpretation of thermodynamics. Gibbs had avoided molecular hypotheses as much as possible in his *Equilibrium*, focusing on the macroscopic energy and entropy concepts and on derived quantities such as the chemical potential and on the function we call Gibbs energy. Having created this macroscopic view in the 1870s, Gibbs decided that it was time to continue his search for the "rational foundations" of thermodynamics at the microscopic or molecular level. This was a description of molecular mechanics, necessarily made statistical because of the stupendous numbers of molecules involved in thermodynamic systems; Gibbs called it "statistical mechanics." The energy and entropy concepts were again of central importance, but now they were calculated as average values, and entropy was interpreted with the probability connection Gibbs had hinted at much earlier, in his *Equilibrium*. He unified the work of his predecessors, Maxwell and Boltzmann, and helped pave the way for the conceptual upheaval called quantum theory just arriving (mostly unnoticed) when Gibbs published his *Elementary Principles in Statistical Mechanics* in 1902.

Gibbs was not a mathematician, but like other great theorists mentioned in this book (for example, Maxwell, Einstein, and Feynman), he knew how to make mathematical methods serve his purposes in the simplest, most direct way. Whenever he approached a physical problem he thought as much about the mathematical language as about the physics.

Among Gibbs's teaching responsibilities was a course in the theory of electricity and magnetism, based on Maxwell's *Treatise on Electricity and Magnetism*. In this subject and others, notably mechanics, the mathematical description treats physical quantities that have direction in space as well as magnitude. Both attributes are built into a mathematical entity called a "vector," which can have three components, corresponding to the three spatial dimensions. Gibbs devised a new method that provided a convenient mathematical setting for the manipulation of vectors.

Gibbs's method of handling vectors was a departure from that of Maxwell and his British colleagues, who relied on the method of "quaternions," formulated by the Irish mathematician and physicist Rowan Hamilton. (William Rowan Hamilton was Ireland's greatest gift to mathematics and physics. While in his twenties, he invented a unified theory of ray optics and particle dynamics that influenced Erwin Schrödinger in his development of wave mechanics, almost a full century later. After the work on optics and dynamics, which was completed when he was twenty-seven, Hamilton's creative genius failed him, or rather strangely misled him. For many years he struggled to rewrite physics with his new quaternion method of mathematics. Hamilton believed he would write a new *Principia;* quaternions were to be as important as Newton's fluxions. But this work was never successful. Hamilton died a recluse, living in a chaotic dreamworld.) Gibbs found quaternions an unnecessary mathematical appendage in physical applications, and demonstrated the advantages of his own method in five papers on electromagnetic theory. For his students, he had a pamphlet printed with the title *Vector Analysis.*

Hamilton had his disciples and partisans. Prominent among them was P. G. Tait, playing his favorite role as polemicist. When news of Gibbs's vector analysis reached Tait, he promptly drew (a reluctant) Gibbs into a prolonged debate in the pages of the British journal *Nature.* Tait labeled Gibbs "one of the retarders of the quaternionic progress," and his vector analysis as "a sort of hermaphrodite monster compounded of the notation of Hamilton and Grassmann." (Hermann Grassmann was a nineteenth-century mathematician and linguist who was one of the first to propose a geometry embracing more than three dimensions.) Gibbs was no match for Tait in polemics, but he knew how to respond without the epithets:

> It seems to be assumed that a departure from quaternionic usage in the treatment of vectors is an enormity. If this assumption is true, it is an important truth; if not, it would be unfortunate if it should remain unchallenged, especially when supported by so high an authority. The criticism relates particularly to notations, but I believe that there is a deeper question of notions underlying that of notations. Indeed, if my offence had been solely in the matter of notation, it would have been less accurate to describe my production as a monstrosity, than to characterize its dress as uncouth.

Gibbs was confident that his method, uncouth or not, served "the *first* duty of the vector analyst . . . to present the subject in such a form as to be most easily acquired, and most useful when acquired." In practice, Gibbs was the clear winner in the debate. Gibbs's biographer Lynde Phelps Wheeler writes that "there has been a steady increase in the use of the vectorial methods of Gibbs through the years until now [the 1950s] they may be said to be practically universal."

A Gibbs Sketch

Josiah Willard Gibbs was born in 1839. His father, also Josiah Willard Gibbs, was a prominent philologist and professor of sacred literature at Yale University. (To the family and contemporaries, the father was "Josiah" and the son "Willard.") Son and father followed different intellectual paths, but they had much in common. One of Willard Gibbs's biographers, Muriel Rukeyser, describes Josiah Gibbs as "the most thoroughly equipped scholar of his college generation," and notes that "the two weapons on which he relied were accurate knowledge and precise statement. He loved [the work of the philologist], this sorting, and tagging, and comparing, this detective work among the clues left by the words of man."

Rukeyser pictures Willard: "A mild, frail child growing up in the Gibbs home with its simple manners, and its little Latin books—its primers and his father's Bible stories, and his mother's soft insistence on mildness." With four sisters, "he was a child, he was a little boy in a house of women. The family was presided over by the long, sympathetic face of the mother and the teaching schedule of the father."

He learned the lessons of death and responsibility as an adolescent. His youngest sister died when he was ten, and shortly after he entered Yale College at age fifteen his mother's health began slowly to decline. Willard's oldest sister, Anna, "more and more took her place as she grew weaker," Rukeyser relates, "and the boy grew up rapidly as the relations of the family shifted. His long face looks out from the early daguerreotypes, with its strong eyes, hostile one moment, and then suddenly soft and perceptive. He takes stillness with him."

After graduating from Yale in 1858 with prizes in Latin and mathematics, Gibbs entered the new Yale graduate school and earned the first Ph.D. in engineering in the United States, and then served Yale for three years as a tutor. By that time (the middle 1860s), his interest in the broader world of science was aroused and he traveled to Europe for three years of study at the scientific centers of Paris, Berlin, and Heidelberg. (But these travels did not take him to Bonn, where his scientific benefactor, Clausius, lived.) He returned to New Haven in 1869 and was appointed professor of mathematical physics at Yale (without salary) in 1871.

Gibbs was unpretentious, friendly, often humorous, and accommodating. According to Crowther, "he instantly laid aside without question any profound work when called upon to perform minor tasks. He never evaded the most trivial college duties, or withheld any of his valuable time from students who sought his instruction." His powers of concentration were so extraordinary that he could probably do the chore, even talk with a student, and hardly interrupt his train of thought.

Gibbs never married, in Wheeler's view because of "his inherited family responsibilities early in life, coupled with uncertainty of his health throughout the period when most young men have thoughts of founding a family of their own." In addition, a close relationship with his oldest sister Anna, who also remained unmarried, may have been important. Anna had "an especially retiring personality, accentuated by poor health," writes Wheeler. "It was said that she and

Willard could be silent together better than anyone else," Rukeyser tells us. "There seems to have been complete understanding between them. Most people deserved her silence, but living persons still remember long days spent in wonderful conversation with Gibbs and his sister—on trains, or in the country."

Gibbs was kind to children. A cousin, Margaret Whitney, remembered special treats when Gibbs took the Whitney children for a sleigh ride:

> He would turn to tuck us in and see that we were all right with a smile so friendly and re-assuring to the little girl beside him that she felt at once at ease with him. My best memory is driving with him in the winter in a cutter, a rare treat for me. The impression of standing beside the sleigh in the snow, waiting to be lifted in, snow all around, crisp air, sleighbells jingling by, all the world in swift motion, and I to be one of them, this sensation stayed with me and can always be evoked. It is well worth a tribute to the kind man who gave it to me.

His health was damaged by scarlet fever when he was a child, and minor illnesses were a problem throughout his life. He had a slight build, but was well coordinated and had athletic ability. Margaret Whitney recalled an encounter between Gibbs and a nervous horse: "He was on horseback, returning to the hotel [on vacation in Keene Valley, New York,] and the horse was misbehaving badly. But so firm was his hand on the rein and so good his seat that although they thought the horse might throw him any minute, he was able to control him and bring him quietly to a halt."

Gibbs accomplished so much in thermodynamics, and at a time when others seemed to be contributing more confusion than progress, that one wonders about his sources of inspiration. Why could Gibbs, so much more clearly than his contemporaries, see the fundamental importance of the Clausius equation? How could he be so certain that adding chemical potential terms to the Clausius equation would make it the master equation it is in modern thermodynamics? Addition of these terms is easily done mathematically, but mathematical ease does not guarantee physical meaning.

We can look at his working habits, which were extraordinarily internal. He lectured and wrote his papers without notes (in contrast to Newton, who could not think without a pen in his hand). He never discussed his researches informally with students or colleagues. Even without such prompting and checking devices, his papers contain few, if any, significant errors.

To an extent perhaps unexcelled in the annals of science, Gibbs was a natural theorist. It may also have been important that he was isolated in his new-world setting from contemporary scientific activity. It is not always true that isolation is an important creative influence in scientific effort, but in cases where established scientific workers are divided into warring camps it may have been an advantage to be as uncommitted and unprejudiced as Gibbs was.

"He expected nothing; *nothing* from outside," writes Rukeyser. "He was sure of himself, and trusted himself." Maxwell's support must have helped bolster that assurance. But even with Maxwell actively promoting his interests, Gibbs was hardly known outside the world of theoretical physics. J. J. Thomson, the discoverer of the electron and one of Maxwell's successors at Cambridge, tells of a conversation with a president on a faculty-recruiting mission from a newly formed American university. "He came to Cambridge," Thomson writes, "and asked me if I could tell him of anyone who could make a good Professor of

Molecular Physics." Thomson told him that one of the greatest molecular physicists in the world was Willard Gibbs, and he lived in America. The president responded that Thomson probably meant *Wolcott* Gibbs, a Harvard chemist. Thomson was emphatic that he did mean *Willard* Gibbs, and he tried to convince his visitor that Gibbs was indeed a great scientist. "He sat thinking for a minute or two," Thomson continues, "and then said, 'I'd like you to give me another name. Willard Gibbs can't be a man of much personal magnetism or I should have heard of him.' "

Another essential clue concerning Gibbs's inspiration is revealed in a comment made by Wheeler, who had considered writing his biography of Gibbs in two volumes, one concerned with Gibbs's scientific work and the other with the nonscientific events of his life, but it soon became clear that the two volumes had to be one: "I came to realize that to an unusual degree Gibbs' scientific work *was* Gibbs, and that really to understand him one must to a certain extent at least understand his work; as his life and work were so largely one, so must his story be."

To many people, including academic dignitaries trading on "personal magnetism," Gibbs seemed inhibited. Yet his friends were impressed by his calm equanimity. Wheeler quotes the daughter of Gibbs's close friend Hubert Newton. Josephine Newton found Gibbs "the happiest man" she ever knew. "This cheerfulness was, I think, due partly to an excellent sense of proportion which enabled him to estimate things at their true value, and partly to the uniformly good digestion which he enjoyed." Why *shouldn't* this man have been happy (and blessed with good digestion)? He was doing profoundly important creative work—and he knew it.

10

The Last Law

Walther Nernst

1864–1941

The Devil and Walther Nernst

According to a story current in Berlin in the early 1900s, God decided one day to create a superman. He worked first on the brain, fashioning a "most perfect and subtle mind." But he had other business, and the job had to be put aside. The Archangel Gabriel saw this marvelous brain and could not resist the temptation to try to create the complete man. He overestimated his abilities, however, and succeeded only in creating a "rather unimpressive looking little man." Discouraged by his failure, he left his creation inanimate. The devil came along, looked with satisfaction upon this unique, but lifeless, being and breathed life into it. "That was Walther Nernst."

This tale is told by Kurt Mendelssohn in a fine biography of Nernst. Mendelssohn also supplies us with a more authentic picture of Nernst: "There is no record of hereditary genius [in Nernst's family] or even of outstanding enterprise. It seemed that Walther owed his brilliance to a lucky throw of the genetic dice." At one time, Nernst considered becoming an actor and "he realized this ambition to some extent by wearing throughout his life the mask of a trusting and credulous little man. His favorite expression of innocent astonishment could be underlined by a twitch of the nose, which removed [his] pince-nez. There was always a note of astonishment in his voice and the outrageous and sarcastic comment of which he was the master was never accompanied by a change in his voice or a smile. He remained genuinely serious and mildly surprised."

Leipzig, Göttingen, and Berlin

As a student, Nernst traveled, according to the nineteenth-century custom, among the universities where the great men of science lived and taught. His educational journey took him to Zurich, Berlin (where Helmholtz lectured on thermodynamics), back to Zurich, then to Graz (to study under Ludwig Boltzmann), and finally to Würzburg (where work with Friedrich Kohlrausch inspired a lifelong interest in electrochemistry). He paused long enough in Graz to write a doctoral dissertation and learn lessons in "irritation physics" from Albert von Ettinghausen, a former student of Boltzmann's and Nernst's collaborator in his dissertation research. Nernst, who could never conquer his impatience, had endless admiration for Ettinghausen's easy acceptance of experimental frustrations. After a dismal failure of an experiment, Ettinghausen might say calmly, "Well, the experiment was not successful, at least not entirely."

Nernst's professional career was a story of almost unmitigated success. In the late 1880s, while at Würzburg, he met Wilhelm Ostwald and became his assistant when Ostwald accepted a professorship at Leipzig. Nernst lost no time in finding occupation for his talents in Ostwald's endeavors, all concerned with building foundations for the new discipline of physical chemistry. Nernst's first publication from Leipzig became one of the classics of the literature of electrochemistry; it presented to the world an equation that came to be known to generations of physical chemistry students as the "Nernst equation."

In 1891, Nernst was appointed assistant professor of physical chemistry at the University of Göttingen. Two years later he published one of the first physical chemistry textbooks—the second text in the field after Ostwald's *Lehrbuch der allgemeinen Chemie* (Textbook of General Chemistry). Nernst's text had the title *Theoretische Chemie* (Theoretical Chemistry), and it was dedicated to Ettinghausen. Nernst constructed his view of physical chemistry on thermodynamic foundations laid by Helmholtz, and on the molecular hypothesis ("Avogadro's hypothesis") advocated by Boltzmann (and strenuously opposed at the time by Ostwald). Still in use thirty years later, in its fifteenth edition, the Nernst text was the most influential in the field.

In three more years, Nernst had so impressed the Ministries of Education, not only in Prussia (where Göttingen is located), but also in Bavaria, that he was offered the professorship of theoretical physics at the University of Munich as Boltzmann's successor. The Prussian minister, Friedrich Althoff, was not to be outdone, however. Mendelssohn tells of the further bureaucratic bargaining, masterfully manipulated by Nernst:

> If Althoff wanted to keep Nernst in Prussia, he now had to make an effort that would go a bit beyond his own departmental responsibility. Nernst's price was the creation of a new chair of physical chemistry at Göttingen, and to go with it an electrochemical laboratory. Althoff could produce the new chair from the funds at his disposal, but for the laboratory he had to get money from the Minister of Finance—and that would take time. Nernst, who was certain he held the whip hand and always knew how to drive a hard bargain, forced Althoff into an unheard of act. It was the promise, to be given in writing, that should the laboratory in Göttingen not materialize, Nernst would get a chair in physics at Berlin. Althoff yielded, possibly because he had every reason to believe the Minister of Finance would play, as indeed he did. That was 1894 and Berlin would have to wait another eleven years.

Nernst's scientific talent extended to applied problems, especially those that had economic possibilities. While at Göttingen, he invented an electric lamp, which he hoped would compete with the Edison lamp, then not fully developed. Nernst's design was an application of his studies of ionic conduction. He first tried to persuade Siemens, an established German electric firm, to buy the patent on the invention. Siemens was not interested, either in the technical possibilities of the lamp or in Nernst's financial demands.

Nernst next offered the patent to Allgemeine Elektrizitäts Gesellschaft (A.E.G.), a newer and more adventurous company. After extended bargaining, in which Nernst demanded a lump sum and refused royalties, he got what he wanted: a million marks, enough to make him a wealthy man. Although it was ingeniously developed by A.E.G., with much of the technical work done by two of Nernst's students, the Nernst lamp finally lost in the competition with other designs. This financial failure for A.E.G. seems not to have discouraged its confidence in Nernst's technical abilities. Emil Rathenau, the A.E.G. chairman, remained friendly with Nernst for the rest of his life.

Although he had acquired wealth, become influential with those highly placed in the political and business worlds, and reached a position of eminence in the new science of physical chemistry, Nernst had not quite reached the pinnacle of success. There was one more academic world to conquer. His next move took him, in Mendelssohn's words, from his Göttingen "place in the sun" to an academic and scientific "summit" at the University of Berlin. In the spring of 1905, Nernst drove his family from Göttingen to Berlin in an open motorcar, accompanied by his favorite mechanic in case of breakdowns. That same year Nernst found the clue he needed to formulate his statement of what is now called the third law of thermodynamics.

Chemical Equilibrium

We might pause here, with Nernst about to make his great discovery, and look more closely at one of Nernst's major research interests, high-temperature chemical reactions involving gaseous components. In the early 1900s, such reactions were of great industrial importance. Franz Simon, a colleague of Nernst's in the 1920s, tells of the prevailing concern with gaseous reactions that inspired Nernst's work: "Fifty years ago [Simon's remarks were written in 1956] there was an intense interest in chemical gas reactions, partly because of the relative simplicity of the problem involved, which seemed to lend itself to treatment by physical methods, and partly because of the economic possibilities. Gas reactions had already played an important role in the growth of chemical heavy industry, and it was realized that ammonia synthesis in particular had become very important indeed for the German economy in peace and war."

The ammonia synthesis reaction is like the water reaction mentioned in the previous chapter, except that it replaces oxygen with nitrogen (N_2) and forms ammonia (NH_3),

$$N_2 + 3\,H_2 \rightarrow 2\,NH_3.$$

Ammonia can be used as a fertilizer or converted to nitrates for the manufacture of explosives. In the industrial process, nitrogen is obtained from air, and hydrogen from a reaction between coal and steam; a high temperature, a high pressure, and a catalyst are required. A chemical engineer might design the process so it begins with nitrogen and hydrogen, and if the temperature is high enough and the catalyst is active, the reaction rapidly forms ammonia. But complete conversion of the reactants nitrogen and hydrogen to ammonia is not possible because the reacting system proceeds ultimately to an equilibrium condition with only partial conversion of the reactants; the reaction goes no further because at equilibrium all change ceases. The yield of ammonia at equilibrium is the maximum attainable.

An engineer would want to know what equilibrium yield of ammonia to expect at various pressures and temperatures in order to design the process for optimal performance. Thermodynamics supplies an efficient parameter for that engineering purpose. It is called an "equilibrium constant," always represented by the symbol *K*, and for the ammonia synthesis reaction is defined

$$K = \frac{p_{NH_3} \times p_{NH_3}}{p_{N_2} \times p_{H_2} \times p_{H_2} \times p_{H_2}},$$

in which the *p*s are pressures of the chemical components indicated. The engineer designs for the largest feasible ammonia pressure p_{NH3}, and therefore benefits from large values of the equilibrium constant *K*. Compare this recipe for the equilibrium constant with the statement of the reaction: corresponding to the two molecules of the product NH_3 are two multiplied factors ($p_{NH3} \times p_{NH3}$) in the numerator, and in the denominator are one p_{N2} factor corresponding to one molecule of the reactant N_2 and three p_{H2} factors for the three molecules of the reactant H_2. Equilibrium constants are defined similarly for other gaseous reactions—a multiplied *p* factor for each component in the reaction, with chemical product terms in the numerator and reactant terms in the denominator.

Nernst's pragmatic goal was to develop methods for calculating equilibrium constants of gaseous reactions at any temperature and total pressure chosen by engineers. At the turn of the century, the principal experimental tool for studies in thermodynamics was the calorimeter. A calorimeter is a well-insulated container like a thermos bottle that keeps coffee hot in the winter and lemonade cold in the summer. (Some wag has wondered how the thermos knows it should keep the lemonade cold and the coffee hot, and not the lemonade hot and the coffee cold.) In the laboratory, the calorimeter is supplied with an efficient stirrer to eliminate nonuniformities and a sensitive thermometer to detect temperature changes (recall Joule's calorimeters and his remarkable thermometers).

By the time Nernst began his investigations, it was clear from calorimetric studies of chemical reactions—done by practitioners of "thermochemistry"—that most reactions are "exothermic." Any such reaction releases thermal energy as it proceeds, causing a temperature rise in the calorimeter, and making it possible to measure the "heat of reaction." Nernst soon found, however, that heats of reaction and the ordinary theory and practice of thermochemistry did not provide all the tools he needed in his studies of gaseous reactions. To solve those problems, he first had to tackle a much broader problem.

Chemical Affinity

This was a matter of long standing, as Nernst noted in his textbook:

> The question of the nature of the forces which come into play in the chemical union or decomposition of substances was discussed long before a scientific chemistry existed. The Greek philosophers themselves spoke of the "love and hate" of atoms of matter. . . . We retain anthropomorphic views like the an-

> cients, changing the names only when we seek the cause of chemical changes in the changing *affinity* of the atoms.
>
> To be sure, attempts to form more definite ideas have never been wanting. All gradations of opinions are found, from the crude notions of Borelli and Lemery, who regarded the tendency of the atoms to unite firmly with each other as being due to their hook-shaped structure . . . to the well-conceived ideas of Newton, Bergman and Berthollet, who saw in the chemical process phenomena of attraction comparable with the fall of a stone to Earth.
>
> It is not too much to say that there is no discovery of any physical action between substances that has not been used by some speculative brain in the explanation of the chemical process; but up to the present the results are not at all commensurate with the ingenuity displayed.

Two of Nernst's more recent predecessors in the study of chemical affinity were the pioneering thermochemists Julius Thomsen and Marcellin Berthelot, who believed that chemical affinities were measured by heats of reactions. The affinity principle asserted by Thomsen, for example, was that "every simple or complex action of a purely chemical nature is accompanied by an evolution of heat." In other words, all spontaneous chemical reactions had to be of the exothermic kind. The Thomsen-Berthelot principle was criticized by Gibbs, Helmholtz, and Boltzmann, who cited instances of spontaneous "endothermic" reactions, which displayed cooling effects, rather than heating effects, in a calorimeter.

The Thomsen-Berthelot principle was not entirely worthless, however. It did agree with experimental observations in a large number of cases. Nernst appreciated these successes and thought they might be as important as the failures: "It would be as absurd to give [the Thomsen-Bertholet principle] complete neglect, as to give it absolute recognition. . . . It is never to be doubted in the investigation of nature, that a rule that holds good in many cases, but which fails in a few cases, contains a genuine kernel of truth—a kernel which has not yet been 'shelled' from its enclosing hull." Nernst was particularly cognizant that the Thomsen-Berthelot principle was most likely to be successful when it was applied to reactions involving solid components.

Nernst's solution to the chemical affinity problem, as it is now practiced, turns to the Gibbs energy, defined

$$G = U + PV - TS. \tag{1}$$

This is often shortened by introducing an internal-external energy called "enthalpy," represented with the symbol H, and defined

$$H = U + PV. \tag{2}$$

This is a composite of the internal energy U with the potential energy PV a system of volume V has by virtue of its existence at the pressure P. Substituting H for $U + PV$ in equation (1), the Gibbs energy equation becomes

$$G = H - TS. \tag{3}$$

Chemical reactions are characterized by Gibbs energy changes $\Delta_r G$. Taking the ammonia synthesis reaction,

$$N_2 + 3\ H_2 \rightarrow 2\ NH_3$$

as an example again, the Gibbs energy change tells us that the synthesis reaction proceeds if $\Delta_r G < 0$. On the other hand, if $\Delta_r G > 0$, the synthesis reaction is impossible, but ammonia *decomposition*, the reverse reaction,

$$2\ NH_3 \rightarrow N_2 + 3\ H_2$$

is possible. Thus, depending on whether $\Delta_r G$ is positive or negative, the reaction can go either way. No matter which direction the reaction chooses, synthesis or decomposition, its ultimate destiny is the equilibrium condition defined by $\Delta_r G = 0$, in which all chemical change ceases. The Gibbs energy change is, in other words, a faithful measure of chemical affinity, the force driving the reaction toward ammonia synthesis or decomposition, or not at all after equilibrium has been reached. Gibbs energy changes give similar accounts of chemical affinities for other reactions.

Chemical reactions are also characterized by enthalpy changes $\Delta_r H$ and entropy changes $\Delta_r S$, which are related to the reaction Gibbs energy change $\Delta_r G$ as dictated by equation (3),

$$\Delta_r G = \Delta_r H - T\Delta_r S. \tag{4}$$

Here we have a calorimetric route to chemical affinity, as measured by $\Delta_r G$, if $\Delta_r H$ and $\Delta_r S$ can be measured calorimetrically. That is a simple matter for the reaction enthalpy $\Delta_r H$. If the reaction proceeds at a fixed pressure, the entire enthalpy change $\Delta_r H$ is converted to thermal energy and is detected as a heat of reaction in the calorimeter.

Unfortunately—and this was the crux of Nernst's problem—reaction entropy changes $\Delta_r S$, unlike the enthalpy changes, are not directly measurable by calorimetry. We can, however, use calorimetric data to calculate $\Delta_r S$ at any temperature we choose, if we know $\Delta_r S$ at any particular temperature. Nernst had the insight to realize what that particular temperature had to be. According to Simon, "Nernst had a hunch that [nature] could reveal her intentions only at absolute zero, the one point of special significance in the whole range of temperature." Nernst surmised from low-temperature data for reactions involving solids that, in effect, for all such reactions the entropy change is equal to zero at absolute zero:

$$\Delta_r S = 0 \text{ when } T = \text{absolute zero}. \tag{5}$$

We say "in effect" because Nernst did not believe in the entropy concept (Gibbs's lessons had not yet been learned). His working equations did not include the entropy S, but instead the mathematical equivalent involving the Gibbs energy G (Nernst used Helmholtz's term, "free energy"). At first, Nernst called equation (5) his "heat theorem." It solved Nernst's chemical affinity problem for reactions involving solids by making it possible to calculate $\Delta_r S$ with calorimetric data, to combine these reaction entropies with reaction enthalpies, and finally to determine chemical affinities, measured as $\Delta_r G$.

Chemical Constants

Nernst's heat theorem, published in 1906, was a major accomplishment. Together with subsequent work in thermochemistry, it earned him a Nobel Prize in chemistry in 1920. But the theorem had little to say about Nernst's original problem, the calculation of equilibrium constants for gaseous reactions. He now had to find his way back to the equilibrium constants.

He had an equation that partly satisfied his needs. It was a differential equation that had been introduced by J. H. van't Hoff in the 1880s. (Jacobus Henricus van't Hoff was a modest, silent, hardworking Dutchman who was one of the founders of physical chemistry. He became the first Nobel laureate in chemistry in 1901. By the time Nernst published his heat theorem, he and van't Hoff were colleagues at the University of Berlin.) The mathematical form of van't Hoff's equation is:

$$\frac{1}{K}\frac{dK}{dT} = f(T),$$

with K an equilibrium constant and $f(T)$ some function of temperature obtainable in calorimetric experiments. Nernst needed an equation for K. He could get it by integrating van't Hoff's differential equation, but that introduced an unknown constant. Passage from a differential equation to an integrated equation always requires an "integration constant." In the differential equation, the constant disappears because the derivative of a constant equals zero, but it cannot be ignored in the integrated equation.

The matter of the integration constant, simple enough mathematically, proved a stubborn problem in the physical context. Nernst discovered that he could dissect the integration constant he needed, call it I, into separate constants for the components involved in the reaction. Consider the "water gas reaction," which forms carbon monoxide (CO) and steam (H_2O) from carbon dioxide (CO_2) and hydrogen (H_2),

$$CO_2 + H_2 \rightarrow CO + H_2O.$$

By Nernst's formula, the integration constant I for this reaction is divided into four separate terms, one for each component, with reactant terms subtracted from product terms,

$$I = i_{CO} + i_{H_2O} - i_{CO_2} - i_{H_2}.$$

Nernst called the i terms "chemical constants." Each one depends only on the physical properties of the component indicated, and is valid not only for the one reaction but for all others in which the component participates. For each component, a separate constant could be calculated and tabulated once and for all.

Nernst's study of gaseous chemical equilibria is a demonstration, if any is needed, that the paths of theoretical research can be devious. At the outset,

Nernst had been diverted to the general problem of chemical affinity, and found the advantages of turning to the little-known thermodynamics of reactions involving solids at low temperatures. While on this tangent, he had uncovered the experimental basis for his heat theorem, with implications reaching beyond the realm of gas equilibria. Returning to the gaseous reactions, and formulating the problem in terms of integration of van't Hoff's differential equation, he had found that he could express the necessary integration constants as summations of separate chemical constants, one for each reaction component.

But Nernst's task was still not complete. The data required for accurate calculation of the chemical constants were not available when Nernst formulated his theory in 1906. He soon embarked on one of the first experimental programs aimed at obtaining the necessary low-temperature data. As an interim measure, he developed formulas for estimating the chemical constants.

Nernst prepared a table of values for his estimated chemical constants and used it to calculate approximate equilibrium constants. "Surveying the whole material available at the time," Simon writes, "he showed that results of his calculations agreed with experimental facts within a rather generous limit of error." In a typical application, his empirical method calculated $K = 1.82$ for the water gas reaction at the high temperature 800°C; the observed value for the equilibrium constant at this temperature was $K = 0.93$.

To some of Nernst's critics, this kind of agreement was not impressive. Gilbert Lewis, a former student of Nernst's, and one of his most important successors in the development of the methods of chemical thermodynamics, credited Nernst's efforts to obtain the low-temperature data for accurate calculations, but deplored "the rapidly growing use of [estimated] chemical constants." Lewis found dismaying "the various efforts which have been made to square the calculations based on these constants with the results of measurements. . . . [They] constitute a regrettable episode in the history of chemistry."

Lewis would note with approval that chemical constants are nowhere to be found in the modern literature of chemical thermodynamics. But Simon reminds us that the "generous limit of error" with which Nernst measured his success "was inifintely preferable to the complete ignorance that existed before. . . . [Nernst's] approximations were very useful to chemical industry, making it possible to get very quickly a rough idea which reactions were thermodynamically feasible in complex reaction patterns."

The Theorem Is a Law

Nernst made it clear that his heat theorem was fundamentally a "law" and not just another formula or mathematical recipe. He insisted not only that his theorem belonged with the two established laws of thermodynamics—as the "third law"—but that there could never be another. This conclusion followed from an extrapolation: the discoverers were three (Mayer, Clausius, and Helmholtz) for the first law, two (Carnot and Clausius) for the second, and just one (W. Nernst) for the third. With no one to discover it, a fourth law of thermodynamics could not exist. The third law was the last law.

With all his immodesty, self-glorification, and sarcastic wit, Nernst continued to expand his influence, not only among the high and mighty, but also within the intimate circle of his graduate students. James Partington, an Englishman who worked in Nernst's Berlin laboratory, writes of Nernst's kind attention to a "very young man, with little experience." Unlike scientific potentates then (and now), Nernst did not ignore the daily labors of his research students, leaving them to sink or swim. Partington found his research difficult, but Nernst's presence was an incentive: "one felt that he could do the work easily himself, and that perseverance would remove lack of skill, a fault which could be cured by application. . . . His true kindness is something I remember with gratitude."

At least one visitor to Berlin in the 1930s had initial reservations about Nernst and his unusual manner. Hendrik Casimir (known for his studies of low-temperature superconductivity phenomena) gives his impressions of Nernst and his research colloquium:

> Mendelssohn has described this institution [the colloquium] in enthusiastic terms as the place where the most prominent physicists of the day pronounced on the most recent developments. It did not strike me that way at all. . . . Discussions were both formal and perfunctory. In a fairly soft, yet penetrating, rather high-pitched voice [Nernst] could proclaim that he already said some of the things presented at the colloquium in his book, and complain that people did not recognize that as a publication. He struck me at the time as a ridiculous figure. . . . Later, I realized that some of the remarks had contained a rather subtle point. In 1964, the centenary of his birth was celebrated at Göttingen and I was invited to give the main talk. On that occasion, I studied his published work more closely and was impressed. True, there were some irritating mannerisms and his mathematics was shaky, but his work shows throughout a remarkably clear and often prophetic vision. And so I had an opportunity to atone in public for an error of judgment I had never voiced.

Joy and Sorrow

Nernst's private life was almost as extraordinarily fortunate as his professional career. His wife Emma was, among many other things, a paragon of domestic efficiency and hard work. She customarily arose at 6 A.M. and kept the Nernst household, which was never simple or quiet, in order. Not long after their arrival in Berlin, the Nernsts were known as the most hospitable family in Berlin. There were five children, three daughters and two sons, and family life was an important part of Nernst's existence.

But not even Nernst could escape the tragedies of two world wars. In the first war, both Nernst sons were killed. Long before the armistice, Nernst could see that Germany was beaten and nearly ruined. In vain, he tried to use his connections—with the kaiser, among others—to prevent further devastation. In 1917, he found escape in the peaceful realm of science by gathering in a monograph his work on the heat theorem. The opening sentences of this book tell of the solace he found in science: "In times of trouble and distress, many of the old Greeks and Romans sought consolation in philosophy, and found it. Today we may as well say there is hardly any science so well adapted as theoretical physics to divert the mind from the mournful present."

Peace finally came, and miraculously Germany began to recover. For a time, there was political and economic chaos, but German science emerged as strong and active as ever; Nernst and his Berlin colleagues could reconstruct scientifically. Now there were conceptual revolutions to be fought. Both the quantum theory and relativity had come over the scientific horizon. Nernst did not contribute to these endeavors, although he understood and appreciated what was happening. More recognition came his way; he was offered (but declined) an ambassadorship to the United States. He was elected rector of the University of Berlin, became a Helmholtz successor as professor of physics, and won a Nobel Prize in 1920.

But Germany had not completely recovered from the political ruin brought on by the first war. In the 1930s, the Nazi influence began to spread; then suddenly and irrevocably the Nazis were in power. Nernst was opposed to the Nazi policies, but lacked the energy and influence to act. Mercifully, he retired to his country home and found a measure of peace in the last seven years of his life. During his final days, Emma sat with him and recorded his last words. "True to his whole character," Mendelssohn writes, Nernst told Emma just before he died, "I have already been to Heaven. It was quite nice there, but I told them they could have it even better."

iii

ELECTROMAGNETISM

Historical Synopsis

Our story must now follow a more zigzag chronology. (Those more comfortable with linear timelines may want to consult the chronology at the end of the book.) Part 2 followed the development of thermodynamics from Carnot in the 1820s to Nernst in the 1930s. The history now returns to the 1820s and 1830s, with the same scientific scenery that inspired the thermodynamicists, the topic of the day being the mysterious and intriguing matter of conversion processes. It was plain to the scientists of the early nineteenth century that the many interconvertible effects—thermal, mechanical, chemical, electrical, and magnetic—demanded unifying principles. Thermodynamicists concentrated at first on thermal and mechanical effects, and from them refined the concepts of energy and entropy and three great physical laws. Eventually, by the end of the nineteenth century, thermodynamicists had discovered that the language of their science encompassed all macroscopic effects—indeed, the entire universe.

There were other unities to be discovered at the same time. In 1820, Oersted observed that a wire carrying an electric current slightly disturbed the magnetic needle of a nearby compass: an electric effect produced a magnetic effect. Oersted's colleagues were not impressed, but an ambitious young laboratory assistant at the Royal Institute in London was; his name was Michael Faraday. In a string of brilliantly designed experiments, Faraday discovered many more "electromagnetic" effects, including those that make possible modern electric motors and generators. In one of the last and most difficult of these experiments, Faraday made the stunning discovery that polarized light is affected by a magnetic field. With that observation he brought light into the domain of electromagnetic phenomena.

Faraday was guided by his superb skill in the laboratory—he was the greatest experimentalist of the nineteenth century—and also by a revolutionary theory. He believed that magnetic, electric, and electromagnetic effects were transmitted through space along "lines of force," which collectively defined a "field." Once it was generated, the field could exist anywhere, even in otherwise empty space. Faraday's associates believed his experiments but not his theory, which was radically at odds with the version of Newtonianism then popular.

But Faraday was joined by two young dissenters who also believed fervently in the field concept. One was William Thomson, and the other a Scotsman who would become the greatest theorist of the nineteenth century, James Clerk Maxwell. Thomson fashioned a limited mathematical theory of Faraday's electric lines of force. Maxwell went much further. Over a period of almost two decades, he constructed a great theoretical edifice beginning with Faraday's field concept. The theory comprised a set of differential equations for the electric and magnetic components of the field and their sources, which condensed into a few lines the theory of all electric, magnetic, and electromagnetic phenomena, including Faraday's experimental demonstration of the electromagnetic nature of light.

The scope and utility of Maxwell's equations are vast. Their physical interpretation has changed over the years. We now consider that the electric field originates in electric charges and the magnetic field in electric currents. Maxwell regarded the electric charge as a product of the field, and could see only an indirect connection between the magnetic field and electric currents. But the equations themselves are valid on a cosmic scale. Like Newton's laws of dynamics and universal gravitation, and the laws of thermodynamics, Maxwell's equations have a reach that extends to the corners of the universe.

11

A Force of Nature

Michael Faraday

Doing Without

The scientists in these chapters are a diverse group. One would look in vain to find particular aspects of their backgrounds or characters that guaranteed their success in science. Some were introverted and solitary, others extroverted and gregarious. Some were neurotic, while others were well adjusted. They could be friendly and agreeable, or unfriendly and contentious. Their marriages were usually happy, but some were disastrous. Their educations were both formal and informal. Some had mentors, others did not. Some founded schools to carry on their work, and others worked alone.

But these outstanding scientists had at least two things in common: they all worked hard, sometimes obsessively, and with only a few exceptions, they came from middle-class backgrounds. The tendency to workaholism is a trait found in most people who achieve outstanding success. More interesting is the rule of middle-class origins. Our physicists led lives in social worlds that covered the full middle-class range, from lower to upper, but rarely found themselves above or below these stations. By far the most prominent exception is the subject of this chapter, Michael Faraday, born in a London slum.

Faraday's father, James, was a blacksmith with a debilitating illness, who could barely support his family. Late in his life, Faraday recalled that in 1801, when economic times were bad, his weekly food allotment was a loaf of bread. His education, he told his friend and biographer, Henry Bence Jones, "was of the most ordinary description, consisting of little more than the rudiments of reading, writing, and arithmetic at a common day-school. My hours out of school were passed at home and in the streets."

But the misfortunes of poverty were balanced by a secure family life. Michael's mother, Margaret, "was the mainstay of the family," writes Faraday's most recent biographer, Pearce Williams. "She made do with what she had for material needs, but offered her younger son that emotional security which gave him the strength in later life to reject all social and political distinctions as irrelevant to his own sense of dignity." No doubt she also deserves credit for the close friendship of the three siblings, Michael, his younger sister, Margaret, and his older brother, Robert.

Faraday had a long climb from the streets of London to his ultimate position of eminence as one of the greatest scientists of his time. Family support helped him take the first steps in that climb, and so did his religious faith in the Sandemanian church. The Sandemanians are a fundamentalist Protestant Christian sect, who teach the essential importance of love, discipline, and community without proselytizing or fiery preaching. Faraday drew daily strength from his religion throughout his life. One of his colleagues, John Tyndall, noted in his diary: "I think that a good deal of Faraday's week-day strength and persistency might be referred to his Sunday Exercises. He drinks from a fount on Sunday which refreshes."

But for all the tenacity and purpose built into Faraday's character by adversity, family support, and religious faith, he would not have a place in our history without two more advantages: extraordinary good luck at several key points in his life, and a personality of enormous intensity. In his first piece of good luck, he was apprenticed to George Riebau, a bookbinder and bookseller. Riebau, a French refugee, liked his lively apprentice, and encouraged him to take advantage of the many books that passed through the shop. "Whilst an apprentice," Faraday told Bence Jones, "I loved to read the scientific books which were under my hands, and, amongst them, delighted in [Jane] Marcet's *Conversations in Chemistry* and the electrical treatises in the *Encyclopaedia Britannica.*" Such was the haphazard beginning of Faraday's education in science.

The books were crucial, but not enough. Faraday began to attend evening lectures, including four given by Humphry Davy at the Royal Institution in London. Davy was one of the most famous scientists of the day and an immensely popular lecturer. As Faraday related to one of Davy's biographers, he was finding the book trade "vicious and selfish," and thought of entering "the service of Science, which I imagined made its pursuers amiable and liberal." He naïvely wrote to Davy asking for a position, and in Faraday's greatest piece of good luck, Davy hired him, first as an amanuensis, and later as assistant in the laboratory at the Royal Institution. Faraday remained at the institution for his entire career and eventually succeeded Davy as the main attraction in the institution's laboratory and lecture theater.

Church, family, friendships made during his bookbinding apprenticeship, and the patronage of Humphry Davy were the external strengths that gave Faraday his opportunities. No less important was his extraordinary internal strength. Tyndall wrote, "Underneath his sweetness and gentleness was the heat of a volcano. He was a man of excitable and fiery nature; but through his high-discipline he converted the fire into a central glow and motive force of life, instead of permitting it to waste itself in useless passion." It took no less than a controlled volcano of energy for Faraday to make his long, strenuous, and hazardous ascent. He chose to study the forces of nature in his research. He was a force of nature himself.

Faraday and Davy

Among major scientists there has probably never been one so handsome, charming, and publicly popular as Humphry Davy. At the time he employed Faraday, he was at the peak of his celebrity. He made his headquarters at the Royal Institution, which had recently been founded by Count Rumford (Benjamin Thompson) for the "teaching, by regular courses and philosophical lectures and experiments, the applications of the new discoveries in science to the improvement of arts and manufactures, and in facilitating the means of procuring the comforts and conveniences of life." During Davy's tenure as professor of chemistry at the institution, this social purpose became secondary to the professor's chemical research and famous scientific lectures.

Davy's lower-middle-class background was not far removed from Faraday's lower-class origins. His father was a wood carver, with a small farm in Penzance, Cornwall. He attended a good grammar school, but his formal education went no further. He found his interest in chemistry as an apprentice to a Penzance apothecary. Thomas Beddoes, a doctor in Clifton, Bristol, gave Davy his first scientific opportunity. Beddoes appointed Davy as the superintendent of experiments at his Medical Pneumatic Institution in Clifton. Davy's experiments with gases, particularly his descriptions of the effects of breathing "laughing gas" (nitrous oxide)—"a sensation analogous to gentle pressure on all the muscles, attended by a highly pleasurable thrilling, particularly in the chest and extremities"—quickly became famous.

Davy's daring experiments and speculations caught Rumford's attention, and in 1799 Rumford appointed Davy to his first position at the Royal Institution, which became a platform for his aspirations in both science and society. In 1812, he married a wealthy and attractive widow, Jane Apreece. "Her passion for rank was as intense as Davy's," writes one of Davy's biographers, J. G. Crowther. "The two social hunters allied in the attack on the aristocratic stockade." For Davy, "the pursuit of science was rapidly subordinated to the pursuit of snobbery."

Soon after Faraday started his scientific apprenticeship with Davy in 1813, he had another fortunate opportunity. The Davys, now Sir Humphry and Lady Davy, embarked on a tour of Europe, accompanied by Faraday as Davy's "assistant in experiments and writing." The European tour was another essential part of Faraday's education, scientific and otherwise. Davy's fame opened doors everywhere in France and Italy, and Faraday met many of the leading scientists of the time. Davy himself was part of the education. He and his eager assistant freely discoursed on topics covering the scientific map and beyond. Lady Davy was a different matter. She talked too much and insisted on treating Faraday as a servant. "She is haughty and proud to an excessive degree and delights in making inferiors feel her power," Faraday wrote to his friend Benjamin Abbott.

Faraday's European experience was as important as any other in his life. Williams tells us that "the young man who landed on English soil in the spring of 1815 was quite different from the youth who had left it in 1813. He had seen a good part of the world, realized its complexity and diversity, and gained a good deal of insight into the ways of men. He had met some of the foremost scientists of the day and had both impressed and been impressed by them."

Discoverer

Most of Faraday's many biographers have portrayed him as a peerless discoverer of experimental facts. This image is certainly accurate as far as it goes, but it neglects another, equally important, side of his genius: his remarkable achievements as a theorist, or "philosopher," as he preferred to be called. We first see him playing the familiar role of the experimentalist.

Faraday's early work at the Royal Institution, while Davy was still active in the institution's affairs, was mainly as a chemist. His first scientific paper, "Analysis of Native Caustic Lime of Tuscany," was published in 1816 when he was twenty-five. By 1820, he had become a journeyman chemist, in demand for his services as an analytical chemist. During the 1820s, he helped keep the institution afloat financially by doing hundreds of chemical analyses. Also in the 1820s, Faraday turned to the topics of his major research, electricity and electromagnetism. Here we find him becoming the outstanding experimentalist of the nineteenth century.

The event that inspired Faraday's interest in electricity and magnetism was a discovery in 1820 by the Danish scientist, Hans Christian Oersted. The experiment was first performed as a demonstration before an audience of scientists. As Oersted described it later (using the third person to refer to himself),

> The plan of the first experiment was, to make the [electrical] current of a little galvanic trough apparatus [a battery], commonly used in his lectures, pass through a very thin platina wire, which was placed over a compass covered with glass. The preparations for the experiment were made, but some accident having hindered him from trying it before the lecture, he intended to defer it to another opportunity; yet during the lecture, the probability of its success appeared stronger, so that he made the first experiment in the presence of the audience. The magnetical needle [the compass], though included in a box, was disturbed; but as the effect was very feeble . . . the experiment made no strong impression on the audience.

Faraday and others were more impressed. Oersted's experiment was a major event in the inauguration of the science of electromagnetism, which would ultimately lead to some of the technologies that are most familiar in our own lives. Faraday paid particular attention to Oersted's demonstration, later in 1820, that a current-carrying wire is surrounded by a *circular* magnetic effect, which forces the compass needle to point in a direction perpendicular to the wire.

Faraday guessed that a current-carrying wire could keep a magnet revolving in *continuous* circular motion around the wire's axis, and he designed the experiment illustrated in figure 11.1 to demonstrate this "electromagnetic rotation." The left side of the figure shows a mercury-filled cup with a stationary electric current carrying wire dipping into it. A small, powerful magnet was placed next to the wire and tethered to the bottom of the cup by a thread. When an electric current was passed through the wire (and the mercury in the cup), the upper pole of the magnet rotated around the wire. The right side of the figure shows a similar experiment in which the magnet was fixed and the current-carrying wire rotated.

These experiments were reported in October 1821 and the paper "thrust Faraday into the first rank of European scientists," writes Williams. "In every laboratory throughout Europe copies of Faraday's rotation apparatus were made and the strange nature of motive force contemplated." Faraday's device had obvious practical possibilities: he had invented the electric motor. He did not pursue these applications, or any others made possible by his inventions. But others did.

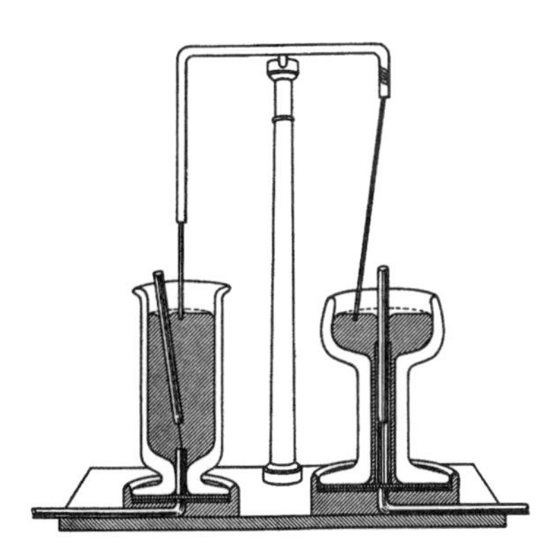

Figure 11.1. Faraday's experiments demonstrating electromagnetic rotation. From plate IV of Michael Faraday, *Experimental Researches in Electricity* (London: Taylor and Francis, 1839), vol. 2.

By the 1830s, the performance of practical "electromagnetic engines" was being studied by James Joule, among others.

The 1821 paper and its immediate reception were an occasion for celebration—and as it turned out, for a plagiarism charge against Faraday. The controversy concerned earlier unsuccessful and unpublished attempts by William Wollaston and Davy, to make a current-carrying wire rotate around its own axis when influenced by a magnet. This was not the same as Faraday's experiment, but it was similar enough that Faraday, who was familiar with the Wollaston-Davy effort, should have acknowledged it. In his haste to publish, he did not, and suspicions were aroused. Wollaston eventually allowed the storm to blow over, but Davy was not so magnanimous. When Faraday was proposed for election as a fellow of the Royal Society three years later, he had Wollaston's support but not that of Davy. Faraday was elected with one vote in opposition, no doubt Davy's vote. The master had broken with the pupil, evidently motivated to some degree by jealously and vanity.

Oersted's experiment was not the first to display an electromagnetic phenomenon. Earlier, François Arago and André Marie Ampère had demonstrated that a helical coil of wire carrying an electric current becomes a magnet, an "electromagnet." In a series of experiments reported in 1831, Faraday investigated this connection between electricity and magnetism mediated by a coil of wire. He discovered the effect he eventually called "electromagnetic (or magneto-electric) induction." The induction took place between two coils of wire wound around an iron ring, one coil carrying an electric current and serving as an electromagnet and the other connected to a copper wire that passed over a compass needle. Here is Faraday's typically meticulous description of the experiment, as recorded in his laboratory notebook:

> I have had an iron ring made (soft iron), iron round and 7/8ths of an inch thick, and ring six inches in external diameter. Wound many coils round, one half of the coils being separated by twine and calico; there were three lengths of wire, each about twenty-four feet long, and they could be connected as one length, or used as separate lengths. By trials with a trough [a voltaic battery,] each was insulated from the other. Will call this side of the ring *A* [see fig. 11.2]. On the other side, but separated by an interval, was wound wire in two pieces, together amounting to about sixty feet in length, the direction being as with the former coils. This side call *B*.
>
> Charged a battery of ten pairs of plates four inches square. Made the coil on

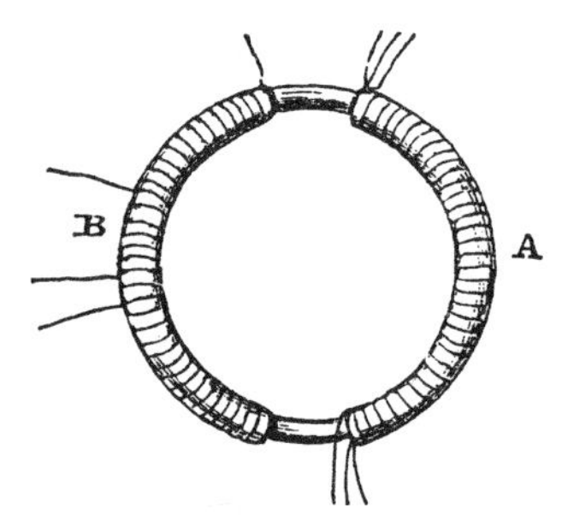

Figure 11.2. Faraday's first electromagnetic induction experiment. From Henry Bence Jones, *The Life and Letters of Faraday* (London: Longmans, Green, 1870), 2:2.

> *B* side one coil, and connected its extremities by a copper wire passing to a distance, and just over a magnetic needle (three feet from wire ring), then connected the ends of one of the pieces on *A* side with battery: immediately a sensible effect on needle. It oscillated and settled at last in original position. On breaking connection of *A* side with battery, again a disturbance of the needle.

When it was connected to the battery, coil *A* became an electromagnet whose magnetic effect induced an electric current in coil *B*, as indicated by the magnetic needle (a compass). Faraday's key discovery, which had been missed for years by Faraday himself and many others, was that the induced electric current was *transient*: it lasted only for a short time after coil *A* was connected. In other words, the induction was in effect only while the magnetic effect was *changing*. Another transient current was induced in coil *B* when coil *A* was disconnected from the battery.

Oersted's experiment displayed a magnetic effect caused by an electric effect. Faraday's first induction experiment demonstrated the inverse, an electric effect caused by a magnetic effect, with the latter originating in an electromagnet. In another induction experiment, Faraday got a similar result by replacing the electromagnet with a permanent magnet. He wound a helical coil of wire around a hollow pasteboard cylinder, connected the coil to a galvanometer (for measuring electrical currents), and rapidly thrust a cylindrical permanent magnet into the cylinder. While the magnet was in motion—but only while it was in motion—the galvanometer indicated that an electric current was induced in the coil.

Like the electromagnetic rotation experiments of 1821, Faraday's 1831 electromagnetic induction experiments had some obvious practical implications, and as usual, Faraday did not exploit them. The induction experiments showed that all one needed to produce electricity was a magnet and a coil of wire. The machines we now call dynamos or electric generators are based on this principle.

The 1830s were prolific years for Faraday. Soon after completing the electromagnetic induction experiments, he embarked on another profoundly important series of experiments, this one focusing on electrochemical decomposition, an interest he had inherited from Davy. In the prototype experiment of this kind, an electric current from a voltaic battery is passed through water, and the gases hydrogen and oxygen are evolved at the two wires making the electrical connection to the water. The chemical reaction promoted by the current is the decomposition of water (H_2O) into hydrogen (H_2) and oxygen (O_2),

$$2\,H_2O \rightarrow 2\,H_2 + O_2.$$

This effect was first observed in 1800, and by the time Faraday turned to electrochemistry in the 1830s, many more electrochemical decompositions had been observed.

Faraday first concluded that in all cases the amount of chemical decomposition produced was proportional to the amount of electricity producing the effect. He also observed that the masses of elements liberated by a definite quantity of electricity were proportional to their chemical equivalent weights. (The equivalent weight of an element is about equal to the mass that combines with one gram of hydrogen. For example, the equivalent weight of oxygen in H_2O is eight grams.)

From the second observation, Faraday concluded that "the equivalent weights of bodies are simply those quantities of them which contain the same quantity of electricity, or have naturally equal electrical powers; it being the *electricity* which *determines* the equivalent [weight], *because* it determines the combining force." These statements, written in 1834, were astonishingly prophetic. They anticipated by more than fifty years the theories developed by physical chemists at the end of the nineteenth century based on the notion that in solution many chemical substances can dissociate into electrically charged components, each with its own equivalent weight.

With even greater prescience, Faraday continued: "Or, if we adopt the atomic theory or phraseology, then the atoms of bodies which are equivalents to each other in their ordinary chemical action, have equal quantities of electricity naturally associated with them." Here he formulated the concept of charged particles in aqueous solutions, the "ions" of modern solution theory. But, like many of his contemporaries, he had reservations about atomism: "I must confess I am jealous of the term *atom;* for though it is very easy to talk of atoms, it is very difficult to form a clear idea of their nature, especially when compound bodies are under consideration." With the gift of hindsight, we wonder why Faraday was not bold enough to believe in charged atoms, and even in an atom of electricity (the electron). This was a step he could not take because it violated his dictum that a postulate is not a truth unless it has the support of (many) experimental facts. Nothing was more important to Faraday than that.

We can credit Faraday with the founding of the science of electrochemistry. He not only proposed the two fundamental laws of electrochemistry mentioned above, but also introduced the language of electrochemistry, such terms as "electrolyte," "electrode," "cathode," "anode," "cation," "anion," and "ion." Faraday had help in the invention of these terms from William Whewell of Trinity College, Cambridge. As Crowther remarks, "The famous terminology of [electrochemistry] was chiefly due to Whewell's excellent etymological taste."

From electrochemistry, Faraday turned in 1837 to electrostatics. He had the idea, which he could see confirmed in the evidence of electrochemistry, that when two electrically charged bodies influence each other the effect depends not only on the charge itself but also on the medium between the two bodies. He designed a device called a "capacitor" in modern terminology. It consisted of two concentric brass spheres separated electrically by shellac insulation. The device could be opened, and the space between the two spheres filled with different insulating materials, gases, liquids, or solids.

Faraday had two precisely identical capacitors of this design made. In a typical experiment, he filled one capacitor with air and the other with another substance, such as glass, sulfur, or turpentine. He then charged one capacitor electrically and connected it with the other, thus dividing the charge between the two capacitors. Finally, with an electrometer he measured the charges on the capacitors. He found that the capacitor filled with a solid material always held a higher charge than the one with air. This was clear evidence that the electrical interaction between two charged bodies involved not only the charge and the distance between the two bodies, but also the medium—the "dielectric," as Faraday called it—occupying the space between the bodies. If the dielectric was solid, some of the charge was induced in the dielectric itself. For Faraday, these "electrostatic induction" experiments illustrated an intimate reciprocal connection between electric forces and the medium in which they were effective: the forces altered the medium, and the medium propagated the forces.

Faraday was always strong physically. In the mountains, he could easily walk thirty miles in a day, and he was incessantly busy at the Royal Institution. His tragic weakness was recurring "ill health connected with my head," as he put it. Even as a young man, he had memory problems, and as he grew older he suffered from bouts of depression and headaches. "When dull and dispirited, as sometimes he was to an extreme degree," his niece Constance Reid recalled, "my aunt used to carry him off to Brighton, or somewhere, for a few days, and they generally came back refreshed and invigorated."

These symptoms increased in severity and frequency until, in 1840, at age forty-nine, Faraday had a major nervous breakdown. Brighton vacations were no longer curative, and for four years he avoided most of his research activities. One can glimpse his desperate condition in a letter to his friend Christian Schönbein, in 1843: "I must begin to write you a letter, though feeling, as I do, in the midst of one of my low nervous attacks, with memory so treacherous, that I cannot remember the beginning of a sentence to the end—hand disobedient to the will, that I cannot form the letters, bent with a certain crampness, so I hardly know whether I shall bring to a close with consistency."

Nevertheless, he came back. In 1845, he was again in his laboratory and closing in on what was to be one of his crowning achievements. This work was initiated by a suggestion in a letter from William Thomson, then a Cambridge undergraduate. Thomson mentioned the effects of electricity on dielectrics, already familiar to Faraday, and then offered the speculation that the electrical constraint of a transparent dielectric might have an effect on polarized light passing through the dielectric.

The phenomenon of light polarization had been known for many years. It was observed particularly in reflected light, and understood as a process that confined the vibrations constituting light waves to a certain plane. About a decade before Thomson's letter, Faraday had tried to detect a change in the plane of polarization of a light beam passing through a dielectric strained by electric charge. He got only negative results then. Faraday replied to Thomson, "Still I firmly believe that the dielectric is in a peculiar state whilst induction is taking place across it." He was again inspired to search for the elusive effect, but modifications of the earlier search for electrical effects on polarized light were no more successful. It then occurred to him that a strong magnet might strain a solid dielectric sufficiently to affect the passage of a beam of polarized light.

In 1845, Faraday began a series of experiments based on this surmise. For the solids passing the polarized light, he tried flint glass, rock crystal, and calcareous spar; he varied the current supplied to his electromagnet, and the placements of the poles: still no success. He then tried a piece of lead glass he had prepared fifteen years earlier—and at last the eureka moment arrived:

> A piece of heavy glass . . . which was 2 inches by 1.8 inches, and 0.5 of an inch thick, being a silico borate of lead, and polished on the two shortest edges, was experimented with. It gave no effects when the *same magnetic poles* or the *contrary* poles were on opposite sides (as respects the course of the polarized ray)—nor when the same poles were on the same side, either with the constant or intermitting current—*BUT*, when contrary magnetic poles were on the same side, there *was an effect produced on the polarized ray*, and thus magnetic force and light were proved to have a relation to each other. This fact will most likely prove exceedingly fertile and of great value in the investigations of both conditions of natural forces.

Indeed. He had demonstrated a link between light and magnetism, the first step along the path that would lead to one of the greatest theoretical accomplishments of the nineteenth century, an electromagnetic theory of light, finally achieved by Maxwell building on Faraday's foundations.

Faraday never tired of telling his readers, correspondents, and audiences about the irreducible importance of tangible experimental facts. "I was never able to make a fact my own without seeing it," he wrote to a friend toward the end of his career. In a letter to his colleague Auguste de la Rive, he recalled, "In early life I was a very lively imaginative person, who could believe in the 'Arabian Nights' as easily as in the 'Encyclopaedia,' but facts were important to me, and saved me. I could trust a fact." The facts were the gifts of the experiments. "Without experiment I am nothing," he said. And there was no end to the experiments: "But still try, for who knows what is possible?" To a lecture audience he said, "I am no poet, but if you think for yourselves, as I proceed, the facts will form a poem in your minds." In the poetry, we find the other side of Faraday's genius.

Philosopher

The twentieth-century philosopher and historian Isaiah Berlin wrote a famous essay, "The Hedgehog and the Fox," in which he classified thinkers as foxes or hedgehogs: foxes know many things, while hedgehogs know one big thing. Faraday was both. As an experimentalist, he learned all the things mentioned and a lot more (Bence Jones lists twenty-two topics pursued by Faraday in his electrical researches alone). But as a theorist, he learned and taught one great thing: that the forces of nature are all interconnected. "We cannot say that any one is the cause of the others, but only that they all are connected and due to a common cause," he said in a lecture at the Royal Institution in 1834. In the 1845 paper reporting his discovery of the effect of magnetism on light, he wrote, "I have long held an opinion almost amounting to conviction . . . that the various forms under which the forces of matter are made manifest have one common origin; or, in other words, are so directly related and mutually dependent, that they are convertible, as it were, one into another, and possess equivalents of power in their action." And in 1849 he said, "The exertions in physical science of late years have been directed to ascertain not merely the natural powers, but the manner in which they are linked together, the universality of each in its action, and their probable *unity in one*."

These were not vague generalities. Faraday's experiments had given him a clear picture of natural forces. Magnetic forces could actually be mapped in the space surrounding a magnet by sprinkling iron filings on a piece of paper placed over the magnet. The filings aligned themselves along the magnetic "lines of force" (see fig. 11.3). Faraday assumed that the force between the magnetic poles was propagated along these lines. In 1849, Thomson introduced the now indispensable term "field of force," or just "field," for an entire network of Faraday's lines of force.

The iron filings showed that magnetic lines of force could be curved. In Oersted's experiment, they followed circles with the current-carrying wire at the center. The lines of force also determined the laws of electromagnetic induction. The rule was that if a wire cut through magnetic lines of force, an electric current was induced in the wire, and the magnitude of the current depended on the rate of cutting of the lines of force. Induction by this mode was particularly evident in Faraday's experiment with the helical coil of wire and the inserted magnet: as the magnet moved, the wire in the coil cut the lines of force carried by the magnet, and a current was induced.

Faraday generalized what he saw in the iron filings responding to a magnetic field to electric and gravitational fields. He had no device like the iron filings for developing an image of these lines of force, but he assumed that they were there, occupying the space—even an otherwise empty space—between interacting bodies.

All of this was a drastic departure from theoretical physics as Faraday found it in the early nineteenth century, which was based largely on a version of Newtonian physics. At the turn of the century, Newtonianism was unchallenged. The world was seen as a system of particles acting under forces that were manifested in the phenomena of electricity, magnetism, and gravitation. Each such force was transmitted instantaneously from one body to another without any mediating influence, and was determined mathematically by Newton's three laws. Theories of light were in a different category, but also reliant on particle models.

The first blow to the prevailing Newtonian view was struck by Thomas Young and Augustin Fresnel, who had by the 1830s demolished the particle theory of light and replaced it with a wave theory. The waves brought another problem. They were conceived as vibrations, but vibrations of what? To answer this question, theorists invented a strange kind of weightless matter called "ether" with some surprising properties: it could pass through ordinary matter completely without friction, and yet when called upon, it could support the extremely high frequencies of the vibrations of light waves.

The ether hypothesis was not the only theoretical device of the time to rely on weightless matter. A weightless fluid called "caloric" was popular in theories of heat, and electricity and magnetism were also treated as weightless fluids.

Faraday had little sympathy for any of these theoretical contrivances. He re-

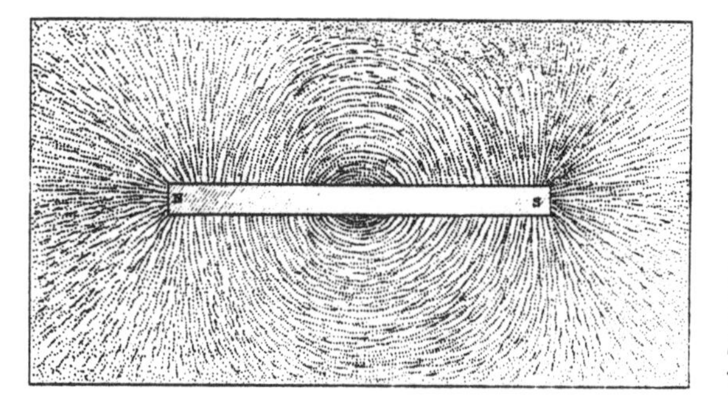

Figure 11.3. Magnetic lines of force traced by fine iron filings. From plate IV of Michael Faraday, *Experimental Researches in Electricity* (London: Taylor and Francis, 1855), vol. 3.

jected the ether concept and all weightless fluids, and refused to accept the Newtonian "action-at-distance" principle, which stated that the effect of a force, electric, magnetic, or gravitational, could reach from one body to another through empty space. In Faraday's worldview, space was occupied by fields comprising lines of force—an electric field was generated by an electric charge, a magnetic field by the poles of a magnet, and a gravitational field by a massive object. Another body could respond to one of these fields, but not at a distance. The response was local, to the condition of the field where the second body was located. John Wheeler, a contemporary theoretical physicist, gives us a picture of a gravitational field that would meet with Faraday's approval:

> The Sun, for instance, can be said to create a gravitational field, which spreads outward through space, its intensity diminishing as the inverse square of the distance from the Sun. Earth "feels" this gravitational field locally—right where Earth is—and reacts to it by accelerating toward the Sun. The Sun, according to this description, sends its attractive message to Earth via a field rather than reaching out to influence Earth at a distance through empty space. Earth doesn't have to "know" that there is a sun out there, 93 million miles distant. It only "knows" that there is a gravitational field at its own location. The field, though nearly as ethereal as the ether itself, can be said to have physical reality. It occupies space. It contains energy. Its presence eliminates a true vacuum. We must then be content to define the vacuum of everyday discourse as a region free of matter, but not free of field.

Faraday's theories were heretical and not popular with his contemporaries. "The reaction to the concept of the line of force was not merely one of indifference," writes Williams, "it was downright hostile, especially when Faraday tried to extend it to gravitation. . . . The *Athenaeum* suggested that he go back to the Royal Institution and work up his sixth form mathematics before he ventured again into the deep seas of Laplacian physics." In 1855, when he was sixty-four, Faraday said to his niece, Constance Reid, "How few understand the physical line of force! They will not see them, yet all the researches on the subject tend to confirm the views I put forth many years since. Thomson of Glasgow seems almost the only one who understands them. He is perhaps the nearest to understanding what I meant. I am content to wait, convinced as I am of the truth of my views."

Faraday's theories were opposed because they were revolutionary, always sufficient reason to stir opposition, and also because Faraday did not speak the sophisticated mathematical language his fellow theorists expected to hear. Beyond rudimentary arithmetic, Faraday had no mathematics; his mathematical methods were about the same as those of Galileo. In Faraday's time, that may actually have been an advantage for creativity. The field concept was the product of "a highly original mind, a mind which never got stuck on formulas," wrote a great twentieth-century field theorist, Albert Einstein. But for Faraday's audience theoretical physics had to be mathematical physics.

Faraday's great fortune was that he had two young followers who—apparently alone—believed in the concepts of lines of force and field, and possessed all the equipment needed to build field theories in the requisite mathematical language. One of these mathematical physicists was William Thomson, as Faraday told his niece. To Faraday's great delight, Thomson formulated a mathematical theory of electric lines of force in 1845, when he was just twenty-one. The other mathematical physicist with his eyes on field theory was James Clerk Maxwell, who later, shortly before Faraday's death, created his great electromagnetic theory of light. Maxwell explained the genesis of his theory, and acknowledged his debt to Faraday and Thomson, in the preface to his *Treatise on Electricity and Magnetism*:

> I was aware that there was supposed to be a difference between Faraday's way of conceiving phenomena and that of the mathematicians, so that neither he nor they were satisfied with each other's language. I had also the conviction that this discrepancy did not arise from either party being wrong. I was first convinced of this by Sir William Thomson, to whose advice and assistance, as well as to his published papers, I owe most of what I have learned on the subject.
>
> As I proceeded with the study of Faraday, I perceived that his method of conceiving the phenomena was also a mathematical one, though not exhibited in the conventional form of mathematical symbols. I also found that these methods were capable of being expressed in the ordinary mathematical forms, and thus compared with those of the professed mathematicians.

Maxwell added that he deliberately read Faraday's *Experimental Researches in Electricity* before reading "any mathematics on the subject."

Our account may give the impression that Faraday was first an experimentalist and then a theorist, in separate scientific lives, so to speak. But there was only one scientific life, a highly creative interplay between the experiments and the theoretical speculations. The experiments suggested the theories, and the theories guided the experiments. Neither endeavor would have succeeded without the other. This ability to work in the theoretical and experimental realms simultaneously and creatively is a rare gift. Only a few of the physicists in this book, perhaps only Newton and Fermi in addition to Faraday, had it. Einstein, Gibbs, Maxwell, Boltzmann, and Feynman were in the first rank of theorists, but not creative experimentalists.

At Home

Faraday's wife, Sarah, was in some ways as remarkable as her husband. More than anyone else, she was the steadying influence that kept the Faraday volcano of energy under control. Williams gives us this picture of her indispensable role in Faraday's life: "Sarah Barnard was a perfect mate for Faraday. From his accounts and from accounts of others, she emerges as a warm and charming person. She was filled with maternal feelings which, in the absence of children of her own, she lavished upon her nieces and upon Faraday himself. This was precisely what Faraday needed. Oftentimes he would become so absorbed in his work in the laboratory that he would forget his meals. Quietly Mrs. Faraday would serve him and see that his health did not suffer."

Wisely, she did not attempt to follow her husband's work. She told her niece that science was already "so absorbing and exciting to him that it often deprives him of his sleep and I am quite content to be the pillow of his mind." In 1838, when he was forty-seven, Faraday wrote to Sarah from Liverpool, "Nothing rests me so much as communion with you. I feel it even now as I write, and I catch myself saying the words aloud as I write them, as if you were here within hearing." Much later, in 1863, when his health was failing, he wrote to Sarah, "My head is full, and my heart also, but my recollection rapidly fails, even as regards the friends that are in the room with me. You will have to resume your old function of being a pillow to my mind, and a rest, a happy-making wife."

The Faradays were childless but immensely fond of children. Two nieces, Constance Reid and Jane Barnard, often filled the void. They enjoyed Faraday's company as much as he did theirs. "A visit to the laboratory used to be a treat when the busy time of the day was over," Constance Reid wrote in her diary. "We often found him hard at work on experiments with his researches, his apron full of holes. If very busy he would merely give a nod, and aunt would sit down quietly with me in the distance, till presently he would make a note on his slate and turn round to us for a talk; or perhaps he would agree to come upstairs to finish the evening with a game of bagatelle, stipulating for half an hour's quiet work first to finish the experiment. He was fond of all ingenious games, and he always excelled in them." With some young visitors, he romped through the institution's lecture theater in a game of hide-and-seek, and then entertained them with tuning forks and resounding glasses. "He was," as one biographer observed, "still a child himself."

Discourses

"Faraday was admirably suited to the Royal Institution, and the Royal Institution admirably suited Faraday; indeed there was probably no other place in British science where Faraday could have flourished. In the same building he could play out both his private and public roles," writes a recent biographer, Geoffrey Cantor. Faraday occupied three spaces at the institution: upstairs, downstairs, and basement. Upstairs was the Faradays' apartment, which they occupied until Faraday retired in 1862. Downstairs were the public rooms, the library, and the lecture theater; and in the basement was the laboratory. We have seen Faraday in his laboratory and at home upstairs in the apartment. Now we find him downstairs, performing as a lecturer and a teacher.

From 1826 until his retirement, Faraday gave a series of lectures for lay audiences that he called Friday Evening Discourses. He took these lectures seriously: he rehearsed them, worried about them, and prepared cards to improve his timing. They were popular, and the income they provided helped alleviate the institution's perennial financial problems. In these lectures and others, Faraday broke his own rule that "lectures which *really teach* will never be popular; lectures which are popular will never *really teach*."

An even greater boon for the institution was the Christmas Lectures given by Faraday for children. They quickly attracted an audience from the upper social strata of London, including Albert Edward, Prince of Wales. The most famous of the Christmas lecture series, called "The Chemical History of a Candle," was published and has gone through innumerable editions in many languages. It shows Faraday in a charming dialogue with his young audience and also with nature. "There is no better, there is no more open door by which you can enter into the study of natural philosophy than by considering the physical phenomena of a candle," he begins. With many enthusiastic asides, he shows how candles are made, how they burn, demonstrates the thermal and chemical structure of the flame, identifies the chemical reactions of combustion, and finally leads his audience into the mysteries of electrochemistry, respiration, and the chemistry of the atmosphere. From the humble candle, he evokes a world of science, for himself as much as for his youthful audience.

Sandemanian

In response to a correspondent who asked about the influence of his religion on his natural philosophy, Faraday wrote, "There is no philosophy in my religion. I am of a very small and despised sect of Christians, known, if at all, as *Sandemanians*, and our hope is founded in the faith that is in Christ."

The Sandemanians originated in Scotland, where they were called Glasites, and later spread to Yorkshire and other parts of England. In Faraday's time, the membership was about one hundred in London and around six hundred in total. "Sandemanianism makes great demands on its members," writes Cantor. "It is not for the half-hearted or for those who wish to practice Christianity only on Sundays. Indeed it is a way of life. In making his confession of faith in 1821 at the age of 29, Faraday solemnly vowed to live according to the precepts laid down in the Bible and in imitation of Christ's perfect example. Sandemanians live strictly by the laws laid down in the Bible, and the sect's stern disciplinary code ensures that any backslider is either brought back into the fold or is excluded—'put away' to use the conventional euphemism."

The beliefs and practices of the Sandemanians are far from the British religious mainstream, and predictably that has brought hostility from followers of other religions. Sandemanians see themselves as a despised sect, as Faraday told his correspondent, and accept that fate because Christ himself was isolated and despised by his contemporaries.

Faraday was elected an elder of the church in 1840, an event of great importance in his life. About four years later, for reasons that are still obscure, he was excluded from the church for a short time. According to one of his biographers, J. H. Gladstone (father of Margaret Gladstone, who gave us some charming glimpses of William Thomson in chapter 7), Faraday accepted an invitation from the queen for a visit on a Sunday early in 1844, and consequently did not appear at the church meetinghouse that day. When he was asked to justify his absence, he did so by insisting that in his mind the queen's command took precedence. That was not the expected repentance.

Cantor disputes this account, pointing out that at the same time Faraday was excluded, so were others, around 20 percent of the membership, including his brother, sister-in-law, and father-in-law. Moreover, Cantor reports that he could find no evidence that Faraday actually visited the queen on the day in question. In any case, Faraday suffered the exclusion, and wrote to Schönbein that it left him "low in health and spirit." He was soon reinstated, but was not reelected an elder for sixteen years.

Faraday's faith was certainly deeply rooted, and despite his protestation that there was no religion in his philosophy, it must have guided his metaphysics, and the metaphysics his physics. He believed that the universe was a divinely inspired edifice. It was less than that if it did not manifest patterns of unity and symmetry. He searched for those patterns in natural forces, sometimes spending years on a single quest. When he succeeded, as he did many times, his religious faith was confirmed and deepened.

Later Life

Faraday crossed a divide in his life during the years from 1841 to 1845, while he was recovering from his breakdown. This was a period of rest—"head-rest," really, for otherwise he was active as ever. In 1841, he and Sarah traveled to Switzerland, where Faraday hiked the mountain trails and roads, sometimes at a phenomenal pace. On one occasion, he walked forty-five miles over rough terrain in ten-and-a-half hours. "I felt a little stiff," he recorded in his journal at the end of this excursion, "and only felt conscious of one small blister," but added: "I would gladly give half this strength for as much memory, but what have I to do with that? Be thankful."

He was out of the laboratory, but continued his dialogue with nature. Here he records in his journal delight in one of his favorite natural events, a thunderstorm: "the morning was sunny and beautiful and the afternoon was stormy, and equally beautiful; so beautiful I never saw the like. A storm came on, and the deep darkness of one part of the mountains, the bright sunshine of another part, the emerald lights of the distant forests and glades under the edge of the cloud were magnificent. Then came lightning, and the Alp thunder rolling beautifully; and to finish all, a flash struck the church, which is a little way from us, and set it on fire, but no serious harm resulted, as it was soon put out." Here he marvels at the sound and fury of an avalanche on the Jungfrau:

> Every now and then thundering avalanches. The sound of these avalanches is exceedingly fine and solemn. . . . To the sight the avalanche is at this distance not terrible but beautiful. Rarely is it seen at the commencement, but the ear tells first of something strange happening, and then looking, the eye sees a falling cloud of snow, or else what was a moment before a cataract of water changed into a tumultuous and heavily waving rush of snow, ice, and fluid, which as it descends through the air, looks like water thickened, but as it runs over the inclined surfaces of the heaps below, moves like paste, stopping and going as the mass behind accumulates or is dispersed.

And here he enjoys an alpine display of another of his favorite natural delights, sky effects: "a succession of exceedingly fine cloud effects came on, the blue sky appearing in places most strangely mixed with snow-peaks and the clouds. To my mind no scenery equals in grandeur the fine sky-effects of such an evening as this. We even had the rose-tint on the snow tops in the highest perfection for a short time."

By 1845, he was partially recovered and back in his laboratory, pursuing magnetic effects on light. At about this same time, he started a series of researches that ended in failure, but this failure was as interesting as other people's successes. Inspired by the connections he had found among other natural forces, he hoped to include gravity in these correlations. In his laboratory notebook, he wrote: "Gravity. Surely this force must be capable of an experimental relation to electricity, magnetism, and the other forces, so as to bind up with them in reciprocal action and equivalent effect." He proposed a sequence of experiments and began to have doubts, but they were dispelled: "ALL THIS IS A DREAM [he wrote in his laboratory notebook]. Still examine it by a few experiments. Nothing is too wonderful to be true, if it be consistent with the laws of nature; and in such things as these, experiment is the best test of consistency."

But the gravitational force refused to "bind up" with the other forces. "The results are negative," he wrote at the end of the paper reporting the work, but added, "they do not shake my strong feeling of the existence of a relation between gravity and electricity." Ten years later he tried again, and closed his last paper with almost the same words.

Faraday was the first in a long line of preeminent physicists who have searched for a theory that unifies gravity with other forces. For many years, Einstein attempted, and ultimately failed, to build a unified field theory that included both gravity and electromagnetism. More recently, the goal has been to find a quantum theory of gravity. That effort, too, has so far failed, but Faraday would note with approval that theorists are still dreaming.

Faraday in later life was not much different from the young man hired by Davy decades earlier. He was now Professor Faraday, D.C.L. (Oxford), fellow of the Royal Society, with medals and dozens of other honors from academies and scientific societies; but he was still unpretentious, sincere, and satisfied with a humble lifestyle. Except for help from his assistant, Charles Anderson—whose contribution was "blind obedience," according to Bence Jones—he worked alone. "I do not think I could work in company, or think aloud, or explain my thoughts," he said late in his life. "I never could work, as some professors do most extensively, by students or pupils. All the work had to be my own."

As he grew older, he suppressed his natural tendency to be gregarious, and became increasingly asocial. "He became more and more selective about the invitations he would accept," writes Williams. "By the mid-1830s the rejection of invitations had become almost complete. He would attend the anniversary dinners of the Royal Society and a *very* few other events. The apartment in the Royal Institution and the laboratory in the basement provided everything he needed for his personal happiness." He called himself "an anchorite." Yet his lectures at the Royal Institution were famous, and he was a celebrity. Driven by his religion, obsessive work habits, and recurring ill health, he resisted the charms of social activities. The public met him in the lecture hall, in correspondence, or not at all.

Faraday never fully recovered from his breakdown in the 1840s. Although he returned to his research in 1845, he was still plagued by periods of memory loss, headaches, giddiness, and depression. He tells about his struggle against increasing mental frailty in letters written to his colleague and close friend Christian Schönbein. In these letters, as nowhere else, Faraday reveals his affliction. Here are some extracts, written between 1845 and 1862, in chronological order:

> My head has been so giddy that my doctors have absolutely forbidden me the privilege and pleasure of working or thinking for a while, and so I am constrained to go out of town, be a hermit, and take absolute rest.
>
> My dear friend, *do you remember* that I *forget*, and that I can no more help it than a sieve can help water running out of it.
>
> I have been trying to think a little philosophy (magnetical) for a week or two, and it has made my head ache, turned me sleepy in the day-time as

> well at nights, and, instead of being a pleasure, has for the present nauseated me.
>
> Even if I go away for a little general health, I am glad to return home for rest in the company of my dear wife and niece . . . my time is to be quiet and look on, which I am able to do with great content and satisfaction.

In his last letter to Schönbein, in 1862, he said good-bye: "Again and again I tear up my letters, for I write nonsense. I cannot spell or write a line continuously. Whether I shall recover—this confusion—do not know. I will not write any more. My love to you."

12

The Scientist as Magician

James Clerk Maxwell 1831–1879

Heart, Head, and Fingers

"There are three ways of learning props [propositions]—the heart, the head, and the fingers; of these the fingers is the thing for examinations, but it requires constant thought. Nevertheless the fingers have fully better retention of methods than the heart has. The head method requires about a mustard seed of thought, which, of course, is expensive, but then it takes away all anxiety. The heart method is full of anxiety, but dispenses with the thought, and the finger method requires great labor and constant practice, but dispenses with thought and anxiety together." This is James Clerk Maxwell offering advice, characteristically concise, cryptic, and profound, to his young cousin Charles Cay. We can translate by identifying the fingers as memory and technique, the head as reason, and the heart as intuition.

Maxwell himself was skilled in all three methods. He demonstrated the competence of his fingers as an outstanding student at Cambridge University; and he built his theories by complex reasoning from physical and mathematical models. But the principal source of his genius was his mastery of the heart method. The brilliance of his scientific intuition and insight puts him in a class with Newton and Einstein.

In the construction of his theory of electromagnetism, the main concern in this chapter, Maxwell's intellectual tool was analogy. "In order to obtain physical ideas without adopting a physical theory," he wrote in the introduction to his first paper on electromagnetism, "we must make ourselves familiar with the existence of physical analogies. By a physical analogy I mean that partial similarity between the laws of one science and those of another which makes each of them illustrate the other." On the road to his theory of electromagnetism, Maxwell invented two successive mechanical analogies. Neither was a theory: about the first he wrote, "I do not think it contains even a shadow of a true theory." But in each he intuitively recognized elements of the truth, which he built into his evolving theory. In the end, he took away the mechanical models, like the removal of a scaffolding, and what was left were mathematical statements, the now-celebrated "Maxwell's equations."

To Maxwell's associates, this reliance on a series of provisional arguments, and their ultimate abandonment to the abstractions of differential equations, seemed like the conjuring trick of a magician. One colleague remarked that Maxwell's world of electromagnetic theory seemed like an enchanted fairyland; he never knew what was coming next. And it didn't help that during Maxwell's lifetime his theory had little experimental support. To be a Maxwellian, you had to subscribe to Maxwell's insights, which could seem decidedly quirky, with few verifying experiments.

One of Maxwell's biographers, C. W. F. Everitt, points to another vital aspect of his genius by comparing him with his two mentors, Michael Faraday and William Thomson. Everitt characterizes Faraday as an "accumulative thinker," Thomson as an "inspirational thinker," and Maxwell as an "architectural thinker." Faraday accumulated the facts of electricity and magnetism by designing and executing experiments. His rule was to "work, finish, publish," and move on. Thomson was the virtuoso; he had inspired answers to all kinds of problems, but rarely wove them into a finished theory. Maxwell had the patience and tenacity that Thomson lacked. "Maxwell's great papers," Everitt writes, "are in total contrast to Thomson's. Seventy or eighty pages long (and tersely written at that), each is evidently the result of prolonged thinking, and each in its own way presents a complete view of its subject." Like Newton, another great architectural thinker, Maxwell developed his major ideas gradually; he started his theory of electromagnetism in 1855 and finished it almost twenty years later, in 1873, with long pauses between papers. He felt that part of the slow evolution of his theories was subconscious. In a letter to a friend, he wrote: "I believe there is a department of mind conducted independent of consciousness, where things are fermented and decocted, so that when they are run off they come clear."

What the architect erected was one of the great intellectual edifices of the nineteenth century. It unified all electric and magnetic phenomena, revealed the electromagnetic wave nature of light, and opened the door to the style and substance of twentieth-century physics. Maxwell did it with head and heart, thought and anxiety, and with an ingredient of the mind that can well be called magical.

Dafty

James Clerk Maxwell was born in Edinburgh, Scotland, in 1831. His father, John Clerk, added the name Maxwell to satisfy some legal conditions that allowed him to inherit a small country estate in Middlebie, Galloway (southwestern Scotland). John Clerk Maxwell was sensitive, cautious, and unconventional. He married Frances Cay, who was practical like her husband, but more decisive and blunt. Their personalities were complementary, and their only son had the good fortune to inherit some of the finer features of both parents.

When the Clerk Maxwells took possession of their Middlebie property, it was mostly undeveloped, not even including a house. With skill and enthusiasm, John Clerk Maxwell supervised every detail of the construction of a house, which he called "Glenlair." The son became as devoted to Glenlair as the father; through childhood, adolescence, and maturity Glenlair was his refuge.

When he was eight, James's idyllic family life at Glenlair was tragically disrupted by the painful death of his mother at age forty-eight from abdominal cancer—apparently the same cancer that killed Maxwell himself at the same early age. The boy's reaction to the tragedy was remarkably detached from his private loss: "Oh, I'm so glad! Now she'll have no more pain." John Clerk Maxwell was a doting father, and more so after his wife's death, but he could be blind to some of his son's most urgent needs. He entrusted James's formal education to a tutor whose pedagogy was, to say the least, uninspired. When his pupil obstinately objected to drills in Latin grammar, the tutor beat him. Lewis Campbell, Maxwell's principal biographer, felt that this harsh treatment had lasting psychological effects, "not in any bitterness," he writes, "though to be smitten on the head with a ruler and have one's ears pulled till they bled might naturally have operated in that direction—but in a certain hesitance of manner and obliquity of reply, which Maxwell was long in getting over, if, indeed, he ever quite got over them." The boy was stoic, the father inattentive, and the tutor remained until a visit by another important figure in Maxwell's early life, his maternal aunt, Jane Cay. She sized up the tutor situation and persuaded the father to send the boy to Edinburgh, where he could join the household of Isabella Wedderburn, his paternal aunt, and attend the Edinburgh Academy.

James's initial experiences at the academy were no happier than those inflicted by the tutor. He appeared the first day dressed in the sensible clothes he wore at Glenlair, designed by his father with little thought of appearance or fashion. The country clothes, and an accompanying Gallovidian accent, made him an easy target for a tormenting gang of schoolmates. But he gave as good as he got, and returned that day to the Wedderburns with his once neat customized clothes in tatters. He seemed "excessively amused by his experiences, and showing not the smallest sign of irritation," reports Campbell. "It may be questioned, however, whether something had not passed within him, of which neither those at home nor his schoolfellows ever knew." His attackers gave him the nickname "Dafty," meaning "strange rather than silly; 'weirdo' might be closest to the modern idiom," Everitt tells us.

The academy did little to subdue the boy's spirit or discourage his unconventional behavior. He made lasting friendships—with, among others, Lewis Campbell, who was to become his biographer, and Peter Guthrie Tait, later professor of natural philosophy at the University of Edinburgh and polemicist par excellence. Cheerful letters to "his papaship" back at Glenlair brought the news from Edinburgh, elaborated with puns, mirror writing, misspellings, riddles, and hoaxes. Here is a sample:

> MY DEAR MR. MAXWELL—I saw your son today, when he told me that you could not make out his riddles. Now, if you mean the Greek jokes, I have another for you. A simpleton wishing to swim was nearly drowned. As soon as he got out he swore that he would never touch water till he learned to swim; but if you mean the curious letters on the last page, they are at Glenlair.
>
> Your aff. Nephew JAMES CLERK MAXWELL

He often signed his letters "Jas. Alex. M'Merkwell" (an anagram), and included in the address "Postyknowswhere."

Maxwell was not a prodigy; unlike Thomson, he did not show early signs of mathematical genius. No doubt the sensitive father, an enthusiastic amateur in all matters of science and technology, deserves major credit for developing his son's talents. Father and son often attended meetings of the Edinburgh Society of Arts and the Edinburgh Royal Society. At age fourteen, displaying a geometrical imagination that would serve him well throughout his career, Maxwell wrote a paper describing a novel method for constructing ovals. John Clerk Maxwell saw to it that James Forbes, a professor of natural philosophy at Edinburgh, read the work. Forbes found it "very remarkable for [the author's] years," and communicated the paper to the Edinburgh Royal Society.

With the ovals, Maxwell's scientific career was launched. After the Edinburgh Academy, he studied with Forbes and William Hamilton (not to be confused with William Rowan Hamilton, the great Irish mathematician and physicist) at the University of Edinburgh. Forbes and Hamilton were at opposite poles in all university matters, and sworn enemies. Forbes was a skilled experimentalist and gave Maxwell free access to his laboratory. Hamilton was a philosopher who forcefully taught that knowledge is not absolute but relative to, and shaped by, the limitations of human senses; to get at the truth, imperfect logical devices such as models and analogies are necessary. The two adversaries agreed on one thing: that young Clerk Maxwell deserved special attention. They gave it—Forbes in the laboratory and Hamilton in the lessons of metaphysics—and their influence was lasting.

But for a student with Maxwell's mathematical talents, Edinburgh was not enough. The next step was Cambridge, the acknowledged center for training "head and fingers" in the methods of mathematics and physics. The centerpiece at Cambridge was the Tripos Syllabus, which prepared students for a punishing series of examinations. Training of the examinees was in the hands of private tutors. The most illustrious of these was William Hopkins, who had coached many Tripos winners, known for some reason as "wranglers." Maxwell joined Hopkins's team and triumphed as a wrangler, but with less than the normal amount of drudgery, as Tait, another wrangler, tells us in this reminiscence:

> [He] brought to Cambridge, in the autumn in 1850, a mass of knowledge which was really immense for so young a man, but in a state of disorder appalling to his methodical tutor. Though that tutor was William Hopkins, the pupil to a great extent took his own way, and it may safely be said that no high wrangler of recent years ever entered the Senate-House [where the Tripos examinations were given] more imperfectly trained to produce "paying" work than did Clerk Maxwell. But by sheer strength of intellect, though with the very minimum of knowledge how to use it to advantage under the conditions of the examination, he obtained the position of Second Wrangler and was bracketed equal with the Senior Wrangler in the higher ordeal of the Smith's Prizes [another competition].

"Second Wrangler" was second place in the competition, but it was an impressive performance for one unprepared for "paying" work. Hopkins said of Maxwell that he was "unquestionably the most extraordinary man he [had] met with in the whole range of [my] experience." It appeared "impossible for Maxwell to think incorrectly on physical subjects."

He was still confirmed in his unconventional ways, but at Cambridge eccentricities, if they were entertaining, were an advantage. "He tried some odd experiments in the arrangement of his hours of work and sleep," writes Campbell. "From 2 to 2:30 A.M. he took exercise by running along the upper corridor, *down* the stairs, along the lower corridor, then *up* the stairs and so on, until the inhabitants of the rooms along his track got up and lay *perdus* behind their sporting-doors to have shots at him with boots, hair-brushes, etc., as he passed." Tait gives this account of further Maxwellian antics: "He used to go up on the pollard at the bathing-shed, throw himself *flat on his face* in the water, dive and cross, then ascend the pollard on the other side, project himself *flat on his back* in the water. He said it stimulated the circulation!"

Maxwell's Tripos performance earned him a scholarship and then a fellowship at Trinity College. During this peaceful time, he started his research on electromagnetism and fell in love with his teenaged cousin, Elizabeth Cay, "a girl of great beauty and intelligence," according to Everitt. The romance did not last, however, because of family concern with "the perils of consanguinity in a family already inbred."

Two years as a Cambridge don left Maxwell restless for a less cloistered existence. "The sooner I get into regular work the better," he wrote to his father. Forbes reported that a professorship of natural philosophy was available at Marischal College, Aberdeen, Scotland. Maxwell applied for the position, complaining about the process of testimonials. One reason for considering Aberdeen was to be nearer to his father, whose health was declining. In the spring of 1856, John Clerk Maxwell died, and a few weeks later Maxwell learned from Forbes that he had the Aberdeen appointment.

Aberdeen, London, Glenlair

Maxwell, like many creative scientists, was not successful as a teacher. While lecturing, his thoughts were so complex and rapid that he could not slow to the mental pace of his students. He was sometimes thrown into a kind of panic by student audiences, as Campbell relates:

> [A] hindrance lay in the very richness of his imagination and the swiftness of his wit. The ideas with which his mind was teeming were perpetually intersecting, and their interferences, like those of waves of light, made "dark bands". . . . Illustrations of *ignotum per ignotius* [the unknown through the more unknown], or of the abstruse by some unobserved property of the familiar, were multiplied with dazzling rapidity. Then the spirit of indirectness and paradox, though he was aware of its dangers, would often take possession of him against his will, and either from shyness, or momentary excitement, or the despair of making himself understood, would land him in "chaotic statements," breaking off with some quirk of ironical humor.

Yet his written style—his papers, formal lectures, and books—were models of clarity. This strange conflict between Maxwell's verbal and written expression impressed one of his Aberdeen students, David Gill, who became an accomplished astronomer:

> Maxwell's lectures were, as a rule, most carefully arranged and written out—practically in a form fit for printing—and we were allowed to copy them. In lecturing he would begin reading his manuscript, but at the end of five minutes or so he would stop, remarking, "Perhaps I might explain this," and then he would run off after some idea which had just flashed upon his mind, thinking

> aloud as he covered the blackboard with figures and symbols, and generally outrunning the comprehension of the best of us. Then he would return to his manuscript, but by this time the lecture hour was nearly over and the remainder of the subject was dropped or carried over to another day. Perhaps there were a few experimental illustrations—and they very often failed—and to many it seemed that Clerk Maxwell was not a very good professor. But to those who could catch a few of the sparks that flashed as he thought aloud at the blackboard in lecture, or when he twinkled with wit and suggestion in after lecture conversation, Maxwell was supreme as an inspiration.

Maxwell completed his first paper on electromagnetism during his four years in Aberdeen. He was also occupied at the time with courting Katherine Dewar, daughter of the principal of Marischal College, and they were married in 1858. Campbell is mostly silent about the marriage, and he seems to say something by his omission. Katherine was seven years older than Maxwell, in constant ill health, and at least in later life, neurotic. If the gossip of Maxwell's friends is to be believed, she resented her husband's scientific activities. Perhaps so, but she skillfully assisted him in several series of experiments. Neither husband nor wife brought passion to the marriage, but it is clear from their correspondence that they were deeply devoted to each other.

Maxwell was left redundant and without a job in 1860 when Aberdeen merged its two colleges, Marischal and King's. He probably had few regrets; Aberdeen was not his social element. He had written earlier to Campbell: "Society is pretty steady in this latitude—plenty of diversity, but little of great merit or demerit—honest on the whole, and not vulgar. . . . No jokes of any kind are understood here. I have not made one for two months, and if I feel one coming I shall bite my tongue."

His next move was to King's College in London, where he was appointed to the professorship of natural philosophy. Maxwell's five years in London were the most creative in his life. He brought his dynamical theory of the electromagnetic field to maturity during that time. In addition, he advanced his theories of gas behavior and color vision, and produced the world's first color photograph. These accomplishments, particularly the electromagnetic theory, were fundamentally important, and he knew it. Although he rarely said so in his letters, this work must have given him great satisfaction. His heavy burden of teaching was less congenial, as he broadly hinted in a letter to Campbell: "I hope you enjoy the absence of pupils. I find that the division of them into smaller classes is a great help to me and to them: but the total oblivion of them for definite intervals is a necessary condition of doing them justice at the proper time."

Finally he concluded that he did not need an academic appointment, with all its accompanying duties for which he was not well suited, to continue his researches. He had comfortable independent means and all the professional contacts he needed to communicate his findings to the scientific world. What he really wanted was more time at Glenlair, "to stroll in the fields and fraternize with the young frogs and old water-rats," as he had done earlier. So in 1865 he resigned from King's College and took up permanent residence at Glenlair.

Campbell sketches Maxwell's Glenlair activities and recreations:

> Both now and afterwards, his favorite exercise—as that in which his wife could most readily share—was riding, in which he showed great skill. [A neighbor]

> remembers him in 1874, on his new black horse, "Dizzy," which had been the despair of previous owners, "riding the ring," for the amusement of the children of Kilquhanity, throwing up his whip and catching it, leaping over bars, etc.
>
> A considerable portion of the evening would often be devoted to Chaucer, Spenser, Milton, or a play by Shakespeare, which he would read aloud to Mrs. Maxwell.
>
> On Sundays, after returning from the kirk, he would bury himself in the works of the old divines. For in theology, as in literature, his sympathies went largely with the past.
>
> [He had] kindly relations with his neighbors and with their children. . . . [He] used occasionally to visit any sick person in the village, and read and pray with them in cases where such ministrations were welcome.
>
> One who visited at Glenlair between 1865 and 1869 was particularly struck with the manner in which the daily prayers were conducted by the master of the household. The prayer, which seemed *extempore*, was most impressive and full of meaning.

Maxwell as laird: a role he clearly enjoyed.

We now turn from personal to scientific biography, and that means first some easy lessons on the mathematical language of electromagnetism.

Vector Lessons

Maxwell's electromagnetic theory is a story of electric and magnetic fields of forces. These forces, like all others, not only have a certain magnitude but also a direction. In addition to force, velocity, momentum, and acceleration are also directional. Nondirectional quantities, called "scalars," are equally important in physics; energy, temperature, and volume are examples. All directional physical quantities are represented mathematically as "vectors," and are distinguished from scalar quantities by their boldface symbols. A force vector might be represented by **F**, a velocity vector by **v**, and a momentum vector by **p**.

Directions of vectors are conveniently specified by resolving their components in three mutually perpendicular directions, which one can picture as east-west, north-south, and up-down axes. The abstract symbols x, y, and z conventionally label these axes, and the vector components measured along the axes are given corresponding labels. The velocity vector **v**, for example, has components designated v_x, v_y, and v_z along the x, y, and z axes. An airplane climbing with a speed of 500 miles per hour at an angle of 30° and in a southeast direction has the velocity components $v_x = v_y = 306$ miles per hour (southeast) and $v_z = 250$ miles per hour (up). See figure 12.1 for a visualization of this vector. Entire equations can be expressed in this vectorial language. Newton's second law of motion, for example, connects the force vector **F** with the rate of change in the momentum vector **p**,

$$\mathbf{F} = \frac{d\mathbf{p}}{dt}$$

Maxwell eventually put all of his electromagnetic field equations in vectorial format, and they are still seen that way. The electric and magnetic fields are represented by the vectors **E** and **B**, and Maxwell's equations relate these vectors to the electric charges and currents always associated with an electromagnetic

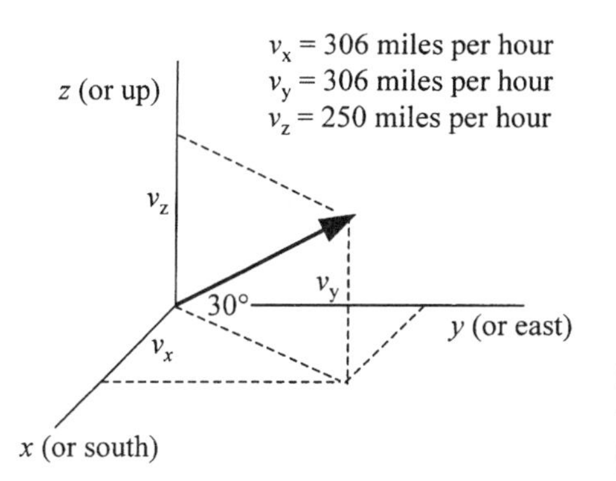

Figure 12.1. Picture of a velocity vector **v** (represented by the arrow) for an airplane headed southeast at 500 miles per hour and climbing at an angle of 30°.

field. Maxwell relied on two key mathematical operations for analyzing a field to reveal its charge and current structure. Both lead to differential equations and were borrowed from the dynamics of fluid motion. One operation, applied at a point in the field, measured what Maxwell called the "convergence," that is, the extent to which the field was aimed at the point. The second operation measured the rotational character of the field at the point. For this, Maxwell eventually settled on the term "curl," after discarding "rotation," "whirl," "twist," and "twirl." See figure 12.2 for Maxwell's illustrations of the convergence and curl operations. In later usage, it was found more convenient to switch the sign and direction of Maxwell's convergence operation and make it into "divergence."

Great Guns

Maxwell's first paper on electromagnetism, published while he was at Aberdeen and twenty-four years old, had the title *On Faraday's Lines of Force*. It was aimed at giving mathematical form to Faraday's field concept. Maxwell was following Thomson, who had earlier composed a mathematical theory of Faraday's concept of electric lines of force. When he started his work, Maxwell wrote to Thomson warning him to expect some "poaching": "I do not know the Game-laws & Patent-laws of science. Perhaps the [British] Association may do something to fix them but I certainly intend to poach among your images, and as for the hints you have dropped about the 'higher' electricity, I intend to take them." Thomson cheerfully opened the gates to his "electrical preserves," wishing Maxwell good hunting.

And Maxwell found it. His theory delved deeper than Thomson's; it concerned

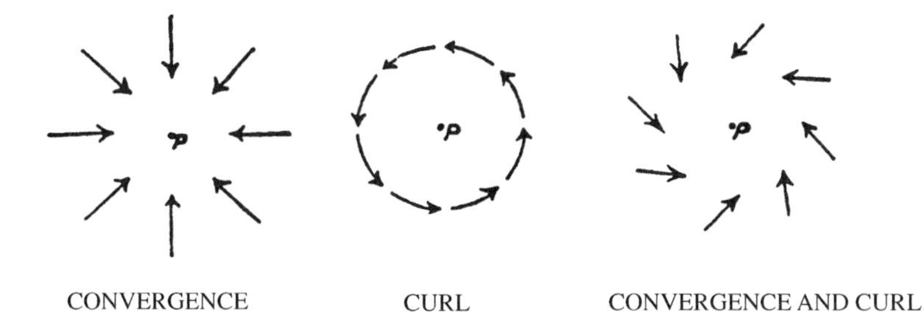

Figure 12.2. Maxwell's representations of convergence and curl operations at a point in a field. From *The Scientific Papers of James Clerk Maxwell*, ed. W. D. Niven (New York: Dover, 1952), 2:265.

magnetic fields as well as electric fields, and showed mathematically how they were interconnected. He found his mathematical ideas in an analogy between Faraday's lines of force and the lines of flow in a fictitious, weightless, incompressible fluid. Like all of the analogies evoked by Maxwell, this one did not constitute a complete physical theory. The gift of the analogy was a short list of equations that accounted for many of the observed phenomena of electricity, magnetism, and electromagnetism. The ingredients of the equations were five vectors, which we now write **A, B, E, H**, and **J**. (The vector notation was not fully developed until later by Oliver Heaviside and Willard Gibbs, but the anachronism violates only the letter, not the spirit, of Maxwell's equations.) The electric field was represented by the **E** vector, and **J** described electric current. For the magnetic field two vectors were required, **B** and **H**. **H** was generated by the currents **J**, as observed in Oersted's experiment. The second magnetic vector, **B**, was equal to **H** in a vacuum but differed from it in a material medium.

The four vectors **B, E, H**, and **J** and their equations unified in concise mathematical form the phenomena observed by Faraday, Ampère, and Oersted. The fifth vector, **A**, was pure Maxwellian speculation. It stood for what Faraday had originally called the "electrotonic state," the special condition created in a wire by a magnet, such that when the wire was moved an electric current was induced. Faraday had changed his mind, however, and eventually abandoned the idea of the electrotonic condition. Maxwell resurrected the concept by introducing his vector **A**, which he called the "electrotonic intensity," and showing in one of his equations that the electric field vector **E** was equal to the rate of change of **A**; that equation was a direct statement of Faraday's law of magnetic induction.

The further history of Maxwell's seemingly innocent vector **A** is interesting. Maxwell changed its name twice, from the original "electrotonic intensity" to "electromagnetic momentum," and then to "vector potential." Maxwell's immediate successors found **A** offensive and wrote it out of the equations. The next generation brought it back, and in 1959 David Bohm and Yakir Aharanov gave the elusive **A** a secure place in electromagnetic theory by showing that without it the field is not fully specified.

After reading an offprint of Maxwell's paper sent courtesy of the author, Faraday responded in a letter that deserves a place in any collection of great scientific correspondence. Faraday expressed his gratitude, apologized for his mathematical innocence, and then made an astonishing suggestion:

> MY DEAR SIR—I received your paper, and thank you for it. I do not venture to thank you for what you have said about "Lines of Force," because I know you have done it for the interests of philosophical truth, but you must suppose it is a work grateful to me, and gives me much encouragement to think on. I was at first almost frightened when I saw such mathematical force made to bear on the subject, and then wondered to see the subject stood it so well. I send by this post another paper to you; I wonder what you will say of it. I hope however, that bold as the thoughts may be, you may perhaps find reason to bear with them. I hope this summer to make some experiments on the *time* [speed] of magnetic action, or rather on the time required for the assumption of the state round a wire carrying a current, that may help the subject on. The time must probably be short as the time of light; but the greatness of the result, if affirmative, makes me not despair. Perhaps I had better have said nothing about it, for I am often long in realizing my intentions, and a failing memory is against me.—Ever yours most truly, M. Faraday.

This was Faraday, nearing the end of his career, communicating with Maxwell, age twenty-six and in his second year at Aberdeen. Maxwell's paper was lengthy and full of equations, and Faraday understood little of the mathematical language. Yet he divined Maxwell's message, and was reminded of his own conjecture, that magnetic (and presumably electric) effects were transmitted in a finite time, not instantaneously. That time was indeed very short, and Faraday's experiments were not successful. But for Maxwell, the theorist, here was a grand revelation. "The idea of the *time* of magnetic action . . . seems to have struck Maxwell like a bolt out of the blue," writes Martin Goldman, a Maxwell biographer. "If electromagnetic effects were not instantaneous that would of course be marvelous ammunition for lines of force, for what could a force be in transit, having left its source but not yet arrived at its target, if not some sort of traveling fluctuation along the lines of force?"

Maxwell's next paper on electromagnetism matched Faraday's conjecture with another. This paper came from London in 1861 and 1862 with the title *On Physical Lines of Force.* It worked the Maxwellian wizardry with a new analogy, this one between the medium through which electric and magnetic forces were transmitted—called the "ether" by Victorian scientists—and the complicated honeycomb-like system of vortex motion shown in figure 12.3. Each cell in the honeycomb represented a vortex, with its axis parallel to the magnetic lines of force. The circles between the cells depicted small particles of electricity that rolled between the vortices like ball bearings and carried electric currents. Maxwell cautioned that this mechanical ether model, like the analogy he used in his previous paper, was to be used with care: "I do not bring it forward as a mode of connection existing in nature, or even as that which I would willingly assent to as an electrical hypothesis. It is, however, a mode of connection which is mechanically conceivable, and easily investigated, and it serves to bring out the actual mechanical connections between the known electromagnetic phenomena; so that I venture to say that anyone who understands the provisional and temporary character of this hypothesis, will find himself helped rather than hindered by it in his search after the true interpretation of the phenomena."

Maxwell endowed his vortices—and the real ether—with a physical property that was crucial in the further evolution of his theory: they were elastic. He knew

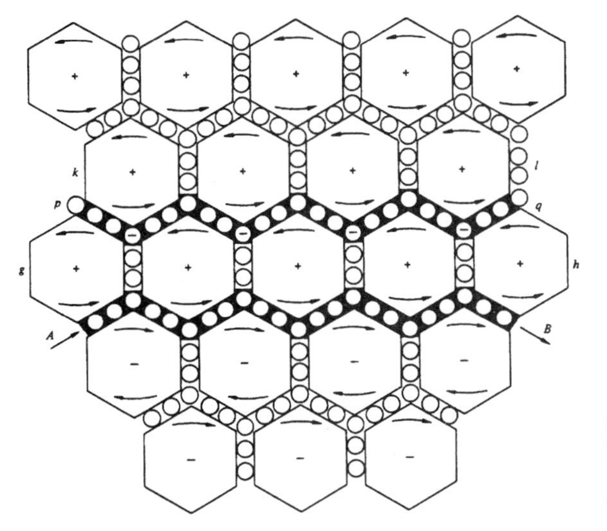

Figure 12.3. Maxwell's vortex model of the ether. From *The Scientific Papers of James Clerk Maxwell*, ed. W. D. Niven (New York: Dover, 1952), vol. 1, plate VIII, fig. 2.

that elastic media of all kinds support transverse wave motion ("transverse" here means perpendicular to the direction of wave propagation), and that the speed of the wave depends on a certain elasticity parameter of the medium. It happened that Maxwell could calculate a value for that parameter from his ether model, and from that the speed of the electromagnetic waves he imagined were propagated through the elastic medium. He did the calculation, and found to his amazement that the result was almost identical to the speed of light that had been measured in Germany by Wilhelm Weber and Rudolph Kohlrausch. In uncharacteristic italics, Maxwell announced his conclusion: "We can scarcely avoid the conclusion that *light consists of the transverse undulations of the same medium which is the cause of electric and magnetic phenomena.*"

Light as traveling electromagnetic waves: it was a simple idea, yet its implications for science and technology were still being realized a hundred years later. Maxwell had brought together under the great umbrella of his equations two great sciences, electromagnetism and optics, previously thought to be unrelated; now Maxwell claimed they were close relatives.

Like most revolutionary developments in science, Maxwell's concept of electromagnetic waves was slow to catch on. Eventually, two decades after Maxwell's *Physical Lines of Force* paper, experimentalists began to think about how to generate, detect, and use electromagnetic waves. At first, they tried to make "electromagnetic light," and that effort failed. Then they looked for electromagnetic waves of a greatly different kind and succeeded spectacularly. The hero in that work was Heinrich Hertz, the Mozart of physics, a man who had immense talent and a short life. We will come to his story later.

Hertz's waves were what we now call radio waves and microwaves. In a primitive form, radio communication, and its offspring, television, were born in Hertz's laboratory. As we now recognize, radio waves, microwaves, and light waves are "colors" in a vast continuous electromagnetic rainbow. Distinguished by their wavelengths, radio waves and microwaves are long, and light waves short. Between are the electromagnetic "colors" we call infrared radiation. On the short-wavelength side of visible light are ultraviolet radiation, x rays, and gamma rays. Wavelengths of radio waves and gamma rays differ by an astronomical ten orders of magnitude. These discoveries came to light during the last decade of the nineteenth century and the first two of the twentieth, sadly too late for Maxwell and Hertz to witness.

Maxwell extracted another, more subtle, conclusion from the elasticity property of his ether model. When the vortices were stretched or compressed in a changing electric field, the particles of electricity between the vortices were displaced, and their movement constituted what Maxwell called a "displacement current." Like any other current, it could generate a magnetic field à la Oersted, and Maxwell incorporated this possibility into his equations. With that addition, Maxwell's list of equations, although different in form, told the same mathematical story as the "Maxwell equations" found in modern textbooks.

In early 1865, Maxwell wrote to his cousin Charles Cay (later in the same year the recipient of Maxwell's "heart, head, and fingers" advice): "I have also a paper afloat, with an electromagnetic theory of light, which till I am convinced to the contrary, I hold to be great guns." This was his third offering on electromagnetism, *A Dynamical Theory of the Electromagnetic Field*, considered by most commentators to be his crowning achievement. He explained the title this way: "The theory I propose may . . . be called a theory of the *Electromagnetic Field*, because it has to do with the space in the neighborhood of the electric or magnetic bodies, and it may be called a *Dynamical* theory, because it assumes that in that space there is matter in motion, by which the observed electromagnetic phenomena are produced."

The "matter in motion" was, as before in his *Lines of Force* papers, the ether, but he now treated it without the mechanistic trappings. Gone were the fluids, vortices, and particles of electricity. In their place was an abstract analytical method introduced in the eighteenth century by Joseph Lagrange as a generalization of Newton's system of mechanics. The great advantage of Lagrange's approach was that it did its work above and beyond the world of hidden mechanisms. The mechanisms might actually be there (for example, in the ether), but the Lagrangian theorist had no obligation to worry about them.

Thomson and P. G. Tait, in their comprehensive *Treatise on Natural Philosophy*, had made abundant use of Lagrange's analytical mechanics, and in a review of the *Treatise* Maxwell explained that Lagrange's method was a "mathematical illustration of the scientific principle that in the study of any complex object, we must fix our attention on those elements of it which we are able to observe and to cause to vary, and ignore those which we can neither observe nor cause to vary." And for the mystified he offered a metaphor: "In an ordinary belfry, each bell has one rope which comes down through a hole in the floor to the bellringer's room. But suppose that each rope, instead of acting on one bell, contributes to the motion of many pieces of machinery, and that the motion of each piece is determined not by the motion of one rope alone, but that of several, and suppose, further, that all this machinery is silent and utterly unknown to the men at the ropes, who can only see as far as the holes in the floor." Each of the bellringer's ropes supplies its own information, and the ropes can be manipulated to obtain the potential energy and kinetic energy of the complex system of bells. Applying Lagrange's methods, "these data are sufficient to determine the motion of every one of the ropes when it and all the others are acted on by any given forces. This is all that the men at the ropes can ever know. If the machinery above has more degrees of freedom than there are in the ropes, the coordinates which express these degrees of freedom must be ignored. There is no help for it."

We will explore a different version of this subtle philosophy in chapter 19, where it provides escape from some otherwise weird predictions of quantum theory. Quantum theorists do not practice Lagrangian mechanics, but for different reasons they see themselves as Maxwellian bellringers. If there is a hidden world beneath their essentially statistical description, they are obliged to omit it from their deliberations, and "there is no help for it."

Maxwell used the Lagrangian method to derive all of the mathematical equipment he had obtained earlier in his *Lines of Force* papers, and then went further to identify his vector A as a measure of "electromagnetic momentum," and to calculate the energy of the electromagnetic field. With these additional elements, his theory of electromagnetism was complete. About a decade later, in 1873, Maxwell summarized his theory, and many other aspects of electromagnetism, in a difficult two-volume work called *A Treatise on Electromagnetism*, which has been called (not entirely as a compliment) the *Principia* of electromagnetism. In the *Treatise*, Maxwell's equations are found almost in the modern vectorial format.

Maxwell pursued numerous topics besides electromagnetism in his researches, including gas theory, thermodynamics, Saturn's rings, and color vision.

His molecular theory of gases, another "dynamical" theory, ranks a close second in importance to his theory of the electromagnetic field. It brought another revolutionary development to physics, the first use of statistical methods to describe macroscopic systems of molecules. In the hands of first Boltzmann and then Gibbs, Maxwell's statistical approach became the fine theoretical tool now called "statistical mechanics."

Symbols to Objects

The last chapter in Maxwell's story began late in the year 1870, when he heard from Glenlair of a new professorship in physics to be established at Cambridge. The university had belatedly realized that it was lagging behind Scottish and German universities, and even Oxford, in science education. Particularly urgent was a need for student and research laboratory facilities. The customary commission was appointed, which recommended a considerable expenditure, and that brought opposition from the nonscience faculty. The matter would have ended there but for the munificence of the chancellor of the university, the seventh duke of Devonshire, who offered to foot the bill for a new laboratory. Devonshire's family name was Cavendish, and he was related to Henry Cavendish, a reclusive, aristocratic, eighteenth-century physicist and chemist, who had conducted pioneering experiments in electricity. It also happened that Devonshire, like Maxwell, had been a Second Wrangler and Smith's Prizeman at Cambridge.

The duke's offer was accepted, and a chair of experimental physics was created for the director of the new facility, to be called the Cavendish Laboratory. The post was first offered to Thomson, but he was well planted in Glasgow and could not imagine leaving. Thomson was then asked to sound out Helmholtz, and that effort also failed; Helmholtz had just been appointed professor of physics in Berlin and director of a new physics institute. The third choice was Maxwell, who was happy and still creative at Glenlair, and not enticed. He could not deny a sense of duty, however, and he offered to stand for the post with the proviso that he might change his mind at the end of the first year. There was no opposition, he was elected, and without realizing it, Cambridge got the greatest of the three candidates.

Maxwell remained, and construction of the new laboratory went forward under his conscientious and expert supervision. His genius was for theoretical work, but he was also a competent experimentalist. The design of the Cavendish was practical and clever, and it served the needs of physics at Cambridge for more than a century. But for two years, as the planning, conferring, and building slowly progressed, Maxwell was left without a professional home: "I have no place to erect my chair [he wrote to Campbell], but move about like a cuckoo, depositing my notions in the chemical lecture-room 1st term; in the Botanical in Lent, and Comparative Anatomy in Easter."

As a newly installed professor, Maxwell was expected to deliver an inaugural lecture, and he obliged without fanfare. In fact, the affair was so casual that most of the Cambridge faculty missed it. Then, in a move that was evidently not entirely innocent, Maxwell issued a formal announcement of his first academic lecture. The esteemed scientists and mathematicians, now in attendance, were treated to a detailed explanation of the Celsius and Fahrenheit temperature scales.

The inaugural lecture survived, however. Maxwell had it printed, and it is a first-rate source of Maxwellian wisdom. It teaches lessons about the interplay between experimental and theoretical science that are still being learned. "In every experiment," he told his (presumably sparse) audience, "we have first to make our senses familiar with the phenomenon, but we must not stop here, we must find out which of its features are capable of measurement, and what measurements are required in order to make a complete specification of the phenomenon. We must make these measurements, and deduce from them the result which we require to find."

He emphasized that the processes of measurement and refinement are subtle and complex. Regrettably, he said, "the opinion seems to have got abroad, that in a few years all the great physical constants will have been approximately estimated, and that the only occupation which will then be left to men of science will be to carry on these measurements to another place of decimal."

But great scientific discoverers do not meet dreary dead ends like this: "We have no right to think thus of the unsearchable riches of creation, or of the untried fertility of those fresh minds into which these riches will be poured." On the contrary:

> The history of science shows that even during that phase of her progress in which she devotes herself to improving the accuracy of the numerical measurement of quantities with which she has long been familiar, she is preparing the materials for the subjugation of new regions, which would have remained unknown if she had been contented with the rough methods of her early pioneers. I might bring forward instances gathered from every branch of science, showing how the labor of careful measurement has been rewarded by the discovery of new fields of research, and by the development of new scientific ideas.

(See the discussion in chapter 25 of quantum electrodynamics; its great success hinged on some very refined measurements.)

Maxwell's projected program of experimental physics seemed worlds apart from the Cambridge Tripos tradition, based on intensive training of theoretical reasoning. But there must not be antagonism, Maxwell said: "There is no more powerful method for introducing knowledge into the mind than of presenting it in as many different ways as we can. When the ideas, after entering through different gateways, effect a junction in the citadel of the mind, the position they occupy becomes impregnable."

The problem for both teacher and student is to "bring the theoretical part of our training into contact with the practical," and to conquer "the full effect of what Faraday has called 'mental inertia,' not only the difficulty of recognizing, among concrete objects before us, the abstract relation which we have learned from books, but the distracting pain of wrenching the mind away from the symbols to the objects, and from the objects back to the symbols. This . . . is the price we have to pay. But when we have overcome the difficulties, and successfully bridged the gulf between the abstract and the concrete, it is not a mere piece of knowledge that we have obtained: we have acquired the rudiment of a permanent mental endowment."

Maxwell as Cavendish Professor in the 1870s was remarkably like Maxwell the student in the 1850s. One of his Cambridge friends who knew him at both stages gave Campbell this sketch:

> My intercourse with Maxwell dropped when we left Cambridge. When I returned in 1872, after an absence of fifteen years, he had lately been installed at the new Cavendish Laboratory, and I had the happiness of looking forward to a renewal of friendship with him. I found him, as was natural, a graver man than of old; but as warm of heart and fresh of mind as ever. . . . The old peculiarities of his manner of speaking remained virtually unchanged. It was still no easy matter to read the course of his thoughts through the humorous veil which they wove for themselves; and still the obscurity would now and then be lit up by some radiant explosion.

Maxwell had research students, but it was not his style to mold them into a team with a common purpose. Arthur Schuster, among the first Cavendish students, recalled that in Maxwell's view it was "best both for the advance of science, and for the training of the student's mind, that everyone should follow his own path. [Maxwell's] sympathy with all scientific inquiries, whether they touched points of fundamental importance or minor details, seemed inexhaustible; he was always encouraging, even when he thought the student was on the wrong path. 'I never try to dissuade a man from trying an experiment,' he once told me; 'if he does not find what he wants, he may find out something else.' "

Maxwell's lecture audiences were minuscule. Ambitious students trained with private tutors for the Tripos examinations; Maxwell's courses, as well as those of other university professors, were of little "paying" value for aspiring wranglers. John Fleming, another one of the early Cavendish students, reported that "Maxwell's lectures were rarely attended by more than a half-dozen students, but for those who could follow his original and often paradoxical mode of presenting truths, his teaching was a rare intellectual treat." It is said that during his tenure as Lucasian Professor at Cambridge, Newton often "lectured to the walls."

Reluctantly at first, the Cavendish Professor took on the huge task of editing the papers of Henry Cavendish, who had performed some remarkable electrical researches in the eighteenth century. Maxwell soon found more enthusiasm for the project, as much with Cavendish the man as with his work. "Cavendish cared more for investigation than for publication," Maxwell wrote in the introduction to *The Electrical Researches of the Honorable Henry Cavendish*. "He would undertake the most laborious researches in order to clear up a difficulty which no one but himself could appreciate, or was even aware of." Here was a purity of purpose and indifference to recognition that Maxwell could appreciate.

Maxwell, as editor of the Cavendish papers, was indulging his deep fascination with science history. He had said in his inaugural lecture:

> It is true that the history of science is very different from the science of history. We are not studying or attempting to study the working of those blind forces which, we are told, are operating on crowds of obscure people, shaking principalities and powers, and compelling reasonable men to bring events to pass in an order laid down by philosophers.
>
> The men whose names are found in the history of science are not mere hypothetical constituents of a crowd, to be reasoned upon only in masses. We recognize them as men like ourselves, and their thoughts, being more free from influence of passion, and recorded more accurately than those of other men, are all the better materials for the study of the calmer parts of human nature.

But the history of science is not restricted to the enumeration of successful investigations. It has to tell of unsuccessful inquiries, and to explain why some of the ablest men have failed to find the key of knowledge, and how the reputation of others has only given a firmer footing to the errors into which they fell.

Heinrich Hertz

When Maxwell died in 1879, his theory of the electromagnetic field and its amazing progeny, electromagnetic waves, had little experimental support, just the indirect evidence that Maxwell's calculated speed of the electromagnetic waves matched the speed of light.

Maxwell's immediate successors thought about electromagnetic waves, but at first could find no feasible way to study them in the laboratory. The turning point, in both the study of electromagnetic waves and the fortunes of Maxwell's theory, came with the force of an intellectual earthquake in a series of experiments brilliantly carried out by Heinrich Hertz.

The year was 1887, and Hertz had recently arrived at the Karlsruhe Technische Hochschule, Baden, Germany. He was only thirty years old, but already well known, and rising rapidly in the academic world. He had been Helmholtz's star research student in Berlin, then briefly *Privatdozent* (instructor) at the University of Kiel, and was now a full professor at Karlsruhe.

When he took up his work in Karlsruhe, Hertz was familiar with Maxwell's theory but not committed to it. Earlier, Helmholtz had tried to interest him in the problem of creating experiments to test Maxwell's assumptions (with a prestigious prize attached), but Hertz had tactfully declined. His aim now was to assemble an apparatus for studying electrical discharges in gases. One item of equipment in the Karlsruhe laboratory was a spark generator called a Rühmkorff coil (distantly related to the ignition coil that generates sparks in a gasoline engine). He tinkered with the coil and was intrigued by its performance in the configuration depicted in figure 12.4. The coil A was connected to two small brass spheres B separated by ¾ centimeter, and also to two straight lengths of thick copper wires 3 meters in length terminating in two metallic spheres, 30 centimeters in diameter. When the coil was activated, sparks were repeatedly generated in the gap at B.

Hertz found that he could connect the coil circuit electrically to a wire loop, as shown in figure 12.5, and with careful adjustment of the size of the loop, obtain observable sparks across the gap M. He then discovered that he could obtain sparks in the wire loop with the connecting wire removed (fig. 12.6).

If there was no wire linking the two circuits electrically, how were they communicating with each other? At this point, Hertz began to realize that his device was generating and detecting electromagnetic waves. The origin of the waves

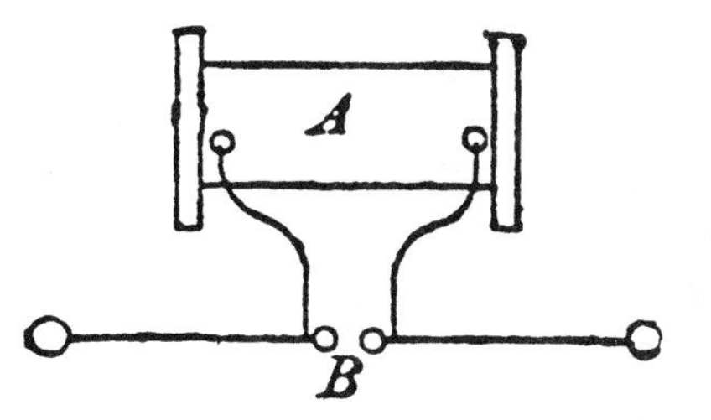

Figure 12.4. Hertz's coil circuit. This figure and the two following are adapted from Heinrich Hertz, "On Very Fast Electric Oscillations," in Wiedemann's *Annalen der Physik und Chemie* 31 (1887): 431.

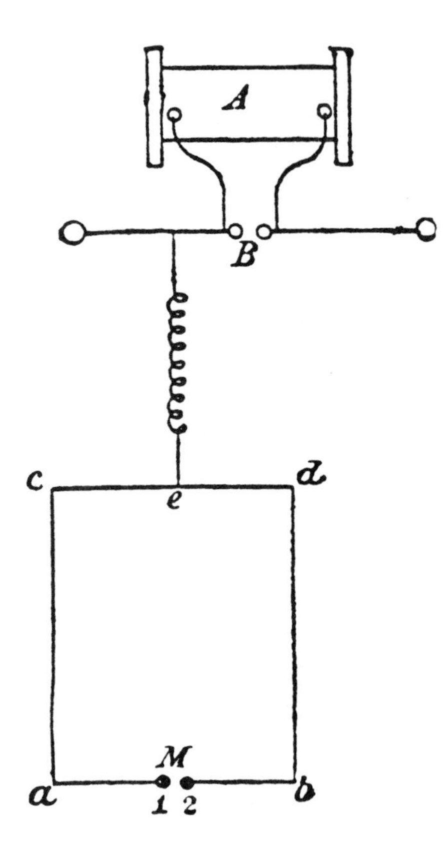

Figure 12.5. Hertz's coil circuit connected to a wire loop with a spark gap.

was a sequence of electrical oscillations initiated by each spark in the coil circuit. The waves were propagated along a wire—or even through free space—to the wire loop, and their presence revealed by the observed sparks at the gap in the loop.

Waves of all kinds have three fundamental characteristics: a wavelength, the distance from one wave crest to the next; a frequency, a count of the number of wave cycles passing a certain point in a unit of time; and a speed of propagation, the distance traveled by a wave crest in a unit of time. Hertz was, above all, interested in the speed of his waves. Was that speed finite? If so, Maxwell's theory was strongly supported against its competitors, based on the concept of action at a distance and an infinite speed of propagation. Hertz soon found a route to that crucial determination. He relied on a simple equation, valid for all kinds of waves, that connects the speed s, frequency v, and wavelength λ,

$$s = \lambda v. \quad (1)$$

The frequency v in this equation could be calculated from the length of the wire and the diameters of the metallic spheres in the coil circuit, using a formula derived earlier by Thomson. Hertz found the calculated frequencies to be exceptionally high, around one hundred million cycles per second. To measure the wavelength λ, Hertz ingeniously reconfigured his apparatus so it generated "standing waves" (as in a violin string), either along a straight wire or in free space, and using one of the wire-loop detectors, he located crests and troughs of the waves; the distance from one measured crest to the next was the wavelength.

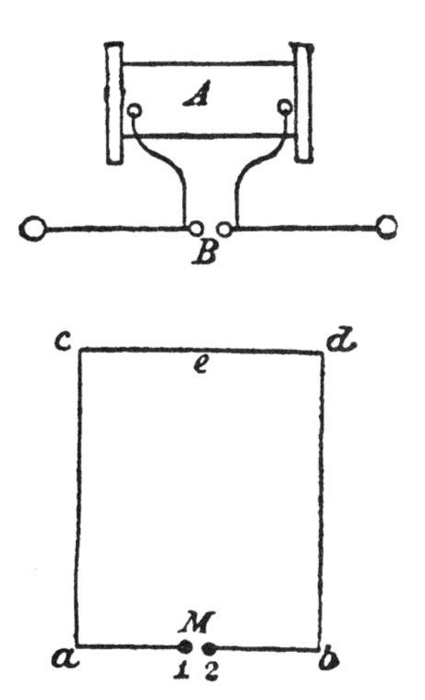

Figure 12.6. Hertz's coil circuit and wire loop disconnected. Sparks are still produced at the gap in the loop.

There were stubborn problems: he made an embarrassing calculational mistake, and the waves were distorted by an iron stove and other objects in the laboratory. But eventually equation (1) told the story Hertz was expecting: his electromagnetic waves had a finite speed, in fact, the speed of light, known to be three hundred million meters per second.

Although he had already made a strong case for Maxwell's theory, Hertz went much further, displaying a thoroughness and ingenuity that would have impressed Faraday. He demonstrated that his electromagnetic waves could be reflected, focused, refracted, diffracted, and polarized—that they were, in every sense but frequency and wavelength, the same as light waves.

In about one "miraculous year" of experiments, Hertz had closed the great debate between the Maxwellians and the proponents of action at a distance. Not surprisingly, Hertz's work was quickly recognized in Britain, and more slowly in Germany, where action-at-a-distance sentiment was strongest. The joke was that Germans learned about Hertz by way of the British. But by the summer of 1889, Hertz's triumph was complete; at a meeting in Heidelberg, he was celebrated by Germany's great men of science. In the same year, he was appointed Clausius's successor in Bonn.

Tragically, Hertz was as unfortunate with his health as he was fortunate with his talent. The first sign of trouble was a series of toothaches in 1888, which led to removal of all his teeth in 1889. By 1892, he was suffering from pains in his nose and throat, and was often depressed. His doctors could give him no satisfactory diagnosis. Several operations failed to provide permanent relief. By December 1893, he knew he would not recover, and in a letter he asked his parents "not to mourn . . . rather you must be a little proud and consider that I am among the especially elect destined to live for only a short while and yet to live enough. I did not choose this fate, but since it has overtaken me, I must be content; and if the choice had been left to me, perhaps I should have chosen it myself."

Hertz died of blood poisoning on New Year's Day, 1894; he was thirty-six years old.

"Maxwell's Equations"

Once a scientific theory has been created, it becomes public property. Friends and enemies of the theory (and the theorist) are licensed to argue for changes in both content and form as they see fit. If the theory has been successful, its content is likely to be more or less permanent. But the form—the mathematical form of a physical theory—may not be so durable. Newton's geometrical mathematical language in the *Principia* did not last, nor did Clausius's mathematically elaborate version of entropy theory. The physical content of Newton's laws of motion and the entropy concept are, however, still with us.

Maxwell's theory met a similar fate. Most of the physical assertions Maxwell made in his *Treatise on Electricity and Magnetism* are permanent fixtures. His equations, on the other hand, have been reshaped by other hands. As we find them in the *Treatise*, the equations are a dozen in number. Maxwell's successors, particularly Hertz and the most gifted of the British Maxwellians, Oliver Heaviside, wanted more "purity" in the equations. They suppressed auxiliary equations, eliminated the vector potential **A** and a companion scalar potential Ψ, and boiled the original dozen down to just four differential equations.

These "Maxwell's equations" have as their mathematical ingredients the electric field vector **E**, the magnetic field vector **B**, the electric current vector **J**, and the density of electric charge ρ. There are two divergence equations and two curl equations, one each for **E** and **B**:

$$\text{div } \mathbf{E} = \rho \tag{2}$$

$$\text{curl } \mathbf{E} = -\frac{1}{c}\frac{\partial \mathbf{B}}{\partial t} \tag{3}$$

$$\text{div } \mathbf{B} = 0 \tag{4}$$

$$\text{curl } \mathbf{B} = \frac{\mathbf{J}}{c} + \frac{1}{c}\frac{\partial \mathbf{E}}{\partial t}. \tag{5}$$

The abbreviation "div" stands for divergence, c is the speed of light, and the derivatives written with the "∂" notation are calculated just for changes in the time t, holding all other variables constant. (Mathematicians call these "partial" derivatives and the equations "partial differential equations.") To put the equations in their most symmetrical—and mathematically least excruciating—form, I have assumed that the electromagnetic field is propagated in a vacuum.

In its modern interpretation, the divergence equation (2) simply states that an electric field **E** is produced by electric charges (included in the density of electric charge ρ). The companion curl equation (3) for **E** tells us what Faraday observed: that a rotational electric field is generated in a changing magnetic field **B**.

The divergence equation (4) for the magnetic field parallels equation (2) for the electric field, except that there is no magnetic counterpart of the electric charge density ρ. Here we see a fundamental difference between electricity and magnetism. One of the two kinds of electricity, positive or negative, can dominate, making the net charge density ρ positive or negative. But a magnetic field cannot be divided this way: every north pole in the field is exactly balanced by a south pole, there is no observable "magnetic charge density," and a zero is required on the right side of the divergence equation (4).

The second curl equation (5) asserts in its first two terms (curl $\mathbf{B} = \frac{\mathbf{J}}{c}$) what Oersted observed: that a rotational magnetic field **B** is generated by an electric current **J**. The third term in equation (5), ($\frac{1}{c}\frac{\partial \mathbf{E}}{\partial t}$), has special significance. Maxwell proved that without it the equation disobeys the fundamental law of electricity that electric charge, like energy, is conserved: it cannot be created or destroyed. The third term in equation (5) saves charge conservation and it represents a ubiquitous kind of electric current (Maxwell's "displacement current"), found even in free space.

Maxwell's field theory, embodied in his equations, closed the book on the nineteenth-century, or "classical," theory of electromagnetism. It also had a long reach into the twentieth century. Einstein first found in Maxwell's equations the clue he needed drastically to revamp the concepts of space and time in his special theory of relativity (chapter 14), and then he followed Maxwell's electromagnetic field theory with his own gravitation field theory. More recently, quantum field theory has become the mainstay of particle physics.

In an appreciation of Maxwell, Einstein wrote: "Before Maxwell people thought of physical reality—in so far as it represented events in nature—as material points, whose changes consist only in motions which are subject to total differential equations [that is, no partial derivatives]. After Maxwell they thought of physical reality as represented by continuous fields, not mechanically explicable, which are subject to partial differential equations [partial derivatives included]. This change in the conception of reality is the most profound and the most fruitful that physics has experienced since Newton."

Partaker of Infinity

Maxwell's contemporaries may have found him difficult to understand, but beneath his eccentricities they always saw generosity, a complete lack of selfishness, and a deep sense of duty. Of all the scientists who populate these chapters, Maxwell and Gibbs were probably the least selfish and self-centered.

Maxwell's selfless devotion to his wife Katherine was particularly strong. Her health was always frail, and Maxwell guarded it with great care, even when his own health was failing. Campbell reports that at one point Maxwell sat by Katherine's bed through the night for three weeks, and tended to the affairs of the Cavendish Laboratory during the day. He was always considerate of colleagues, especially those who had not received much attention. He was the first to recognize (and promote) the importance of Gibbs's work on thermodynamics. His generous comments on the doctoral thesis of a young Dutch physicist, Johannes van der Waals, are typical: "The molecular theory of the continuity of the liquid and gaseous states forms the subject of an exceedingly ingenious thesis by Mr. Johannes Diderick van der Waals, a graduate of Leyden. . . . His attack on this difficult question is so able and so brave, that it cannot fail to give a notable impulse to molecular science. It has certainly directed the attention of more than one inquirer to the study of the Low-Dutch language in which it is written."

Maxwell's referee's reports on papers of young colleagues sometimes offered more insights than the papers themselves. His report to William Crookes concerning research on electrical discharges in gases dropped some hints that could have been (but were not) followed to the discovery of the electron. As Bruce Hunt remarks, Maxwell's referee's report in 1879 to George Fitzgerald "stands as perhaps the clearest marker of the point at which Maxwell's theory passed from his own hands into those of a new generation." Fitzgerald, who became a leading Maxwellian, was then a newcomer to electromagnetism, and he gratefully accepted Maxwell's pointers.

In the 1850s, while he was at Aberdeen, Maxwell wrote in a letter, "I wish to say that it is in personal union with my friends that I hope to escape the despair which belongs to the contemplation of the outward aspect of things with human eyes. Either be a machine and see nothing but the 'phenomena,' or else try to be a man, feeling your life interwoven, as it is, with many others, and strengthened by them whether in life or death."

Much later he expressed to a friend "a favorite thought," the mystical belief, "that the relation of parts to wholes pervades the invisible no less than the visible world, and that beneath the individuality which accompanies our personal life there lies hidden a deeper community of being as well as of feeling and action." Campbell marveled that while Maxwell "was continually striving to reduce to greater definiteness men's conceptions of leading physical laws, he seemed habitually to live in a sort of mystical communion with the infinite."

In the 1850s and 1860s, Maxwell taught evening classes for working-class people, first in Cambridge, and then in Aberdeen and London. Several biographers have remarked that Maxwell was, in a paternal way, "feudal" in his treatment of artisans and the servants and tenants at Glenlair. J. G. Crowther remarks that those biographers "cannot be satisfied with his role of gentleman-farmer or laird in the middle of the nineteenth century. The modernity of Maxwell's science, and the antiquity of his sociology and religion appear incongruous. But it may be noted that though his views on sociology were antique, they were superior to those of nearly all his scientific contemporaries. He at least thought about these problems."

Maxwell's religious views were conventional, at least up to a point. His mother was a Presbyterian and his father an Episcopalian. As a child in Edinburgh, he attended services in both churches. He could recall long passages from the Bible, and his letters to Katherine were full of pious biblical references and quotations. His private faith probably went deeper than that, but he chose not to advertise it. He responded to an invitation to join an organization dedicated to reconciling science with religion with a refusal and this explanation: "I think that the results which each man arrives at in his attempts to harmonize his science with his Christianity ought not to be regarded as having any significance except to the man himself, and to him only for a time, and should not receive the stamp of society."

Yet occasionally in his writings Maxwell did reveal something about the religious and other metaphysical underpinnings of his science. In his inaugural lecture at Aberdeen, he said:

> But as physical science advances we see more and more that the laws of nature are not mere arbitrary and unconnected decisions of Omnipotence, but that they are essential parts of one universal system in which infinite Power serves only to reveal unsearchable Wisdom and eternal Truth. When we examine the truths of science and find that we can not only say "This is so" but "This must be so,

> for otherwise it would not be consistent with the first principles of truth"—or even when we can only say "This ought to be so according to the analogy of nature" we should think what a great thing we are saying, when we pronounce a sentence on the laws of creation, and say they are true, or right, when judged by the principles of reason. Is it not wonderful that man's reason should be a judge over God's works, and should measure, and weigh, and calculate, and say at last "I understand and I have discovered—It is right and true."

While he was still a student at Cambridge, he wrote in an essay: "Happy is the man who can recognize in the work of Today a connected portion of the work of life, and an embodiment of the work of Eternity. The foundations of his confidence are unchangeable, for he has been made a partaker of Infinity." Much later, when he was dying, he said to a friend: "My interest is always in things rather than in persons. I cannot help thinking about the immediate circumstances which have brought a thing to pass, rather than about any will setting them in motion. What is done by what is called myself is, I feel, done by something greater than myself in me. My interest in things has always made me care much more for theology than for anthropology; states of the will only puzzle me."

iv

STATISTICAL MECHANICS

Historical Synopsis

In the first three parts of the book, the themes have been mechanics, thermodynamics, and electromagnetism, which can be grouped under the broader heading of "macrophysics"—that is, the physics of objects of ordinary size and larger. This fourth part of the book addresses for the first time the vastly different realm of "microphysics." As used here, the term means the physics of molecules, atoms, and subatomic particles. Microphysics will be a major theme in the book from now on, particularly here in part 4, and then in parts 6 (quantum mechanics), 7 (nuclear physics), and 8 (particle physics).

Molecules (and the atoms they contain) are very small, incredibly large in number, chaotic in their motion, and difficult to isolate and study as individuals. But *populations* of molecules, like human populations, can be described by statistical methods. The strategy is to focus on average, rather than individual, behavior. Insurance companies do their business this way, and so do molecular physicists. The insurance company statistician might calculate the average life span for an urban population of males in a certain income bracket. The physicist might seek an average energy for a population of gas molecules occupying a certain volume at a certain temperature. The method works well enough for the insurance company to make a profit, and even better for the physicist because molecules are far more numerous and predictable than human beings. By determining energy, or an average value for some other mechanical property of molecules, the physicist practices what Gibbs called "statistical mechanics."

The single chapter in this part of the book introduces the man who did the most to define, develop, and defend statistical mechanics. He was Ludwig Boltzmann, who wrote his most important papers on statistical mechanics in the 1870s. For Boltzmann, statistical mechanics was most profitable in discussions of the entropy concept. He found a molecular basis for the second law of thermodynamics, and made the entropy concept accessible by linking entropy with disorder.

Boltzmann built on foundations laid by Maxwell, who had in turn been inspired by Clausius. In the late 1850s, Clausius showed how to calculate average values for molecular speeds and distances traveled by molecules between collisions with other molecules. He recognized that the molecules of a population have different speeds distributed above and below the average, but his statistical mechanics supplied no way to determine the distribution. In two papers, written in 1858 and 1866, Maxwell defined the missing molecular distribution law and applied it in many different ways to the theory of gas behavior. The line of development of statistical mechanics from Clausius to Maxwell to Boltzmann continued to Gibbs. A masterful treatise published by Gibbs in 1901 gave statistical mechanics the formal structure it still has today, even after the intervening upheaval brought by quantum theory.

To believe in statistical mechanics, one must believe in molecules. At the beginning of the twenty-first century, we don't have to be persuaded, but late in the nineteenth century Boltzmann had influential and obstinate opponents who could not accept the reality of molecules. Boltzmann enthusiastically engaged his adversaries in friendly and unfriendly debates, but they outlasted him. Albert Einstein then took up the debate and showed how to make molecules real and visible.

13

Molecules and Entropy
Ludwig Boltzmann 1844–1906

Peregrinations

Restlessness was the story of his life and work. Ludwig Boltzmann saw the physical world as a perpetually agitated molecular chaos; and, like the molecules, he never found rest himself. He moved from one academic post to another seven times during his career of almost forty years. The chronology goes like this: two years (1867–69) at the University of Vienna as an assistant professor; four years (1869–73) as an assistant professor of mathematical physics at the University of Graz; back to Vienna for three years (1873–76) as a professor of mathematics; to Graz again for fourteen years (1876–90) as a professor of experimental physics; four years (1890–94) as a professor of theoretical physics at the University of Munich; a second return to Vienna for four years (1894–1900), this time as a professor of theoretical physics; two years in Leipzig (1900–1902) as a professor of theoretical physics; and a third and final return to Vienna to succeed himself in the chair still unoccupied since his departure two years earlier.

These were not forced departures. From the early 1870s on, Boltzmann was famous in the scientific world and much in demand. To entice him to return from Munich to Vienna, the Austrian minister of culture had to offer him the highest salary then paid to any Austrian university professor. Competing faculties described him as the "uncontested first representative" of theoretical physics "recognized as such by all nations," and "the most important physicist in Germany and beyond." In this job market, Boltzmann was not above some hard bargaining with the appropriate ministries. Late in his life he was negotiating for his next move soon after he had completed the last one. The Vienna authorities finally decided enough was enough: they would take him back (for the third time) only if he would give his word that he would never take another job outside Austria.

But Boltzmann's restlessness was driven by more than salaries and the other things he complained about in his correspondence, such as the quality of the students and German cooking. He moved incessantly because the polar opposites of his personality would give him no peace. He joked that these polarities were determined on the night of his birth between Shrove Tuesday and Ash Wednesday—Carnival and Lent. The modern diagnosis would be bipolar disorder or manic depression. His health was troubled in other ways—he had asthma, migraine headaches, poor eyesight, and angina pains—but the periods of depression were far worse, and finally intolerable. Traveling and relocating would lift him from one depression but not prevent the next. The move to Leipzig, for example, brought relief, but not for long. Within a year he was suffering again and driven to an unsuccessful attempt at suicide.

When he was not gripped by the deep melancholy of his depressions, Boltzmann was, in a word, brilliant. "I am a theoretician from head to toe," he said. "The idea that fills my thoughts and deeds [is] the development of theory. To glorify it no sacrifice is too great for me: since theory is the content of my entire life." Among nineteenth-century theorists, he was in a class with Gibbs; only Maxwell ranked higher.

Boltzmann was famous not only for his theories but also, and perhaps more so, for his superb ability as a teacher and lecturer. Lise Meitner, who attended Boltzmann's cycle of lectures on theoretical physics in Vienna just after the turn of the century (and later was a codiscoverer of uranium fission), left this appreciation:

> He gave a course that lasted four years. It included mechanics, hydrodynamics, elasticity theory, electrodynamics, and the [molecular] theory of gases. He used to write the main equations on a very large blackboard. By the side he had two smaller blackboards, where he wrote the intermediate steps. Everything was written in a clear and well-organized form. I frequently had the impression that one might reconstruct the entire lecture from what was on the blackboard. After each lecture it seemed to us as if we had been introduced to a new and wonderful world, such was the enthusiasm that he put into what he taught.

In spite of his many psychological tensions, Boltzmann was open and informal with his students and sensitive to their needs. "He never exhibited his superiority," writes Fritz Hasenöhrl, who succeeded Boltzmann at the University of Vienna. "Anybody was free to put him questions and even criticize him. The conversation took place quietly and the student was treated as a peer. Only later one realized how much he had learned from him. He did not measure others with the yardstick of his own greatness. He also judged more modest achievements with goodwill, so long as they gave evidence of serious and honest effort."

Ernst Mach, Boltzmann's perennial opponent in debates on atomism, seemed to be offended by all this informality: "Boltzmann is not malicious," Mach wrote in a letter, "but incredibly naïve and casual . . . he simply does not know where to draw the line." During his second tenure in Graz, Boltzmann accepted and then quickly rejected an appointment as Gustav Kirchhoff's successor at the University of Berlin. It is said that a factor in his decision was a haughty remark from Frau Helmholtz: "Professor Boltzmann I am afraid you will not feel at ease here in Berlin."

Boltzmann married Henriette von Aigentler, a handsome young woman with luxuriant blond hair and blue eyes. Although it was considered quite inappropriate at the time, she took a strong interest in her husband's work and had his encouragement. "It seems to me," he wrote in his letter proposing marriage, "that a constant love cannot endure if the wife has no understanding, no enthusiasm for the endeavors of the husband, but is merely his housekeeper rather than a companion in his struggles." The couple had five children, three daughters and two sons, whom they adored. It is recorded that Boltzmann bought two pet rabbits for the youngest daughter, Elsa, over the objections of Henriette. The animals lived in Boltzmann's study, outside Henriette's domain. Boltzmann's biographers do not say much about Henriette, but we can be sure that she was a strong and resourceful woman, if for no other reason than that she lived with, and survived, her husband's neuroses.

If Boltzmann had not succeeded as a physicist, he might have been a humorist. He was Mark Twain in reverse, a European who traveled to America. During the summer of 1905, he gave a series of lectures at the "University of Berkeley," and back in Vienna, reported on the incredible ways of Californians in a piece called "A German Professor's Journey into Eldorado." "The University of Berkeley," he writes, "is the most beautiful place imaginable. A park a kilometer square, with trees which must be centuries old, or is it millennia? Who can tell at a moment's notice? In the park there are splendid modern buildings, obviously too small already; new ones are under construction, since both space and money are available."

But it was a spoiled paradise: "Berkeley is teetotal: to drink or retail beer and wine is strictly forbidden." Berkeley water was not a good alternative: "My stomach rebelled," and more drastic measures were called for:

> I ventured to ask a colleague about the location of a wine merchant. The effect my question produced reminded me of a scene in the smoking-car of a train between Sacramento and Oakland. An Indian had joined us, who asked quite naïvely for the address of a . . . well, as he was an Indian, let's say the address of a house with bayaderes [Hindu dancing girls] in San Francisco. Most of the people in the smoker were from San Francisco and there are certainly girls there with the motto: "Give me money, I give you honey," but everyone was startled and embarrassed. My colleague reacted in exactly the same way when I asked about the wine merchant. He looked about anxiously in case someone was listening, sized me up to see if he could really trust me and eventually came out with the name of an excellent shop selling California wine in Oakland. I managed to smuggle in a whole battery of wine bottles and from then on the road to Oakland became very familiar.

Among the bizarre culinary habits of the Californians was oatmeal. Boltzmann was offered some by his hostess, Mrs. Hearst, the mother of the newspaper tycoon William Randolph Hearst. It was "an indescribable paste on which people might fatten geese in Vienna—then again, perhaps not, since I doubt whether Viennese geese would be willing to eat it." The after-dinner entertainment compensated for such lapses, however. The Hearst music room was comparable to "any of the smaller Viennese concert halls." Boltzmann, an accomplished pianist, played a Schubert sonata, and was enchanted by the piano, "a Steinway from the most expensive price-range." He had heard such pianos but never touched one. "[At] first I found the mechanics strange, but how quickly one becomes accustomed to good things. The second part of the first movement went well and in the second movement, an Andante, I forgot myself completely: I was not playing the melody, it was guiding my fingers. I had to hold myself back forcibly from playing the Allegro as well, which was fortunate because there my technique would have faltered." In Mrs. Hearst's grand music room, Boltzmann found his Eldorado: "If the hardships which beset my Californian visit had ever made me regret, from then on they ceased to do so."

Log Lessons

Before we turn to the work of Boltzmann and his great contemporaries, Clausius, Maxwell, and Gibbs, on statistical mechanics, we need to take a brief detour into mathematical territory on the subjects of logarithms and exponential functions.

The 2 in 10^2 (= 100) is an exponent or power. This notation is invaluable for expressing very large numbers: it is much easier to write 10^{23} than 1 followed by 23 zeros. The exponent need not be a fixed number; it can be a variable with any value, such as x in 10^x.

Exponents are convenient in that they are added in multiplication and subtracted in division. For example,

$$10^3 \times 10^2 = 10^{3+2} = 100000,$$

and

$$\frac{10^3}{10^2} = 10^{3-2} = 10.$$

With variables as exponents, the corresponding statements are

$$10^x \times 10^y = 10^{x+y},$$

and

$$\frac{10^x}{10^y} = 10^{x-y}.$$

These algebraic properties make it possible to convert a multiplication into an addition and a division into a subtraction. To multiply two numbers this way, we first convert them into powers of ten—that is, find values of the exponents x and y in the above equations; then add x and y to obtain $x + y$, and the product as 10^{x+y}, or $x - y$, and the quotient as 10^{x-y}.

The powers of ten (the x and y) in these recipes are called "logarithms (logs) to the base 10." They are tabulated to make it convenient to convert any number into a power of ten, and vice versa. Until the advent of hand calculators, "log tables" were an indispensable computational tool. Logarithmic functions are still standard equipment in algebra. The notation "log" denotes a power of ten, as in

$$\log 10^x = x,\ \log 10^y = y, \text{ and } \log 10^{x+y} = x + y.$$

The general concept of logarithms was invented early in the seventeenth century by John Napier, a Scotsman, and independently by Joost Bürgi, a Swiss. Napier and Henry Briggs devised the computational scheme I have described involving powers of ten.

The number ten is convenient as a "base" in logarithmic calculations, but any other number can serve the same purpose. When he was still an undergraduate at Cambridge, Newton discovered that "natural" logarithms to a special base, now denoted with the symbol e, could be calculated by accumulating added terms in a series. Newton's formula for $\ln(1 + x)$, with "ln" representing a natural logarithm to the base e, is

$$\ln(1 + x) = x - \frac{x^2}{2} + \frac{x^3}{3} - \frac{x^4}{4} + \ldots,$$

in which ". . . " means that the series continues forever (the next two terms are $+ \frac{x^5}{5}$ and $- \frac{x^6}{6}$). If x is less than one, however, only the first few terms may be needed, because later terms are small enough to be negligible.

Regardless of the base—10, e, or any other number—logarithms can be positive, negative, or equal to zero. The rules are worth noting (and proving). For a logarithmic function $\ln x$,

$$\begin{aligned} \ln x &> 0 \text{ if } x > 1 \\ &= 0 \text{ if } x = 1 \\ &< 0 \text{ if } x < 1. \end{aligned}$$

The Story of *e*

Functions containing e are ubiquitous in the equations of physics. Prototypes are the "exponential functions" e^x and e^{-x}. Both are plotted in figure 13.1, showing that e^x increases rapidly as x increases ("exponential increase" is a popular phrase), and e^{-x} decreases rapidly. The constant e may seem mystifying. Where does it come from? Why is it important?

Mathematicians include e in their pantheon of fundamental numbers, along with 0, 1, π, and i. The use of e as a base for natural logarithms dates back to the seventeenth century. Eli Maor speculates that the definition of e evolved somewhat earlier from the formulas used for millennia by moneylenders. One of these calculates the balance B from the principal P at the interest rate r for a period of t years compounded n times a year,

$$B = P\left(1 + \frac{r}{100n}\right)^{nt} \qquad (1)$$

For example, if we invest P = \$1000 at the interest rate r = 5% with interest compounded quarterly ($n = 4$), our balance after $t = 20$ years is

$$B = (\$1000)\left(1 + \frac{5}{(100)(4)}\right)^{(4)(20)} = \$2701.48.$$

Equation (1) has some surprising features that could well have been noticed by an early-seventeenth-century mathematician. Suppose we simplify the formula by considering a principal of P = \$1, a period of t = 1 year, and an interest rate of r = 100% (here we part with reality), so the right side of the formula is

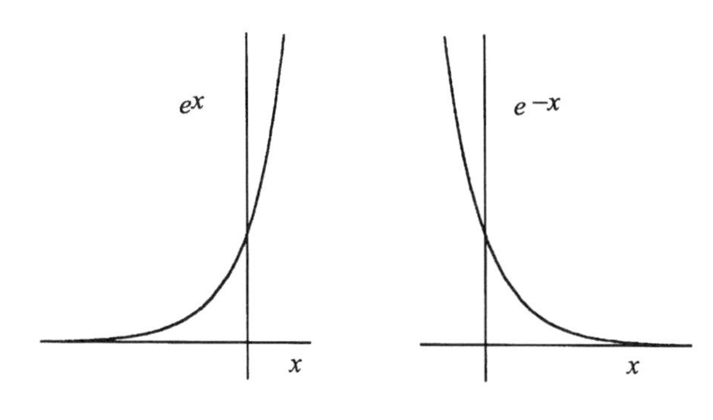

Figure 13.1. Typical exponential functions: the increasing e^x and the decreasing e^{-x}.

simply $\left(1 + \frac{1}{n}\right)^n$. Our seventeenth-century mathematician might have amused himself by laboriously calculating according to this recipe with n given larger and larger values (easily done with a calculator by using the y^x key to calculate the powers). Table 13.1 lists some results. The trend is clear: the effect of increasing n becomes more minuscule as n gets larger, and as $\left(1 - \frac{1}{n}\right)^n$ approaches a definite value, which is 2.71828 if six digits are sufficient. (For more accuracy, make n larger.) The "limit" approached when n is given an *infinite* value is the mathematical definition of the number e. Mathematicians write the definition

$$e = \lim_{n \to \infty} \left(1 + \frac{1}{n}\right)^n. \tag{2}$$

Physicists, chemists, engineers, and economists find many uses for exponential functions of the forms e^x and e^{-x}. Here are a few of them:

1. When a radioactive material decays, its mass decreases exponentially according to

$$m = m_0 e^{-at},$$

in which m is the mass at time t, m_0 is the initial mass at time $t = 0$, and a is a constant that depends on the rate of decay of the radioactive material. The exponential factor e^{-at} is rapidly decreasing (a is large) for short-lived radioactive materials, and slowly decreasing (a is small) for long-lived materials.

2. A hot object initially at the temperature T_0 in an environment kept at the lower, constant temperature T_1 cools at a rate given by

$$T = T_1 + (T_0 - T_1)\, e^{-at}.$$

3. When a light beam passes through a material medium, its intensity decreases exponentially according to

$$I = I_0 e^{-ax},$$

Table 13.1

N	$(1 + 1/n)^n$
1	2
2	2.25
5	2.48832
10	2.59374
100	2.70481
10000	2.71815
1000000	2.71828
10000000	2.71828

in which I is the intensity of the beam after passing through the thickness x of the medium, I_0 is the intensity of the incident beam, and a is a constant depending on the transparency of the medium.

4. Explosions usually take place at exponentially increasing rates expressed by a factor of the form e^{at}, with a a positive constant depending on the physical and chemical mechanism of the explosion.

5. If a bank could be persuaded to compound interest not annually, semiannually, or quarterly, but *instantaneously*, one's balance B would increase exponentially according to

$$B = Pe^{rt/100},$$

where P is the principle, r is the annual interest rate, and t is the time in years.

Brickbats and Molecules

The story of statistical mechanics has an unlikely beginning with a topic that has fascinated scientists since Galileo's time: Saturn's rings. In the eighteenth century, Pierre-Simon Laplace developed a mechanical theory of the rings and surmised that they owed their stability to irregularities in mass distribution. The biennial Adams mathematical prize at Cambridge had as its subject in 1855 "The Motions of Saturn's Rings." The prize examiners asked contestants to evaluate Laplace's work and to determine the dynamical stability of the rings modeled as solid, fluid, or "masses of matter not mutually coherent." Maxwell entered the competition, and while he was at Aberdeen, devoted much of his time to it.

He first disposed of the solid and fluid models, showing that they were not stable or not flat as observed. He then turned to the remaining model, picturing it as a "flight of brickbats" in orbit around the planet. In a letter to Thomson, he said he saw it as "a great stratum of rubbish jostling and jumbling around Saturn without hope of rest or agreement in itself, till it falls piecemeal and grinds a fiery ring round Saturn's equator, leaving a wide tract of lava and dust and blocks on each side and the western side of every hill buttered with hot rocks. . . . As for the men of Saturn I should recommend them to go by tunnel when they cross the 'line.' " In this chaos of "rubbish jostling and jumbling" Maxwell found a solution to the problem that earned the prize.

This success with Saturn's chaos of orbiting and colliding rocks inspired Maxwell to think about the chaos of speeding and colliding molecules in gases. At first, this problem seemed too complex for theoretical analysis. But in 1859, just as he was completing his paper on the rings, he read two papers by Clausius that gave him hope. Clausius had brought order to the molecular chaos by making his calculations with an *average* dynamical property, specifically the average value of v^2, the square of the molecular velocity. Clausius wrote this average quantity $\overline{v^2}$ and used it in the equation

$$PV = \frac{1}{3}Nm\overline{v^2} \quad (3)$$

to calculate the pressure P produced by N molecules of mass m randomly bombarding the walls of a container whose volume is V.

In Clausius's treatment, the molecules move at high speeds but follow extremely tortuous paths because of incessant collisions with other molecules. With all the diversions, it takes the molecules of a gas a long time to travel even a few meters. As Maxwell put it: "If you go 17 miles per minute and take a totally new course [after each collision] 1,700,000,000 times in a second where will you be in an hour?"

Maxwell's first paper on the dynamics of molecules in gases in 1860 took a major step beyond Clausius's method. Maxwell showed what Clausius recognized but did not include in his theory: that the molecules in a gas at a certain temperature have many different speeds covering a broad range above and below the average value. His reasoning was severely abstract and puzzling to his contemporaries, who were looking for more-mechanical details. As Maxwell said later in a different context, he did not make "personal enquiries [concerning the molecules], which would only get me in trouble."

Maxwell asked his readers to consider the number of molecules dN with velocity components that lie in the specific narrow ranges between v_x and $v_x + dv_x$, v_y and $v_y + dv_y$, v_z and $v_z + dv_z$. That count depends on N, the total number of molecules; on dv_x, dv_y, and dv_z; and on three functions of v_x, v_y, and v_z, call them $f(v_x)$, $f(v_y)$, and $f(v_z)$, expressing which velocity components are important and which unimportant. If, for example, $v_x = 10$ meters per second is unlikely while $v_x = 500$ meters per second is likely, then $f(v_x)$ for the second value of v_x is larger than it is for the first value. Maxwell's equation for dN was

$$dN = Nf(v_x)f(v_y)f(v_z)dv_x dv_y dv_z. \quad (4)$$

Maxwell argued that on the average in an ideal gas the three directions x, y, and z, used to construct the velocity components v_x, v_y, and v_z, should all have the same weight; there is no reason to prefer one direction over the others. Thus the three functions $f(v_x)$, $f(v_y)$, and $f(v_z)$ should all have the same mathematical form. From this conclusion, and the further condition that the total number of molecules N is finite, he derived

$$f(v_x) = \frac{N}{\alpha\sqrt{\pi}}e^{-v_x^2/\alpha^2} \quad (5)$$

with α a parameter depending on the temperature and the mass of the molecules. This is one version of Maxwell's "distribution function."

A more useful result, expressing the distribution of speeds v, regardless of direction, follows from this one,

$$g(v) = \frac{4N}{\alpha^3\sqrt{\pi}} v^2 e^{-\frac{v^2}{\alpha^2}}. \tag{6}$$

The function, $g(v)$, another distribution function, assesses the relative importance of the speed v. Its physical meaning is conveyed in figure 13.2, where $g(v)$ is plotted for speeds of carbon dioxide molecules ranging from 0 to 1,400 meters per second; the temperature is assumed to be 500 on the absolute scale (227°C). We can see from the plot that very high and very low speeds are unlikely, and that the most probable speed at the maximum point on the curve is about 430 meters per second (= 16 miles per minute).

Maxwell needed only one page in his 1860 paper to derive the fundamental equations (5) and (6) as solutions to the proposition "To find the average number of particles [molecules] whose velocities lie between given limits, after a great number of collisions among a great number of equal particles." The language—calculation of an "average" for a "great number" of molecules and collisions—prescribes a purely *statistical* description, and that is what Maxwell supplied in his distribution functions.

Thus, without "personal enquiries" into the individual histories of molecules, Maxwell defined their statistical behavior instead, and this, he demonstrated in his 1860 paper, had many uses. Statistically speaking, he could calculate for a gas its viscosity, ability to conduct heat, molecular collision rate, and rate of diffusion. This was the beginning "of a new epoch in physics," C. W. F. Everitt writes. "Statistical methods had long been used for analyzing observations, both in physics and in the social sciences, but Maxwell's ideas of describing actual physical processes by a statistical function [e.g., $g(v)$ in equation (6)] was an extraordinary novelty."

Maxwell's theory predicted, surprisingly, that the viscosity parameter for gases

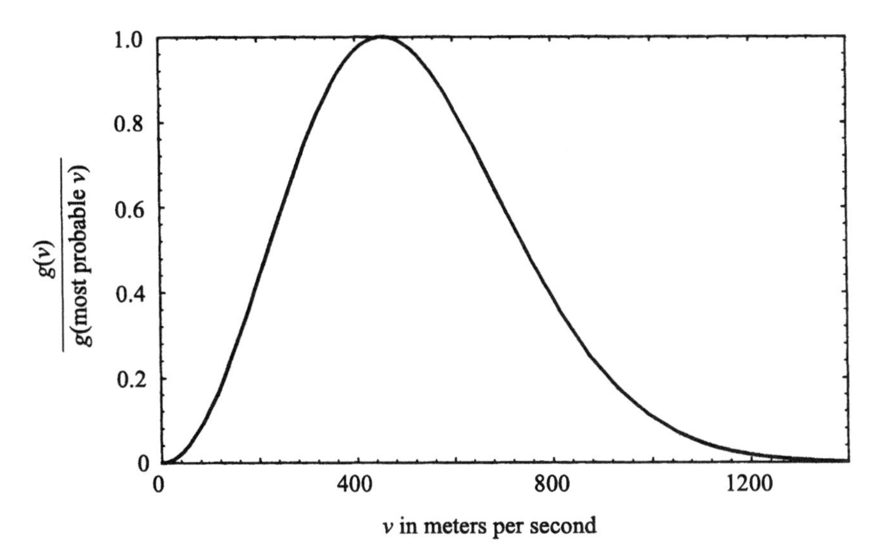

Figure 13.2. Maxwell's distribution function $g(v)$. The plot is "normalized" by dividing each value of $g(v)$ by the value obtained with v given its most probable value.

should be independent of the pressure of the gas. "Such a consequence of a mathematical theory is very startling," Maxwell wrote, "and the only experiment I have met with on the subject does not seem to confirm it." His convictions were with the theory, however, and several years later, ably assisted by his wife Katherine, Maxell demonstrated the pressure independence experimentally. Once more, the scientific community was impressed by Maxwellian wizardry.

But the theory could not explain some equally puzzling data on specific heats. A specific heat measures the heat input required to raise the temperature of one unit, say, one kilogram, of a material one degree. Measurements of specific heats can be done for constant-pressure and constant-volume conditions, with the former always larger than the latter.

Maxwell, like many of his contemporaries, believed that heat resides in molecular motion, and therefore that a specific heat reflects the number of modes of molecular motion activated when a material is heated. Maxwell's theory supported a principle called the "equipartition theorem," which asserts that the thermal energy of a material is equally divided among all the modes of motion belonging to the molecules. Given the number of modes per molecule, the theory could calculate the constant-pressure and constant-volume specific heats and the ratio between the two. If the molecules were spherical, they could move in straight lines and also rotate. Assuming three (x, y, and z) components for both rotation and straight-line motion, the tally for the equipartition theorem was six, and the prediction for the specific-heat ratio was 1.333. The observed average for several gases was 1.408.

Maxwell never resolved this problem, and it bothered him throughout the 1870s. In the end, his advice was to regard the problem as "thoroughly conscious ignorance," and expect that it would be a "prelude to [a] real advance in knowledge." It was indeed. Specific-heat theory remained a puzzle for another twenty years, until quantum theory finally explained the mysterious failings of the equipartition theorem.

Maxwell's Demon(s)

The statistical method opened another door for Maxwell, into the realm of the second law of thermodynamics. In his idiomatic way, he amused himself by imagining a bizarre scheme for violating the second-law axiom that heat always passes from hot to cold. "Let A & B be two vessels divided by a diaphragm," he wrote in a letter to P. G. Tait, "and let them contain elastic molecules in a state of agitation which strike each other and the sides. Let the number of particles be equal in A and B but those in A have the greatest energy of motion [that is, A is at a higher temperature than B]." If the diaphragm has a small hole in it, molecules will go through it and transfer their energy from one vessel to the other.

"Now conceive a finite being who knows the paths and velocities of all the molecules by simple inspection, but who can do no work except open and close [the] hole in the diaphragm by means of a slide without mass." The task of this "being" is to open the hole and allow molecules to pass from B to A if they have greater than the average speed in A, and from A to B if they have less than the average speed in B. The "being" keeps the two-way molecular traffic balanced, so the number of molecules in A and B does not change. The result of these maneuvers is that the molecules in A become more energetic than they were originally, and those in B less energetic. This amounts to wrong-way heat flow, an infringement of the second law: "[The] hot system has got hotter and the cold colder and yet no work has been done, only the intelligence of a very observant and neat-fingered being has been employed." When Tait told Thomson about Maxwell's talented "being," Thomson promoted it to the status of a "sorting demon."

Had Maxwell actually defeated the second law? He did not claim victory: his "neat-fingered" demon was an imposter. If we could design a demon that controls molecular traffic without doing any work, Maxwell argued, then we could actually violate the second law. "Only we can't," he concluded, "not being clever enough." The demon fails in its assignment, and on the average—statistically speaking—more hot molecules pass through the hole from A (where the temperature is higher) to B, than from B to A. This is the normal direction of heat flow permitted by the second law. Maxwell's message is that the basis for the second law is the statistical behavior of vast numbers of molecules, and no amount of technical ingenuity can reverse these statistical patterns. As he put it to John Strutt (later Lord Rayleigh) in 1870: "The 2nd law of thermodynamics has the same degree of truth as the statement that if you throw a tumblerful of water into the sea, you cannot get the same tumblerful out again."

Commentary on Maxwell's demon has become a minor industry among physicists. The demon has been the subject of countless papers and even a few books. Some of these authors have apparently not trusted Maxwell's sagacity, and tried to invent a better demon. They have been clever, but not clever enough: Maxwell and the second law have been upheld.

Entropy and Disorder

We come now to Boltzmann's role in the development of statistical mechanics, supporting and greatly extending the work already done by Clausius and Maxwell. Boltzmann's first major contribution, in the late 1860s, was to broaden Maxwell's concept of a molecular distribution function. He established that the factor for determining the probability that a system of molecules has a certain total (kinetic + potential) energy E is proportional to e^{-hE}, with h a parameter that depends only on temperature. This "Boltzmann factor" has become a fixture in all kinds of calculations that depend on molecular distributions, not only for physicists but also for chemists, biologists, geologists, and meteorologists.

Boltzmann assumed that his statistical factor operates in a vast "phase space" spanning all the coordinates and all the velocity components in the system. Each point in the phase space represents a possible state of the system in terms of the locations of the molecules and their velocities. As the system evolves, it follows a path from one of these points to another.

Boltzmann constructed his statistical theory by imagining a small element, call it $d\omega$, centered on a point in phase space, and then assuming that the probability dP for the system to be in a state represented by points within an element is proportional to the statistical factor e^{-hE} multiplied by the element $d\omega$:

$$dP \propto e^{-hE} d\omega,$$

or

$$dP = Ae^{-hE} d\omega, \qquad (7)$$

where A is a proportionality constant. Probabilities are always defined so that when they are added for all possible events they total one. Doing the addition of the above dPs with an integration, we have

$$\int dP = 1,$$

so integration of both sides of equation (7),

$$\int dP = \int Ae^{-hE}d\omega = A\int e^{-hE}d\omega,$$

leads to

$$1 = A\int e^{-hE}d\omega,$$

and this evaluates the proportionality constant A,

$$A = \frac{1}{\int e^{-hE}d\omega} \tag{8}$$

Substituting in equation (7), we have

$$dP = \frac{e^{-hE}d\omega}{\int e^{-hE}d\omega}. \tag{9}$$

This is an abstract description, but it is also useful. If we can express any physical quantity, say the entropy S, as a function of the molecular coordinates and velocities, then we can calculate the average entropy $\overline{S}$ statistically by simply multiplying each possible value of S by its corresponding probability dP, and adding by means of an integration,

$$\begin{aligned} \overline{S} &= \int SdP \\ &= \frac{\int Se^{-hE}d\omega}{\int e^{-hE}d\omega}. \end{aligned} \tag{10}$$

So far, Boltzmann's statistical treatment was limited in that it concerned reversible processes only. In a lengthy and difficult paper published in 1872, Boltzmann went further by building a molecular theory of irreversible processes. He began by introducing a molecular velocity distribution function f that resembled Maxwell's function of the same name and symbol, but that was different in the important respect that Boltzmann's version of f could evolve: it could change with time.

Boltzmann firmly believed that chaotic collisions among molecules are responsible for irreversible changes in gaseous systems. Taking advantage of a mathematical technique developed earlier by Maxwell, he derived a complicated equation that expresses the rate of change in f resulting from molecular collisions. The equation, now known as the Boltzmann equation, justifies two great propositions. First, it shows that when f has the Maxwellian form seen in equation (5) its rate of change equals zero. In this sense, Maxwell's function expresses a static or *equilibrium* distribution.

Second, Boltzmann's equation justifies the conclusion that Maxwell's distribution function is the only one allowed at equilibrium. To make this point, he introduced a time-dependent function, which he later labeled H,

$$H = \int (f \ln f) d\sigma, \tag{11}$$

in which $d\sigma = dv_x dv_y dv_z$. The H-function, teamed with the Boltzmann equation, leads to Boltzmann's "H-theorem," according to which H can never evolve in an increasing direction: the rate of change in H, that is, the derivative $\frac{dH}{dt}$, is either negative (H decreasing) or zero (at equilibrium),

$$\frac{dH}{dt} \leq 0. \tag{12}$$

Thus the H-function follows the irreversible evolution of a gaseous system, always decreasing until the system stops changing at equilibrium, and there Boltzmann could prove that f necessarily has the Maxwellian form.

As Boltzmann's H-function goes, so goes the entropy of an isolated system according to the second law, except that H always decreases, while the entropy S always increases. We allow for that difference with a minus sign attached to H and conclude that

$$S \propto -H. \tag{13}$$

In this way, Boltzmann's elaborate argument provided a molecular analogue of both the entropy concept and the second law.

There was, however, an apparent problem. Boltzmann's argument seemed to be entirely mechanical in nature, and in the end to be strictly reliant on the Newtonian equations of mechanics or their equivalent. One of Boltzmann's Vienna colleagues, Joseph Loschmidt, pointed out (in a friendly criticism) that the equations of mechanics have the peculiarity that they do not change when time is reversed: replace the time variable t with $-t$, and the equations are unchanged. In Loschmidt's view, this meant that physical processes could go backward or forward with equal probability in any mechanical system, including Boltzmann's assemblage of colliding molecules. One could, for example, allow the molecules of a perfume to escape from a bottle into a room, and then expect to see all of the molecules turn around and spontaneously crowd back into the bottle. This was completely contrary to experience and the second law. Loschmidt concluded that Boltzmann's molecular interpretation of the second law, with its mechanical foundations, was in doubt.

Boltzmann replied in 1877 that his argument was not based entirely on mechanics: of equal importance were the laws of probability. The perfume molecules *could* return to the bottle, but only against stupendously unfavorable odds. He made this point with an argument that turned out to be cleverer than he ever had an opportunity to realize. He proposed that the probability for a certain physical state of a system is proportional to a count of "the number of ways the inside [of the system] can be arranged so that from the outside it looks the same," as Richard Feynman put it. To illustrate what this means, imagine two vessels like those guarded by Maxwell's demon. The entire system, including both vessels, contains two kinds of gaseous molecules, A and B. We obtain Boltzmann's count by systematically enumerating the number of possible arrangements of molecules between the vessels, within the restrictions that the total number of molecules does not change, and the numbers of molecules in the two vessels are always the same.

The pattern of the calculation is easy to see by doing it first for a ridiculously small number of molecules, and then extrapolating with the rules of "combinatorial" mathematics to systems of realistic size. Suppose, then, we have just eight noninteracting molecules, four As and four Bs, with four molecules (either A or B) in each vessel. One possibility is to have all the As in vessel 1 and all the Bs in vessel 2. This allocation can also be reversed: four As in vessel 2 and four Bs in vessel 1. Two more possibilities are to have three As and one B in vessel 1, together with one A and three Bs in vessel 2, and then the reverse of this allocation. The fifth, and last, allocation is two As and two Bs in both vessels; reversing this allocation produces nothing new. These five allocations are listed in Table 13.2 in the first two columns.

Boltzmann asks us to calculate the number of molecular arrangements allowed by each of these allocations. If we ignore rearrangements within the vessels, the first two allocations are each counted as one arrangement. The third allocation has more arrangements because the single B molecule in vessel 1 can be any one of the four Bs, and the single A molecule in vessel 2 can be any of the four As. The total number of arrangements for this allocation is $4 \times 4 = 16$. Arrangements for the fourth allocation are counted similarly. The tally for the fifth allocation (omitting the details) is 36. Thus for our small system the total "number of ways the inside can be arranged so that from the outside it looks the same," a quantity we will call W, is

$$W = 1 + 1 + 16 + 16 + 36 = 70.$$

Table 13.2

Vessel 1	Vessel 2	Number of arrangements
4A	4B	1
4B	4A	1
3A + B	A + 3B	16
A + 3B	3A + B	16
2A + 2B	2A + 2B	36
		70

This result can be obtained more abstractly, but with much less trouble, by taking advantage of a formula from combinatorial mathematics,

$$W = \frac{N!}{N_A\,!\,N_B\,!}, \tag{14}$$

where $N = N_A + N_B$ and the "factorial" notation ! denotes a sequence of products such as $4! = 1{\cdot}2{\cdot}3{\cdot}4$. For the example, $N_A = 4$, $N_B = 4$, $N = N_A + N_B = 8$, and

$$W = \frac{8!}{4!4!} = \frac{(1{\cdot}2{\cdot}3{\cdot}4{\cdot}5{\cdot}6{\cdot}7{\cdot}8)}{(1{\cdot}2{\cdot}3{\cdot}4)(1{\cdot}2{\cdot}3{\cdot}4)} = 70.$$

In Boltzmann's statistical picture, our very small system wanders from one of the seventy arrangements to another, with each arrangement equally probable. About half the time, the system chooses the fifth allocation, in which the As and Bs are completely mixed, but there are two chances in seventy that the system will completely unmix by choosing the first or the second allocation.

An astonishing thing happens if we increase the size of our system. Suppose we double the size, so $N_A = 8$, $N_B = 8$, $N = N_A + N_B = 16$, and

$$W = \frac{16!}{8!8!} = 12870.$$

There are now many more arrangements possible. We can say that the system has become much more "disordered." As is now customary, we will use the term "disorder" for Boltzmann's W.

The beauty of the combinatorial formula (14) is that it applies to a system of any size, from microscopic to macroscopic. We can make the stupendous leap from an N absurdly small to an N realistically large and still trust the simple combinatorial calculation. Suppose a molar amount of a gas is involved, so $N = 6 \times 10^{23}$, $N_A = 3 \times 10^{23}$, $N_B = 3 \times 10^{23}$, and

$$W = \frac{(6 \times 10^{23})!}{(3 \times 10^{23})!(3 \times 10^{23})!}.$$

The factorials are now enormous numbers, and impossible to calculate directly. But an extraordinarily useful approximation invented by James Stirling in the eighteenth century comes to the rescue: if N is very large (as it certainly is in our application) then

$$\ln N! = N \ln N - N.$$

Applying this shortcut to the above calculation of the disorder W, we arrive at

$$\ln W = 4 \times 10^{23}$$

or

$$W = e^{4 \times 10^{23}}.$$

This is a *fantastically* large number; its *exponent* is 4×10^{23}. We cannot even do it justice by calling it astronomical. This is the disorder—that is, the total number of arrangements—in a system consisting of ½ mole of one gas thoroughly mixed with ½ mole of another gas. In one arrangement out of this incomprehensibly large number, the gases are completely unmixed. In other words, we have one chance in $e^{4 \times 10^{23}}$ to observe the unmixing. There is no point in expecting *that* to happen. Now it is clear why, according to Boltzmann, the perfume molecules do not voluntarily unmix from the air in the room and go back into the bottle.

Boltzmann found a way to apply his statistical counting method to the distribution of energy to gas molecules. Here he was faced with a special problem: he could enumerate the molecules themselves easily enough, but there seemed to be no natural way to count the "molecules" of energy. His solution was to assume as a handy fiction that energy was parceled out in discrete bundles, later called energy "quanta," all carrying the same very small amount of energy. Then, again following the combinatorial route, he analyzed the statistics of a certain number of molecules competing for a certain number of energy quanta. He found that a particular energy distribution, the one dictated by his exponential factor e^{-hE}, overwhelmingly dominates all the others. This is, by an immense margin, the most probable energy distribution, although others are possible.

Boltzmann also made the profound discovery that when he allowed his energy quanta to diminish to zero size, the logarithm of his disorder count W was proportional to his H-function inverted with a minus sign, that is,

$$\ln W \propto -H \tag{15}$$

Then, in view of the connection between the H-function and entropy (the proportionality (13)), he arrived at a simple connection between entropy and disorder,

$$S \propto \ln W. \tag{16}$$

Boltzmann's theoretical argument may seem abstract and difficult to follow, but his major conclusion, the entropy-disorder connection, is easy to comprehend, at least in a qualitative sense. Order and disorder are familiar parts of our lives, and consequently so is entropy. Water molecules in steam are more disordered than those in liquid water (at the same temperature), and water molecules in the liquid are in turn more disordered than those in ice. As a result, steam has a larger entropy than liquid water, which has a larger entropy than ice (if all are at the same temperature). When gasoline burns, the order and low entropy of large molecules such as octane are converted to the disorder and higher entropy of smaller molecules, such as carbon dioxide and water, at high temperatures. A pack of cards has order and low entropy if the cards are sorted, and disorder and higher entropy if they are shuffled. Our homes, our desks, even our thoughts have order or disorder. And entropy is there too, rising with disorder, and falling with order.

Entropy and Probability

We can surmise that Gibbs developed his ideas on statistical mechanics more or less in parallel with Boltzmann, although in his deliberate way Gibbs had little to say about the subject until he published his masterpiece, *Elementary Principles in Statistical Mechanics*, in 1901. That he was thinking about the statistical interpretation of entropy much earlier is clear from his incidental remark that "an uncompensated decrease of entropy seems to be reduced to improbability." As mentioned, Gibbs wrote this in 1875 in connection with a discussion of the mixing and unmixing of gases. Gibbs's speculation may have helped put Boltzmann on the road to his statistical view of entropy; at any rate, Boltzmann included the Gibbs quotation as an epigraph to part 2 of his *Lectures on Gas Theory*, written in the late 1890s.

Gibbs's *Elementary Principles* brought unity to the "gas theory" that had been developed by Boltzmann, Maxwell, and Clausius, supplied it with the more elegant name "statistical mechanics," and gave it a mathematical style that is preferred by today's theorists. His starting point was the "ensemble" concept, which Maxwell had touched on in 1879 in one of his last papers. The general idea is that averaging among the many states of a molecular system can be done conveniently by imagining a large collection—an ensemble—of replicas of the system, with the replicas all exactly the same except for some key physical properties. Gibbs proposed ensembles of several different kinds; the one I will emphasize he called "canonical." All of the replicas in a canonical ensemble have the same volume and temperature and contain the same number of molecules, but may have different energies.

Averaging over a canonical ensemble is similar to Boltzmann's averaging procedure. Gibbs introduced a probability P for finding a replica in a certain state, and then, as in Boltzmann's equation (7), he calculated the probability dP that the system is located in an element of phase space,

$$dP = Pd\omega. \tag{17}$$

These probabilities must total one when they are added by integration,

$$\int dP = 1. \tag{18}$$

Gibbs, like Boltzmann, was motivated by a desire to compose a statistical molecular analogy with thermodynamics. Most importantly, he sought a statistical entropy analogue. He found that the simplest way to get what he wanted from a canonical ensemble was to focus on the *logarithm* of the probability, and he introduced

$$S = -\,k \ln P \tag{19}$$

for the entropy of one of the replicas belonging to a canonical ensemble. The constant k (Gibbs wrote it $1/K$) is a very small number with the magnitude 1.3807 $\times\ 10^{-23}$ if the energy unit named after Joule is used. It is now known as "Boltzmann's constant" (although Boltzmann did not use it, and Max Planck was the first to recognize its importance).

In Gibbs's scheme, as in Boltzmann's, the probability P is a mathematical tool for averaging. To calculate an average energy $\overline{E}$ we simply multiply each energy

E found in a replica by the corresponding probability $dP = Pd\omega$ and add by integrating

$$\bar{E} = \int PEd\omega. \tag{20}$$

The corresponding entropy calculation averages the entropies for the replicas given by equation (19),

$$\begin{aligned}\bar{S} &= \int P(-k \ln P)d\omega \\ &= -k \int P \ln Pd\omega.\end{aligned} \tag{21}$$

The constant k is pervasive in statistical mechanics. It not only serves in Gibbs's fundamental entropy equations, but it is also the constant that makes Boltzmann's entropy proportionality (16) into one of the most famous equations in physics,

$$S = k \ln W. \tag{22}$$

The equation is carved on Boltzmann's grave in Vienna's Central Cemetery (in spite of the anachronistic k).

We now have two statistical entropy analogues, Gibbs's and Boltzmann's, expressed in equations (21) and (22). The two equations are obviously not mathematically the same. Yet they apparently calculate the same thing, entropy. One difference is that Gibbs used probabilities P, quantities that are always less than one, while Boltzmann based his calculation on the disorder W, which is larger than one (usually much larger). It can be proved (in more space than we have here) that the two equations are equivalent if the system of interest has a single energy.

Gibbs proved that a system represented by his canonical ensemble does have a fixed energy to an extremely good approximation. He calculated the extent to which the energy fluctuates from its average value. For the average of the square of this energy fluctuation he found kT^2C_v, where k is again Boltzmann's constant, T is the absolute temperature, and C_v is the heat capacity (the energy required to increase one mole of the material in the system by one degree) measured at constant volume. Neither T nor C_v is very large, but k is very small, so the energy fluctuation is also very small. A similar calculation of the entropy fluctuation gave kC_v, also very small.

Thus the statistical analysis, either Gibbs's or Boltzmann's, arrives at an energy and entropy that are, in effect, constant. They are, as Gibbs put it, "rational foundations" for the energy and entropy concepts of the first and second laws of thermodynamics.

Note one more important use of the ever-present Boltzmann constant k. Gibbs proved that the h factor appearing in Boltzmann's statistical factor e^{-hE} is related to the absolute temperature T by $h = \frac{1}{kT}$, so the Boltzmann factor, including the temperature, is $e^{-\frac{E}{kT}}$.

Boltzmann and Gibbs Updated

Boltzmann and Gibbs gave us what is now called the "classical" version of statistical mechanics. With the advent of quantum theory in the early 1900s, some changes had to be made. The main problem was that in the view of nineteenth-century physics, molecules exist in a continuum of mechanical states, while quantum theory was founded on the principle that molecules are allowed only certain discrete states and no others. This means that the energy variable E, which can have continuous values in Boltzmann's and Gibbs's equations, must be replaced by the particular values E_1, E_2, etc., one for each quantum state allowed to the molecular system.

The necessary repairs to the classical equations are remarkably easy to make. The classical Boltzmann factor $e^{-\frac{E}{kT}}$ becomes $e^{-\frac{E_i}{kT}}$ in quantum theory, with $i =$ 1,2, etc. The *integral* of Boltzmann factors, $\int e^{-\frac{E}{kT}}\,d\omega$ in equation (9) (remember that $h = \frac{1}{kT}$), becomes a *sum* of Boltzmann factors $\sum e^{-\frac{E_i}{kT}}$ covering all of the system's accessible quantum states. This summation plays a leading role in modern statistical mechanics. It is called a "partition function," and is represented with the symbol Z,

$$Z = \sum e^{-\frac{Ei}{kT}}. \tag{23}$$

Thus Boltzmann's classical probability equation (9) adapted for quantum theory so it calculates the probability P_i for the ith quantum state is

$$P_i = \frac{e^{-\frac{Ei}{kT}}}{Z}. \tag{24}$$

The entropy S_i of the ith quantum state, adapted from the classical equation (19), is

$$S_i = -k\ln P_i \tag{25}$$

and the average entropy for a canonical ensemble is calculated with an adaptation of equation (21),

$$\overline{S} = -k\sum P_i \ln P_i. \tag{26}$$

The average-energy calculation, an adaptation of equation (20), is

$$\overline{E} = \sum P_i E_i. \tag{27}$$

The keys to this version of statistical thermodynamics are equations (24) and (25), and they in turn require the energies E_i. Each system has its own hierarchical set of energies specified by the energy equations of quantum theory and the precise numerical data of molecular spectroscopy.

Combatants

When Boltzmann published the second volume of his *Lectures on Gas Theory* in 1898, he was not optimistic about its reception. The molecular basis for his theory was being attacked by eminent and not-so-eminent critics. "I am convinced that these attacks are merely based on a misunderstanding," he wrote in his forward to the *Lectures*, "and that the role of [molecular] gas theory has not yet been played out. . . . In my opinion it would be a great tragedy for science if the theory of gases were temporarily thrown into oblivion because of a momentary hostile attitude toward it, as was for example the wave theory [of light] because of Newton's authority."

Boltzmann's most prominent adversaries were Ernst Mach, Wilhelm Ostwald, and Georg Helm. Ostwald believed that a grand scheme could be formulated that encompassed all of the fields of science, beginning with the energy concept as a unifying principle. He became convinced that energy fluxes and transformations determined the laws of physics and chemistry. Molecules and atoms were figments of the mathematics; energy in all its forms was the universal reality. Helm also adhered to Ostwald's school of "energetics."

Mach, the most able and obstinate of Boltzmann's opponents, did not subscribe to energetics, but he was an ardent antiatomist. He could not accept atoms and molecules because he could find no direct evidence for their existence. If atoms could not be seen, Mach argued, "we have as little right to expect from them, as from the symbols of algebra, more than we put into them, and certainly not more enlightenment than from experience itself." Boltzmann's explanation of the second law as a consequence of the molecular chaos was superficial. "In my opinion," Mach wrote, "the roots of this (entropy) law lie much deeper, and if success were achieved in bringing about agreement between the molecular hypothesis and the entropy law this would be fortunate for the hypothesis, but not for the entropy law."

At about the same time Mach made these remarks, Boltzmann published his *Principles of Mechanics*, which began with the epigraph,

> Bring forward what is true
> Write it so that it's clear
> Defend it to your last breath!

This could have been Boltzmann's battle cry in the war against the antiatomists. Arnold Sommerfeld, a student at the time, and later a prominent quantum physicist, witnessed a skirmish in the war at an 1895 conference of natural scientists in Lübeck, and recorded this picture of Boltzmann in combat: "The paper on *Energetik* [energetics] was given by Georg Helm from Dresden: behind him stood Wilhelm Ostwald, behind both the philosophy of Ernst Mach, who was not present. The opponent was Boltzmann, seconded by Felix Klein. Both externally and internally, the battle between Boltzmann and Ostwald resembled the bull with the supple fighter. However, this time the bull was victorious over the torero in spite of the latter's artful combat. The arguments of Boltzmann carried the day. We, the young mathematicians of that time, were all on Boltzmann's side."

Boltzmann had the sympathies of the "young mathematicians," but the Mach-Ostwald forces prevailed through the turn of the century. Then, in 1905, while Boltzmann was dining with Mrs. Hearst in California and being seduced by her piano, a twenty-six-year-old Albert Einstein wrote a theoretical paper that brought the beginning of the end to the war against molecules. Einstein argued that molecules of a certain kind could actually be seen, counted, and tracked. He had in mind "colloidal" particles, which can be dispersed in an aqueous or other liquid medium and remain suspended there permanently, like oxygen and nitrogen molecules in Earth's atmosphere. In size they can be some five orders of magnitude larger than ordinary molecules, but Einstein proved with Boltzmann's method that all molecules, large and small, display the same kind of statistical behavior. He derived an equation for the average straight-line distance λ a colloidal particle travels in its random motion during a period of time t. Assuming that the particles are all spherical and prepared with the same radius r, Einstein obtained

$$\lambda^2 = \frac{RT}{3\pi N_A r\eta} t,$$

where R is a constant that appears in the equation relating the pressure, volume, and temperature of an ideal gas, the "gas constant," T is the absolute temperature, N_A is Avogadro's number, and η is a coefficient that measures the viscosity of the liquid medium in which the particles move.

Einstein appreciated that colloidal particles are large enough to be seen with a microscope (at least by scattered light in an "ultramicroscope"), and he surmised that his equation could be subjected to a direct experimental test, which would decide the contentious issue of the reality of atoms and molecules. The man who had the patience and skill to make the crucial experimental test, Jean Perrin, was at first unaware of Einstein's theory. But, like Einstein, he believed that colloidal particles behave like mega-molecules. In a series of experiments started in 1906 he carefully demonstrated an analogy between the equilibrium distribution of colloidal particles in resin suspensions and the distribution of gas molecules in the atmosphere. Einstein's equation finally came to Perrin's attention in 1909, and he proved its validity by calculating Avogadro's number N_A, beginning with measured values of the other parameters in the equation. His result, $N_A = 7 \times 10^{23}$, was in reasonable agreement with other determinations of N_A, including one ($N_A = 6 \times 10^{23}$) obtained by Max Planck in 1900 with a completely different theoretical and experimental basis.

Einstein's theory, Perrin's meticulous experiments, and other experiments, such as J. J. Thomson's discovery of the electron and Ernest Rutherford's investigations of radioactivity, finally left no doubt about the reality of molecules. In 1909, Ostwald surrendered: "I am now convinced [he wrote in the preface to his *Outlines of General Chemistry*] that we have recently become possessed of experimental evidence of the discrete or grained nature of matter, which the atomic hypothesis sought in vain for hundreds and thousands of years." Mach was apparently never persuaded.

When Boltzmann returned to Vienna from California in 1905, he was unaware of Einstein's paper, and Perrin's experiments were a few years in the future. No doubt he thought about engaging the enemy once again, but he did not have the chance. Sometime in early 1906 he met his final, inescapable depression. During the spring and summer of 1906, his mental state grew steadily worse. "Boltzmann had announced lectures for the summer semester," Mach wrote later, "but he had to cancel them because of his nervous condition. In informed circles one knew that Boltzmann would most probably never be able to exercise his professorship again. One spoke of how necessary it was to keep him under surveillance, for he had already made attempts at suicide."

Boltzmann, Henriette, and Elsa went to the resort town of Duino near Trieste for a summer holiday. A few days before they were to return to Vienna Boltzmann committed suicide by hanging himself.

V RELATIVITY

Historical Synopsis

Relativity begins with a modest question: How does your physics relate to my physics if we are moving relative to each other? Galileo gave one answer: We find exactly the same laws of mechanics if our relative speed is constant. Newton said the same thing but more elaborately by referring all motion—yours, mine, and everyone else's—to an absolute frame of reference in space and time. Nineteenth-century theorists found Newton's absolute frame a convenient place to locate the hypothetical medium they called the ether, which propagated light and other electromagnetic waves.

Ether physics was a prominent endeavor among Victorian scientists, but it had fatal flaws. For one thing, ether physicists could never agree on a standard model for the mechanical structure of the ether. Also questionable was the concept of motion through an ether anchored in Newton's absolute frame of reference. A series of experiments performed by Albert Michelson and Edward Morley in the 1880s that aimed at detecting Earth's motion relative to an "ether sea" was an impressive failure. The stubborn fact, always observed, is that the speed of light in empty space is the same regardless of the speed and direction of the light source.

A young patent examiner in Bern, Switzerland, named Albert Einstein published a paper in 1905 that resolved the ether problem by simply ignoring it. Einstein postulated two empirical principles that could not be denied: constancy of the speed of light, and a generalization of Galileo's relativity principle to include electromagnetic and optical phenomena. Beginning with these two principles, and without recourse to the ether concept, he proved that, for observers moving relative to each other at constant speeds, length and time measurements are different, perhaps drastically different if the speed is close to the speed of light. For example, if a stationary observer watches a clock moving at high speed he or she sees it ticking more slowly than an observer traveling with the clock. In addition to this "time dilation," Einstein's 1905 paper insisted that the length dimension of the clock, or of anything else, is contracted in the direction of motion for the stationary observer.

Einstein designed his 1905 "special" theory of relativity with two limitations: it focused on "inertial" systems, those moving at constant relative speeds; and although the theory was compatible with Maxwell's equations for the electromagnetic field, its scope did not include another great theory from the past, Newton's gravitation theory. Einstein soon realized that a "general" theory of relativity must recognize both gravitational effects and noninertial systems—that is, those accelerating relative to each other. His first step in that direction, later called the "equivalence principle," asserted Einstein's "happiest thought," that acceleration and gravitation are intimately related to each other: where there is acceleration there are artificial gravitational effects that are indistinguishable from the real thing.

As he proceeded with the equivalence principle as his guide, Einstein became aware that space and time are peculiarly warped in accelerating systems; Euclidean formulas such as the calculation of the circumference-to-diameter ratio for a circle as π are slightly in error. This gave him the vital clue that a general theory of relativity had to be based on non-Euclidean geometry. As it happened, a complete theory of non-Euclidean spaces, developed in the 1850s by Bernhard Riemann, provided just the right mathematical tools for Einstein to construct a theoretical edifice that linked geometry and gravitation. At the same time, he found a generalized equation of motion that was also determined in the Riemann manner by the geometry. His motto, physics as geometry, was taken up by many of his successors.

14

Adventure in Thought

Albert Einstein

Like Columbus

Modern theoretical physicists like to think of themselves as intellectual explorers, and the greatest of them have indeed discovered new and exotic physical worlds, both microscopic and macroscopic. Travel in these intellectually distant realms has proved hazardous because it takes the explorer far from the world of ordinary experience. Werner Heisenberg, one of the generation of theorists who found the way to the quantum realm, the strangest of all the physical worlds, likened the intellectual expeditions of modern physics to the voyage of Columbus. Heisenberg found Columbus's feat remarkable not because Columbus tried to reach the East by sailing west, nor because he handled his ships masterfully, but because he decided to "leave the known regions of the world and sail westward, far beyond the point from which his provisions could have got him back home again." The man who ranks above all others as an intellectual Columbus is Albert Einstein. He took such expeditions far beyond "the safe anchorage of established doctrine" into treacherous, uncharted seas. Not only was he a pioneer in the quantum realm; he discovered and explored much of the territory of modern physics.

These great explorations were started, and to a large extent completed, when Einstein was in his twenties and working in a quiet corner of the scientific world, the Swiss Patent Office in Bern. Life in the patent office, as Einstein found it, was a "kind of salvation." The work was interesting, and not demanding; without the pressures of an academic job, he was free to exploit his marvelous ability "to scent out that which was able to lead to fundamentals and to turn aside from everything else, from the multitude of things which clutter up the mind and divert it from the essential."

Einstein had tried to place himself higher professionally, but his prospects after graduating from the Zürich Polytechnic Institute (since 1911 known as the Swiss Technical University or ETH) were not brilliant. He had disliked and opposed most of his formal education. The teachers in his Munich gymnasium said he would never amount to anything, and deplored his disrespectful attitude. The gymnasium experience aroused in Einstein a profound distrust of authority, particularly the kind wielded by Prussian educators. Ronald Clark, one of Einstein's biographers, describes the Luitpold Gymnasium Einstein attended in Munich as probably "no better and no worse than most establishments of its kind: It is true that it put as great a premium on a thick skin as any British public school but there is no reason to suppose that it was particularly ogreish. Behind what might be regarded as not more than normal discipline it held, in reserve, the ultimate weapon of appeal to the unquestionable Prussian god of authority. Yet boys, and even sensitive boys, have survived as much."

Einstein's father, Hermann, was a cheerful optimist—"exceedingly friendly, mild and wise," as Einstein recalled him—but prone to business failures. One of these drove the family from Munich to Milan, with Einstein left behind to complete his gymnasium courses. He had few friends among his classmates, and now with his family gone, he could no longer bear life in Munich, or anywhere else in Germany. He abruptly joined his family members in Italy and informed them that he planned to surrender his German citizenship. That meant no gymnasium diploma, but Einstein planned to do the necessary studying himself to prepare for the Zürich Poly entrance examination. Life in Italy, and later in Switzerland, was free and promising again, and it "transformed the quiet boy into a communicative young man," writes Abraham Pais, a recent Einstein biographer. For a few happy months, Einstein celebrated his release from a dismal future by roaming northern Italy.

A temporary setback, failing marks in the Poly admission examination, proved to be a blessing. To prepare for a second try, Einstein attended a Swiss cantonal school in Aarau, where the educational process was, for a change, a joy. In Aarau, Einstein lived with the Winteler family. Jost Winteler was the head of the school, and "a somewhat casual teacher," writes Clark, "as ready to discuss work or politics with his pupils as his fellow teachers. [He] was friendly and liberal-minded, an ornithologist never happier than when he was taking his students and his own children for walks in the nearby mountains." Even in old age, Einstein recalled vividly his year in Aarau: "This school left an indelible impression on me because of its liberal spirit and the unaffected thoughtfulness of the teachers, who in no way relied on external authority."

In early 1896, Einstein paid a fee of three marks and was issued a document declaring that he was no longer a German citizen; he would be a stateless person for the next five years. Later in the year he passed the Zürich Poly examination with good marks and began the four-year preparation of a *fachlehrer*, a specialized high-school teacher. Hermann had suffered another business disaster, so Einstein's means were now limited—a monthly allowance of one hundred Swiss francs, from which he saved twenty francs to pay for his Swiss naturalization papers. But there was nothing meager about his vision of the future. In a letter to Frau Winteler, he wrote, "Strenuous labor and the contemplation of God's nature are the angels which, reconciling, fortifying and yet ceaselessly severe, will guide me through the tumult of life."

On the whole, Einstein did not respond with much enthusiasm to his course work at the Zürich Poly. He recognized that some of the mathematics courses were excellent—one of his mathematics professors, Hermann Minkowski, later made vital contributions to the mathematical foundations of the theory of relativity—but the courses in experimental and theoretical physics were uninspiring.

At first he was fascinated by laboratory work, but his experimental projects rarely met with the approval of his professor, Heinrich Weber. In exasperation, Weber finally told his pupil, "You are a smart boy, Einstein, a very smart boy. But you have one great fault: you do not let yourself be told anything."

Einstein responded by simply staying away from classes and reading in his rooms the great nineteenth-century theorists, Kirchhoff, Helmholtz, Hertz, Maxwell, Hendrik Lorentz, and Boltzmann. Fortunately, the liberal Zürich program allowed such independence. "In all there were only two examinations," Einstein writes in his autobiographical notes, "aside from these, one could just about do as one pleased. . . . This gave one freedom in the choice of pursuits until a few months before the examination, a freedom which I enjoyed and have gladly taken into the bargain the bad conscience connected with it as by far the lesser evil."

The punishment appears to have been more than a bad conscience, however. Preparation for the final examination was a nightmare, and the outcome successful largely due to the help of a friend, Marcel Grossmann, who had a talent for taking impeccable lecture notes. Einstein tells us, again in his autobiographical notes, that the pressure of that examination "had such a deterring effect [on me] that, after I had passed . . . I found consideration of any scientific problems distasteful for an entire year." And he adds this thought concerning the heavy hand the educational system lays on a student's developing intellectual interests: "It is, in fact, nothing short of a miracle that the modern methods of instruction have not yet entirely strangled the holy curiosity of inquiry; for this delicate little plant, aside from stimulation, stands mainly in need of freedom."

Einstein graduated from the Poly in the fall of 1900, and a few months later passed two important milestones in his life: he published his first paper—in volume 4 of the *Annalen der Physik*, which contained, just forty pages later, Max Planck's inaugural paper on quantum theory—and he received his long-awaited Swiss citizenship. Although he was to leave Switzerland nine years later, and did not return to settle, Einstein never lost his affection for the humane, democratic Swiss and their splendid country, "the most beautiful corner on Earth I know."

He was now job hunting. An expected assistantship at the Zürich Poly under Weber never materialized. ("Weber . . . played a dishonest game with me," Einstein wrote to a friend.) Two temporary teaching positions followed, and then with the help of Marcel Grossmann's father, Einstein was appointed technical expert third class at the Bern Patent Office in 1902.

Now that he had steady employment, Einstein thought of marriage, and a year later he and Mileva Maric, a classmate at the Zürich Poly, were married. Mileva came from a Slavic-Serbian background. She was pretty, tiny in stature, and slightly crippled from tuberculosis in childhood. She had hoped to follow a career in science, and went to Zürich because Switzerland was the only German-speaking country at the time admitting women to university studies. The couple became lovers soon after both entered the program at the Zürich Poly. By 1901, the affair had deepened: Mileva was pregnant. In 1902, a daughter, Liserl, was born at Mileva's parents' home in Novi Sad. When she returned to Zürich, Mileva did not bring the baby, and in 1903, shortly after Einstein and Mileva were married, the girl was apparently given up for adoption.

The marriage was never a success. After the trials of her pregnancy, a difficult birth, and the loss of the child, Mileva's career plans collapsed. She was jealous of Einstein's freewheeling friends, and prone to periods of depression. On his side, Einstein was not a sensitive husband; too much of his intellectual and emotional strength was spent on his work to make a difficult marriage succeed. In old age, Einstein recalled that he had entered the marriage with a "sense of duty." He had, he said, "with an inner resistance, embarked on something that simply exceeded my strength."

Precursors

In 1905, when he was twenty-six, happily employed in the Bern Patent Office, and yet to make the acquaintance of (another) theoretical physicist, Einstein published three papers in the *Annalen der Physik*. This was volume 17 of that journal, and it was, as Max Born remarks, "one of the most remarkable volumes in the whole scientific literature. It contains three papers by Einstein, each dealing with a different subject and each today acknowledged to be a masterpiece."

The first of the 1905 papers was a contribution to quantum theory, which developed a theory of the photoelectric effect by picturing light beams as showers of particles, or "quanta." I will have more to say about that revolutionary paper in chapter 15. The second paper, on the reality of molecules observed as colloidal particles, was mentioned in chapter 13 above. Our concern now is with the third paper, which presented Einstein's version of the theory of relativity.

By the time Einstein entered the field, relativity theory had a long and distinguished history. Einstein counted among his precursors some of the giants: Galileo, Newton, Maxwell, and Lorentz. Galileo stated the relativity principle applied to mechanics in his usual vividly observed style:

> Shut yourself up with some friend below decks on some large ship, and have with you some flies, butterflies, and other flying animals. Have a large bowl of water with some fish in it; hang up a bottle that empties drop by drop into a wide vessel beneath it. With the ship standing still, observe carefully how the little animals fly with equal speeds to all sides of the cabin. The fish swim indifferently in all directions; the drop falls into the vessel beneath; and in throwing something to your friend, you need to throw it no more strongly in one direction than another, the distances being equal; jumping with your feet together, you pass equal spaces in every direction. When you have observed all these things carefully (though there is no doubt that when the ship is standing still everything must happen this way), have the ship proceed with any speed you like, so long as the motion is uniform and not fluctuating this way and that [not accelerating]. You will discover not the least change in all the effects named, nor could you tell from any of them whether the ship was moving or standing still.

Galileo's ship, or any other system moving at constant speed, is called in modern terminology an "inertial frame of reference," or just an "inertial frame," because in it Galileo's law of inertia is preserved. Galileo's relativity principle, generalized, tells us that the laws of mechanics are exactly the same in any inertial frame ("nor could you tell from any of the [observed effects] whether the ship was moving or standing still").

Newton's statement of the relativity principle, which he derived from his three laws of motion, was similar, except that it raised the later contentious issue of "space at rest": "The motions of bodies included in a given space are the same among themselves, whether that space is at rest, or moves uniformly forwards in a right [straight] line without any circular motion." At rest with respect to what? Newton believed in the concept of absolute space relative to which all motion, or lack of motion, could be referred. In the same vein, he adopted an absolute time frame in which all motion could be measured; one time frame served all observers.

Maxwell and his contemporaries accepted Newton's concept of absolute space, and they filled it with the all-pervading medium they called ether. The principal role of the ether for nineteenth-century theorists was to provide a mechanism for the propagation of light and other electromagnetic fields through otherwise empty space. The ether proved to be a versatile theoretical tool—too versatile. British and Continental theoreticians could never reach a consensus concerning which of the many ether models was the standard one.

The man who saw ether physics and its connections with field theory most clearly, and at the same time helped Einstein find his way, was Hendrik Lorentz, professor of theoretical physics at the University of Leiden from 1877 to 1912. Lorentz was revered by generations of young physicists for his remarkable ability to play the dual roles of creative theorist and sympathetic critic. Like Maxwell and Gibbs, it was not his style to gather a school of research students, yet physicists from all over the world attended his lectures on electrodynamics. After the turn of the century, he was recognized by one and all as the leader of the international physics community. Beginning in 1911, he acted as president of the Solvay Conferences in Brussels, named after Ernest Solvay, an industrial chemist with formidable wealth and an amateur's interest in physics, who paid the bill for the participants' elegant accommodations at the conferences. No one but Lorentz could bring harmony to these international gatherings, which Einstein liked to call "Witches' Sabbaths." "Everyone remarked on [Lorentz's] unsurpassed knowledge, his great tact, his ability to summarize lucidly the most tangled arguments, and above all his matchless linguistic skill," writes one of Lorentz's biographers, Russell McCormmach. After attending the first Solvay Conference, Einstein wrote to a friend, "Lorentz is a marvel of intelligence and exquisite tact. A living work of art! In my opinion he was the most intelligent of the theoreticians present."

As a theorist, Lorentz's principal goal was to unify at the molecular level the physics of matter with Maxwell's physics of electromagnetic fields. One of the foundations of Lorentz's theory was the concept that the seat of electric and magnetic fields was an absolutely stationary ether, which permeated all matter with no measurable resistance. Another cornerstone provided the assumption that (to some degree) matter consisted of very small charged particles, which Lorentz eventually identified with the particles called "electrons" discovered in 1897 by J. J. Thomson in cathode rays. The electrons generated the electric and magnetic fields, and the fields, in turn, guided the electrons through the immobile ether. Lorentz used Maxwell's equations, written for the ether's stationary frame of reference, to describe the fields, and he accepted the message of the equations that in that frame the speed of light was the same regardless of the speed and the direction of the light source.

To summarize, and bring the story around to Einstein's point of view, imagine two observers, the first at rest in the ether, and the second at rest in a room moving at constant speed with respect to the ether. The room carries a fixed light source, and the two observers compare notes concerning the light signals generated by the source. According to Lorentz's theory, the first observer finds that the speed of a light beam is independent of its direction. But the second observer sees things differently: suppose one of the walls of his or her room moves away from a light beam after it is generated, while the opposite wall moves toward it. If the light source is fixed in the center of the room, a light beam directed toward a wall retreating from the beam will seem to be slower than a beam directed to a wall approaching the beam. Thus for the second observer the speed of light is not the same in all directions.

To take this argument beyond a thought experiment, we can picture Earth as a "room" moving through the ether and conclude that for us, the occupants of the room, the speed of light should be different when it is propagated in different directions. We anticipate that if we can observe this directional effect it will define Earth's motion with respect to the ether. Several experiments, designed and executed in the late nineteenth century, had this motivation. The most refined of these was performed by Albert Michelson and Edward Morley in 1887. Their conclusion, probably the most famous negative result in the history of physics, was that the speed of light (in empty space) has no dependence whatever on the motion, direction, or location of the light source.

This was a damaging, but not quite fatal, blow to Lorentz's electron theory. He found that he could explain the Michelson-Morley result by assuming that moving material objects contract slightly in their direction of motion, *just* enough to frustrate the Michelson-Morley experiment and other attempts to define Earth's motion through the ether by measuring changes in the speed of light. The cause of this contraction, as Lorentz saw it, was a very slight alteration of molecular forces in the direction of the motion.

Now it is time to bring Technical Expert Third Class Einstein on stage and follow his creation of what came to be called the special theory of relativity. He is acquainted with Galileo's relativity principle. He is aware of Newton's concept of absolute space and time. He has read Lorentz carefully, and he is impressed that experimentalists can find no way to detect Earth's motion relative to the ether by measuring changes in the speed of light.

Doctrine of Space and Time

For Einstein, there were two important kinds of theories. "Most of them are constructive," he wrote. "They attempt to build up a picture of the more complex phenomena out of the materials of a relatively simple formal scheme from which they start out." As an example, he cited the molecular theory of gases. It begins with the hypothesis of molecular motion, and builds from that to account for a wide variety of mechanical, thermal, and diffusional properties of gases. "When we say that we have succeeded in understanding a group of natural processes," Einstein continued, "we invariably mean that a constructive theory has been found which covers the processes in question."

Theorists since Galileo and Newton have also created what Einstein called "principle theories." These are theories that "employ the analytic, not the synthetic, method. The elements which form their basis and starting point are not hypothetically constructed but empirically discovered ones, general characteristics of natural processes, principles that give rise to mathematically formulated criteria which the separate processes or the theoretical representations of them have to satisfy." The supreme example of a principle theory, Einstein pointed out, is thermodynamics, based on the energy and entropy principles called the first and second laws of thermodynamics.

Einstein saw relativity as a principle theory. He began his 1905 paper on relativity by postulating two empirical principles on which his theory, with all its startling conclusions, would rest. The first principle generalized Galileo's relativity principle by asserting that (as Einstein put it several years later),

> *The laws of nature are independent of the state of motion of the frame of reference, as long as the latter is acceleration free [that is, inertial].*

The phrase "laws of nature" is all-inclusive; it encompasses the laws of electromagnetic and optical, as well as mechanical, origin. This is a grandly democratic principle: *all* inertial frames of reference are equal; *none* is different or preferred.

The second of Einstein's principles gives formal recognition to the constancy of the speed of light:

> *Light in empty space always propagates with a definite [speed], independent of the state of motion of the emitting body.*

Whereas Lorentz had struggled to *explain* the invariance of the speed of light with a constructive theory that hypothesized motion-dependent molecular forces, Einstein bypassed all the complications by simply promoting the constancy to a postulate. For Lorentz and his contemporaries, it was a problem, for Einstein a principle.

Einstein's two principles led him to conclude that the speed of light in free space is the only measure of space and time that is reliably constant from one observer to another. All else is relative. Different observers cannot express their physical laws in a shared, absolute frame of reference, as Newton taught. Observers in different inertial frames find that their physical worlds are different according to a new "doctrine of space and time," as Einstein put it.

We can follow the rudiments of Einstein's argument by first considering the elementary question of time measurements. Imagine a timing device recommended by Einstein, called a "light clock"; figure 14.1 displays the light clock as it is seen by an observer who travels along with it. Light flashes are generated by the source S; they travel to the mirror M, and are reflected back to the detector D. The short time for one flash to make the round trip from S to M to D represents one "tick" of the clock. If c is the speed of light, and L_0 is the distance from S and D to M, this time, call it Δt_0, is equal to $\frac{L_0}{c}$ for the trip from S to M, and also for the return trip from M to D, so

$$\Delta t_0 = \frac{2L_0}{c}. \quad (1)$$

Now, keeping in mind Einstein's principle of the constancy of the speed of light, we look at the light clock from the point of view of a second observer, who sees the clock in an inertial frame moving at the constant speed v. Figure 14.2 shows the path of a light flash as seen by this observer. The clock is shown in

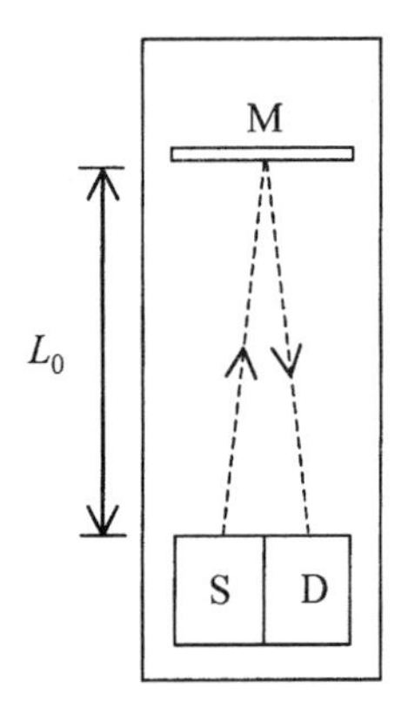

Figure 14.1. Einstein's light clock, as seen by an observer traveling with the clock. The distance between the source and the detector is exaggerated. This figure, the one that follows, and fig. 14.4 are adapted with permission from Robert Resnick, David Halliday, and Kenneth Krane, *Physics*, 4th ed. (New York: Wiley, 1992), 470.

three positions: at A when the light flash leaves the source, at B when it is reflected by the mirror, and at C when it reaches the detector. (The inertial frame containing the light clock is moving extremely fast: Galileo's ship has become a spaceship.) The time representing one tick of the clock is now Δt, and the clock moves the distance $v\Delta t$ in that time. The corresponding distance traveled by the light flash is $2L$, and that is clearly greater than the distance $2L_0$ the light flash travels for the first observer during one tick of the clock. The speed of light is exactly the same for both observers, Einstein's principle insists, so the time Δt for one tick according the second observer is greater than the time Δt_0 for one tick according to the first observer. In other words, the two observers perceive the clock ticking at different rates; it goes slower for the observer who sees the clock moving.

The mathematical connection between Δt and Δt_0 follows from the geometry of figure 14.2. The time interval for the light to travel the distance is $2L$ is

$$\Delta t = \frac{2L}{c}, \tag{2}$$

and as shown in the diagram in figure 14.3, abstracted from figure 14.2,

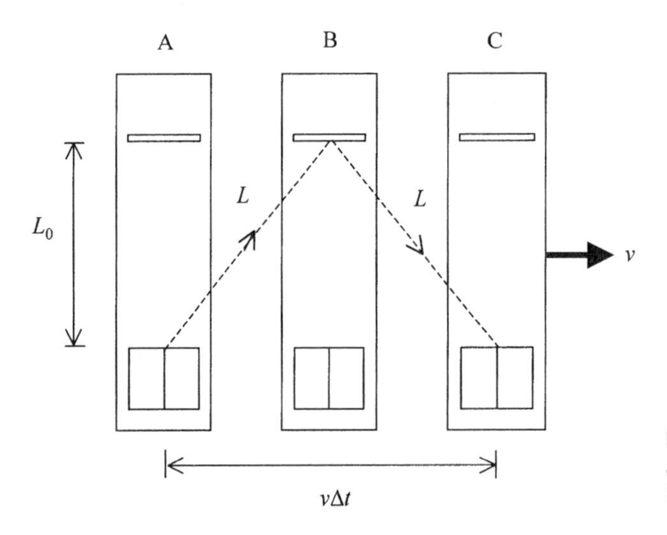

Figure 14.2. Einstein's light clock as seen by an observer who observes the clock moving at constant speed *v*.

$$2L = 2\sqrt{L_0^2 + (v\Delta t\ /\ 2)^2}.$$

Therefore,

$$\Delta t = \frac{2\sqrt{L_0^2 + (v\Delta t\ /\ 2)^2}}{c}.$$

Substituting for L_0 from equation (1) and solving for Δt, we arrive at

$$\Delta t = \frac{\Delta t_0}{\sqrt{1 - v^2\ /\ c^2}}. \tag{3}$$

If the speed v has any ordinary value, that is, much less than the speed of light c, the ratio v/c is very small, the denominator in equation (3) is nearly equal to one, $\Delta t = \Delta t_0$, and time measurements are not appreciably affected. As v approaches c, however, the denominator becomes less than one, Δt is greater than Δt_0 and time measurements are different for the two observers.

Equation (3) places limitations on the speed v: with $v = c$ the equation generates a physically questionable infinite value for Δt, and with $v > c$ the square root becomes an "imaginary" number in mathematical parlance, and even more unacceptable physically. We will find this prohibition on any speed equal to or larger than the speed of light to be a general feature of Einstein's theory.

The relativistic calculation of time intervals expressed by equation (3) speaks of real physical effects, not just artifacts of the mathematics. Light clocks, and all other physical aspects of time, including aging, are really seen differently by different observers moving (at high speed) relative to each other. In fact, if we can boost ourselves to a speed comparable to c relative to Earth—which is possible and not dangerous if we accelerate to the high speed slowly, as in a spaceship accelerating at the rate of Earth's gravitational acceleration g—we can enter a time machine and age decades while Earth and its inhabitants age millennia.

In company with this slowing or dilation of time in a moving inertial frame, Einstein's principles also demand a contraction of length measurements. This, too, can be demonstrated with the handy light clock. As in figure 14.2, we see the clock moving at a constant speed v, but this time parallel to its length, as shown in figure 14.4.

We again imagine that a light flash is produced by the source S, and that the flash is reflected by the mirror M back to the detector D. Let Δt_1 be the time interval for the light to travel from the source to the mirror. During this time, the mirror moves through the distance $v\Delta t_1$, so the light flash must travel $L + v\Delta t_1$ to

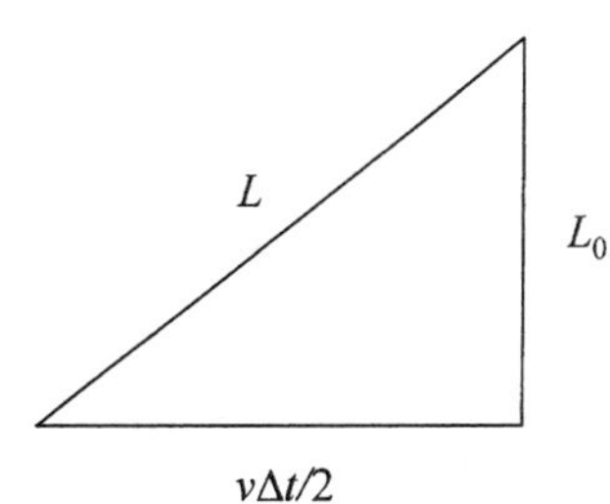

Figure 14.3. A right triangle constructed from the distances shown in fig. 14.2, demonstrating that according to the Pythagorean theorem $L^2 = L_0^2 + (v\Delta t\ /\ 2)^2$ or $L = \sqrt{L_0^2 + (v\Delta t\ /\ 2)^2}$.

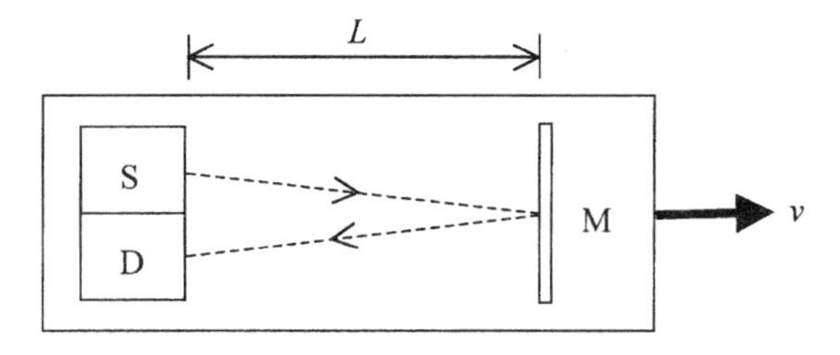

Figure 14.4. Einstein's light clock again, as perceived by an observer who sees the clock moving parallel to the clock's length at constant speed v.

reach the mirror. Noting that the speed of light is c, as always, we can calculate this same distance as $c\Delta t_1$, and write the equation

$$c\Delta t_1 = L + v\Delta t_1$$

or

$$\Delta t_1 = \frac{L}{c - v}. \tag{4}$$

Now follow the light flash on its return from the mirror to the detector, supposing that this trip requires the time interval Δt_2. The light begins at the mirror, located at $L + v\Delta t_1$, and finishes at the detector, which has traveled the distance $v\Delta t_2$ in the first interval and $v\Delta t_2$ in the second. Thus the distance traveled by the light on its backward return trip is

$$(L + v\Delta t_1) - (v\Delta t_1 + v\Delta t_2) = L - v\Delta t_2.$$

↑ Light begins here at the mirror

↑ . . . and finishes here at the detector.

The light still has the speed c, so we can also calculate this distance as $c\Delta t_2$ and obtain

$$c\Delta t_2 = L - v\Delta t_2$$

or

$$\Delta t_2 = \frac{L}{c + v}. \tag{5}$$

The total time interval Δt, the time for one click of the clock, is the sum of Δt_1 and Δt_2, calculated in equations (4) and (5),

$$\Delta t = \Delta t_1 + \Delta t_2 = \frac{L}{c - v} + \frac{L}{c + v}.$$

Two algebraic maneuvers (forming a common denominator and then dividing numerator and denominator by c^2) convert this to

$$\Delta t = \frac{2L}{c}\frac{1}{1 - v^2 / c^2}. \tag{6}$$

This equation reveals the length contraction when it is compared with equation (3) combined with equation (1),

$$\Delta t = \frac{2L_0}{c} \frac{1}{\sqrt{1 - v^2 / c^2}}.$$

If the two calculations of Δt are compatible, we must have

$$L = L_o\sqrt{1 - v^2 / c^2}, \quad (7)$$

which tells us that the length L_o found by an observer traveling with the clock is contracted to $L_0 \sqrt{1 - v^2 / c^2}$ for an observer watching the clock move at the constant speed v. This is the same equation that Lorentz had concluded earlier was necessary to account for Michelson and Morley's frustrated attempts to detect Earth's motion through the ether.

Equations (3) and (7), expressing the relativity of time and length, embody Einstein's new doctrine of space and time. They cover what physicists call "kinematics"—that is, physics without the energy concept. Einstein's next step was to broaden his theory into a "dynamics," with energy included. He began to construct the dynamics in another brilliant 1905 paper, where he reached "a very interesting conclusion": "The mass of a body is a measure of its energy content." He thought about this proposition for several years. In 1906, it occurred to him that "the conservation of mass is a special case of the law of conservation of energy." A year later he concluded that, "With respect to inertia, mass m is equivalent to energy content of magnitude mc^2." This is a verbal statement of the equation that is now the world's most famous: $E = mc^2$.

Underlying this energy equation is the concept that mass, like time and length, is relative. Both time and length depend on the relative speed of the object observed, and so does mass. The relevant equation, which calculates the mass m of an object moving at the constant speed v, is (with no proof this time)

$$m = \frac{m_0}{\sqrt{1 - v^2 / c^2}}, \quad (8)$$

resembling equation (3) for time intervals. At rest ($v = 0$), the object has its lowest mass m_0; in motion, the mass of the object increases, but only slightly at ordinary speeds much less than c.

Equation (8) equips us with some clues concerning Einstein's celebrated mc^2. Multiplied by c^2 the equation calculates mc^2,

$$mc^2 = \frac{m_0 c^2}{\sqrt{1 - v^2 / c^2}}. \quad (9)$$

In the physics of the familiar world, v/c in this equation is very small, v^2 / c^2 is even smaller, and we can take advantage of the mathematical fact that

$$\frac{1}{\sqrt{1 - x}} = 1 + \frac{x}{2}$$

if x is very small. We apply this approximation to equation (9) with $x = v^2 / c^2$, and arrive at

$$mc^2 = m_0c^2 + \frac{m_0v^2}{2}.$$

Recognizing with Einstein that $E = mc^2$, we have

$$E = m_0c^2 + \frac{m_0v^2}{2}. \qquad (10)$$

This divides the total energy E into two parts. One term, $\frac{m_0v^2}{2}$, is the familiar kinetic energy carried by an object of mass m_0. The second term, m_0c^2, unlocks the secret. Einstein understood this quantity, as we do today, to be a kind of potential energy possibly obtainable from the "rest mass" m_0. Because c^2 has an immense magnitude, this mass-equivalent energy is also immense. From a mass of one kilogram (2.2 pounds), complete conversion of mass to energy would generate energy equivalent to the daily oil consumption per day in the entire United States (fifteen million barrels).

Ordinary chemical reactions convert mass to energy, but on a minuscule scale; formation of one kilogram of H_2O in the reaction

$$2\ H_2 + O_2 \rightarrow 2\ H_2O$$

converts about 1.5×10^{-10} kilogram of mass to energy. Nuclear reactions are more efficient; they convert a few tenths of a percent of the mass entering the reaction to energy. When matter meets antimatter, the conversion is complete. In his 1905 paper, Einstein suggested that radioactive materials such as radium might lose measurable amounts of mass as they decay, but for many years he could see no practical consequences of the mass-energy equivalence. (In 1934, the *Pittsburgh Gazette* headlined a story reporting an Einstein lecture with "Atom Energy Hope Is Spiked by Einstein. Efforts at Loosing Vast Force [Are] Called Fruitless.") The full lesson of $E = mc^2$ was learned in the 1940s and 1950s with the advent of nuclear physics, nuclear weapons, nuclear reactors, and nuclear anxiety.

A further accomplishment of Einstein's relativity theory was that it brought a permanent end to the ether concept by simply depriving the ether of any good reason to exist. If there were an ether, it would provide an absolute and preferred frame of reference, contrary to Einstein's first principle, and motion through the ether would be manifested by variations in the speed of light, contradicting the second principle. An ether obituary was written by Einstein and Leopold Infeld in their estimable book for the lay reader, *The Evolution of Physics*: "It [the ether] revealed neither its mechanical construction nor absolute motion. Nothing remained of all its properties except that for which it was invented, i.e., its ability to transmit electromagnetic waves."

Berlin

Our narrative returns to Einstein's life now, and follows his odyssey into the scientific world and beyond. Einstein's accomplishments during his seven years in the Bern Patent Office were unique in their creative brilliance. Inevitably, recognition came, and suddenly, in just five years, he reached the pinnacle of the scientific and academic world.

In 1909, when he was thirty, and still unacquainted with a "real physicist," Einstein left the patent office and took a position as associate professor at the University of Zürich. He was Clausius's successor: "There had been no professor of theoretical physics or mathematical physics," Abraham Pais, Einstein's biographer, notes, "since Clausius had left the university, in 1867." Pais also paints this picture of Einstein as a sometimes unenthusiastic teacher: "He appeared in class in somewhat shabby attire, wearing pants that were too short and carrying with him a slip of paper the size of a visiting card on which he had sketched his lecture notes." "He enjoyed explaining his ideas," Ernst Straus, one of Einstein's assistants, remarks, "and was exceptionally good at it because of his own way of thinking in intuitive and informal terms. What he presumably found irksome was the need to prepare and present material that was not at the moment at the center of his interest. Thus the preparation of lectures would interfere with his own thoughts."

In Zürich, Einstein was already beginning to show signs of the restlessness that was hard to understand in a man who always said he wanted to do nothing but think about theoretical physics. In five years, he would live in three countries and hold academic positions in four universities. In another five years, he would be immersed in various political matters, including pacifism, Zionism, and international government. "In his sixties," Pais explains, "[Einstein] once commented that he had sold himself body and soul to science, being in flight from the 'I' and 'we' to the 'it.' Yet he did not seek distance between himself and other people. The detachment lay within and enabled him to walk through life immersed in thought. What was so uncommon about this man is that at the same time he was neither out of touch with the world nor aloof."

His next move, in 1911, was from Zürich to Prague, where he was appointed full professor at the Karl-Ferdinand (or German) University. In Prague, he felt isolated intellectually and culturally. There were few scientific colleagues with whom he could discuss his work, and he had little in common with either the Czech or the German community. Sixteen months later he was on the move again, back to Zürich, this time to the Swiss Technical University (ETH, previously the Zürich Poly).

A little more than a year later, in the spring of 1913, Max Planck and Walther Nernst arrived in Zürich with their wives for the some sightseeing—and to entice Einstein to go to Berlin. Their offer included membership in the Prussian Academy of Sciences with a handsome salary, a chair at the University of Berlin (with no obligation to teach), and the directorship of a physics institute to be established. This was a great opportunity, but Einstein was ambivalent. He had turned his back on Germany seventeen years earlier, and he was no less distrustful of the Prussian character now than he was then. But for Einstein there was always one consideration above all others. "He had had enough of teaching. All he wanted to do was think," as Pais puts it. His decision probably came quickly, but to Planck and Nernst, symbols of the Prussian scientific establishment, he said he needed to consider the offer. He told them that when they saw him again they would know his decision: he would carry a rose, red if his answer was yes, and white if no.

The letter Planck and Nernst wrote to the Prussian Ministry of Education in support of Einstein's appointment tells a lot about where Einstein's reputation stood in 1913, alleged failures included:

> [Einstein's] interpretation of the time concept has had sweeping repercussions on the whole of physics, especially mechanics and even epistemology. . . . Although this idea of Einstein's has proved itself so fundamental for the development of physical principles, its application still lies for the moment on the frontier of the measurable. . . . Far more important for practical physics is his penetration of other questions on which, for the moment, interest is focused. Thus he was the first man to show the importance of the quantum theory for the energy of atomic and molecular movements, and from this he produced the formula for specific heats of solids. . . . He also linked the quantum hypothesis with the photoelectric and photochemical effects. . . . All in all, one can say that among the great problems, so abundant in modern physics, there is hardly one to which Einstein has not brought some outstanding contribution. That he may sometimes have missed the target in his speculations, as, for example, in his theory of light quanta [now called "photons" and indispensable as a member of the family of elementary particles], cannot be held against him. For in the most exact of natural sciences every innovation entails risk. At the moment he is working on a new theory of gravitation, with what success only the future will tell.

The rose was red, and Einstein moved to Berlin, delighted that he would be free of lecturing, but with misgivings concerning his end of the bargain. "The gentlemen in Berlin are gambling on me as if I were a prize hen," he told a friend before leaving Zürich. "As for myself I don't even know whether I'm going to lay another egg."

An imposing measure of the dominion of science is that it brought together on amicable terms two men as totally dissimilar as Einstein and Planck. Einstein avoided all formality and ceremony, detested the Prussian traditions of discipline, militarism, and nationalism, and for most of his life was a pacifist. Yet, this casual, untidy, anti-Prussian pacifist had a deep respect for Max Planck, the formal, impeccably dressed servant of the Prussian state. What Einstein saw and appreciated in Planck was the strength of his integrity and the depth of his commitment to science. Einstein always had admiration—and sometimes friendship—for anyone who could match the intensity of his own devotion to physics.

The move to Berlin was the final blow to an already slipping marriage. Soon after her arrival in Berlin, Mileva returned to Zürich with her two sons and remained there. Her subsequent life was not happy. She could not accept the separation, or the divorce that came in 1919. Her means were modest, even after Einstein transmitted to her his Nobel Prize money, received in 1921. The younger son, Eduard, was mentally unstable for much of his life and died a schizophrenic in a Zürich psychiatric hospital.

Einstein was now a bachelor, and under the "loving care" of a "cousine," who he claimed "drew me to Berlin." This was Elsa Einstein Löwenthal, both a first and second cousin to Einstein (their mothers were sisters, and their grandfathers brothers), and a friend since childhood. She had married young, was now divorced, and was living with her two daughters, Margot and Ilse, in Berlin when Einstein arrived. In 1917, Einstein suffered a serious breakdown of his health, and Elsa was on hand to supervise his care and feeding. The patient recovered and two years later married the nurse. Although Einstein rarely expressed his appreciation, he must have realized that Elsa was indispensable. Like some of the other wives mentioned in these profiles, she became an efficient manager of her husband's nonscientific affairs, and allowed him to get on with his main business, thinking about theoretical physics.

Pais gives us this sketch of Elsa: "gentle, warm, motherly, and prototypically bourgeoisie, [she] loved to take care of her Albertle. She gloried in his fame." Charlie Chaplin, who entertained the Einsteins in California, described Elsa this way: "She was a square-framed woman with abundant vitality; she frankly enjoyed being the wife of a great man and made no attempt to hide the fact; her enthusiasm was endearing."

Hardly any chapter from the Einstein story is conventional or predictable, but the most bizarre episode by far was the public reaction to Einstein's elaboration of his 1905 "special" theory of relativity to a "general" theory of relativity in 1915. It was not the theory itself, which few people understood, but the announcement that one of the theory's predictions had been confirmed, that brought the avalanche of attention.

Einstein had used his general theory to show that a gravitational field has a bending effect on light rays, and he had calculated the expected effect of the Sun's gravity on light originating from stars and passing near the Sun before reaching telescopes on Earth. The effect was small, but measurable if the observations could be made during a solar eclipse. After failures, delays, and much political interference—the First World War was in progress at the time—two British expeditions, one under Arthur Eddington to the island of Principe off the coast of West Africa, and another led by Andrew Crommelin to Sobral in northern Brazil, observed the eclipse of 1919, and succeeded in confirming Einstein's predictions.

Overnight, Einstein became the most famous scientist in the world. He was besieged by distinguished and not-so-distinguished colleagues, learned societies, reporters, and plain people. "Since the flood of newspaper articles," he wrote to a friend, "I have been so swamped with questions, invitations, challenges, that I dream I am burning in Hell and the postman is the Devil eternally roaring at me, throwing new bundles of letters at my head because I have not answered the old ones." It is all but impossible to understand what prompted this reaction to what was after all an esoteric and theoretical effort. The mathematician and philosopher Alfred Whitehead expressed public sentiment on the more rational side: "a great adventure in thought [has] at length come to safe shore."

Spacetime

The new doctrine of space and time brought by Einstein's 1905 special theory demanded relativity of time as well as relativity of length and space. If an observer in an inertial frame describes some event with the coordinates x, y, z, and the time t, another observer in a different inertial frame uses different coordinates, call them x', y', z', and a different time t', to express the physics of the event. The time variable is not separate from the spatial variables, as it is in Newtonian physics. It enters the Einstein picture seemingly on the same footing as the spatial variables. This point of view was taken by one of Einstein's former mathematics professors, Hermann Minkowski, and developed into a mathematical structure that would eventually be indispensable to Einstein as he ventured beyond special relativity to general relativity. Minkowski put forward his program at the beginning of an address delivered in 1908: "The views of space and time which I wish to lay before you have sprung from the soil of experimental physics, and therein lies their strength. They are radical. Henceforth space by itself, and time by itself, are doomed to fade away into mere shadows, and only a kind of union of the two will preserve an independent reality."

Physics is about events in space and time. We locate each event in space in a reference frame equipped with a coordinate system. For example, two events are located in two spatial dimensions with the coordinate pairs x_1, y_1 and x_2, y_2 (fig. 14.5), and the spatial interval l between them is calculated by constructing a right triangle (fig. 14.6) and applying the Pythagorean theorem:

$$l^2 = \Delta x^2 + \Delta y^2 \text{ or } l = \sqrt{\Delta x^2 + \Delta y^2}.$$

In three spatial dimensions, these equations add a term Δz^2 for the third dimension:

$$l^2 = \Delta x^2 + \Delta y^2 + \Delta z^2 \text{ or } l = \sqrt{\Delta x^2 + \Delta y^2 + \Delta z^2}. \tag{11}$$

In the spirit of field theory, we treat space as a continuum and make the calculations for two neighboring events separated by the very small interval dl. That calculation follows the same recipe as equation (11), with l replaced by the much smaller dl, and Δx, Δy, Δz replaced by the smaller dx, dy, dz,

$$dl^2 = dx^2 + dy^2 + dz^2 \text{ or } dl = \sqrt{dx^2 + dy^2 + dz^2}. \tag{12}$$

Minkowski asks us to replace this three-dimensional picture with a four-dimensional one, each physical event being located by a "world point" with *four* coordinates, the three spatial coordinates x, y, z and the time coordinate t. How do we calculate an interval in this four-dimensional picture of space and time—or better, "spacetime"—comparable to dl in three dimensions? The rules of mathematical physics do not allow a simple addition of spatial and time terms, as in $dt^2 + dx^2 + dy^2 + dz^2$, because dx, dy, dz measure one thing (length) and dt another (time). If two terms are added in a physical equation, they must measure the same thing and have the same units.

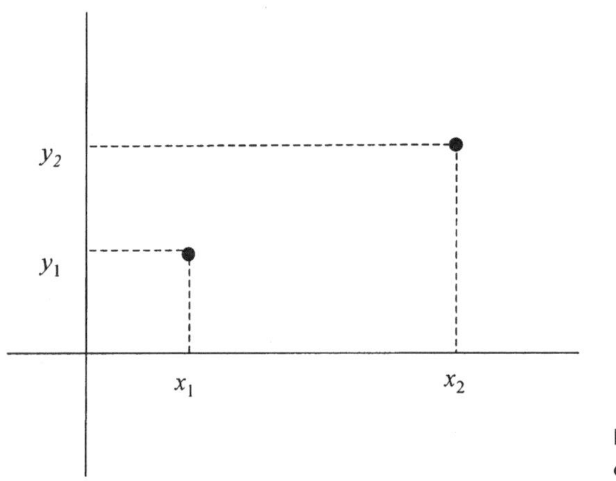

Figure 14.5. Location of two events in two spatial dimensions at the two points x_1,y_1 and x_2,y_2.

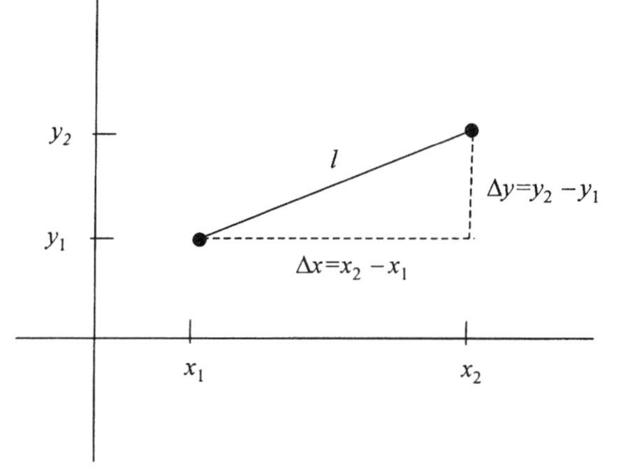

Figure 14.6. Calculation of the spatial interval l between the two events of fig. 14.5.

The simplest way to approach connections and intervals in spacetime is to imagine two events joined by that most reliable of measuring devices, a light ray. Suppose a flash of light is generated at one world point x_1, y_1, z_1, t_1, and then later detected at the world point x_2, y_2, z_2, t_2. The distance traveled by the light flash is $\sqrt{\Delta x^2 + \Delta y^2 + \Delta z^2}$ with $\Delta x = x_2 - x_1$, $\Delta y = y_2 - y_1$, and $\Delta z = z_2 - z_1$, as before. The same distance is calculated by multiplying the speed of light c by Δt, the time elapsed between the two events, that is,

$$\sqrt{\Delta x^2 + \Delta y^2 + \Delta z^2} = c\Delta t.$$

This equation is better suited to later discussions if it is rearranged slightly. Square both sides of the equation and move all terms to one side,

$$c^2\Delta t^2 - \Delta x^2 - \Delta y^2 - \Delta z^2 = 0.$$

For neighboring events in the spacetime continuum, Δx, Δy, Δz, Δt become dx, dy, dz, dt, and the equation is

$$c^2dt^2 - dx^2 - dy^2 - dz^2 = 0. \tag{13}$$

The quantity calculated is the square of a spacetime interval, and in relativity theory it is represented with ds^2,

$$ds^2 = c^2dt^2 - dx^2 - dy^2 - dz^2. \tag{14}$$

Physicists call ds a "world line element." It is a fundamental entity in relativity theory.

However it is calculated—equation (14) is only one of many possibilities—the world line element ds shows how, in the four-dimensional world of spacetime, physical events are connected. For the light flash we have been discussing, $ds^2 = 0$, according to equations (13) and (14), and the events joined by ds are said to be "lightlike." The square ds^2 can also be positive or negative: if positive, the events connected are "timelike," and if negative the events are "spacelike."

Minkowski emphasized that Einstein's world of spacetime events has a fundamental symmetry that makes the line element ds invariant in all inertial frames. If you measure ds in one frame, where the coordinates are x, y, z, t, and then ds' in another frame whose coordinates are x', y', z', t', the two measurements must be equal, $ds = ds'$, no matter what kinds of events are connected by the line element—lightlike, timelike, or spacelike. From the simple condition $ds = ds'$, Minkowski extracted the four equations that express the relativity of the two sets of coordinates x, y, z, t and x', y', z', t'. These equations, which Lorentz had previously derived in the different context of his own theory, are now called the "Lorentz transformation."

Einstein was not at first impressed by Minkowski's mathematical recasting of special relativity theory. He found it "banal" and called it "superfluous erudition." But later, as he explored the mathematically more complicated world of general relativity, he found Minkowski's concepts indispensable. He had to admit that, without Minkowski, relativity theory "might have remained stuck in its diapers."

Physics as Geometry

Einstein's 1905 theory "in diapers" had made a powerful statement about the physical world, but Einstein knew immediately that there was room for improvement. For one thing, the theory seemed to be restricted to inertial systems. For another, it was compatible with Maxwell's electromagnetic theory, but not with another great theory inherited by Einstein, Newton's gravitation theory. To realize its potential, the theory had to recognize noninertial systems, those accelerating relative to each other, and at the same time extend its scope to gravitation.

The first step Einstein took in this direction killed both of these birds with one stone. As he explained later, "I was sitting in a chair in the patent office at Bern when all of a sudden a thought occurred to me: 'If a person falls freely he will not feel his own weight.' I was startled. This simple thought made a deep impression on me. It impelled me toward a theory of gravitation." This was Einstein's first mental image of what he would later call "the equivalence principle." The central idea is that gravitation is relative. The person in free fall, locked inside a falling elevator, let's say, finds no evidence of gravity: *everything* in the elevator seems to be at rest and without weight. An outside observer, on the other hand, sees the elevator accelerating in the grip of a gravitational field.

The elevator inhabitants have the opposite experience if the elevator is removed from the gravitational field and accelerated at a constant rate upward with an attached rope (fig. 14.7). Now the outside observer sees no gravitational field, while the inside observer and all his or her belongings are held to the floor of the elevator exactly as if they were in a gravitational field. The "equivalence"

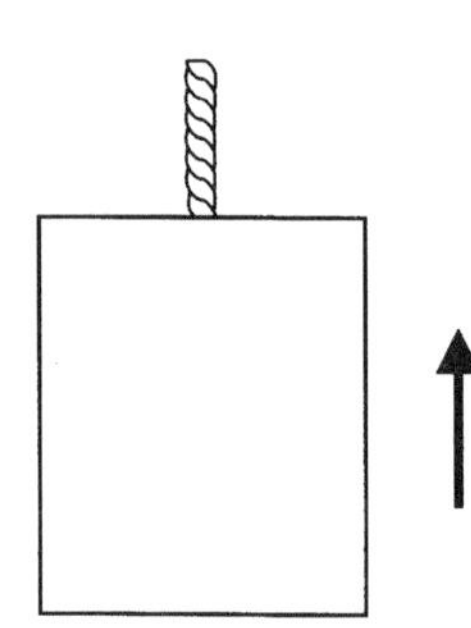

Figure 14.7. An elevator on a rope accelerated upward at a constant rate, as seen by an outside observer.

here is between an accelerating system in field-free space and an inertial system in a gravitational field. Reasoning this way, Einstein began to see how both gravitation and acceleration could be introduced into relativity theory.

The elevator-on-a-rope image (developed later by Einstein and Infeld) shows how the equivalence principle justifies an initial version of Einstein's prediction of light rays bent by gravity, which ten years later would bring the world clamoring to his door. Picture the elevator on a rope with a light ray traveling across the elevator from left to right. The outside observer sees elevator and light ray as shown in figure 14.8. Because the light ray takes a finite time to travel from wall to wall, and the elevator is accelerated upward during that time, the outside observer sees the light ray traveling the slightly curved path shown. The inside observer also sees the light ray bent, but is not aware of the acceleration and attributes the effect to the equivalent gravitational field that holds that observer to the floor of the elevator. The inside observer believes that the light ray *should* respond to a gravitational field because it has energy, and therefore, by the $E = mc^2$ prescription, also has mass. Like any other object with mass, the light ray responds to a gravitational field.

With the equivalence principle as his guide, Einstein began in 1907 to generalize his relativity theory so that it encompassed gravity and acceleration. As he proceeded, he became increasingly convinced that he was dealing with a problem in a strange kind of geometry. Even in special relativity there are hints that acceleration and the equivalent gravitation spell violations of some of Euclid's theorems, such as the rule that the ratio of a circle's circumference to its diameter is equal to the number π. Einstein could, for example, argue from special relativity that the measured circumference-to-diameter ratio of a rapidly rotating disk had to be slightly larger than π.

By 1912, when he returned from Prague to Zürich, Einstein was hoping to find salvation in the mathematics of non-Euclidean geometry. He got some crucial help from his invaluable friend Marcel Grossmann, now professor of mathematics at the Zürich ETH, who advised him to read the work of Bernhard Riemann on differential geometry. In the 1850s, Riemann had made a general study of non-Euclidean spaces by defining the "curvature" of lines drawn in those spaces.

To calculate curvatures, Riemann used the mathematical tool that Minkowski would borrow sixty years later, the squared line element ds^2. As mathematicians will, Riemann imagined a completely general version of the line element equation involving any number of dimensions and including all possible quadratic terms. Consider, for example, two-dimensional Euclidean geometry with the line element

$$ds^2 = dx^2 + dy^2. \tag{15}$$

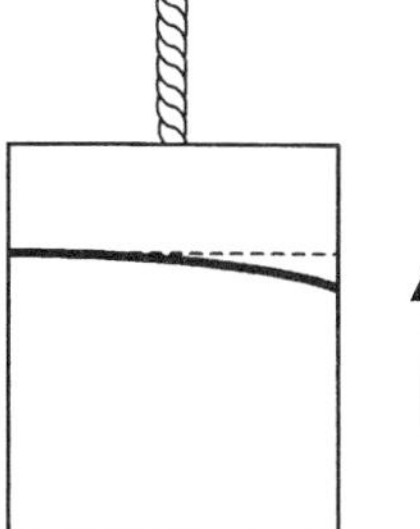

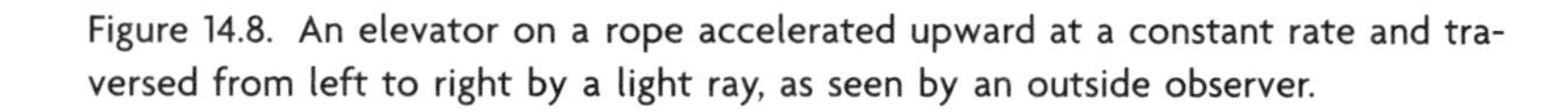

Figure 14.8. An elevator on a rope accelerated upward at a constant rate and traversed from left to right by a light ray, as seen by an outside observer.

In Riemann's scheme, we expand this to include terms in the other two mathematically possible quadratic factors, $dxdy$ and $dydx$,

$$ds^2 = (1)dx^2 + (0)dxdy + (0)dydx + (1)dy^2$$

The added terms are multiplied by zero coefficients because they do not actually appear in the ds^2 equation; the other two terms have coefficients of one, as in equation (15). All we need to know about two-dimensional Euclidean geometry in Riemann's analysis is the four coefficients in parentheses in the last equation. We collect them in a 2×2 table represented by g,

$$g = \begin{pmatrix} 1 & 0 \\ 0 & 1 \end{pmatrix},$$

called a "metric tensor."

In three-dimensional Euclidean space the line element is

$$ds^2 = dx^2 + dy^2 + dz^2,$$

and by the same conventions the metric tensor is the 3×3 table

$$g = \begin{pmatrix} 1 & 0 & 0 \\ 0 & 1 & 0 \\ 0 & 0 & 1 \end{pmatrix}.$$

In four-dimensional Minkowski spacetime, with the line element of equation (14), the metric tensor is represented by the 4×4 table

$$g = \begin{pmatrix} c^2 & 0 & 0 & 0 \\ 0 & -1 & 0 & 0 \\ 0 & 0 & -1 & 0 \\ 0 & 0 & 0 & -1 \end{pmatrix}.$$

The mathematical raw material for Riemann's curvature calculation is contained in the metric tensor g for the geometry in question: given a geometry defined by its metric tensor, Riemann shows how to calculate the curvature. The three metric tensors quoted happen to yield zero curvature: they specify "flat" geometries. But many other geometries have curvature and are thus non-Euclidean, as their metric tensors reveal in Riemann's analysis.

After several years of mistakes and false starts (mercifully, not part of our story), Einstein finally realized in 1915 that with Riemann's mathematical tools he could derive a field equation that intimately links gravity and geometry. His equation, reduced to its simplest form, is

$$\mathbf{G} = \frac{8\pi G}{c^4}\mathbf{T}, \tag{16}$$

in which G is the Newtonian gravitational constant, and **G** and **T** are "tensors," meaning that they are specially defined so that the equation has exactly the same mathematical form in all frames of reference, inertial or noninertial. (Note that G and $\mathbf{G}$ have different meanings.)

The tensor $\mathbf{G}$ is Einstein's adaptation of Riemann's curvature calculation; it depends entirely on the relevant spacetime metric tensor g and its derivatives. The tensor $\mathbf{T}$ supplies all the necessary information on the gravitation source by specifying the energy and matter distribution. Thus the field equation (16) says geometry on the left side and gravity on the right. Propose a gravitation source ($\mathbf{T}$) and the equation gives the Einstein tensor $\mathbf{G}$, and ultimately the geometry in terms of the spacetime metric tensor g.

Gravity determines geometry in Einstein's field equations, and not surprisingly, geometry determines motion. Einstein continued with his physical argument by deriving a generalized equation of motion whose principal mathematical ingredient is the indispensable spacetime metric tensor g. Thus the sequence of the entire calculation is

$$\text{Gravitation source} \rightarrow \text{Curvature} \rightarrow \text{Metric tensor } g \rightarrow \text{Equation of motion.}$$

The gravitation source is expressed by $\mathbf{T}$, the curvature by $\mathbf{G}$, the metric tensor is extracted from $\mathbf{G}$, and the equation of motion is defined by g. This, in a nutshell, is one way to tell the story of Einstein's general theory of relativity. Notice that no *forces* are mentioned: *geometry* is the intermediary between gravitation and motion. The title of the story is "Physics As Geometry."

Geometry as revealed by Einstein's field equation (16) always means spatial curvature or non-Euclidean geometry if gravitation is present. But, except in extreme cases (for example, black holes), the extent of the curvature is extremely small. Richard Feynmann uses Einstein's theory to estimate that the Euclidean formula $4\pi r^2$ for calculating the surface area of a sphere from the radius r is in error by 1.3 parts per million in the intense gravitational field at the surface of the Sun.

Einstein offered two applications of his general theory as tests of its validity. One was the calculation of the bending of light rays near the Sun, later to be confirmed in the famous Eddington and Crommelin expeditions. The other was a calculation of the orbit of Mercury, showing that the orbit is not fixed, as demanded by Newtonian theory, but slowly changes its orientation at the rate of 42.9 seconds of arc per century. This effect had been observed and measured as 43.5 seconds. When he saw this success of his theory, Einstein was euphoric. "For some days I was beyond myself with excitement," he wrote to a friend. As Pais puts it, "From that time he knew: Nature had spoken; he had to be right."

Destiny, or God Is Subtle

It is an inescapable and mostly unfathomable aspect of scientific creativity that it simply does not last. Einstein once wrote to a friend, "Anything really new is invented only in one's youth. Later one becomes more experienced, more famous—and more stupid." Most of the scientists whose stories are told in this book did their important work when they were young, in their twenties or thirties. Some, notably Planck and Schrödinger, were approaching middle age when they did their best work. But with the exception of Gibbs, Feynman, and Chandrasekhar, none did outstanding work toward the end of his or her life.

Although unique in most other respects, Einstein's creative genius was only a little less ephemeral. According to Pais, Einstein's creativity began to decline after 1924, when he was forty-five. Pais sketches Einstein's career after 1913, the year he arrived in Berlin: "With the formulation of the field equations of gravitation in November 1915, classical physics (that is, nonquantum physics) reached its perfection and Einstein's scientific career its high point. . . . Despite much illness, his years from 1916 to 1920 were productive and fruitful, both in relativity and quantum theory. A gentle decline begins after 1920. There is a resurgence toward the end of 1924. . . . After that, the creative period ceases abruptly, though scientific efforts continued unremittingly for another thirty years."

After about 1920, Einstein became more a part of the world of politics, and no doubt that drew on his time and energy. He traveled a lot and made many public appearances. He despised the publicity, but at the same time it cannot be denied that he enjoyed performing before an audience. His older son, Hans Albert, tells us that he was "a great ham." The social life in Berlin was an attraction; the Einsteins counted among their acquaintances well-known intellectuals, statesmen, and educators. And Einstein had at least several extramarital romantic attachments during the 1920s and 1930s.

So Einstein the extrovert weakened the creative spirit that belonged to Einstein the introvert. But that only partly explains the decline. Two other factors may have been more important. In 1925 and 1926, the methods of quantum mechanics made their appearance and dominated developments in theoretical physics for many years. Einstein quickly accepted the utility of quantum mechanics, but to the end quarreled with its interpretation. Most physicists became reconciled to the peculiar brand of indeterminism that quantum mechanics seems to demand, but Einstein would not have it. As a second generation of quantum physicists introduced and exploited the revolutionary new methods, Einstein became the conserver. He hoped to see beyond what he felt was the incompleteness of quantum theory, without breaking with some of the great traditions of physics that were more important to him than temporary successes. He never found what he was looking for, although he searched for many years. "The more one chases after quanta, the better they hide themselves," he wrote in a letter. In his stubbornness, he became isolated from most of his younger colleagues.

Einstein's tenacity—certainly one of his strongest personality traits—brought another grand failure. In the late 1920s, he began work on a "unified field theory," an attempt to unite the theories of gravitation, electromagnetism, and perhaps, quanta. He was fascinated—one might say obsessed—by this effort for the rest of his life. The drama of Einstein struggling furiously with this theoretical problem should answer any claim that great scientists do their work as thinking machines without passionate commitment. In 1939, he wrote to Queen Elizabeth of Belgium, with whom he corresponded for many years, "I have hit upon a hopeful trail, which I follow painfully but steadfastly in company with a few youthful fellow workers. Whether it will lead to truth or fallacy—this I may be unable to establish with any certainty in the brief time left to me. But I am grateful to destiny for having made my life into an exciting experience."

And a few years later, in a letter to a friend: "I am an old man known as a crank who doesn't wear socks. But I am working at a more fantastic rate than ever, and I still hope to solve my problem of the unified physical field. . . . It is no more than a hope, as every variant entails tremendous mathematical difficulties. . . . I am in an agony of mathematical torment from which I am unable to escape."

He must have been tired, and at times discouraged. After one approach led to still another dead end, he told an assistant he would publish, "to save another fool from wasting six months on the same idea." Perhaps the most famous of Einstein's many quotable sayings is "God is subtle, but not malicious," by which he meant "Nature conceals her mystery by her essential grandeur, but not by her cunning." After many futile years devoted to the search for the unifying field theory, he said to Hermann Weyl, "Who knows, perhaps he is a little malicious."

Yet the miracle of Einstein's creative spirit was that if he felt despair, it was never lasting. One of Einstein's most recent biographers, Albrecht Fölsing, tells us that "he was capable of pursuing a theoretical concept, with great enthusiasm for months and even years at a stretch; but when grievous flaws emerged—which invariably happened in the end—he would drop it instantly at the moment of truth, without sentimentality or disappointment over the time and effort wasted. The following morning, or a few days later at the most, he would have taken up a new idea and would pursue that with the same enthusiasm." "After all," Einstein wrote to a friend, "to despair makes even less sense than to strive for an unattainable goal."

Letters

Einstein received an enormous volume of mail, from all kinds of people on all kinds of subjects. When he was not overwhelmed by them, he enjoyed these letters and answered them. Excerpts from his responses give us fragments of the personal autobiography he never wrote.

To members of the "Sixth Form Society" of an English grammar school, who had elected him as their rector, he wrote: "As an old schoolmaster I received with great joy and pride the nomination to the Office of Rectorship of your society. Despite my being an old gypsy there is a tendency to respectability in old age—so with me. I have to tell you, though, that I am a little (but not too much) bewildered that this nomination was made independent of my consent."

Einstein was asked many times about his religion. He was a "deeply religious nonbeliever," he wrote to a friend, and he explained to a sixth grader, "Every one who is seriously involved in the pursuit of science becomes convinced that a spirit is manifest in the laws of the Universe—a spirit vastly superior to that of man, and one in the face of which we with our modest powers must feel humble. In this way the pursuit of science leads to a religious feeling of a special sort, which is indeed quite different from the religiosity of someone more naïve." "I do not believe in a personal God and I have never denied this but have expressed it clearly," he wrote to an admirer. "If something is in me which can be called religious, then it is the unbounded admiration for the structure of the world so far as science can reveal it." His religion did not include morality: "Morality is of the highest importance—but for us, not God." In response to an evangelical letter from a Baptist minister he wrote, "I do not believe in the immortality of the individual, and I consider ethics to be an exclusively human concern with no superhuman authority behind it."

Einstein detested militarism and nationalism. "That a man can take pleasure in marching to the strains of a band is enough to make me despise him," he wrote. He believed that Gandhi's strategy of civil disobedience offered hope: "I believe that serious progress can be achieved only when men become organized on an international scale and refuse, as a body, to enter military or war service."

His commitment to pacifism was at first unmitigated. In an interview, he said "I am not only a pacifist but a militant pacifist. I am willing to fight for peace. . . . Is it not better for a man to die for a cause in which he believes, such as peace, than to suffer for a cause in which he does not believe, such as war?"

But the horrors of Nazi anti-Semitism converted him from an "absolute" to a "dedicated" pacifist: "This means that I am opposed to the use of force under any circumstances except when confronted by an enemy who pursues the destruction of life as an *end in itself.*"

Many of his correspondents wanted to know what it was like to live a life in physics. He explained that, for him, there was a detachment: "My scientific work is motivated by an irresistible longing to understand the secrets of nature and by no other feelings. My love for justice and the striving to contribute towards the improvement of human conditions are quite independent from my scientific interests."

And in the detachment he found another part of the motivation: "Measured objectively, what a man can wrest from Truth by passionate striving is utterly infinitesimal. But the striving frees us from the bonds of the self and makes us comrades of those who are the best and the greatest."

Bird of Passage

It was Einstein's fate to roam and never settle in a place he could comfortably call home. Switzerland was his favorite place, but he did not stay there long after leaving the patent office. Berlin kept him for almost twenty years, and for a time left him in peace. But in the 1920s the Nazis became influential and brought with them the three scourges of nationalism, militarism, and anti-Semitism. We have already seen the devastating effects of Nazi policies in Nernst's time, and the Nazi destruction of the German scientific establishment will continue to be a morbid theme in later chapters. Anti-Semitism had been evident throughout the 1920s, but for Einstein at least not a threat. That was no longer the case in the early 1930s when the Nazis came to power.

After short stays in Belgium, England, and California, Einstein relocated to Princeton, where he joined the newly founded Institute for Advanced Study. Compared to that of Berlin, the intellectual climate in Princeton was less than exciting. "Princeton is a wonderful little spot," he wrote to Queen Elizabeth, "a quaint ceremonious village of puny demigods on stilts." But it served his main purpose: "By ignoring certain special conventions, I have been able to create for myself an atmosphere conducive to study and free from distraction."

In Princeton, Einstein ended his flight and returned to his routine. As always, he was in touch with world affairs. In the 1940s, the Manhattan Project, aimed at developing a nuclear bomb, was organized, and Einstein's influence helped in the initial stages. The "pet project," unified field theory, was his major concern in Princeton, however. More than ever, he became the "artist in science," searching endlessly for the unified theory with the mathematical simplicity and beauty that would satisfy his intuition and aesthetic sense.

Abraham Pais, whose biography of Einstein is the best of the many written, leaves us with this glimpse of Einstein about three months before he died in 1955. He had been ill and unable to work in his office at the institute. Pais visited him at home and went upstairs and knocked at the door of [his] study. There was a gentle "Come." As I entered, he was seated in his armchair, a blanket over his knees, a pad on the blanket. He was working. He put his pad aside at once and greeted me. We spent a pleasant half hour or so; I do not recall what was discussed. Then I told him I should not stay any longer. We shook hands, and I said goodbye. I walked to the door of the study, not more than four or five steps away. I turned around as I opened the door. I saw him in his chair, his pad on his lap, a pencil in his hand, oblivious to his surroundings. He was back at work.

vi QUANTUM MECHANICS

Historical Synopsis

Our story has so far been a tale of five great scientific revolutions. The first, initiated by Galileo and largely completed by Newton, brought mechanics and the concept of universal gravitation. The second, pioneered by Carnot and carried on by Mayer, Joule, Helmholtz, Thomson, Clausius, Gibbs, and Nernst, gave us thermodynamics. In the third, Faraday and Maxwell introduced the field concept and constructed a theory of electromagnetism. The work of Clausius, Maxwell, Boltzmann, and Gibbs in the fourth revolution, called statistical mechanics, opened the door to molecular physics. And the fifth revolution, Einstein's relativity theory, rebuilt our view of space, time, and gravitation.

This part of the book starts one more account of scientific revolution. The story begins conveniently in 1900 and twenty-five years later arrives at a new science, now called "quantum theory," "quantum mechanics," or "quantum physics," which probes further the microworld of molecules, atoms, and subatomic particles. A usage note: to distinguish pre- and postquantum physics I will now use the adjectives "classical" for the former and "quantum" for the latter, as in "classical mechanics" and "quantum mechanics," and "classical physics" and "quantum physics." ("Quantal" would be a better partner for "classical," but that term is rarely used.)

In its early stages, the quantum revolution had three great leaders: Max Planck, whose disciplined insights gave the first glimpse of what was coming; Albert Einstein, who became as deeply committed to this intellectual adventure as to relativity theory; and Niels Bohr, who brought the revolution to its greatest crisis. Each of the three pioneers first faced the task of reconciling classical physics with the strange conclusions forced by the new physics, and each in his own way failed. Planck and Bohr tried to dispel the mysteries by building the new physics partially into the framework of the old. Einstein quickly accepted the most drastic features of the new physics, and then stubbornly probed for a deeper level of physical reality.

None of these efforts succeeded entirely, and it took another generation of quantum theorists to complete the revolution. In this second generation were Werner Heisenberg, Erwin Schrödinger, Wolfgang Pauli, and Louis de Broglie. Their legacy is quantum mechanics, a brand of physics that is perhaps intellectually more challenging than any other. In the microworld it explores, the laws of quantum physics are, to us in our macroworld, strange and mysterious. Schrödinger gives us this warning:

> As our mental eye penetrates into smaller and smaller distances and shorter and shorter times, we find nature behaving so entirely differently from what we observe in visible and palpable bodies of our surroundings that *no* model shaped after our large-scale experiences can ever be "true." A completely satisfactory model *of this type* is not only practically inaccessible, but not even thinkable. Or, to be more precise, we can, of course, think it, but however we think it, it is wrong; not perhaps quite as meaningless as a "triangular circle," but much more so than a "winged lion."

On that cautionary note, we begin our story of quantum theory and quantum mechanics.

15

Reluctant Revolutionary

Max Planck

Physics Is Finished

The first of the revolutionary quantum theorists we meet, Max Planck, would not have succeeded in revolutions of the other kind. Planck was born into the conservative society of nineteenth-century Prussia, and in his formal, disciplined way, he remained committed to the Prussian traditions, even in his scientific work, it seemed, throughout his life. Planck's life was devoted to an intense, sometimes desperate search—a "hunger of the soul," in Einstein's words—for what was absolute and fundamental. "It is of paramount importance," Planck wrote in his scientific autobiography, "that the outside world is something independent from man, something absolute, and the quest for the laws which apply to this absolute appeared to me as the most sublime scientific pursuit in life." His faith in physics, ideally rooted in the principles of classical physics, as a manifestation of the absolute principles had the intensity of a religious belief. His intellectual strength and integrity, Einstein tells us, grew from an "emotional condition . . . more like that of a deeply religious man or a man in love; the daily effort is not dictated by either a purpose or a program, but by an immediate need."

In one of those ironies that seems part of a trite novel, Planck was advised in 1875 when he was seventeen not to make a career in physics, particularly theoretical physics, because the significant work was finished except for the details. Planck took his own advice, however, and eventually made his way to Berlin, where he studied under two of Germany's most famous physicists, Hermann Helmholtz and Gustav Kirchhoff. The great scientists were less than inspiring in the lecture hall—Helmholtz's lectures were poorly prepared, and Kirchhoff's were "dry and monotonous"—but in their writings, and in the principles of their subject, thermodynamics, Planck found what he sought, "something absolute."

By 1890, Planck had fully developed his ideas on thermodynamics and suffered some setbacks. Possibly because he chose to emphasize the then new concept of entropy, Planck found it nearly impossible at first to make a favorable impression, or any impression at all, on Germany's great thermodynamicists. Kirchhoff only found fault with Planck's papers, and Helmholtz did not bother to read them. Even Rudolf Clausius, who was responsible for the entropy concept that Planck used and refined, had no time for Planck or his papers. Another disappointment came when Planck discovered that much of his work on entropy theory had been anticipated by Willard Gibbs in America. Finally, in 1895, with help from his father, Planck received an academic appointment "as a message of deliverance" from the University of Kiel.

A few years later, Planck was still seeking broader professional recognition. He found it by entering a competition sponsored by the University of Göttingen, and, on a point related to electrical theory, innocently siding with Helmholtz against an antagonistic viewpoint held by Wilhelm Weber of Göttingen. Predictably, Planck's entry was refused first prize in the Göttingen competition, but with his work belatedly recognized by Helmholtz, Planck was in luck. In 1889, with Helmholtz supporting his candidacy, Planck was appointed as Kirchhoff's successor at the University of Berlin.

Blackbody Radiation

Max Planck's story as an unenthusiastic revolutionary began in about 1895 in Berlin, with Planck established as a theoretical physicist and concerned with the theory of the light and heat radiation emitted by special high-temperature ovens known in physical parlance as "blackbodies." Formally, a blackbody is an object that emits its own radiation when heated, but does not reflect incident radiation. These simplifying features can be built into an oven enclosure by completely surrounding it with thick walls except for a small hole through which radiation escapes and is observed.

The color of radiation emitted by blackbody (and other) ovens depends in a familiar way on how hot the oven is: at 550°C it appears dark red, at 750°C bright red, at 900°C orange, at 1000°C yellow, and at 1200°C and beyond, white. This radiation has a remarkably universal character: in a blackbody oven whose walls are equilibrated with the radiation they contain, the spectrum of the color depends exclusively on the oven's temperature. No matter what is in the oven, a uniform color is emitted that changes only if the oven's temperature is changed. A theory that partly accounted for these fundamental observations had been derived by Kirchhoff in 1859.

To Planck there were unmistakable signs here of "something absolute," that sublime presence he had pursued in his thermodynamic studies. The blackbody oven embodied an idealized, yet experimentally accessible, instance of radiation interacting with matter. Blackbody theoretical work had been advancing rapidly because the experimental methods for analyzing blackbody spectra—that is, the rainbow of emitted colors—had been improving rapidly. The theory visualized a balanced process of energy conversions between the thermal energy of the blackbody oven's walls and radiation energy contained in the oven's interior. By the time Planck started his research, the blackbody radiation problem had developed into a theoretical tree with some obviously ripening plums.

Planck first did what theoreticians usually do when they are handed accurate experimental data: he derived an empirical equation to fit the data. His guide in this effort was a thermodynamic connection between the entropy and the energy of the blackbody radiation field. He defined two limiting and extreme versions of the energy-entropy relation, and then guessed that the general connection was a certain linear combination of the two extremes. In this remarkably simple way, Planck arrived at a radiation formula that did everything he wanted. The formula so accurately reproduced the blackbody data gathered by his friends Heinrich Rubens and Ferdinand Kurlbaum that it was more accurate than the spectral data themselves: "The finer the methods of measurement used," Planck tells us, "the more accurate the formula was found to be."

The Unfortunate *h*

Max Born, one of the generation of theoretical physicists that followed Planck and helped build the modern edifice of quantum theory on Planck's foundations, looked on the deceptively simple maneuvers that led Planck to his radiation formula as "one of the most fateful and significant interpolations ever made in the history of physics; it reveals an almost uncanny physical intuition." Not only was the formula a simple and accurate empirical one, useful for checking and correlating spectral data; it was, in Planck's mind, something more than that. It was not just *a* radiation formula, it was *the* radiation formula, the final authoritative *law* governing blackbody radiation. And as such it could be used as the basis for a theory—even, as it turned out, a revolutionary one. Without hesitation, Planck set out in pursuit of that theory: "On the very day when I formulated the [radiation law]," he writes, "I began to devote myself to the task of investing it with true physical meaning."

As he approached this problem, Planck was once again inspired by "the muse entropy," as the science historian Martin Klein puts it. "If there is a single concept that unifies the long and fruitful scientific career of Max Planck," Klein continues, "it is the concept of entropy." Planck had devoted years to studies of entropy and the second law of thermodynamics, and a fundamental entropy-energy relationship had been crucial in the derivation of his radiation law. His more ambitious aim now was to find a *theoretical* entropy-energy connection applicable to the blackbody problem.

As mentioned in chapter 13, Ludwig Boltzmann interpreted the second law of thermodynamics as a "probability law." If the relative probability or disorder for the state of a system was W, he concluded, then the entropy S of the system in that state was proportional to the logarithm of W,

$$S \propto \ln W.$$

In a deft mathematical stroke, Planck applied this relationship to the blackbody problem by writing

$$S = k \ln W \tag{1}$$

for the total entropy of the vibrating molecules—Planck called them "resonators"—in the blackbody oven's walls; k is a universal constant and W measures disorder. Although Boltzmann is often credited with inventing the entropy equation (1), and k is now called "Boltzmann's constant," Planck was the first to recognize the fundamental importance of both the equation and the constant.

Planck came to this equation with reluctance. It treated entropy in the statistical manner that had been developed by Boltzmann. Boltzmann's theory taught the lesson that conceivably—but against astronomically unfavorable odds—any macroscopic process can reverse and run in the unnatural, entropy-decreasing direction, contradicting the second law of thermodynamics. Boltzmann's quantitative techniques even showed how to calculate the incredibly unfavorable odds. Boltzmann's conclusions seemed fantastic to Planck, but by 1900 he was becoming increasingly desperate, even reckless, in his search for an acceptable way to calculate the entropy of the blackbody resonators. He had taken several wrong directions, made a fundamental error in interpretation, and exhausted his theoretical repertoire. No theoretical path of his previous acquaintance led where he was certain he had to arrive eventually—at a derivation of his empirical radiation law. As a last resort, he now sided with Boltzmann and accepted the probabilistic version of entropy and the second law.

For Planck, this was an "act of desperation," as he wrote later to a colleague. "By nature I am peacefully inclined and reject all doubtful adventures," he wrote, "but by then I had been wrestling unsuccessfully for six years (since 1894) with this problem of equilibrium between radiation and matter and I knew that this problem was of fundamental importance to physics; I also knew the formula that expresses the energy distribution in normal spectra [his empirical radiation law]. A theoretical interpretation *had* to be found at any cost, no matter how high."

The counting procedure Planck used to calculate the disorder W in equation (1) was borrowed from another one of Boltzmann's theoretical techniques. He considered—at least as a temporary measure—that the total energy of the resonators was made up of small *indivisible* "elements," each one of magnitude ε. It was then possible to evaluate W as a count of the number of ways a certain number of energy elements could be distributed to a certain number of resonators, a simple combinatorial calculation long familiar to mathematicians.

The entropy equation (1), the counting procedure based on the device of the energy elements, and a standard entropy-energy equation from thermodynamics, brought Planck almost—but not quite—to his goal, a theoretical derivation of his radiation law. One more step had to be taken. His argument would not succeed unless he assumed that the energy ε of the elements was proportional to the frequency with which the resonators vibrated, $\varepsilon \propto \nu$, or

$$\varepsilon = h\nu, \tag{2}$$

with h a proportionality constant. If he expressed the sizes of the energy elements this way, Planck could at last derive his radiation law and use the blackbody data to calculate accurate numerical values for his two theoretical constants h and k.

This was Planck's theoretical route to his radiation law, summarized in a brief report to the German Physical Society in late 1900. Planck hoped that he had in hand at last the theoretical plum he had been struggling for, a general theory of the interaction of radiation with matter. But he was painfully aware that to reach the plum he had ventured far out on a none-too-sturdy theoretical limb. He had made use of Boltzmann's statistical entropy calculation—an approach that was still being questioned. And he had modified the Boltzmann technique in ways that modern commentators have found questionable. Abraham Pais, one of the best of the recent chroniclers of the history of quantum theory, says that Planck's adaptation of the Boltzmann method "was wild."

Even wilder was Planck's use of the energy elements ε in his development of the statistical argument. His procedure required the assumption that energy, at least the thermal energy possessed by the material resonators, had an inherent and irreducible graininess embodied in the ε quantities. Nothing in the universally accepted literature of classical physics gave the slightest credence to this idea. The established doctrine—to which Planck had previously adhered as faithfully as anyone—was that energy of all kinds existed in a continuum. If a resonator or anything else changed its energy, it did so through continuous values, not in discontinuous packets, as Planck's picture suggested.

In Boltzmann's hands, the technique of allocating energy in small particle-like elements was simply a calculational trick for finding probabilities. In the end, Boltzmann managed to restore the continuum by assuming that the energy elements were very small. Naturally, Planck hoped to avoid conflict with the classical continuum doctrine by taking advantage of the same strategy. But to his amazement, his theory would not allow the assumption that the elements were arbitrarily small; the constant h in equation (2) could not be given a zero value.

Planck hoped that the unfortunate h, and the energy structure it implied, were unnecessary artifacts of his mathematical argument, and that further theoretical work would lead to the result he wanted with less drastic assumptions. For about eight years, Planck persisted in the belief that the classical viewpoint would eventually triumph. He tried to "weld the [constant] h somehow into the framework of the classical theory. But in the face of all such attempts this constant showed itself to be obdurate." Finally Planck realized that his struggles to derive the new physics from the old had, after all, failed. But to Planck this failure was "thorough enlightenment. . . . I now knew for a fact that [the energy elements] . . . played a far more significant part in physics than I had originally been inclined to suspect, and this recognition made me see clearly the need for introduction of totally new methods of analysis and reasoning in the treatment of atomic problems."

The physical meaning of the constant h was concealed, but Planck did not have much trouble extracting important physical results from the companion constant k. By appealing to Boltzmann's statistical calculation of the entropy of an ideal gas, he found a way to use his value of k to calculate Avogadro's number, the number of molecules in a standard or molar quantity of any pure substance. The calculation was a far better evaluation of Avogadro's number than any other available at the time, but that superiority was not recognized until much later. Planck's value for Avogadro's number also permitted him to calculate the electrical charge on an electron, and this result, too, was superior to those derived through contemporary measurements.

These results were as important to Planck as the derivation of his radiation law. They were evidence of the broader significance of his theory, beyond the application to blackbody radiation. "If the theory is at all correct," he wrote at the end of his 1900 paper, "all these relations should be not approximately, but absolutely, valid." In the calculation of Avogadro's number and the electronic charge, Planck could feel that his theory had finally penetrated "to something absolute."

In part because of Planck's own sometimes ambivalent efforts, and in part because of the efforts of a new, less inhibited scientific generation, Planck's theory stood firm, energy discontinuities included. But the road to full acceptance was long and tortuous. Even the terminology was slow to develop. Planck's energy "elements" eventually became energy "quanta," although the Latin word "quantum," meaning quantity, had been used earlier by Planck in another context. Not until about 1910 did Planck's theory, substantially broadened by the work of others, have the distinction of its formal name, "quantum theory."

Another View

The interpretation of Planck's work outlined here has been accepted by science historians for many years. The crucial episode in the story is Planck's arrival at the equation $\varepsilon = h\nu$, which calculates the size of the energy elements distributed to the blackbody resonators. Because the energy elements have a definite size, and are indivisible, a resonator can have the energies $0, \varepsilon, 2\varepsilon, 3\varepsilon, \ldots$, but no others, and any energy change must be discontinuous, because a change of less than one unit is not allowed. The resonator energy is, to use the modern terminology, "quantized."

Did Planck hold this view of the resonators and their energy? Most science historians have assumed that he did, but that notion has been challenged by Thomas Kuhn, who can find little, if any, evidence that Planck recognized the concept of energy discontinuity in his early papers. Kuhn justifies his position by showing how Planck made use of the Boltzmann statistical calculation without sacrificing the classical picture of the resonators changing their energy continuously. Kuhn believes Planck kept this classical view until 1908, when he began to formulate a second theory that included the energy discontinuity as it is recognized today. The following quote from a letter written in 1908 expresses what Kuhn believes to be Planck's first acceptance of the energy discontinuity: "There exists a certain threshold: the resonator does not respond at all to very small excitations; if it responds to larger ones, it does so only in such a way that its energy is an integral multiple of the energy element $h\nu$, so that the instantaneous value of the energy is always represented by such an integral multiple."

Kuhn stresses that his revised reading of Planck's work does not diminish Planck's stature. He is convinced that this view "in no way devalues the contribution due to Planck. On the contrary, Planck's derivation of his famous blackbody distribution law becomes better physics, less sleepwalking, than it has been taken to be in the past."

So we are left with two versions of the tale of Planck's discovery. But perhaps we do not have to make a choice. Both versions emphasize what is important: the intensity of Planck's commitment to both the old physics and the new. Whether he actually recognized the quantization concept in 1900 or eight years later, it is clear that he could not be satisfied with the new theory in any form until he had made every possible effort to reshape it and find a way back to the classical principles. What an irony it was that Planck, the most reluctant of revolutionaries, was given the first glimpse of this alien world. He was not free either to follow the traditional physics of his convictions or to expand and build in the domain of the new physics. He was a conserver by nature, and fate had handed him the rebel's role. The best measure of Planck's intellect and integrity is that he succeeded in that role.

Einstein's Energy Quanta (Photons)

One of the few perceptive readers of Planck's early quantum theory papers was the junior patent examiner in Bern, Albert Einstein. To Einstein, the postulate of the energy elements was vivid and real, if appalling, "as if the ground had been pulled from under one, with no firm foundation seen anywhere upon which one could have built." As it happened, the search for a "firm foundation" occupied Einstein for the rest of his life. But even without finding a satisfying conceptual basis, Einstein managed to discover a powerful principle that carried the quantum theory forward in its next great step after Planck's work. He presented his theory in one of the papers published during his "miraculous year" of 1905.

Planck was cautious in his use of the quantum concept. For good reason, considering its radical implications, he had hesitated to regard the quantum as a real entity. And he was careful not to infer anything concerning the radiation field, partly light and partly heat radiation, contained in the blackbody oven's interior. The energy quanta of which he spoke belonged to his resonator model of the vibrating molecules in the oven's walls. Einstein, in one of his 1905 papers, and in several subsequent papers, presented the "heuristic" viewpoint that real quanta existed and that they were to be found, at least in certain experiments, as constituents of light and other kinds of radiation fields. He stated his position with characteristic clarity and boldness: "In accordance with the assumption to be considered here, the energy of a light ray . . . is not continuously distributed over an increasing space but consists of a finite number of energy quanta which are localized in space, which move without dividing, and which can only be produced and absorbed as complete units."

Although it was hedged with the adjective "heuristic," the picture Einstein presented was attractively simple: the energy contained in radiation fields, particularly light, was not distributed continuously but was localized in particle-like entities. Einstein called these particles of radiation "energy quanta"; in modern usage, complicated by the changing fortunes of Einstein's theory, they are called "photons."

Einstein developed his concept of photons in a variety of short, clever arguments written somewhat in the style of Planck's 1900 paper. The entropy concept and fundamental equations from thermodynamics again opened the door to the quantum realm. Entropy equations for a radiation field make the field look like an ideal gas containing a large but finite number of independent particles. Each of these radiation particles—photons, in modern parlance—carries an amount of energy given by one of Planck's energy elements $h\nu$, with ν now representing a radiation frequency. If there are N photons, the total energy is

$$E = Nh\nu. \tag{3}$$

Einstein drew from this equation the conclusion that the radiation field, like the ideal gas, contains N independent particles, the photons, and that the energy of an individual photon is

$$\varepsilon = \frac{E}{N} = h\nu.$$

No doubt Einstein was convinced by this reasoning, but it is not certain that anyone else in the world shared his convictions. The year was 1905. Planck's quantum postulate was still generally ignored, and Einstein had now applied it to light and other forms of radiation, a step Planck himself was unwilling to take for another ten years. What bothered Planck, and anyone else who read Einstein's 1905 paper, was that the concept of light in particulate form had not been taken seriously by physicists for almost a century. The optical theory prevailing then, and throughout most of the nineteenth century, pictured light as a succession of wave fronts bearing some resemblance to the circular waves made by a pebble dropped into still water. It had been assumed ever since the work of Thomas Young and Augustin Fresnel in the early nineteenth century that light waves accounted for the striking interference pattern of light and dark bands generated when two specially prepared light beams are brought together. Other optical phenomena, particularly refraction and diffraction, were also simply explained by the wave theory of light.

One hundred years after Young's first papers, Albert Einstein was rash enough to suggest that there might be some heuristic value in returning to the observation once proposed by Newton, that light can behave like a shower of particles. Einstein had found particles of light in his peculiar use of the quantum postulate. And, more important, he also showed in one of his 1905 papers that experimental results offered impressive evidence for the existence of particles of light, in astonishing contradiction to previous experiments that stood behind the seemingly impregnable wave theory.

The most important experimental evidence cited by Einstein concerned the "photoelectric effect," in which an electric current is produced by shining ultraviolet light on a fresh metal surface prepared in a vacuum. In the late 1890s and early 1900s, this photoelectric current was studied by Philipp Lenard (the same Lenard who later conceived a rabid, anti-Semitic hatred for Einstein, and worked furiously in the ultimately successful campaign to drive Einstein from Germany). Lenard discovered that the current emitted by the illuminated "target" metal consists of electrons whose kinetic energy can accurately be measured, and that the emitted electrons acquire their energy from the light beam shining on the metallic surface. If the classical viewpoint is taken—that light waves beat on the metallic surface like ocean waves, and that electrons are disturbed like pebbles on a beach—it seems necessary to assume that each electron receives more energy when the illumination is more intense, when the waves strike with more total energy. This, however, is not what Lenard found; in 1902, he discovered that, although the total number of electrons dislodged from the metallic surface per second increases in proportion to the intensity of the illumination, the individual electron energies are independent of the light intensity.

Einstein showed that this puzzling feature of the photoelectric effect is comprehensible once the illumination in the experiment is understood to be a collection of particle-like photons. He proposed a simple mechanism for the transfer of energy from the photons to the electrons of the metal: "According to the concept that the incident light consists of [photons] of magnitude $h\nu$ . . . one can conceive of the ejection of electrons by light in the following way. [Photons] penetrate the surface layer of the body [the metal], and their energy is transformed, at least in part, into kinetic energy of electrons. The simplest way to imagine this is that a [photon] delivers its entire energy to a single electron; we shall assume that is what happens."

Each photon, if it does anything measurable, is captured by one electron and transfers all its energy to that electron. Once an electron captures a photon and carries away as its own kinetic energy the photon's original energy, the electron attempts to work its way out of the metal and contribute to the measured photoelectric current. As an electron edges its way through the crowd of atoms in the metal, it loses energy, so it emerges from the metal surface carrying the captured energy minus whatever energy has been lost in the metal. If the energy that the metal erodes from an electron is labeled P, if the captured photon's original energy, also the energy initially transferred to the electron, is represented with Planck's $h\nu$ (ν is now the frequency of the illuminating ultraviolet light), and if energy is conserved in the photoelectric process, the energy E of an electron emerging from the target can be written

$$E = h\nu - P. \tag{4}$$

Einstein's picture of electrons being bumped out of metal targets in single photon-electron encounters easily explains the anomaly found by Lenard. Each interaction leads to the same photon-to-electron energy transfer, regardless of light intensity. Therefore electrons joining the photoelectric current from some definite part of the metallic target have the same energy whether just one or countless photons strike the metal per second. Although admirably simple, this explanation must have seemed almost as far-fetched to Einstein's skeptical audience as the rest of his arguments. The rule that one photon is captured by one electron "not only prohibits the killing of two birds by one stone," as the British theorist James Jeans remarked, "but also the killing of one bird by two stones."

More than anything else Einstein achieved in physics, his photon theory was treated with distrust and skepticism. Not until 1926 was the now standard term "photon" introduced by Gilbert Lewis. What was obvious to Einstein by simply exercising his imagination and intuition was still being seriously questioned twenty years later. It took something approaching a mountain of evidence to make a permanent place for photons in the world of quantum theory.

While Einstein was beginning his bold explorations of the quantum realm, Planck was becoming the chief critic of his own theory. Planck seems to have had no regrets—perhaps he was pleased—that the work of building quantum theory had passed to Einstein and a new generation. Late in his life he wrote, with no sense of the personal irony, "A new scientific truth does not triumph by convincing its opponents and making them see the light, but rather because . . . a new generation grows up that is familiar with it."

The Greatest Good

Planck lived by his conscience. As John Heilbron, Planck's most recent biographer, puts it, "His clear conscience was the only compass he needed." It guided him through a life of triumph and tragedy lived in many spheres. He was a devoted family man, a skilled lecturer, a talented musician, a tireless mountaineer, a formidable administrator, a mentor venerated by junior colleagues, and an inspiration for all. Einstein, who in personality and background seemed to be almost an anti-Planck, listed for Max Born the pleasures of being in Berlin, concluding with: "But chiefly this: to be near Planck is a joy."

Planck was happiest in the company of his family. "How wonderful it is to set everything else aside," he wrote, "and live entirely within the family." His second wife, Marga, remarked: "He only showed himself in all his human qualities in the family." With his first wife, Marie, who died in 1909, he raised two sons, Karl and Erwin, and twin daughters, Emma and Grete. Lise Meitner, a talented, determined, and shy young woman who went to Berlin in 1907 to pursue a career in physics (an all but impossible goal for a woman at the time), was befriended by Planck and taken into the family. In a reminiscence of Planck she wrote: "Planck loved happy, unaffected company, and his home was a focus for such social gatherings. The more advanced students and physics assistants were regularly invited to Wangenheimstrasse. If the invitations fell during the summer semester, we played tag in the garden, in which Planck participated with almost childish ambition and great agility. It was almost impossible *not* to be caught by him."

But good fortune was never a permanent condition in Planck's life. Karl, the elder son, died of wounds suffered in World War I. A few years later one of the twin daughters, Grete, died shortly after childbirth. The baby survived; the other twin, Emma, went to help care for the child and married the widower, and *she* died in childbirth. Planck was devastated by these losses. After the twins' deaths, he wrote in a letter to Hendrik Lorentz: "Now I mourn both my dearly loved children in bitter sorrow and feel robbed and impoverished. There have been times when I doubted the value of life itself." But he had immense inner and outer resources. He could always escape in his work, not only in the solitary studies of theoretical physics, but also in the public life of the university and the powerful academies, societies, and committees of German science.

He was an accomplished lecturer. Meitner, who had come from Vienna and Boltzmann's exuberant performances in the lecture hall, was at first disappointed by Planck's lectures, but soon came to appreciate the difference between Planck's private and public styles: "Planck's lectures, with their extraordinary clarity, seemed at first somewhat impersonal, almost dry. But I very quickly came to understand how little my first impression had to do with Planck's personality."

For decades Planck was influential in the Berlin Academy, a German counterpart of the British Royal Society; the German Physical Society, custodian of the leading physics journal, *Annalen der Physik;* and the Kaiser-Wilhelm Society, created to funnel private funds into research institutes. In 1930, three years after his "retirement," Planck was elected to the presidency of the Kaiser-Wilhelm Society. Heilbron writes that in this, the most elevated position Planck held, he dealt "with ministers and deputies, with men of commerce, banking and industry, with journalists, diplomats, and foreign dignitaries." He was known as "the voice of German scientific research." At the same time, he was a dominating influence in the Berlin Academy, remained active in the Physical Society, and gave a cycle of lectures at the university. As Heilbron observes, "Planck was evidently an exact economist with his time."

Somehow Planck found time for recreation, but nothing frivolous. He was an excellent pianist; he had even considered a musical career. Music was an emotional experience for him. He found the romantic composers, Schubert, Schumann, and Brahms, preferable to the more intellectual music of Bach (except for parts of the *Saint Matthew Passion*). Musical evenings were a fixture in the Planck household, with Planck accompanying the renowned concert violinist Joseph Joachim, or playing trios with Joachim and Einstein. For physical recreation he chose mountaineering, "without stopping or talking and Alpine accommodation without comfort or privacy," writes Heilbron. A day in the mountains could do as much for Planck's soul as a Brahms symphony.

Planck lived almost ninety years. He witnessed the two world wars, two Reichs, and the Weimar Republic. He saw the great German scientific establishment, which he had helped build, destroyed by Nazi anti-Semitic racial policies and other insanities. He deplored everything the Nazis did, but chose to remain in Germany, with the hope that he could help pick up the pieces after it was all over. He was nearly killed in a bombing raid, and his house in the Berlin suburb of Grünewald was damaged. Through all of this Planck held on to a measure of hope for the future. But there was worse to come.

In February 1944, Grünewald was flattened in a massive air raid. Planck's house was destroyed, and with it his library, correspondence, and diaries. About a year later, Planck's remaining son from his first marriage, Erwin, was executed as a conspirator in a plot against Hitler. "He was a precious part of my being," Planck wrote to a niece and a nephew. "He was my sunshine, my pride, my hope. No words can describe what I have lost with him."

Late in his life, Planck wrote: "The only thing that we may claim for our own with absolute assurance, the greatest good that no power in the world can take from us, and the one that can give us more permanent happiness than anything else, is integrity of soul. And he whom good fortune has permitted to cooperate in the erection of the edifice of exact science, will find his satisfaction and inner happiness, with our great poet Goethe, in the knowledge that he explored the explorable and quietly venerates the inexplorable." An adaptation of the last phrase—"He explored the explorable and quietly venerated the inexplorable"—might have been Max Planck's epitaph.

16

Science by Conversation

Niels Bohr 1885–1962

Hail to Niels Bohr

Quantum theory was not an overnight success. Its reception during the first decade of its history was hesitant, and its practitioners were scarce. By 1910, the Planck postulates were more or less recognized, but they had been applied mostly to problems concerning radiation and the solid state, and hardly at all in the realm of atoms and molecules. There had been no movement toward the formulation of a general quantum physics.

In the summer of 1913, there appeared in the *Philosophical Magazine* the first of a series of papers that began to turn the tide. The author was Niels Bohr, a twenty-eight-year-old Danish physicist with a rare personality. Bohr's theory described the behavior of atoms, particularly hydrogen atoms, with a carefully concocted mixture of the Planck postulates and the classical mechanics of Kepler and Newton. Bohr applied the theory, with spectacular success, to the beautiful spectral patterns emitted by hydrogen gas when it is excited electrically. (The physical apparatus is similar to that used in neon lighting.) This was, to the physicists of the time, an incredible achievement. Spectroscopists, the experimentalists who study the regularities of light wavelengths (spectra) emitted by atoms and molecules, had done their work so long without benefit of a theory that they had despaired of ever finding one. Bohr's papers brought new hope for spectroscopy, and for quantum theory as well.

To some extent, Bohr's role in this was good fortune. Quantum theory loomed large enough in 1913 that its value to atomic physics could not have been missed much longer. Even so, Bohr's task was no simple exercise. It took skill and intuitive sense in large measure to devise a workable mixture of classical and quantum physics. Einstein remarked that he had had similar ideas, "but had no pluck to develop them." To Einstein, Bohr's sensitive application of the "insecure and contradictory foundation" supplied by quantum theory to atomic problems was a marvel, "the highest form of musicality in the sphere of thought."

Bohr did more than create theoretical masterpieces. He also built, almost single-handedly, a great school of theoretical and experimental physics in Copenhagen. The Bohr Institute (officially, the University Institute of Theoretical Physics) was inaugurated on March 3, 1921, and it quickly attracted an extraordinary collection of young German, English, Russian, Dutch, Hungarian, Indian, Swedish, and American physicists. Bohr offered them a place to live and work when academic positions were scarce and theoretical physicists, like artists, were poor.

Activities at the institute were not always what one would expect from a learned gathering: Ping-Pong (played in the library), girl watching, and cowboy movies were favorite pastimes. But a lot of strenuous and brilliant work was done in this seemingly easy atmosphere. Wolfgang Pauli, Werner Heisenberg, Paul Dirac, Lev Landau, Felix Bloch, Edward Teller, George Gamow, and Walter Heitler were all visitors at the Bohr Institute: their names and accomplishments tell a large part of what happened in quantum physics during the crucial years of the 1920s and 1930s. Robert Oppenheimer writes of this period and Bohr's indispensable role in it: "It was a heroic time. It was not the doing of one man; it involved the collaboration of scores of scientists from many different lands, though from first to last the deep creative and critical spirit of Niels Bohr guided, restrained, and finally transmuted the enterprise."

Bohr had few of the characteristics expected of a man of such influence. His lectures were likely to be "neither acoustically nor otherwise completely understandable." Despite a prodigiously thorough effort, his papers and books were frequently repetitious and dense. Anecdotes are told of his unembarrassed questions about matters of common knowledge. His stock of jokes at any time was limited to about six. Yet his personality was forceful and penetrating. Bohr spoke with a gentle directness and sincerity that impressed students, colleagues, and presidents alike. As Léon Rosenfeld, one of Bohr's collaborators, remarked of the stream of visitors to Copenhagen: "They come to the scientist, but they find the man, in the full sense of the word."

Bohr's generosity was repaid in a remarkable way. Apparently, Bohr could not think creatively without human company. Throughout his career he conceived, shaped, and finished his scientific ideas in conversations with small, critical audiences, usually selected from those at hand at the institute. So attuned were his thoughts to a living presence that no part of the creative process could proceed without a human sounding board. Papers and lectures were written in restless, erratic dictating sessions that were sometimes monologues. One of Bohr's assistants, Oskar Klein, gives us a glimpse of Bohr refining a lecture: "With some writing paper and a pencil in front of me I was placed at a table around which Bohr wandered, alternately dictating in English and explaining in Danish, while I tried to get the English on paper. Sometimes there were long interruptions either for pondering what was to follow, or because Bohr had thought about something outside the theme he had to tell me about. . . . Often, also, work was interrupted by short running trips or cycling to the shore together with the family for bathing."

Bohr's energy and tenacity in the perfecting of a paper seemed almost superhuman. Every word, sentence, concept, and equation had to be reviewed and revised. After five or six drafts (the last one probably on a printer's page proofs), with no end in sight, Bohr would retire to some quiet corner of the institute, accompanied by the indispensable amanuensis, and the struggle would continue. Finally, unbelievably, Bohr would be satisfied. Wolfgang Pauli, who was often invited to Copenhagen for his services as a valuable, but not always sympathetic, critic, responded to one invitation with: "If the last proof is sent away, then I will come."

With his relentless insistence on clarity, and his vast gift for coaxing criticism from others in marathon conversations, Bohr managed to penetrate some of the most difficult problems in quantum physics, including those of a conceptual and philosophical nature. His arguments had daring and a thoroughness that was unassailable. His interpretation of quantum theory, particularly its paradoxes, contrasted with and often contradicted Einstein's viewpoints. Beginning in 1927 at a Solvay conference, and continuing for twenty years, Bohr and Einstein carried on a friendly debate concerning the meaning of quantum physics. Einstein could never accept Bohr's conclusion that the microworld of atoms and molecules is ultimately indeterminate, and did his best to break Bohr's defenses. But Bohr always had an answer to Einstein's criticisms, and his arguments prevailed.

Like some of the other physicists whose stories are told in these chapters, Bohr was blessed with an ideal marriage. Margrethe Nørlund Bohr was a lovely, intelligent woman, and a fine manager and hostess. The Bohrs had six sons—two of them did not survive childhood, and the eldest, Christian, was drowned in a sailing accident—and after 1932 they lived in the Carlsberg "House of Honor" for Denmark's first citizen. Rosenfeld tells us about Margrethe's vital place in this complicated existence: "Margrethe's role was not an easy one. Bohr was of a sensitive nature, and constantly needed the stimulus of sympathy and understanding. When children came . . . Bohr took very seriously his duty as paterfamilias. His wife adapted herself without apparent effort to the part of hostess, and evenings at the Bohr home were distinguished by warm cordiality and exhilarating conversation."

Bohr won a Nobel Prize. He advised Presidents Roosevelt and Truman and Prime Minister Churchill, and became known in every corner of the world of physics. His life, personality, and aspirations became legendary. Only Einstein and Marie Curie, among scientists of the twentieth century, reached positions of such eminence. But before all else Bohr's place was with the carefree, yet devoted and gifted, members of the institute, taking and using their criticism, and enjoying their spoofing:

> Hail to Niels Bohr from the worshipful nations!
> You are the master by whom we are led,
> Awed by your cryptic and proud affirmations,
> Each of us, driven half out of his head,
> —Yet remains true to you,
> —Wouldn't say boo to you,
> Swallows your theories from alpha to zed,
> —Even if—(Drink to him,
> —Tankards must clink to him!)
> None of us fathoms a word you have said!

The Bohr-Rutherford Atom

No doubt it is significant that Niels Bohr began his career as a practicing physicist in a laboratory "full of characters from all parts of the world working with joy under the energetic and inspiring influence of the 'great man.' " The "great man" in the laboratory of Bohr's apprenticeship—known as "Papa" or "the Prof" to the inhabitants—was Ernest Rutherford, who gave us the concept of the atomic nucleus. Rutherford, a New Zealander transplanted to England, presided over nuclear physics during its most creative and, one might say, in view of later developments, its most innocent and happiest years.

The atomic nucleus is a particle about 10^{-13} centimeter in diameter that carries a positive charge and most of the atom's mass. It is surrounded by a balancing negative charge to a total atomic radius of about 10^{-8} centimeter. In other words, the atom is a hundred thousand times bigger than its nucleus. In dimension, the nucleus in the atom is like "a fly in a cathedral," according to Ernest Lawrence, who helped build nuclear physics on Rutherford's foundations. (But this is a fantastically heavy fly; it weighs several thousand times more than the cathedral.)

Rutherford drew his atomic model from the evidence of a monumental series of experiments reported in 1913 by Hans Geiger—later of "Geiger counter" fame, and one of the most gifted in Rutherford's group of experimentalists—and Ernest Marsden, a young student. Geiger and Marsden observed the scattering of alpha particles (helium ions produced by radioactive materials) by thin metallic foils. Most of the alpha particles passed through the thin foils with little or no deflection, as expected, but the paths of a few were drastically altered, as if they had collided with something very small and very massive in the metallic foil—the atomic nuclei of Rutherford's model.

Bohr joined Rutherford and his "tribe" at the University of Manchester in 1912, just as the nuclear atom was beginning to emerge. (Bohr had also spent a brief time working for J. J. Thomson at the Cavendish Laboratory, Cambridge. Bohr's rudimentary English, and his not always tactful insistence on critical discussions, seem to have alienated Thomson, who was inclined to be distant on scientific matters anyway.) The Manchester laboratory and its chief were much to Bohr's liking: "Rutherford is a man you can rely on; he comes regularly and enquires how things are going and talks about the smallest details—Rutherford is such an outstanding man and really interested in the work of all the people around him," Bohr wrote to his brother Harald. Although Bohr showed signs of being a theorist, a breed of physicist not always welcome in Rutherford territory, his talent, obvious sincerity, and lack of pretension—and previous fame as a soccer player—seem to have impressed Rutherford immediately: "Bohr's different. He's a football player!"

Bohr was fascinated by the nuclear model of the atom, not only by its impressive successes in accounting for the Geiger-Marsden foil experiments, but also for its most conspicuous failure. It was obvious that no simple version of the nuclear atom could have the infinite stability atoms normally have. For example, it seemed reasonable to picture the negative electricity surrounding the nucleus as electrons moving in planetlike orbits around the nucleus. But electrons circulating in orbits should have behaved like the electrical charge circulating or oscillating in a radio antenna, and therefore an atom containing orbital electrons should have imitated the antenna and continuously radiated energy. Sooner or later, the electrons would have collapsed into the nucleus, thus destroying the atom.

Such was the unnatural fate predicted for Rutherford's nuclear atom by the classical theory of electrodynamics. But this problem of atoms collapsing on themselves was no challenge to the nucleus itself: the Geiger-Marsden foil experiments left no doubt that Rutherford's picture of the nucleus was correct. The mystery to be solved, for which the Geiger-Marsden data offered no clues, concerned the status of the surrounding electrons.

To Bohr—and several others who had thought about the problem before him—it was clear that, however the electrons disposed themselves in atoms, they had to obey physical laws that were in some sense radically different from the laws of radio antennas and other objects from the macroworld. Bohr noted in his first paper on atomic structure, *On the Constitution of Atoms and Molecules*, the "general acknowledgment of the inadequacies of the classical electrodynamics in describing the behavior of systems of atomic size."

But why allow the classical theory, which had been applied and tested only in the macroscopic realm, to create a mystery concerning nonradiating atomic electrons when there was no reason to believe that the classical theory applied? Why adhere to the classical theory and assume that electrons in atoms *should* radiate energy? The greatest accomplishment of Bohr's theory was that it introduced the assumption that electrons have "waiting places" or "stationary states" in which they do *not* radiate, and have constant, stable energies. This postulate, which Bohr restated and reexamined throughout five lengthy papers published between 1913 and 1915, finally emerged as this statement: "An atomic system possesses a number of states in which no emission of energy takes place, even if the particles are in motion relative to each other, and such an emission is to be expected in ordinary electrodynamics. The states are denoted as 'stationary' states of the system under consideration."

How could electrons be described as they moved around in an atom under the restriction of Bohr's stationary states? Bohr, like Planck, felt that classical physics should be retained wherever possible. Although classical *electrodynamics* created the difficulty that orbiting electrons should radiate energy, there appeared to be no reason why the laws of classical *mechanics*, which governed the orbital motion of planets, should be rejected. So Bohr pictured electrons in stationary states moving in circular or elliptical orbits prescribed by the mechanics of Newton and Kepler. On the other hand, when an electron changed from one stationary state to another it did so in a discontinuous "jump," *not* governed by classical mechanics. Bohr stated a second postulate: "The dynamical equilibrium of the systems in the stationary states is governed by the ordinary laws of mechanics, while those laws do not hold for the transition from one state to another."

Bohr found it expedient to characterize an electron occupying one of the stationary states by specifying the electron's "binding energy" E in the orbit—the energy required to remove the electron from the atom that holds it—and its frequency of rotation ω, the number of orbital circuits completed per second. He derived the classical equation

$$E^3 = R\omega^2, \tag{1}$$

relating E and ω, with R a composite of several constants whose values had been accurately measured.

In the version of his theory we are viewing, Bohr deftly committed his theory to the quantum viewpoint by introducing a second energy-frequency connection by way of "extra-mechanical fiat," in the apt phrase of the science historians John Heilbron and Thomas Kuhn. Bohr's second equation was

$$E = n\alpha h\omega, \tag{2}$$

with E, as before, an electron's binding energy, n a positive integer called a "quantum number," and a proportionality factor to be evaluated at a later stage. According to this equation, with $n = 1,2,3,\ldots,$ an atomic electron can have the "quantized" energy values

$$E = \alpha h\omega, 2\alpha h\omega, 3\alpha h\omega, \ldots,$$

and no others. Bohr was asserting here a formal analogy with Planck's rule that atoms in the walls of a blackbody oven can have only the quantized energies

$$E = 0,\ hv, 2hv, 3hv, \ldots.$$

The atom as a dynamic quantum-emitting entity took shape with a second and radically different Bohr frequency rule. This one pictured an atom jumping from one stationary state of higher energy E_1 to another of lower energy E_2, with an energy change $E_1 - E_2$, and emitting radiation whose frequency v is connected with the energy change by Planck's constant h:

$$E_1 - E_2 = hv. \tag{3}$$

The two frequencies, ω and v, the first representing an electron's rotation frequency and the second a radiation frequency, were separated in Bohr's theory. This was a drastic departure from the classical theory, which would have pictured an orbiting electron irradiating at a frequency equal to its rotation frequency.

Although the rotation frequency ω and the radiation frequency v were *generally* separated in Bohr's theory, the theory did allow for the exaggerated case in which an atom was so stretched in size that it became a classical object, behaved like an ordinary radio antenna, and radiated frequencies equivalent to the electron rotation frequencies. In this special case, $\omega = v$, and the quantum-theoretical laws merged into the classical laws.

The theoretical device of connecting the quantum and classical realms—making them "correspond," as Bohr put it—was one of Bohr's most valuable contributions, and one in which he took particular pride. This "correspondence principle" was used by Bohr throughout much of his work on quantum theory, and it finally became a cornerstone in the quantum mechanics created by Werner Heisenberg.

When equations (1) and (2) are combined by eliminating ω, a simple equation results relating the electron binding energy E and the quantum number n,

$$E = \frac{R}{\alpha^2 h^2 n^2}.$$

If this equation is written twice for two states whose energies are E_1 and E_2, and quantum numbers are n_1 and n_2,

$$E_1 = \frac{R}{\alpha^2 h^2 n_1^2} \text{ and } E_2 = \frac{R}{\alpha^2 h^2 n_2^2},$$

and these two results are substituted in equation (3), we have

$$hv = \frac{R}{\alpha^2 h^2}\left(\frac{1}{n_1^2} - \frac{1}{n_2^2}\right),$$

or

$$v = \frac{R}{\alpha^2 h^3}\left(\frac{1}{n_1^2} - \frac{1}{n_2^2}\right). \quad (4)$$

By invoking his correspondence argument, Bohr proved that the constant α had the value ½, and this put his frequency equation (4) in its final form:

$$v = \frac{4R}{h^3}\left(\frac{1}{n_1^2} - \frac{1}{n_2^2}\right). \quad (5)$$

Balmer's Formula

Bohr did not pick these equations out of theoretical thin air. He was guided by the observed patterns in the radiation spectra emitted by elemental substances, particularly atomic hydrogen. If the components of the hydrogen emission spectrum are sorted out by an instrument called a spectroscope, the observed frequencies fall in regular series. One of the hydrogen spectral series had been discovered thirty years before Bohr's work by Johann Balmer, a Swiss schoolteacher accomplished in the art of distilling precise numerical formulas from complex physical data. Balmer discovered that the visible lines in the hydrogen emission spectrum had frequencies that fit a formula such as

$$v = R'\left(\frac{1}{2^2} - \frac{1}{n^2}\right),$$

in which R' represents a parameter whose value is determined by the spectral data, and n here is any integer larger than 2: $n = 3,4,\ldots$. Balmer appreciated that his formula might imply a more general formula such as

$$v = R'\left(\frac{1}{n_1^2} - \frac{1}{n_2^2}\right), \quad (6)$$

with n_1 given the value 2 in his spectral series, but possibly other values in other series.

Balmer's formula, and a variety of other empirical rules of spectroscopy contributed particularly by the Swedish spectroscopist Johannes Rydberg (whose version of Balmer's formula is quoted above), had been known for years without arousing any suspicion that they contained simple clues to atomic structure. Bohr once remarked that the Balmer-Rydberg formula and others like it were regarded in the same light "as the lovely patterns in the wings of butterflies; their beauty

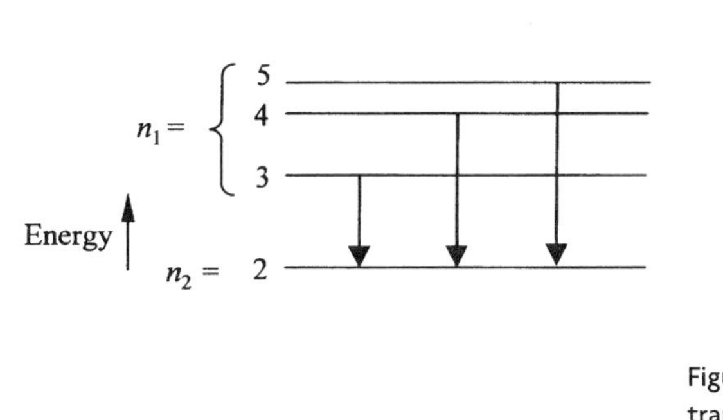

Figure 16.1. An energy-level diagram showing emission transitions for three of the lines in the hydrogen Balmer series.

can be admired, but they are not supposed to reveal any fundamental biological laws."

The "lovely patterns" of the hydrogen emission spectrum were the substance of Bohr's theory. His arguments were pointed, and sometimes forced so that his derived equations would match the observed spectral patterns. Bohr's immediate theoretical aim was accomplished when he derived equation (5), which imitated the Balmer-Rydberg equation (6). The final and crucial test of the theory was passed when the theoretically derived constant $\frac{4R}{h^3}$ in equation (5) was compared with its empirical counterpart R' in equation (6). Calculation of the former from the known fundamental constants (the electronic charge e and mass m and Planck's constant h were involved) came to within a few percent of the measured values of the latter. This was an impressive achievement. Not often in the history of science has a theoretician had such success in bringing theory together with experiment without benefit of those handy numerical devices disrespectful students call "fudge factors."

Bohr's equation (5) is displayed in figure 16.1 as an energy-level diagram. Each horizontal line represents the energy of a stationary state and is labeled with a value of a quantum number, n_1 or n_2. The downward-jumping atomic transitions that produce three of the emitted frequencies in the Balmer series are indicated with arrows.

Images and Connections

Bohr's theory presents an abstract picture: it reveals atoms in modes of behavior that are unrecognizable in the world of ordinary objects. Bohr tells us that atomic electrons are in orbital motion, but that the orbiting electrons have a peculiarly limited, quantized energy, and that they change from one orbit to another in discontinuous jumps that cannot be described completely by the theory.

What does this mean? As Bohr himself recognized, any answers—at least any verbal answers—have limitations. The trouble is that we lack the appropriate language. In a conversation with Heisenberg, Bohr remarked that

> there can be no descriptive account of the structure of the atom; all such accounts must necessarily be based on classical concepts which no longer apply. You see that anyone trying to develop such a theory is really trying the impossible. For we intend to say something about the structure of the atom but lack

> a language in which we can make ourselves understood. We are in much the same position as a sailor, marooned on a remote island where conditions differ radically from anything he has ever known and where, to make things worse, the natives speak a completely alien tongue. He simply must make himself understood, but has no means of doing so. In that sort of situation a theory cannot "explain" anything in the usual strict scientific sense of the word. All it can hope to do is reveal connections and, for the rest, leave us to grope as best we can.

The language of atomic physics, according to Bohr, is something like the language of poetry: "The poet is not nearly so concerned with describing facts as with creating images and establishing connections."

If the substance of Bohr's atomic theory was not descriptive, if it supplied no reliable account of what actually happened within an atom, what was it good for? Why was it so quickly successful? Bohr's theory, like many other aspects of quantum physics, was rooted in the world of experimental findings. The models created by his theory, said Bohr, "have been deduced, or if you prefer guessed, from experiments, not from theoretical foundations." Unlike Einstein, who searched for physical reality in the realm of pure mathematical thought, and often showed indifference to experimental tests of his theories, Bohr was inclined to work backward from fundamental empirical findings to an efficient and reasonable set of postulates. Einstein found his "creative principle" in mathematics. Bohr's creative principle was likely to be a key experimental result. One of Bohr's chief sources of inspiration in the making of his atomic theory was the Balmer-Rydberg formula for the hydrogen spectral lines.

Bohr attended the 1913 meeting of the British Association for the Advancement of Science only a few months after his first papers on atomic theory had appeared, and heard his theory discussed with sympathy and understanding. James Jeans opened the discussion of radiation problems by pointing to Bohr's "ingenious and suggestive, and I think we must add convincing, explanation of spectral series," and assessing the unconventional postulates with the remark, "The only justification at present put forward for these assumptions is the very weighty one of success." When Einstein heard of Bohr's theory in 1913, he was amazed: "Then the frequency of the light does not depend at all on the frequency [of rotation] of the electron. . . . And this is an *enormous achievement*. . . . It is one of the greatest discoveries." But there were others who found the postulates and the correspondence argument forced and unconvincing. Richard Courant, a Göttingen mathematician who defended Bohr against the critics—becoming "a martyr to the Bohr model"—recalls Carl Runge, a Göttingen spectroscopist, saying "Niels, it is true, has made a nice enough impression, but he obviously has done a strange if not crazy stunt with that paper."

The critics were gradually converted, or simply outvoted by the expanding group of Bohr's young, talented disciples. For almost a decade, Bohr's theory, and an elaboration of it developed by the Munich theorist Arnold Sommerfeld, dominated and guided research in atomic physics. As Bohr had intended, his theory began to organize a unified basis for the previously empirical science of spectroscopy. The theory and its achievements had come close to their zenith in 1919 when Sommerfeld wrote this hymn to the beauties of quantum theory applied to atomic spectroscopy: "What we can hear today from the spectra is a veritable atomic music of the spheres, a carillon of perfect whole number relations, an increasing order and harmony in multiplicity."

We Did Not Know It

Beyond its applications to spectroscopy, Bohr's theory performed with distinction the duty of all great theories: it uncovered and unified new fields of experimental and theoretical research. One of the most impressive and surprising experimental confirmations of Bohr's concepts was reported in 1914 by James Franck and Gustav Hertz (a nephew of Heinrich Hertz) from the Kaiser-Wilhelm Institute of Physical Chemistry in Berlin. The Franck-Hertz experiment gave a clear-cut, striking demonstration of the existence of stationary states as intrinsic properties of atoms. Franck and Hertz developed a method for creating electron beams that carried variable, but controlled, amounts of kinetic energy. Atoms of gaseous mercury were placed in the path of such an electron beam so energy could be transferred from electrons to atoms. Franck and Hertz found that when the beam energy reached a certain critical value there was an almost complete transfer of energy from the beam to the mercury atoms, and the beam current abruptly dropped. From the viewpoint of Bohr's theory, electrons with the critical beam energy induced a transition between two of mercury's stationary states.

The plan of the Franck-Hertz experiment follows so directly from Bohr's theoretical suggestions concerning stationary states that one can read the Franck-Hertz paper—and some textbook writers have—and imagine that its authors were advised by Bohr. But the ways of scientific progress are imperfect: Franck and Hertz had not seen Bohr's 1913 paper, and even if they had seen the paper before collecting their own results, they probably would not have believed what they read. Franck's candid remarks on the attitude in Berlin at the time show how dim the light can be that shines on major scientific discoveries (from an interview given by Franck in 1960, quoted by the science historian Gerald Holton):

> It might interest you that when we made the experiments that we did not read the literature well enough—and you know how that happens. On the other hand, one would think that other people would have told us about it. For instance, we had a colloquium at the time in Berlin at which all the important papers were discussed. Nobody discussed Bohr's paper. Why not? The reason is that fifty years ago one was so convinced that nobody would, with the state of knowledge we had at that time, understand spectral line emission, so that if somebody published a paper about it, one assumed "probably it is not right." So we did not know it.

Not Crazy Enough

Jeremy Bernstein, a contemporary theoretical physicist and astute commentator on life in the scientific community, tells a story about a visit to the United States in 1958 by Wolfgang Pauli, who had come with what he thought was a new general theory of particle physics composed by his friend and debating partner, Werner Heisenberg. Pauli presented the theory to an audience at Columbia University that included Bohr.

> After Pauli finished [writes Bernstein], Bohr was called upon to comment. Pauli remarked that at first sight the theory might look "somewhat crazy." Bohr replied that the problem was that it was "not crazy enough." . . . [Then] Pauli and Bohr began stalking each other around the large demonstration table in the front of the lecture hall. When Pauli appeared in the front of the table, he would tell the audience that the theory *was* sufficiently crazy. When it was Bohr's turn he would say it wasn't. It was an uncanny encounter of two giants of modern physics. I kept wondering what in the world a non-physicist visitor would have made of it.

Bohr might have been thinking of his own earlier theories as much as of Pauli's account of Heisenberg's theory. Bohr's atomic theory was certified crazy by more than one of his colleagues, but the theory was not, as it happened, crazy enough. When Bohr began his work on atomic structure, he was unwilling to submit himself intellectually to all the apparent nonsense and contradictions implied by the Planck-Einstein quantum theory. He could rely on the concept of energy quanta, but he had little use for the photon concept and the seemingly irrational wave-particle duality it implied. He could introduce an "extramechanical" postulate that pictured electrons jumping discontinuously from one stationary state to another, but could not part with the classical picture of electrons in continuous orbital motion. What Bohr proposed was only half of a complete atomic theory—a theory that was only half crazy enough.

These comments are made with hindsight, and should not imply that Bohr might have done better. Bohr could hardly have conducted single-handedly a revolution that kept physics in a state of upheaval for twenty-five years. Even Einstein lacked the courage to build an atomic theory on the questionable foundations supplied by the early quantum theory.

By the early 1920s, the Bohr-Sommerfeld atomic theory, and with it most of the rest of quantum theory, was in deep trouble. Although the Bohr method could work wonders with the hydrogen atom, it could do little without excessive difficulty when confronted with atoms more complicated than hydrogen. In the words of the science historian Max Jammer, the quantum theory just prior to 1925 "was, from the methodological point of view, a lamentable hodgepodge of hypotheses, principles, theorems, and computational recipes rather than a logical, consistent theory." Most problems were solved initially with the methods of classical physics and then translated into the language of quantum physics by clever use of the correspondence principle. Frequently the work of translating required more "skillful guessing and intuition than systematic reasoning."

For a time, the community of quantum physicists was struck by an epidemic of theoretician's paralysis. Max Born, whose greatest work was about to come, wrote to Einstein in 1923: "As always, I am thinking hopelessly about quantum theory, trying to find a recipe for calculating helium and other atoms; but I am not succeeding in this either. The quanta are really in a hopeless mess." Pauli thought he would try a different line of work: "Physics is very muddled at the moment; it is much too hard for me anyway, and I wish I were a movie comedian or something like that and had never heard anything about physics." The prevailing mood of dismay was summarized by Hendrik Kramers, Bohr's first assistant and an accomplished theoretician in his own right: "The quantum theory has been very much like other victories; you smile for months; and then weep for years."

But great scientists are blessed with a simple, durable optimism with which they accept the most crushing, disastrous failures as useful steps in the right direction, to be followed sooner or later by new developments and general, evolutionary progress. Planck could struggle eight years in vain to remake his theory in the classical mold and conclude that the entire, seemingly useless effort brought "thorough enlightenment." Einstein could try ninety-nine wrong approaches to a unified field theory and be satisfied that "at least I know 99 ways it won't work." And Bohr, who might have been defending his theory to the last ditch against all rivals, was working as hard as anyone to make a new theory and discard the old one. Ineffectual as his effort was in handling the broader problems of atomic theory, Bohr had faith that it was, like all good theories, at least partly right. Whatever strange concepts were brought by the next theories, those theories could not be made without the connections already seen by Bohr and his great predecessors, Planck and Einstein. Einstein once commented on the "tragedy" of a "deduction killed by a fact": "Every theory is killed sooner or later in that way. But if the theory has good in it, that good is embodied and continued in the next theory."

Peril and Hope

A peculiarity—and potential danger—of scientific work is that it requires the discipline of a detached, objective point of view. For most physicists, detachment is necessary because ordinary human experience is not always a reliable guide to physical principles. The danger is that scientists can become so armored by their objectivity that they fail to anticipate, or perhaps even think about, the consequences of a scientific advance once it is put in a human context.

A prime example of this danger is obvious to us all in the objective principles of the work done before and during the Second World War by nuclear scientists. While they were still detached, nuclear scientists discovered that neutron capture by atoms of a rare uranium isotope, U^{235}, causes the uranium to undergo fission (that is, to split into two fragments of approximately equal mass) with the release of large amounts of energy. Also released are more neutrons, and the tally proves to be more than two neutrons released for each neutron captured.

With that objective discovery, nuclear physics lost its innocence. The possibility of a nuclear chain reaction, in which neutrons produced in one fission event cause more fissioning, was soon recognized. The chain reaction was realized in controlled fashion in nuclear reactors and in uncontrolled fashion in nuclear weapons.

Some of the nuclear scientists who developed the technology of nuclear weapons did their work with a conscience, and some did not. At first, with the prospect of nuclear weapons in the hands of the Nazis, conscience was almost irrelevant. Even Einstein, for most of his life a pacifist, accepted the urgency of the nuclear bomb project. With Leo Szilard and Eugene Wigner, two Hungarian theoretical physicists, Einstein wrote a letter to President Roosevelt in 1939 describing the terrible dangers and the necessity for immediate action. After the war, the threat was the nuclear weapons themselves.

Of all the scientists who struggled with the nuclear threat, the man who stands out today, sixty years later, as the most farsighted and courageous is Niels Bohr. The human consequences were clear to Bohr almost immediately, even before the first nuclear bomb was built and tested. He had the vision to recognize what

Robert Oppenheimer called "not only a great peril but a great hope." Bohr's particular concern was the possibility of an unlimited arms race. He was not alone: after the war he was joined by many others from the scientific community.

Bohr and members of his institute had done important work in nuclear theory during the 1920s and 1930s. In 1939, he and John Wheeler wrote a classic paper on the theory of the fission process, and by 1941 Bohr was convinced that a nuclear explosion was possible with U^{235} if a large enough mass of the isotope could be assembled. At first, the extremely difficult technological task of separating the U^{235} isotope impressed him as an impossibility. But he changed his mind when he saw the huge effort being made in the United States at Los Alamos, New Mexico, and elsewhere, by members of the "Manhattan Project."

Bohr did not make extensive contributions to the development of the nuclear bombs. He spent time at Los Alamos, but his thoughts were more political than technical. Impressive as the bomb project was technologically, Bohr could see that its political ramifications were even more complicated and important. British and American scientists had joined forces, but in 1944, when Bohr began to face the political issues, the Soviet Union knew little or nothing about the bomb project. As Bohr saw it, there was one possibility for avoiding a deadly nuclear arms race between East and West: Stalin should be informed that a nuclear bomb was imminent and offered a share in its control. "The very act of making and accepting such a gesture," Alice Kimball Smith writes, "might . . . produce a radical alteration in the world view of the actors in the drama and create a pattern in international relationships. Only by a policy of true 'openness' could accelerated competition be avoided."

After the postwar atmosphere of nuclear confrontation, Bohr's proposal seems fantastic, but it was, as Smith notes, "based on some highly realistic judgments." Bohr was familiar with the high level of Soviet scientific talent. He knew that the news of a nuclear explosion would prompt a massive Soviet effort that would be successful in at most a few years. Any initial advantage on the side of the West was sure to be temporary, and to think otherwise could be dangerous.

Bohr was persuasive and obstinate enough to convert to his way of thinking some men who were highly placed in the British and American governments. In Britain, he had Sir John Anderson, chancellor of the exchequer, and Lord Cherwell, Churchill's scientific adviser and confidant, on his side. (Cherwell—Frederick Lindemann—was a former student of Nernst's.) In the United States, his most influential ally was Felix Frankfurter, the Supreme Court chief justice and a close friend of Roosevelt's.

Having reached this high level of political influence, Bohr next had the far more formidable task of persuading Churchill and Roosevelt to take his proposal seriously. First, an interview with Churchill was arranged by Cherwell, and it was a fiasco. Churchill seems to have distrusted Bohr almost as much as he did Stalin. Sir Henry Dale, president of the Royal Society, was present at the meeting and saw his fears confirmed that Bohr, with his "mild, philosophical vagueness of expression and in his inarticulate whisper," would not be understood by a "desperately preoccupied Prime Minister." Churchill terminated the meeting before Bohr had an opportunity to present the main points of his proposal. "We did not speak the same language," Bohr said later. Churchill's comment to Cherwell was, "I did not like the man when you showed him to me, with his hair all over his head."

Bohr's discussion with Roosevelt was more civil, but hardly more productive.

Vannevar Bush, Roosevelt's unofficial science adviser, prepared him for the meeting. "Do you think I will be able to understand him?" Roosevelt wanted to know. Bush replied, "No, I do not think you probably will." Roosevelt listened courteously for an hour and a half, and Bohr "went away happy." But, says Bush, "I doubt that the President really understood him at all."

So in the end, Bohr's vision of an open nuclear policy came to nothing, and worse, his reliability was questioned. As Churchill put it to Cherwell: "The President and I are much worried about Professor Bohr. How did he come into the business? He is a great advocate of publicity. He made an unauthorized disclosure to Chief Justice Frankfurter, who startled the President by telling him he knew all the details. . . . What is all this about? It seems to me Bohr ought to be confined or at any rate made to see that he is very near the edge of mortal crimes."

Bohr's hopes were never realized, but his failure no longer matters in the shaping of our judgment of the man. No other scientist has made such a heroic effort to bring the worlds of science and politics together. For Bohr it was not heroism. He simply did what he had always done. Persuasive conversation was his constant method for finding and holding an important position. The conversation could be with a student, an assistant, a colleague, or if necessary, with a preoccupied prime minister or an uninformed president. The scientist's occupational hazard of too much detachment from human problems was never a danger in Bohr's work. For Bohr, scientific problems *were* human problems, no more and no less.

17

The Scientist as Critic
Wolfgang Pauli

What Would Pauli Say?

The modern version of quantum theory—now known as "quantum mechanics"—was born and grew to maturity in just five years, between 1925 and 1930. More was accomplished during those five years than in the preceding twenty-five years, or, for that matter, in the seventy years that have followed. Progress before 1925 was constantly hampered by conceptual doubts. Paradoxes such as the wave-particle duality—the contradiction between the Einstein particle theory of light and the classical wave theory—were disturbing and limiting. But by 1925 these difficulties had, perhaps from familiarity, become less inhibiting. Theorists stopped worrying about the conceptual strangeness of the quantum realm, and began to make a new physics with the strangeness incorporated in it. Once the conceptual barriers were passed, progress was astonishingly rapid. For those who had the vision, it was as if a great fog had lifted. Suddenly it was possible to see in many directions with a clarity no one could have anticipated.

Quantum physicists of the new breed began to practice in the early 1920s. They were mostly second-generation quantum physicists, having been born after Planck read his famous paper to the Berlin Physical Society in 1900. (One might fancy that the appearance of Planck's paper was a signal for the birth of a whole crop of gifted physicists: Wolfgang Pauli, Frédéric Joliot, and George Uhlenbeck in 1900; Werner Heisenberg, Enrico Fermi, and Ernest Lawrence in 1901; Robert Oppenheimer, John von Neumann, and George Gamow in 1904.) One of the most brilliant and influential members of this talented group was Pauli, who not only made major contributions of his own but also, like Bohr, shaped his colleagues' work in long, critical discussions. During the crucial years of the 1920s and 1930s, many quantum physicists felt that their work was not finished until they faced Pauli and his relentless criticism, or lacking the Pauli presence, asked the question, "What would Pauli say?"

One of Pauli's assistants, Rudolf Peierls, tells about Pauli's role as a critic: "To discuss some unfinished work or some new and speculative idea with Pauli was a great experience because of his understanding and his high intellectual honesty, which would never let a slipshod or artificial argument get by." Much of Pauli's effectiveness as a critic was the result of his legendary disregard for his colleagues' pet sensitivities. "Some people have very sensitive corns," he once said, "and the only way to live with them is to step on these corns until they are used to it." A typical Pauli remark, on reading a paper of little significance and less coherence, was, "It is not even wrong." Another comment to a colleague whose papers were not of the highest quality: "I do not mind if you think slowly, but I do object when you publish more quickly than you think."

Pauli found targets for his biting comments on all levels of competence and importance. After a long argument with the Russian theorist Lev Landau, whose work was as brilliant but not so well expressed as his, Pauli responded to Landau's protest that not *everything* he said was nonsense with: "Oh no. Far from it. What you said was so confused that one could not tell whether it was nonsense or not." What may have been Pauli's debut as a belittler of authority was made during his Munich student days. In response to a comment made by Einstein at a colloquium he had this to contribute from the back of a crowded lecture hall: "You know, what Mr. Einstein said is not so stupid."

Of Antimetaphysical Descent

From his youth, Pauli was round in face and body, and physically awkward, in contrast with his lack of intellectual awkwardness. A biographer claims that Pauli managed to pass his driver's test only after taking one hundred driving lessons. One of the most enduring contributions to the Pauli legend was the "Pauli Effect," according to which Pauli could, by his mere presence, cause laboratory accidents and catastrophes of all kinds. Peierls informs us that there are well-documented instances of Pauli's appearance in a laboratory causing machines to break down, vacuum systems to spring leaks, and glass apparatus to shatter. Pauli's destructive spell became so powerful that he was credited with causing an explosion in a Göttingen laboratory the instant his train stopped at the Göttingen station. But none of this misfortune was visited on Pauli himself. That this was a true corollary of the Pauli Effect no one doubted after an elaborate device was contrived to bring a chandelier crashing down when Pauli arrived at a reception. Pauli appeared, a pulley jammed, and the chandelier refused to budge.

Pauli's intellectual inheritance was strong. His father, Wolfgang Joseph, was a professor at the University of Vienna and an expert on the physical chemistry of proteins. His mother, Bertha Schütz, was a newspaper correspondent and the daughter of a singer at the Imperial Opera in Vienna. The father came from a respected Prague Jewish family named Pascheles. He studied medicine at the Charles University in Prague, where one of his classmates was the son of Ernst Mach. At about the time Mach moved to the University of Vienna, Wolfgang Pascheles became a professor there, changed his name to Pauli, and joined the Catholic Church.

The Paulis' only son was born in 1900, and was baptized with the names Wolfgang Ernst Friederich; the second name was for Ernst Mach, who became the child's godfather. At the baptism "[Mach] was a stronger personality than the Catholic priest," Pauli liked to explain when asked about his religion, "and the result seems to be that in this way I [was] baptized 'anti-metaphysical' instead of Catholic. . . . [It] still remains a label which I myself carry, namely: 'of anti-metaphysical descent.' "

Young Wolfgang was a prodigy at all levels of his schooling, not only in mathematics and physics but also in the history of classical antiquity. When the gymnasium classroom activities became boring, he read Einstein's papers on general relativity (only a few years after they were written), and published three papers on relativity that impressed the well-known mathematician and relativist Hermann Weyl.

In company with Werner Heisenberg, who in a few years would initiate the revolution that led to quantum mechanics, Pauli started his career as a research student under Arnold Sommerfeld, a professor at the University of Munich and a renowned teacher of theoretical physics. Pauli liked to joke with Heisenberg about Sommerfeld's martial mustaches and austere manner: "Doesn't he look the typical old Hussar officer?" But the student's respect for the teacher was more lasting than the jokes. "In later years," Peierls writes, "it was surprising when Sommerfeld visited [Pauli], to watch the respect and awe in his attitude to his former teacher, particularly striking in a man who was not normally inclined to be diffident." And Sommerfeld admired his gifted student. He handed the nineteen-year-old Pauli the formidable task of writing an encyclopedia article on relativity. Sommerfeld found the article "simply masterful," and Einstein agreed.

After Munich, Pauli made his brilliant and caustic presence known in Göttingen. In 1921, he became an assistant to Max Born, who had established the University of Göttingen as a center for research in theoretical physics that rivaled Bohr's Copenhagen institute. Born found Pauli "very stimulating." But there were problems: Pauli "liked to sleep in" and did not always appear when he was needed as Born's deputy at 11:00 A.M. lectures. It finally became necessary for the Borns "to send our maid over to him at half past ten, to make sure he got up." Like most of Pauli's associates, Born tolerated this behavior with remarkable good humor. To Born, whose eye for scientific talent was as experienced as Bohr's, Pauli "was undoubtedly a genius of the highest order."

After a year in Göttingen, Pauli moved to Bohr's institute, and one of the most fruitful and lasting partnerships in modern physics was formed. Although Bohr and Pauli never collaborated as authors—perhaps they never agreed—each in his own way had a need for critical conversation. Bohr had perfected the technique of developing his ideas by debating with anyone in sight. Sometimes with students and assistants, the "debate" was simply Bohr thinking aloud. Other times, as in discussions with Einstein and Erwin Schrödinger, the debate became deadlocked over stubborn conceptual problems. But Pauli, with his unsurpassed genius for criticism, was Bohr's favorite partner in debate. Their arguments never ended, but they always progressed, and Bohr became dependent on them. Léon Rosenfeld, one of Bohr's assistants, tells us that if Pauli was not present in person Bohr would focus on his letters: "The arrival of a letter from Pauli was quite an event; Bohr would take it with him when going about his business, and lose no occasion of looking it up again or showing it to those who would be interested in the problem at issue. On the pretext of drafting a reply, he would for days on end pursue with the absent friend an imaginary dialogue almost as vivid as if [Pauli] had been sitting there, listening with his sardonic smile."

Pauli was one of the more itinerant of the quantum physicists. After Munich, Göttingen, and Copenhagen, he went to Hamburg, where he ascended the academic ladder. In 1928, at age twenty-eight, he was appointed to the chair of physics at the Swiss Technical University (ETH) in Zürich. There he remained, except for the five years (1940–45) he spent at the Institute for Advanced Study in Princeton.

Until about 1934, Pauli's personal life was complicated. In 1929, he married a young dancer, Käthe Deppner, who soon left him for a chemist. That annoyed Pauli: "Had she taken a bullfighter I would have understood but an ordinary chemist. . . ." A period of crisis ensued, from which he was rescued by psychoanalysis supervised by Carl Jung, and by a stable marriage in 1934 to Francisca (Franca) Bertram.

The Exclusion Principle

Pauli was first drawn to the frustrations and mysteries of quantum theory as a student listening to Sommerfeld's lectures. He soon became conversant with Sommerfeld's elaborate extension of Bohr's theory and developed a complex application of that theory to the structure of the hydrogen molecule. At the same time, he was critical of the Bohr-Sommerfeld theory, remarking to his fellow student Werner Heisenberg that the whole thing was "atomysticism." To Pauli, with his extraordinarily sensitive ear for the harmonies of formal argument—a sort of mathematical perfect pitch—the quantum theory of the time seemed "muddled." "Everyone is still groping about in a thick mist," Pauli complained to Heisenberg, "and it will probably be quite a few years before it lifts. Sommerfeld hopes that experiments will help us to find some new laws. He believes in numerical links, almost a kind of number mysticism."

Ever since Bohr's first work, it had been known that certain states representing atomic behavior had discrete energies that could be calculated from integers called "quantum numbers," and that when an atom changes its energy it does so in "quantum jumps" between these "stationary states." For about ten years following Bohr's 1913 papers, much of the work on quantum theory focused on the theme of quantum numbers. One of the questions that always had to be answered in the making of atomic models based on quantum numbers was how many quantum numbers were needed for each electronic state to account for the observed physical and chemical behavior of atoms. First there was one quantum number (Bohr's model), then two, then three, and finally, according to Pauli, four.

Pauli found that he could work wonders with a fourfold array of quantum numbers assigned to each state available to an atom's electrons. The key to the model was a set of rules that dictated each electron's choice of quantum numbers. Two rules introduced by Bohr were applicable: the same set of quantum number assignments is available to all electrons in all atoms, and electrons occupy available states lying lowest in energy first. To these Pauli added a broad principle, later called the "exclusion principle" or the "Pauli principle," which did as much to clarify atomic and molecular theory as the more sophisticated theories that followed. Pauli asserted, with a degree of simplicity uncommon in quantum physics, that the set of four quantum numbers describing a state inhabited by an atomic electron must be unique for that electron: no two electrons in a given atom can occupy a state characterized by exactly the same set of values for the four quantum numbers.

Later theory established that the Pauli principle applies to any system of electrons. Wherever electrons gather—in atoms, molecules, or solids—they must organize themselves under the Pauli principle. No two electrons in proximity can be sufficiently alike physically to occupy states carrying exactly the same set of quantum numbers. This often means that electrons simply avoid each other; in atoms they collect in concentric shells.

Spin

That four—and not three—quantum numbers were necessary to make the electron story complete was for a time a deep theoretical puzzle. It had become clear in the earlier theory that the quantum number count for an electronic state is a reflection of the number of dimensions in which an electron moves. An atomic electron in orbital motion around a nucleus moves in three dimensions, and therefore requires three quantum numbers, but only three, for its description. What physical significance could be attached to a fourth quantum number? If analogies to classical physics could be trusted, there was one obvious speculative answer. Electrons, like planets, might have spin motion around an internal axis, in addition to orbital motion.

This idea had occurred to several theorists, including Arthur Compton, Heisenberg, Bohr, and Pauli, but it had problems. For one thing, the ordinary spin of planets and baseballs is rotational motion in three dimensions. If that was the way electrons spun, no fourth quantum number should have been needed. Perhaps, then, spinning electrons were not like spinning baseballs; in some mysterious way, could electron spin be motion outside the familiar three spatial dimensions underlying classical physics? Although he was skeptical about the spin concept, Pauli believed that his fourth quantum number did relate to something "which cannot be described from the classical point of view."

This is where matters stood in late 1925, when, as B. L. van der Waerden puts it, "the spell was broken." What the esteemed theorists feared to do was done quickly and easily by two Dutch graduate students, George Uhlenbeck and Samuel Goudsmit, at the University of Leiden. With Pauli as their inspiration, they arrived at the essentials of the electron spin concept. Uhlenbeck explains their initial reasoning:

> Goudsmit and myself hit upon this idea by studying a paper by Pauli, in which the famous exclusion principle was formulated and in which for the first time, *four* quantum numbers were ascribed to the electron. This was done rather formally; no concrete pictures were connected with it. To us, this was a mystery. We were so conversant with the proposition that every quantum number corresponds to a degree of freedom, and on the other hand with the idea of a point electron [with no three-dimensional structure like that of planets and baseballs], which obviously had [only] three degrees of freedom, that we could not place the fourth quantum number.

The two young graduate students saw immediately the advantages of identifying the fourth quantum number with a special kind of spin motion available to electrons in a realm beyond the usual three spatial dimensions. More slowly they saw the disadvantages. They consulted with their mentor, Paul Ehrenfest, professor of theoretical physics at Leiden. They also got help from the founder of the Leiden school, Hendrik Lorentz (Ehrenfest was his successor), who was interested but not encouraging. After preparing a summary of their findings for Ehrenfest, they thought better of it and told Ehrenfest they had decided not to publish. But Ehrenfest was wiser than they were in the ways scientific careers are made. He said he had already sent the paper to a journal. While better-known theoreticians worried about the peculiar details of the spin concept, Uhlenbeck and Goudsmit had a fine opportunity: "Both of you are young enough to afford a stupidity," Ehrenfest told them.

One of the many who lost out in the competition to write a successful electron spin theory was Pauli's assistant, Ralph Kronig. Several months before the Uhlenbeck-Goudsmit paper reached a journal via Ehrenfest, Kronig arrived at similar conclusions and discussed them with Pauli. But Kronig was not so lucky as his Dutch counterparts. Pauli, the relentless critic, talked him out of publishing. Peierls remarks that in later years, "Pauli did not like to be reminded of this story." Electron spin is certainly one of the seminal ideas of twentieth-century physics and chemistry. Yet Uhlenbeck and Goudsmit did not receive a Nobel Prize for their theory. Kronig's claims possibly explain the omission.

Not only electrons but all of the other elementary particles (for example, protons, neutrons, and positrons) have spin motion, and most of them are allowed just two spin states. The theory dictates that the quantum numbers specifying the spin states are $+\frac{1}{2}$ and $-\frac{1}{2}$. (Most quantum numbers have integer values. Spin quantum numbers, with half-integer values, are exceptional.) The two spin states are pictured roughly with the spin axis oriented "up" for one state and "down" for the other.

In view of what has been said about quantum numbers counting the dimensions in which electrons move, the reader may wonder about the hydrogen atom electron, certainly moving in three dimensions and also endowed with spin motion, yet in Bohr's theory accurately described by the *single* quantum number n. Like all other electrons in other atoms, the hydrogen electron is represented by four quantum numbers. But hydrogen is a special case. In hydrogen, and in no other atoms, the energies of electron states depend to a good approximation only on the single quantum number n, and not on the other three. Bohr was lucky: he could build his model of the hydrogen atom as if it were one-dimensional.

The Critic

Pauli's grasp of physical problems was supreme among his contemporaries, probably not surpassed even by Einstein. Born recalled that "ever since the time he had been my assistant in Göttingen, I had been aware that he was a genius, comparable with Einstein himself. Indeed from the point of view of pure science, he was possibly even greater than Einstein." Pauli's achievements, the enunciation of the exclusion principle and several major contributions in nuclear physics and particle physics, certainly rank among those of the masters of modern physics. Yet his full greatness did not equal that of Einstein, Bohr, or Heisenberg.

To some extent, Pauli was held in check by his own brilliance. At times, he understood physics too well. His critical sense became so refined and broad in scope he could not exercise his creative powers with the imagination and intuitive facility possessed by some of his contemporaries. To Heisenberg, whose reckless departures from the principles of classical physics were soon to be spectacularly successful, Pauli said, "Perhaps it's much easier to find one's way if one isn't too familiar with the magnificent unity of classical physics. You have a decided advantage there." Then he added appreciatively, "Lack of knowledge is no guarantee of success."

But if Pauli's fine critical sense was a personal restraint, it was an inspiration for many of his colleagues. Like a great literary critic, Pauli expressed, for all who had the intelligence to listen, a penetrating, sometimes painfully sharp, yet balanced voice of experience and insight. Much of the best theoretical work in modern physics was done with Pauli attending either in person or in spirit, "sitting there listening with his sardonic smile."

18

Matrix Mechanics

Werner Heisenberg 1901–1976

Twins

The birth of the grand synthesis of quantum theory—now known as "quantum mechanics"—was not the happy event it might have been. To everyone's surprise, what came into the world was not one infant but two—twins. And to make matters worse, the two births were months apart, with different doctors officiating; there were even some ugly rumors about the parentage of the two arrivals. Erwin Schrödinger and his colleagues in Munich and Berlin, who claimed the child they called "wave mechanics," found little to admire in the other child, called "matrix mechanics," claimed by Werner Heisenberg and his friends in Göttingen and Copenhagen. Said Schrödinger about matrix mechanics: "I was discouraged, if not repelled, by what seemed to me a rather difficult method of transcendental algebra, defying any visualization." And Heisenberg had this to say about wave mechanics in a letter to Wolfgang Pauli: "The more I think about the physical portion of the Schrödinger theory, the more repulsive I find it. . . . What Schrödinger writes about visualizability 'is probably not quite right' [one of Bohr's favorite euphemisms], in other words it's crap." For a time, it appeared that physics would have to support two infant versions of quantum mechanics, with an embarrassing rivalry on matters of heritage and title. But fortunately there were some who appreciated and understood both children. All were relieved to find that both twins were healthy and legitimate and deserving of the family name, quantum mechanics.

Wunderkind

Werner Heisenberg, whose skill in the delivery of far-reaching theories brought matrix mechanics into the world (a few months before Schrödinger attended the birth of wave mechanics), was born in Würzburg, Germany, late in 1901. At the time, Werner's father, August, taught ancient languages at the Altes Gymnasium in Würzburg. According to David Cassidy, Heisenberg's most recent biographer,

"August Heisenberg is remembered by his family, superiors, and pupils as a rather stiff, tightly controlled, authoritarian figure. A former student recalled that the schoolmaster demanded 'unbending fulfillment of duty, absolute self-control, and meticulous precision.' " Heisenberg's mother, Annie, was attuned to life in a household that centered on her husband's career. With little assistance, she cared for her two sons and kept her house in fine order. Her formal education was limited—women were excluded from German universities at the time—but advanced enough through self-education and instruction from her father (the rector of a prestigious gymnasium in Munich) for her to add grading of student homework to her many other chores. August Heisenberg was no less driven in his working habits. He carried a course load that many present-day teachers would consider inhuman, participated extensively in political affairs relating to education, and produced a vast scholarly output. His efforts were rewarded. In 1910, he was appointed to the important chair of Greek philology at the University of Munich.

Heisenberg grew up in a family atmosphere that was comfortable but not always secure. One sign of psychological tension was a furious rivalry between Werner and his older brother Erwin, "stoked by August," writes Cassidy. "As boys [Cassidy continues], the two often fought fierce battles with each other. As they grew older, they fought even more frequently and intensely. Finally, after one particularly bloody battle—in which they beat each other with wooden chairs—they called a truce and went their separate ways. After that, they had little to do with each other, except for occasional family visits as adults."

During the formative years of Heisenberg's adolescence, Europe was torn by World War I. In the political and economic chaos that followed the war, Germans young and old were adrift and desperate. "The reins of power had fallen from the hands of a deeply disillusioned older generation," Heisenberg writes in his autobiography, "and the younger one drew together in an attempt to blaze new paths, or at least to discover a new star by which they could guide their steps in the prevailing darkness." Heisenberg found his guiding star in the romantic ideals of the youth movement called the Deutscher Neupfadfinder (German New Boy Scouts). He became the leader of a group of younger boys, who were intimate friends for the rest of his life. They hiked, climbed, camped, and earnestly debated Germany's future.

August Heisenberg contributed to his son's scientific education by introducing him to the speculations of the Greek philosopher-scientists, and the boy found the scientific writings of the Greeks more believable than his textbooks, with their bizarre pictures of molecules containing bonds illustrated with hooks and eyes. While he was still young, Heisenberg, like Boltzmann, Planck, and Einstein, became an accomplished musician. At first, he considered a career as a pianist, but Einstein's creations seemed nearer and more exciting than those of Mozart. So in 1920, at age nineteen, he presented himself to Arnold Sommerfeld at the University of Munich as a prospective student in theoretical physics.

Sommerfeld's stern presence, somewhat like Planck's, was impressive but not intimidating. "The small, squat man with his martial, dark mustache looked rather austere to me," Heisenberg recalled, "but his very first sentences revealed his benevolence, his genuine concern for young people, and in particular for the boy who had come to ask his guidance and advice." Heisenberg, just graduated from the gymnasium, and unimpressed by the difficulty of what he proposed, told Sommerfeld he wanted to explore and extend Einstein's general theory of relativity. Sommerfeld allowed him to attend the advanced seminar, but also prescribed courses from the standard physics curriculum.

As Heisenberg entered Sommerfeld's lecture hall one day, he noticed "a dark-haired student with a somewhat secretive face." This was Wolfgang Pauli, who was to be Heisenberg's close friend, "though often a very severe critic." Heisenberg and Pauli joked about Sommerfeld, and Pauli offered unadmiring opinions of Sommerfeld's elaborate extension of Bohr's atomic theory. It was all a grand "muddle," in Pauli's view.

The high point of Heisenberg's education in physics came during his fourth semester, when Sommerfeld took his bright student to Göttingen to attend a series of lectures on atomic theory given by Niels Bohr, an occasion known to the students as the "Bohr Festival." Heisenberg's recollection of these lectures gives a picture of the almost messianic impression Bohr made:

> I shall never forget the first lecture. The hall was filled to capacity. The great Danish physicist, whose very stature proclaimed a Scandinavian, stood on the platform, his head slightly inclined and a friendly but somewhat embarrassed smile on his lips. Summer light flooded in through the wide-open windows. Bohr spoke fairly softly, with a slight Danish accent. When he explained the individual assumptions of his theory, he chose the words very carefully, much more carefully than Sommerfeld usually did. And each one of his carefully chosen sentences revealed a long chain of underlying thoughts, of philosophical reflections, hinted at but never fully expressed. I found this approach highly exciting; what he said seemed both new and not quite new at the same time. We had all of us learned Bohr's theory from Sommerfeld, and knew what it was about, but it all sounded quite different from Bohr's lips. We could clearly sense that he had reached his results not so much by calculation and by demonstration as by intuition and inspiration.

Young as he was, Heisenberg did not hesitate to speak with Bohr and even argue against some of the work Bohr had reported in his lectures. One discussion was so absorbing it took the master and the enthralled student out of Göttingen to nearby Hainberg Mountain. "This walk was to have profound repercussions on my scientific career," Heisenberg recalls in his autobiography, "or perhaps it is more correct to say that my real scientific career only began that afternoon. . . . Suddenly the future looked full of hope and new possibilities, which I painted to myself in the most glorious colors." About a year later, Heisenberg visited Bohr's institute in Copenhagen and found its occupants awesomely gregarious and full of atomic physics. He soon felt at home, however, and for a few weeks resumed the long, "infinitely instructive" talks and walking tours with Bohr.

Heisenberg's first academic position was in Göttingen. In 1922, he became an assistant to Max Born. Heisenberg's predecessor in Göttingen had been Pauli. Born had been impressed by Pauli's talents, if not his dependability, but his new assistant was even more remarkable: "I had Heisenberg here during the winter (as Sommerfeld was in America)," Born wrote to Einstein. "He is easily as gifted as Pauli but has a more pleasing personality. He also plays the piano very well." To Born, noting contrasts with Pauli, he seemed "like a simple farm boy, with short, fair hair, clear bright eyes, and a charming expression."

Heisenberg, like Bohr ten years earlier, started his career in atomic physics at a critical time, "when the difficulties in quantum theory became more and more embarrassing. Its internal contradictions seemed to become worse and worse, and to force us into a crisis." The Bohr theory had worked its wonders with the problem of the hydrogen atom, and had done all it could do with the theory of multielectron atoms—no insignificant contribution. Most theorists, Bohr included, were struggling to find a new theory. Heisenberg took the first significant step toward a resolution while he was with Born in Göttingen as a *privatdozent* (instructor).

Heisenberg's inspiration was prompted, as great inspirations often are, by an enforced change of scene. "Toward the end of May, 1925," Heisenberg writes,

> I fell so ill with hay fever that I had to ask Born for fourteen days' leave of absence. I made straight for Helgoland [a small island in the North Sea], where I hoped to recover quickly in the bracing sea air, far from blossoms and meadows. On my arrival, I must have looked quite a sight with my swollen face; in any case, my landlady took one look at me, concluded that I had been in a fight and promised to nurse me through the aftereffects. My room was on the second floor, and since the house was built high up on the southern edge of the rocky island, I had a glorious view over the village, and the dunes and the sea beyond. As I sat on my balcony I had ample opportunity to reflect on Bohr's remark that part of infinity seems to lie within the grasp of those who look across the sea. Apart from daily walks and long swims, there was nothing to distract me from my problem, and so I made swifter progress than I would have done in Göttingen.

A New Mechanics

Heisenberg made his breakthrough at almost the same time that Pauli developed his exclusion principle. Recall that in Pauli's view the atomic landscape could be seen ultimately as a fine-grained system of stationary states occupied by electrons according to the dictates of the exclusion principle. Pauli's theory was a major step in the evolution of the concept of quantization. Planck had introduced energy quanta; Einstein had built a theory of radiation quanta or photons; and Bohr had constructed a picture of atoms existing in quantized stationary states. Pauli began to unify these theoretical fragments by enumerating the stationary states with quantum numbers.

However, Pauli's work was itself fragmentary as a theoretical edifice because the fourfold set of quantum numbers he postulated was based as much on empirical knowledge as on theoretical derivation. There was an urgent need for a general theory that deduced the quantum numbers rather than postulating them. Physicists still searched for a grand synthesis that encompassed the entire quantum realm, starting with a few mathematical statements.

Heisenberg took the first confident steps on this theoretical path. He put together the beginnings of a theory that eventually probed deeply into the dynamic workings of atoms. It was an atomic mechanics constructed in parallel to Newton's mechanics, but the resemblance was formal and abstract. Heisenberg shaped his theory with what Léon Rosenfeld called "formal virtuosity." Like Einstein, Heisenberg found his creative principle in mathematics. He once remarked that "it was natural for me to use a formal mathematical view which in some respects was an esthetic judgment."

By simplifying the axiomatic beginnings, and by building along mathematical lines, Heisenberg avoided the pitfalls distressing Bohr's theory. Without committing himself concerning the physical status of individual atomic electrons, he managed to build a dynamics that resembled the mathematical form of Newtonian mechanics and its elaborations. In an efficient, abstract way, he bridged the ordinary world and the atomic world. Bohr had crossed this bridge earlier, but with the difference that he had visualized the inner workings of atoms with some of the attributes of large-scale objects, such as the orbital motion of planets. Heisenberg's bridge to the atomic realm was formal and thoroughly mathematical, and it offered no such convenient images of atomic interiors.

Heisenberg was building in a style of theoretical architecture that was unfamiliar in atomic physics. This was an approach guided by mathematical models that formally resembled the Newtonian equations of motion, but was otherwise based only vaguely, if at all, on classical models or "pictures." The essential attitude, which soon became and remained dominant in quantum theory, was later bluntly summarized by Paul Dirac: "The main object of physical science is not the provision of pictures, but is the formulation of laws governing phenomena and the applications of these laws to the discovery of new phenomena. If a picture exists, so much the better; but whether a picture exists or not is a matter of only secondary importance."

Heisenberg's analysis worked with two fundamental physical ingredients, both of which were simple and observable, although neither helped much in the framing of physical pictures. First was the set of frequencies emitted by an atom when it jumps between stationary states in the manner originally proposed by Bohr. If an atom performs one of these quantum jumps downward from the higher energy E_2 to the lower energy E_1, a spectral "line" is emitted whose frequency, call it ν_{21}, is specified by the Bohr rule,

$$\nu_{21} = \frac{E_2 - E_1}{h}.$$

This concept is generalized to specify any frequency ν_{mn} emitted when an atom jumps between any two stationary states whose energies are E_{m} and E_{n},

$$\nu_{\mathrm{mn}} = \frac{E_{\mathrm{m}} - E_{\mathrm{n}}}{h}.$$

The entire set of frequencies $\{\nu_{\mathrm{mn}}\}$ collects all the lines observable in the atom's emission spectrum.

The second basic ingredient in Heisenberg's analysis evolved from a problem implied but not solved in Bohr's theory. Bohr had used the concept of atoms jumping between stationary states, but he could not cope with the problem of how one knew when and where a particular atom was going to make a particular kind of jump. This was a difficulty Rutherford had immediately spotted when he saw Bohr's first papers. "It seems to me," he wrote to Bohr in 1913, "that you would have to assume that the electron [about to jump] knows beforehand where it is going to stop." Rutherford was asking for a deterministic mechanism like those familiar in classical physics.

Bohr never managed to make his theory work that way, but he later took a valuable hint from a paper written by Einstein in 1916. The idea had occurred to Einstein that atoms making quantum jumps are like disintegrating radioactive atoms. Predictions concerning the when and where of individual radioactive disintegrations had also proved impossible, and in the absence of a better procedure, the laws of radioactivity had for some time been formulated statistically, as predictions of what *probably* would happen to a radioactive atom. From the viewpoint of an individual atom, this is an indeterminate description because the statistical statement says nothing with certainty about individual processes; it is an account of *average* behavior inferred from data taken on a very large number of atoms. Einstein saw that this statistical description could be extended to all atomic change. Among other things, he managed "in an amazingly simple and general way" to arrive at Planck's radiation law by defining probabilities for the occurrence of all possible atomic transitions. Bohr took up this theme and found a place in his own atomic theory for Einstein's "transition probabilities."

So we find Heisenberg in 1925 extending the Einstein-Bohr canon. The second physical constituent in Heisenberg's analytical recipe, accompanying the set of spectral frequencies $\{v_{mn}\}$, was a set of transition probabilities. If the probability for the m-to-n transition, labeled let's say A_{mn}, is large, the transition is likely to occur, and the spectral line whose frequency is v_{mn} is intense. Thus the transition probabilities are theoretical manifestations of the observable spectral line intensities.

Heisenberg found that the transition probabilities A_{mn} and the frequencies v_{mn} could be used in a method of calculation that resembled a well-established technique known as "Fourier analysis" (invented by Joseph Fourier in the early nineteenth century for his analytical theory of heat). For each observable quantity known in Newtonian mechanics, Heisenberg found a quantum counterpart that was recognizable as a "Fourier expansion," formulated with the frequencies and the transition probabilities.

In later developments, the sets of transition probabilities were arranged in square arrays with all the entries concerning state 1 in row 1, entries for state 2 in row 2, and so forth. If a total of three states is involved, the square array has the appearance

$$\begin{pmatrix} A_{11} & A_{12} & A_{13} \\ A_{21} & A_{22} & A_{23} \\ A_{31} & A_{32} & A_{33} \end{pmatrix}.$$

Guided by the Fourier procedure, which was mostly a mathematical technique, and striving for a dynamics that formally resembled Newtonian mechanics when the arrays were replaced by corresponding classical variables, Heisenberg arrived at a workable quantum mechanics.

Inspiration

Once he had managed to "jettison all the mathematical ballast" he brought from Göttingen to his second-floor room on Helgoland, with its partial view of infinity, Heisenberg quickly saw the form of his new mechanics. As it took shape, and he could see that it was physically and mathematically consistent, Heisenberg was distracted by an intense excitement—"I began to make countless errors"—and even by a curious anxiety: "At first, I was deeply alarmed. I had a feeling that, through the surface of the atomic phenomena I was looking at a strangely beautiful interior and felt almost giddy at the thought that I had to probe this wealth of mathematical structures nature had so generously spread out for me." The first successful calculations had been completed by three o'clock one morning. Sleep was impossible: "So, as a new day dawned, I made for the southern tip of the island, where I had been longing to climb a rock jutting out into the sea. I now did so . . . and waited for the sun to rise."

But in the wake of his initial optimism and excitement, Heisenberg began to feel uneasy about his new mechanics, because it worked with a peculiar kind of algebra. Two variables, call them x and y, represented as square arrays in the Heisenberg manner, obeyed a strange multiplication rule: the product xy was not always mathematically equivalent to the product yx with the factors reversed, as in ordinary algebra. "The fact that xy was not equal to yx was very disagreeable to me," Heisenberg writes. "I felt that this was the only point of difficulty in the whole scheme; otherwise I would be perfectly happy." Most of the theory was constructed by June 1925, when Heisenberg received an invitation to lecture at the Cavendish Laboratory in Cambridge. The choice was to complete the work quickly or "throw it into the flames." Pauli, the invaluable critic, read the manuscript, and responded "with jubilation." It gave him "new hope, and a renewed enjoyment of life." Heisenberg presented his paper to Born, but in Cambridge he said nothing about his recent efforts.

Matrix Mechanics

"Heisenberg's latest paper, soon to be published, appears rather mystifying but is certainly true and profound," Born wrote to Einstein in July 1925. To Born, it was clear that a genuine quantum mechanics was at hand, and he began developing a full mathematical statement of the theory. He was particularly intrigued by the remarkable multiplication rule: "Heisenberg's symbolic multiplication rule did not give me rest, and after days of concentrated thinking and testing I recalled an algebraic theory I had learned from my teacher, Rosanes, in Breslau." The algebraic theory concerned "matrices," mathematical arraylike entities, whose algebra had been formulated by Arthur Cayley, with a mathematician's foresight, some seventy years earlier. The peculiar multiplication rule discovered by Heisenberg was strictly analogous to matrix multiplication; the Heisenberg arrays were formally identifiable as matrices. Once Born had this clue, the way was cleared for the development of a quantum "matrix mechanics." That work was started by Born, Heisenberg, and a young matrix expert, Pascual Jordan.

Born and Heisenberg found themselves in an alien mathematical world in which they were not fluent with the language. "I do not even know what a matrix is," Heisenberg complained to Jordan. As it happened, however, the Göttingen physicists were not lacking in good advice on how to handle their mathematical difficulties. The great mathematician David Hilbert also lived in Göttingen, and he, better than anyone in the world, spoke the mathematical language the physicists needed to learn. Edward Condon, an American who was on the Göttingen scene, tells about Hilbert's advice: "Hilbert was having a great laugh on Born and Heisenberg and the Göttingen theoretical physicists because when they first discovered matrix mechanics they were having, of course, the same kind of trouble that everybody else had in trying to solve problems and to manipulate and really do things with matrices. So they went to Hilbert for help."

Hilbert told them that for him matrices were handy devices for bringing out certain formal aspects of problems written in another mathematical idiom, that of differential equations. Because physicists had for many years exploited the language of differential equations to great advantage in other problems, Hilbert suggested that the matrices might be manifestations of more-useful equations of the differential kind. According to Condon, the Göttingen theorists thought that was "a goofy idea and that Hilbert did not know what he was talking about." But Hilbert was rarely wrong. Just six months later, Erwin Schrödinger found the equations Hilbert had prophesied, and demonstrated that they accomplished the same things as matrix mechanics and more—with the familiar methods of differential equations.

War and Aftermath

Heisenberg began his career in the 1920s and 1930s, during a time of great achievement in atomic physics. The work was done by theorists and experimentalists who were young—many of them in their twenties. They came from all over the world and met in Copenhagen, Göttingen, Berlin, and Munich. It was an international community whose citizens swore allegiance as much to science as to their home countries. For the scientists who were lucky enough to participate, it must have been an intellectual's paradise.

But at the same time physicists were thriving on this spirit of internationalism, political forces feeding on the most intense feelings of nationalism were rising in Germany. The National Socialist (Nazi) Party, led by Adolf Hitler, was the focus. By 1933, Hitler and the Nazis were in power and Germany was rapidly becoming isolated from the rest of the world, as many of its most renowned physicists, chemists, and mathematicians were forced to emigrate.

Heisenberg witnessed these grim events from Leipzig, where he had been appointed professor of theoretical physics in 1927. "When I returned to my Leipzig Institute at the beginning of the summer term of 1933," Heisenberg writes in his autobiography, "the rot had begun to spread. Several of my most capable colleagues had left Germany, others were preparing to flee." Heisenberg never belonged to the Nazi Party, and by the early 1930s he had no sympathy for its ideals or tactics. Yet he did not seriously consider emigration; he loved his country, and to that extent he was a nationalist.

Not many of Germany's great physicists and chemists stayed, but a few did. In addition to Heisenberg, there were Otto Hahn, the radiochemist who with Fritz Strassmann did the experiments that led to the discovery of nuclear fission; Max von Laue, best known for his work in x-ray crystallography; and Max Planck. Heisenberg went to see Planck, now an old man but solid as ever in his ideals and integrity: "Planck received me in a somewhat somber but otherwise friendly and old-fashioned living room; all that was missing was an oil lamp over its central table. Planck seemed to have grown a good many years older since our last meeting. His finely chiseled face had developed deep creases, his smile seemed tortured, and he was looking terribly tired."

Planck said that he had recently met with Hitler and had tried to make him understand that he was destroying the German universities: "I had hoped to convince him that he was doing enormous damage . . . by expelling our Jewish colleagues; to show how senseless and utterly immoral it was to victimize men who had always thought of themselves as German, and who had offered up their lives for Germany like everyone else." The effort was futile. "I failed to make myself understood," Planck said. "There is simply no language in which one can talk to such men."

Planck could offer few words of encouragement, but his advice was to stay and hold on to what was now most precious, the students: "You cannot stop the catastrophe, and in order to survive you will be forced to make compromise after compromise. But you can try to band together with others and form islands of constancy. You can gather young people around you, teach them to become good scientists and thus help to preserve the old values . . . for such groups can constitute so many seed crystals from which new forms of life can grow."

Acting on Planck's advice proved to be an excruciating, and often perilous, game. By the middle 1930s, a bogus movement called "Aryan physics" or "German physics," originated and promulgated by two Nobel Prize–winning experimentalists, Johannes Stark and Philipp Lenard, was gaining strength. Their attitude was blatantly anti-Semitic. They aimed to suppress the prevalent theoretical "Jewish physics"—relativity and quantum theory—and promote in its place a more concrete science with transparent empirical foundations. Their original target, Einstein, was now gone from Germany, but "Jewish formalism" persisted in the theories of Einstein's friends, Planck, Laue, and "the theoretical formalist, Heisenberg, spirit of Einstein's spirit."

Stark and company launched a vicious campaign of vilification against Heisenberg, which intensified during 1936 and 1937, and ultimately threatened his academic position and even his safety. It was a period of "unending loneliness" from which he began to emerge when he married Elisabeth Schumacher, a strong young woman who was thirteen years his junior. Shortly after the marriage, Heisenberg made the courageous and risky decision to write to Heinrich Himmler, head of the SS (Schutzstaffel) and in effect the Reich's chief of police, requesting that the charges against him be officially investigated. If he couldn't be cleared, he would resign and volunteer for military service. The SS investigation was prolonged and humiliating, but finally favorable in its judgment of Heisenberg's political reliability. "Heisenberg's character is decent," the SS investigators reported. "Heisenberg is typical of the apolitical academic. . . . Over the course of several years, Heisenberg has allowed himself to be convinced more and more of National Socialism through its successes and is today positive toward it. He is however of the view that political activity is not suitable for a university teacher, save for the occasional participation in indoctrination camps and the like." Even so, Heisenberg had to be careful: he could not mention the names of Jewish physicists to his students or in his papers, and he occasionally had to represent Nazi Germany abroad.

When war broke out in September 1939, Heisenberg was ordered to join the Uranium Project—known as the "Uranium Club" to its members—founded to follow up the possibilities raised by the Hahn-Strassmann nuclear fission experiments. By late 1939, Heisenberg and his colleagues had concluded that a nuclear chain reaction was possible in natural uranium if it was initiated by neutrons whose energy had been reduced to low levels in a "moderator," either heavy water or graphite. They were also convinced that the rare isotope uranium 235 could be used as a nuclear explosive. Intensive work followed, and "toward the end of 1941," Heisenberg writes, "our 'Uranium Club' had, by and large, grasped the physical problems involved in the technical exploitation of atomic energy [not including nuclear bombs]." At this point German nuclear research was perhaps a year ahead of British and American nuclear efforts.

One problem the Uranium Club experts recognized but could not solve was how to separate uranium 235 from the much more abundant uranium 238. Such separations—of isotopes with nearly the same mass—had never been attempted, or hardly imagined. To prepare even a small amount of uranium 235 would require years and vast resources. At the time, Hitler allowed no ordnance development that did not promise results in six months.

To the extent that it was possible for anyone enduring the war years in Nazi Germany, Heisenberg was lucky. Nuclear bomb development, which he evidently feared and knew he had to avoid, was an impossibility in Germany. With no distortion of the facts, the Uranium Club could advise that only one kind of nuclear effort was feasible, the development of a graphite- or heavy-water-moderated nuclear reactor. Work on the heavy-water design began in Berlin at the Kaiser-Wilhelm Physics Institute.

Heavy Allied bombings of Berlin started in 1943, and the reactor research had to be moved to a safer location. The village of Haigerloch in the south near Stuttgart was chosen for the new site. Haigerloch was not only safe, it was hardly of this world. As one of the chroniclers of German wartime nuclear research, Robert Jungk, writes, "In all Germany there were few such operatically romantic sites as Haigerloch. . . . At this spot, which had hardly changed since the middle ages, the most modern German power station was built." Work on the reactor was resumed in a chamber carved in the rock beneath the town's "half-Gothic, half-Baroque church." While the experimental work progressed, Heisenberg would sometimes go up to the church and play Bach fugues on the organ. "It was the most fantastic period of my life," one of Heisenberg's colleagues remarked later.

In a grim way, Heisenberg was indeed a lucky man. Soon after he left Berlin for Haigerloch in 1944, an attempt was made to assassinate Hitler. The plot failed and some of Heisenberg's friends and associates (including Planck's son Erwin) were arrested and executed. Had Heisenberg remained in Berlin, and without the "benefits" of the SS investigation and his compliance with the demands of the regime, his life would certainly have been in great danger.

In April 1945, Allied armies invaded southern Germany. The French were to occupy Haigerloch, but American intelligence officers knew that Heisenberg had relocated his institute there and in nearby Hechingen. A small unit code-named the "Alsos Commission," advised by Samuel Goudsmit of earlier electron-spin fame, raced ahead of the French army to snatch the German scientists, their papers, and their equipment. The prize they most wanted to capture—Heisenberg—could not be found, however. As he had planned, Heisenberg had left Hechingen at the last minute (on a bicycle) to join his family in the Bavarian Alps, where they had been waiting out the last months of the war.

Elisabeth Heisenberg tells of her husband's journey through the chaos left by the defeated and destitute German army: "While all this [the occupation] was taking place in Hechingen, Heisenberg was riding east on his bicycle. He was on the road for three days and nights until he arrived home safe and sound." He evaded "bands of marauding, tattered figures speaking foreign languages, who had been released or had escaped from some prison camp or from forced labor, and were now roaming the countryside plundering." He saw teenagers, drafted into the German army at the end, "now camping along side the road, crying, hungry and lost, not knowing what to do." Everywhere there were soldiers on the move, all "going somewhere, some to the east, others to the west or north, without a plan, exhausted and threatening." At the end of his ride, Heisenberg did not have long to wait before the arrival of the Alsos unit. When he was finally taken prisoner, Heisenberg writes, "I felt like an utterly exhausted swimmer setting foot on firm land."

With the hope that further intelligence could be gathered, Heisenberg and nine other German scientists were kept in internment for six months. Included in the group, in addition to Heisenberg, were Otto Hahn, Max von Laue, Walter Gerlach (who with Otto Stern had designed a classic experiment that demonstrated the existence of spin states), and Carl Friedrich von Weizächer, a colleague and close friend of Heisenberg's. The ten were taken to a large country estate near Cambridge called Farm Hall, and there they were "imprisoned." Hahn describes their pampered existence: "Our life in England was truly luxurious. Breakfast consisted of porridge or cornflakes, bacon and eggs, toast, butter and marmalade. For luncheon and dinner we had rump steaks or a roast, very often with *pommes frites*. It was no wonder we all began to put on weight. Five prisoners of war were detailed to look after us, among them a very good cook. Inside the house and the very large garden, these prisoners were as free as we were."

They exercised, played cards, read Dickens, and held seminars (which revealed by way of secret microphones that Heisenberg's grasp of nuclear bomb physics was primitive); Heisenberg played Beethoven sonatas on a fine piano. If it had not seemed so unreal, and if their families in Germany had not been facing starvation or worse, life in this "prison," which they called the "golden cage," would have been an idyll.

Vision

Released from the golden cage, and back in Germany, Heisenberg could face reality once more, no doubt with relief. He became director of the Max Planck Institute (formerly the Kaiser Wilhelm Institute) in Göttingen. Like Planck a generation earlier, he worked with great energy and vision to raise German science from its postwar devastation. In addition to rebuilding the institute at Göttingen, his voice was influential on science matters, domestic and international, in the West German chancellor's office. As always, he focused his research on the most fundamental theoretical problems. Beginning in the 1950s, he pursued the dream of the unifiers—a generalized field theory. He hoped to find a fundamental wave equation that embodied all of elementary particle physics. When he thought he had a particularly promising possibility, he submitted it to the usual test, Pauli's criticism. Elisabeth Heisenberg describes the fierce exchange of letters between Heisenberg and Pauli while they thrashed out the meaning of the theory: "The letters were harsh and without mercy. It was really like a battle, and each volley was answered by an equally strong one from the other side. This 'battle' turned out well in the first round. [Heisenberg] finally succeeded in convincing Pauli of his ideas."

Pauli became enthusiastic about the direction the theory was taking. In one letter, he wrote: "This is powerful stuff. . . . The cat is out of the bag, and has shown its claws. . . . A very happy New Year. Let us march forward toward it. It's a long way to Tipperary, it's a long way to go." Pauli decided to travel to the United States and lecture on the theory, but Heisenberg was anxious: "I did not like the idea of this encounter between Wolfgang in his present mood of exaltation and the sober American pragmatists, and tried to stop him from going." One of the "sober American pragmatists" was Jeremy Bernstein, whose account of Pauli's appearance at Columbia University before an audience including Bohr was quoted above.

In the end, Pauli agreed with Bohr's assessment that the theory was "not crazy enough." He gave up on it, and wrote to Heisenberg, "You're free to go your own way, but I want nothing more to do with it." Abraham Pais notes in his chronicle of modern theoretical physics, *Inward Bound*, that the theory and its variations "were not influential in the long run."

So the story of Heisenberg's grand theoretical effort had an unsuccessful conclusion. But it was a story of high intellectual adventure, of reaching once more for the creative spirit. "One moonlit night we walked all over Hainberg Mountain [near Göttingen]," Elisabeth Heisenberg writes, "and [Heisenberg] was completely enthralled by the visions he had, trying to explain his newest discovery to me. He talked about the miracle of symmetry as the original archetype of creation, about harmony, about the beauty of simplicity, and its inner truth. It was a high point of our lives."

19

Wave Mechanics
Erwin Schrödinger and Louis de Broglie

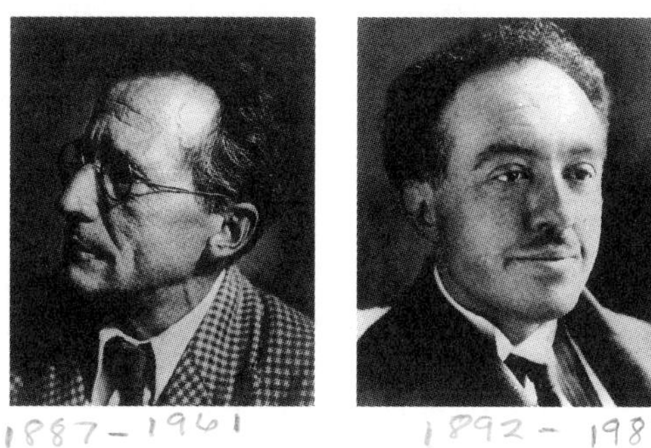

Hopes and Fears

Paul Dirac has offered the opinion that his fellow theorists are guided not only by their hopes, but just as importantly, by their fears. Theoretical researchers find it hard, he says, to ignore fears that their work contains hidden, possibly disastrous, flaws; and their thoughts, influenced by this worry, are not so logical as they might be: "You might think a good research worker would review the situation quite calmly and unemotionally and with a completely logical mind, and proceed to develop whatever ideas he has in an entirely rational way. This is far from being the case. The research worker is only human and, if he has great hopes, he also has great fears. . . . As a result, his course of action is very much disturbed. He is not able to fix his attention on the correct line of development."

If there was a fundamental fear threatening the development of quantum theory during its first two decades, it was the concept of wave-particle duality, demanded because light can appear to be wavelike in certain experiments and particle-like in others. Einstein was among the first to face the duality mystery. In spite of long-established experimental and theoretical evidence for light as waves, Einstein proposed a particle theory of light to explain puzzling features of the photoelectric effect. Einstein's equation $E = h\nu$ for the energy E of a light particle or photon casually introduces the duality theme: the equation combines E, a property of light as a particle, with the frequency ν, a property of light as a wave.

From the logical viewpoint, this was a paradox, which hardly any theoretician but Einstein had the courage to confront. How could light be two essentially different things, wave and particle, at the same time? The duality seemed to be a threat, a "fundamental blemish" that might, if pushed too far, bring the entire theoretical edifice crashing down.

The Brothers de Broglie

The first among theoreticians to follow Einstein's lead in facing the challenge of the wave-particle enigma was a French nobleman, Louis-Victor de Broglie. As a younger son born into an illustrious, wealthy, aristocratic family, Louis de Broglie was entitled to princely status, but not expected to pursue a career as intellectual and sedentary as science. To the old duc de Broglie, Louis's grandfather, science was "an old lady content with the attractions of old men." But Louis's older brother, Maurice, had managed to make a fine reputation in experimental physics while simultaneously pursuing a more traditional naval career. Influenced by his brother, and no doubt protected by him from family opposition, Louis became a theoretical physicist after taking a nonscientific degree in ancient history and paleography.

Beginning in 1913, Maurice de Broglie had done experimental work on x rays, in which, to the experimentalist at least, the wave-particle question was hard to avoid. His first x-ray experiments followed the discovery that beams of x rays interfere with each other to give characteristic bright and dark patterns. Such "diffraction" effects had been observed almost a century earlier in experiments with ordinary light, and explained with a wave theory. One of the discoverers of x-ray diffraction, with its implication that x-ray beams could be understood as processions of waves, was William Bragg, a British experimentalist who had just previously found convincing evidence that x rays have particle properties. Bragg first began to see x rays the other way, as waves, through the eyes of his son, Lawrence Bragg, who invented and applied a famous equation that treats x rays as waves and allows a detailed analysis of x-ray diffraction patterns. Having had concrete experience with x rays as both particles and waves—and, as experimentalists, being unthreatened by the fears of theorists—the Braggs were among the first to recognize that neither the wave nor the particle theory of x rays was adequate in itself. In 1912, the elder Bragg wrote: "The problem becomes not to decide between theories of x rays, but to find one theory which possesses the capacities of both."

By the early 1920s, Maurice de Broglie had seen enough of x-ray behavior to share Bragg's opinion and to pass this viewpoint on to his brother Louis, who by then was realizing his talents as a theorist. Louis de Broglie writes: "My brother considered x rays as a combination of wave and particle, but not being a theoretician, he did not have particularly clear ideas on the subject." For a time, the two brothers worked together on experiments involving the study of "recoil" electrons produced when x rays are scattered by solid materials.

In this experimental work, and in "long discussions with my brother on the interpretation of his beautiful experiments," Louis de Broglie was "led to profound meditations on the need of always associating the aspects of waves with that of particles." He began to look on wave-particle duality as a natural symmetry applicable not only to radiation forms such as light and x rays but also to the elementary constituents of matter, particularly electrons. Ever since the work of J. J. Thomson in the late 1890s, electrons had been understood as tiny particles carrying a definite charge and mass. At the time de Broglie formulated his theory, there was no evidence whatever that electrons could show themselves as anything but particles. Yet, on the basis of his firm belief in wave-particle symmetry, and arguing along the lines suggested mainly by Einstein's special theory of relativity, de Broglie arrived at several crucial results predicting that electrons and the other "particle" constituents of matter should show manifestations of wave behavior.

Matter Waves

De Broglie's argument began with the supposition that "the basic idea of quantum theory is the impossibility of considering an isolated fragment of energy without assigning a certain frequency to it." The particles of radiation—and of matter as well—had a level of existence that was fundamentally a "periodic process." Such was the physical content of the Planck-Einstein equation $E = h\nu$, with its energy term E and its frequency factor ν. De Broglie also pointed out that a photon's wavelength, a wave property, could be related to the photon's momentum, a particle property, by combining $E = h\nu$ with another energy equation,

$$E = mc^2,$$

derived from special relativity. The two equations combined read

$$E = h\nu = mc^2,$$

from which we obtain

$$mc = \frac{h\nu}{c}.$$

Because the photon's speed is c, the term mc in the last equation can be regarded as the photon's momentum p, so

$$p = \frac{h\nu}{c}. \qquad (1)$$

Switching now from the particle viewpoint and momentum considerations to the wave viewpoint, we make use of the equation $\lambda\nu = c$, which connects the wavelength λ, frequency ν, and speed c of light waves, to calculate ν with

$$\nu = \frac{c}{\lambda}.$$

When this way of expressing the frequency is substituted into the momentum equation (1), the result is

$$p = \frac{h}{\lambda}. \qquad (2)$$

This equation still refers to photons, but de Broglie saw no reason why electrons and other particles of matter, since they, too, were "isolated fragments of energy," should not also have associated frequencies and wavelengths. In a derivation of more complexity than the one outlined here—but beginning with the energy equation $mc^2 = h\nu$—de Broglie justified the momentum equation (2) for all kinds of material particles. This was de Broglie's major contribution. It suggested that electrons, and all other particles of matter, not only had momentum and energy attributes, as J. J. Thomson had established several decades earlier, but also a mysterious wavelength.

So de Broglie's momentum-wavelength equation (2) joined the Planck-Einstein energy-frequency connection $E = h\nu$ as another duality equation with a particle quantity (the momentum p) on one side, a wave quantity (the wavelength λ) on the other, and the ever-present Planck's constant h standing between.

Einstein For, Copenhagen Against

To Einstein, at least, de Broglie's theoretical argument was convincing, almost self-evident, by its generality and simplicity. When Einstein heard of de Broglie's work from his friend Paul Langevin (de Broglie presented his theory as a doctoral thesis for Langevin), he replied with Einsteinian eloquence that de Broglie had "lifted a corner of the great veil." Einstein took up the cause of the new "wave mechanics," and the benefit to de Broglie was crucial: "The scientific world of the time hung on every one of Einstein's words, for he was at the peak of his fame. By stressing the importance of wave mechanics, the illustrious scientist had done a great deal to hasten its development."

At first, Einstein was the only physicist of note to take de Broglie's side. De Broglie was no stranger in the scientific community; in Copenhagen and Göttingen, his name and reputation were known, but not favorably. Several unfriendly debates had pitted de Broglie and some of his French colleagues against Bohr and the Copenhageners, with the latter usually coming out ahead. The most famous of these rivalries concerned element 72. In Paris, this element was associated with rare-earth elements and called "celtium," while in Copenhagen it was "hafnium" (a Latinized version of Copenhagen), and on the basis of a suggestion by Bohr, considered to be related to the element zirconium. When Bohr and his colleagues were proved right on this and several other occasions, de Broglie and his allies acquired a reputation for supporting misguided theories. Predictably, de Broglie's radical ideas about electron waves were not taken seriously in Copenhagen and in other places where Bohr's influence was strong.

Electron Waves Observed

The experimental discovery of electron waves predicted by de Broglie's theory was finally reported in 1927, by Clinton Davisson and Lester Germer in the United States, and by G. P. Thomson (the only son of J. J. Thomson, who was the first to see electrons as particles) in England.

The Davisson-Germer experiments, more complete and definitive than those of Thomson, evolved over almost a decade of difficult experimental development. As the experiments were finally and most successfully done, an electron beam of a precisely determined low energy was formed and directed at a specially prepared face of nickel crystal, and scattered portions of the beam were collected by a moveable detector. Experiments with this apparatus showed that electrons were not scattered uniformly in all directions. Instead, under certain conditions, a sharply defined current of electrons was observed in a direction for which the angle of incidence on the crystal's surface was equal to the angle of reflection. If, to the electrons, the nickel surface were entirely smooth and flat, this result would not have been surprising: throw a rubber ball at a smooth wall and it always bounces off at an angle equal to the angle of incidence. But to particle-like electrons, a nickel surface cannot conceivably be smooth: electrons in the form of particles are much smaller and less massive than nickel atoms. Reflection of electron particles from a nickel surface is, in Davisson's apt description, "like imagining a handful of bird shot being regularly reflected by a pile of large cannon balls." The difficulty is that a "surface made up of large cannon balls is much too coarse-grained to serve as a reflector for particles as small as bird shot."

Davisson and Germer successfully analyzed their data by treating the electron beam as if it were a beam of x rays displaying its wave manifestations. Lawrence Bragg had treated the reflection of x-ray waves from crystal planes by imagining the effect produced on individual rays. A "primary" ray was reflected at the crystal face and joined by "secondary" rays reflected by successive layers of atoms in the crystal. The reflected rays, both primary and secondary, formed a concerted and reinforced wave front if all the waves joined in step, crests falling on crests, and troughs on troughs. Bragg's equation guaranteed this condition, and he applied the equation to the determination of crystal structures.

Davisson and Germer found that they could unravel their mysterious data if they discarded the previously accepted picture of an electron beam as a shower of particles and assumed instead that the Bragg equation applied. This was impressive evidence for the theoretical viewpoint expressed by de Broglie at about the same time Davisson and Germer started their experiments. Design and interpretation of the experiments might well have been guided by de Broglie's theory, because with elaborations the theory predicted everything Davisson and Germer observed. But, as we have seen, experimentalists are not always in close touch with theorists, and vice versa. Davisson and Germer did not read de Broglie's paper and then set out on a systematic search for electron waves; their experiments originated in litigation, a famous patent suit.

The principal parties to the suit were the General Electric and Western Electric Companies. General Electric had applied for a basic patent on a three-electrode (triode) vacuum tube that was similar to a design already owned by Western Electric. It was the contention of General Electric that theirs was a high-vacuum device, whereas the Western Electric tube required appreciable air for its operation. According to the General Electric argument, the air molecules formed positive ions that then bombarded the oxide surface of the tube's cathode, releasing electrons the tube needed for its operation. Western Electric hoped to refute this argument by gathering experimental evidence on the effects of positive-ion bombardment on oxide surfaces. The work was started by Germer under Davisson's direction at the Western Electric Laboratories. The General Electric claim was disproved, and the suit was eventually decided in Western Electric's favor.

The bombardment experiments were continued after the settlement of the suit, however, and extended to include bombardment of bare metal surfaces from which the oxide coating was removed. As Germer remarks, it was also possible, "by changing a few potentials on some of the electrodes, to measure emission under electron bombardment." So the work that finally led to an elegant demonstration of electron waves was "undertaken as a sort of sideline." The electron studies were continued for several years, and the data showed an increasingly complex and strange pattern. A major clue was revealed accidentally when a flask of liquid air exploded and shattered the evacuated tube containing the nickel target. Reconstructing the apparatus required cleaning the nickel surface by degassing at high temperatures. This had the unforeseen effect of forming a few large nickel crystals not present in the original target. The complexities were now traced to the crystals, and experiments were started with a *single* nickel crystal whose reflection planes could be oriented in a controlled manner.

Up to this time—it was now 1926—Davisson and Germer were unaware of de Broglie's theory of electron waves. At a meeting of the British Association for the Advancement of Science in Oxford, Davisson heard of the new wave theory and realized that the patterns of the bombardment data, which he and Germer were already finding suggestive of x-ray behavior, actually told a story of wave phenomena. "The experiments were at once guided by the theory," writes Germer, "and were quickly successful." Davisson shared a Nobel Prize in 1937 with the other discoverer of electron waves, G. P. Thomson, who tells us that the inspiration for *his* work with electron waves came while watching another experiment that later gave results that were "quite erroneous and entirely instrumental in origin."

Beauty before Science

We have followed the story of Louis de Broglie's theoretical vision of wave-particle symmetry. We have also seen how de Broglie and his colleagues in Paris had in various ways isolated themselves from the "Copenhagen-Göttingen axis," and made it unlikely that further theoretical work on the mechanics of electron waves would be done by the established practitioners of quantum physics. So it was that Erwin Schrödinger, a scientific loner based in Zürich, became the chief architect of electron wave mechanics, after de Broglie's work on the foundations.

Schrödinger was born in Vienna in 1887. (Schrödinger, Bohr, and Born were about the same age, older than the other founders of quantum mechanics, Heisenberg, Dirac, and Pauli, who were all born around 1900.) His father, Rudolf, not only ran the family linoleum business successfully but maintained an active, near professional interest in botany, chemistry, and Italian painting. One of Schrödinger's biographers, William Scott, writes of the strong tie between father and son: "As friend, teacher and tireless partner in conversation, Rudolf Schrödinger shared his lively intellectual life with his son and only child. Looking back on his childhood, Schrödinger remembered his father as the 'Court of Appeal' for all subjects of interest." Schrödinger's formal education began at the Akademische Gymnasium, where ancient languages and literature were major subjects. From his maternal grandmother, who was English, he acquired proficiency in the English language; in later years he wrote and spoke English with style and fluency. His ability with other modern languages was also remarkable; he lectured and entertained audiences in French and Spanish, as well as in German and English.

Schrödinger entered the University of Vienna shortly after Ludwig Boltzmann's tragic death, but Boltzmann's influence was still alive in the cycle of lectures on theoretical physics given by his successor, Friedrich Hasenöhrl. Many years later, Schrödinger still held Hasenöhrl's lectures as his "supreme model" and regarded the Boltzmann line of thought as his "first love in science. No other has ever thus enraptured me or will ever do so again."

At first, Schrödinger found it difficult to face modern developments in atomic theory: "Its inherent contradictions sounded harsh and crude, when compared with the pure and inexorably clear development of Boltzmann's reasoning. I even, as it were, fled from it for a while." The intensity of Schrödinger's concern for both the philosophical and the mathematical problems of physics impressed his professors and his fellow students. His appearance at a mathematics seminar was pointed out to a new student with a whispered, "Das ist der *Schrödinger.*"

In 1918, after World War I, Schrödinger looked forward to a career as a part-time physicist and a full-time philosopher. A chair at the University of Czernowitz seemed imminent. "I was prepared to do a good job lecturing in theoretical physics . . . but for the rest to devote myself to philosophy." Suddenly, in the aftermath of the war, Czernowitz was no longer part of Austria. "My guardian angel intervened. . . . I had to stick to theoretical physics, and, to my astonishment, something occasionally emerged from it."

For several years, Schrödinger followed the kind of itinerant academic career common in German university life; after short stays in Jena, Stuttgart, and Breslau, he finally settled for six years at the University of Zürich (where Clausius and Einstein had been among his predecessors). This was the most active period of his life, when the great work on wave mechanics was completed. Then, in 1927, Max Planck retired and persuaded Schrödinger to go to Berlin as his successor. For a time, life was pleasant in Berlin: Planck, Einstein, and Max von Laue were there, and Berlin was a major center for theoretical and experimental research.

But then the Nazi nightmare descended, and Schrödinger joined the general exodus of Germany's leading intellectuals. He was not Jewish, and was one of the few German scientists to emigrate without being forced out. Traveling again, he went to Oxford, to Graz, back to Oxford, to Ghent, and to Rome, where he was approached by Eamon de Valera—mathematician, scientist, and prime minister of Ireland. De Valera proposed an Institute for Advanced Studies in Dublin (modeled after the one in Princeton). Funds were short; studies at first were to be confined to two "paper and pencil" schools, a School of Celtic Studies, and a School of Theoretical Physics, which de Valera invited Schrödinger to direct. Schrödinger accepted, and in neutral Ireland he found life peaceful and productive once more. He was a popular lecturer in Dublin, endearing himself to the Irish with his knowledge of Irish music, Celtic design, and the Gaelic language. But the Irish weather did not suit him. In 1956, failing health and a longing for his native Austria took him back to Vienna.

A recent biography of Schrödinger by Walter Moore probes the depths of Schrödinger's complex personality, and tells you all about the man. Moore informs us that Schrödinger found it easy to fall in love, particularly with young women. "Erwin was intensely concerned with sexual experience," writes Moore. "One might say that he was devoted to it as the principal nonscientific occupation of his life. Not only did he enjoy making love, but he also conceived of it as a way to achieve transcendence and to perpetuate himself." Schrödinger's love affairs were numerous and intense, and he had several illegitimate children. But through it all his marriage to Anny (Annemarie) Bertel survived. Anny's attitude concerning the marriage was as remarkable as Schrödinger's was. "She regarded him as a great man in all respects, who was above criticism on any ground," says Moore. "She was willing to tolerate his every Seitensprung [extramarital affair] and acted as an insurance whenever he wished to end one." In love and science, Schrödinger was fascinated by beauty. He wrote to Max Born: "I have no higher aim than to work out the *beauty* of science. I put beauty before science. We are always longing for our neighbor's housewife and for the perfection we are least likely to achieve."

Schrödinger's scientific work was remarkably broad. One of his earliest efforts concerned a theory of color perception. At one time or another, he dealt with nearly all aspects of modern physics: statistical mechanics, x-ray diffraction, general relativity, unified field theory, and the theory of specific heats, as well as the more familiar work on wave mechanics. In 1944, he published a little book entitled *What Is Life?*—one of the first excursions into the realm of molecular biology. (Francis Crick, who with James Watson discovered the double-helix DNA model, tells us that Schrödinger's book was largely responsible for his conversion from physics to molecular biology.) Like Einstein and Bohr, Schrödinger found unity in the diversity of his interests. In the foreword to *What Is Life?* he speaks of the "small number of definite ways of thought that are relevant to [me] and to which [I] therefore return again and again on various occasions."

Schrödinger's Equation

Schrödinger acknowledged that his work on wave mechanics owed debts not only to de Broglie but also to "short but infinitely far-seeing" remarks of Einstein's and to a dualistic mechanics created almost a century earlier by the Irish physicist and mathematician William Rowan Hamilton. Long before any suspicion had been aroused that the physical world was made of wave-particle entities, Hamilton had composed a unified theory of light-ray and particle motion. Carried to its logical conclusion, Hamilton's dynamics implied that any particle should have associated with it a system of waves. Hamilton did not state this conclusion—probably he did not even think of it—because in the 1830s there was no evidence whatever that wave manifestations of particles existed. But Hamilton's dualistic mechanics had a formal, mathematical beauty that kept it alive for the ninety years needed to bring the duality theme back again in the work of de Broglie and Einstein. So it was natural for Schrödinger to turn to Hamilton's theory and broaden it into a more complete wave mechanics.

One basis for Hamilton's theory is an analogy between the optics of a light beam regarded as a ray and the mechanics of a material particle. But this picture is, as Schrödinger noted, an approximation, at least for the light ray, because light is more than a bundle of rays. The rays have a wavelike fine structure, which leads to such phenomena as diffraction and interference. Ray optics says nothing about these effects; it is simply a convenient, but approximate, form of a broader and more refined theory of optics. The more-complete theory, which can be called "wave optics," gives a detailed picture of the wave structure, accounts for diffraction and interference effects, and shows that the rays are fictitious entities constructed perpendicularly to wave fronts.

With analogy as his principal justification, Schrödinger reasoned that this mechanics-optics parallel should hold at all levels—that if ray optics is an approximate form of wave optics, then ordinary mechanics, the analogue of ray optics in Hamilton's scheme, is an approximation for a more fundamental mechanics, a new wave mechanics:

$$\begin{pmatrix}\text{ordinary}\\\text{mechanics}\end{pmatrix} \text{ is to } \begin{pmatrix}\text{wave}\\\text{mechanics}\end{pmatrix} \text{ as } \begin{pmatrix}\text{ray}\\\text{optics}\end{pmatrix} \text{ is to } \begin{pmatrix}\text{wave}\\\text{optics}\end{pmatrix}$$

If wave optics reveals the wavelike structure of light waves, the new mechanics would presumably show the wave structure of material particles such as electrons.

Beginning with these plausible assertions, Schrödinger derived the mathematical aspects of his theory by mixing four ingredients: Hamilton's arguments; the fundamental differential equation of optics; the Planck energy-frequency equation $E = h\nu$; and the de Broglie momentum-wavelength equation $p = \frac{h}{\lambda}$. After several false starts, he arrived at the differential equation now known to students of physics and chemistry as "the Schrödinger equation." The equation was soon successful in an astonishing variety of atomic and molecular problems. Except that he had not found a way to recognize the requirements of Einstein's special theory of relativity, a limitation that is not serious in the theory of atoms and molecules, Schrödinger had, in just six months, put together a complete mathematical quantum theory. His 1926 papers were, in the words of the science historian Max Jammer, "undoubtedly one of the most influential contributions made in the history of science. . . . In fact, the subsequent development of non-relativistic quantum theory was to no small extent merely an elaboration and application of Schrödinger's work."

Mathematically speaking, Schrödinger's equation is unremarkable. It resembles other equations derived to represent other kinds of waves: water waves, electromagnetic waves, light waves, and sound waves. It is also an energy equation, expressing in a special mathematical language that the total energy of the system described, let's say a hydrogen atom, is equal to the atom's kinetic energy plus its potential energy. This is just the quantum mechanical equivalent of the classical principle of conservation of energy. When the equation is solved it yields a "wave function," represented by Schrödinger and ever since with the Greek letter Ψ (uppercase psi).

The wave function is so called because it displays, as expected, wavelike properties. It depends on the location in time and space where it is evaluated, so its mathematical form when it describes a single particle such as an electron is $\Psi(x,y,z,t)$, in which x,y, and z are coordinates defining a point in space, and t is the time variable. An undisturbed atom or molecule does not change with time; in that case t can be omitted from the wave function, and for a single particle written $\psi(x,y,z)$ (ψ is a lowercase psi).

Easy for Beginners, Hard for Experts

Abraham Pais, the best of the chroniclers of twentieth-century physics, has remarked that quantum mechanics is like Vladimir Horowitz's assessment of Mozart's music: "too easy for beginners and too hard for experts." He means that with a superficial grasp of quantum mechanics one can make the calculations—play the notes, so to speak—but to reach a full understanding of what the calculations mean (like Horowitz's mastery of Mozart) is a far more difficult task. The physical interpretation of Schrödinger's equation and its elaborations is still—long after Schrödinger's original papers—a subject for lively controversy.

The first interpretive problem, taken up by Schrödinger, then Born, and then Pauli, was the physical meaning of the wave function. The concept that finally evolved was entirely unexpected, and it fueled years of debate. Born and Pauli concluded that the wave function has an irreducible statistical meaning. For a single electron in a free atom the wave function squared ψ^2 measures the probability of finding the electron at or near a given location: where ψ^2 is large in value, for example near the center of an atom, the electron is likely to be found. Quantum mechanics is, in other words, a kind of statistical mechanics.

But quantum mechanics is profoundly different from the classical statistical mechanics of Clausius, Maxwell, Boltzmann, and Gibbs, which is based on an underlying physical reality comprising molecules. We can view this molecular realm, and see how the molecules generate the statistics. But evidence, both theoretical and experimental, accumulating over many years has most present-day physicists convinced that the statistical picture offered by quantum mechanics has no such underlying interpretation; the ultimate reality in the quantum realm, it seems, is statistical, and that is that.

Heisenberg's Uncertainty Principle

At about the same time as Schrödinger was composing his equation, Heisenberg published a paper that revealed the statistical nature of quantum theory another way. Heisenberg's discovery, his most important achievement, is called the "uncertainty principle," and it has many astonishing ramifications. One of them is the conclusion that if you measure precisely the position of a particle, say an electron in an atom, you inevitably disturb the electron so much that its subsequent behavior is almost completely uncertain.

Put more formally, Heisenberg's principle asserts that if Δx is the uncertainty in the position of an electron in some direction x, and Δp_x is the uncertainty in the momentum in that direction, then Δx and Δp_x are related reciprocally according to

$$\Delta p_x \Delta x \geq \frac{h}{2\pi}, \tag{3}$$

in which h is again Planck's constant. Similar statements hold for the other two spatial directions, y and z. If Δx is small, as it must be after a precise measurement of position, then Heisenberg's principle demands that the momentum uncertainty Δp_x must be large, in order for the product $\Delta p_x \Delta x$ to exceed the value $\frac{h}{2\pi}$ as required by equation (3). Remembering that momentum equals velocity times mass, we see that a precise measurement of position leaves us in almost complete ignorance of the electron's subsequent velocity, concerning both its magnitude and its direction.

Heisenberg made this drastic conclusion more concrete by imagining a position measurement made with a special microscope. He knew (after Bohr reminded him) that the resolution of any microscope depends on the wavelength of the light forming the image: the smaller the wavelength, the greater the resolution. For a precise position measurement of an electron in an atom, a small wavelength is needed, in fact so small that the "light" rays required are actually gamma rays, whose photons are highly energetic. Each gamma-ray photon carries energy far in excess of the energy that holds an electron in an atom. When such a photon collides with an atomic electron and is scattered into Heisenberg's microscope, it is likely to knock the electron right out of the atom, never to return.

The conclusion is that the electron and the atom containing it are so severely damaged in the process of the measurement that they are useless for further measurements. One significant measurement is possible on a particular electron, but no more; and it is certainly impossible to follow continuously the electron's trajectory in the atom or anywhere else. If electron trajectories cannot be measured, say quantum theorists, then they should not be recognized by the theory. Orbital motion of atomic electrons, as pictured by Bohr and Sommerfeld, is out.

If electrons in atoms are as elusive as Heisenberg's argument indicates, how can we hope to form a useful picture of an atom's electronic structure that concedes the uncertainty and still reveals *something* about what goes on electronically in an atom's interior? It is clear that no atomic theory based on individual electrons following definite paths is acceptable. But fortunately we do not need a theory of that kind. It is possible to formulate an atomic theory that deals in probabilities rather than certainties.

Suppose, for example, observations are made on *many* atoms. Because electron-locating measurements are likely to be ruinously disturbing to the atom observed, we must understand that each atom is good for only one observation. If we use the Heisenberg gamma-ray microscope, each measurement does no more than register one possible location of an electron in an atom. Results from many such measurements build a composite, statistical picture of the habitat of atomic electrons.

The Heisenberg microscope has never been realized. It is a "thought experiment" that defies no physical principles, but is not technically feasible. But the well-established methods of x-ray diffraction accomplish the same thing. By analyzing x rays reflected by many atoms in a crystal, one can construct a statistical map that shows where electrons are and are not located in the atoms of the crystal. Good statistical maps of electron densities in atoms are difficult to generate experimentally; but Schrödinger's wave functions tell essentially the same story, and a refined statistical picture of electrons in atoms can be *calculated* using an appropriate formulation of Schrödinger's equation. For a free atom, the equation defines the wave function ψ at any location in the atom, and ψ^2 calculates the probability of finding an electron at that location.

The pattern of Heisenberg's principle extends beyond momentum and position to other dynamic variables linked in the same manner by their reciprocal indeterminacy. The most important of these further connections brings energy and time together. If Δt and ΔE are time and energy uncertainties, then, in analogy with the equality-inequality (3),

$$\Delta E \Delta t \geq \frac{h}{2\pi}. \qquad (4)$$

The Heart of Quantum Mechanics

Schrödinger's equation has a property that is, to the mathematician, routine and unexciting. The equation is "linear," meaning that if it has the solutions Ψ_1 and Ψ_2, then it also has the "superposition" solution $\Psi = \Psi_1 + \Psi_2$. Experimentalists, who make a living testing mathematical pronouncements of theorists, have found

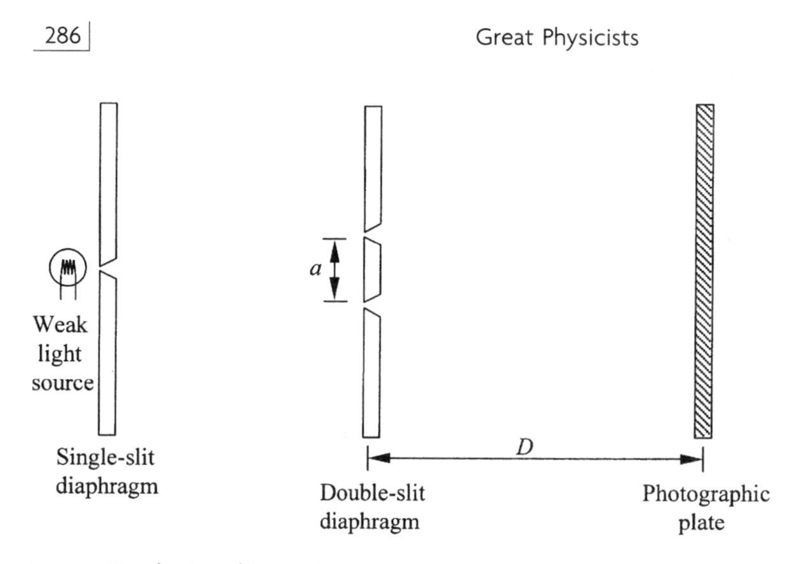

Figure 19.1. The two-slit experiment.

ingenious ways to observe superposition states, and their investigations have led them beyond the mathematics to what Richard Feynman calls "the heart of quantum mechanics."

The prototype of experiments designed to demonstrate superposition states consists of a light source, two diaphragms, one containing a single slit and the other a double slit, and a photographic plate serving as a detector (fig. 19.1). If the wavelength λ of the light is small compared to the distance a between the double slits, bright and dark bands appear on the plate, with centers of the bright bands separated by the distance $\frac{\lambda D}{a}$, D being the distance between the double slit and the photographic plate.

Since the early nineteenth century, bright and dark bands in experiments of this kind have been accepted as evidence of "interference" phenomena. Light is pictured as a wave train, which is "diffracted" (spread out) after passing through a slit. The double slit forms two diffracted wave trains that overlap (fig. 19.2). In the region of the overlap there can be both cancellation, where crests from one wave train fall on troughs from the other, and reinforcement, where wave crests fall on crests and troughs on troughs. Bright bands appear where there is reinforcement and dark bands where there is cancellation. All of this is easily expressed in the mathematical language of Schrödinger's wave mechanics. The separate diffracted wave trains are designated by the two wave functions Ψ_1 and Ψ_2, and the overlapping region, where interference occurs, by the superposition $\Psi_1 + \Psi_2$ (fig. 19.3).

Normally, the two-slit experiment is performed with a strong light source that sends many photons into the apparatus at the same time, but it can also be done with a source so weak that only one photon at a time traverses the space between the double-slit diaphragm and the photographic plate. Even in this situation, if enough time is allowed for many photons to be detected by the plate, the usual interference pattern of bright and dark bands is displayed.

This is, as Feynman remarks, "a phenomenon which is impossible, *absolutely impossible*, to explain in a classical way." The problem is that the experiment confronts us with the spectacle of a *single* photon *interfering with itself*. The

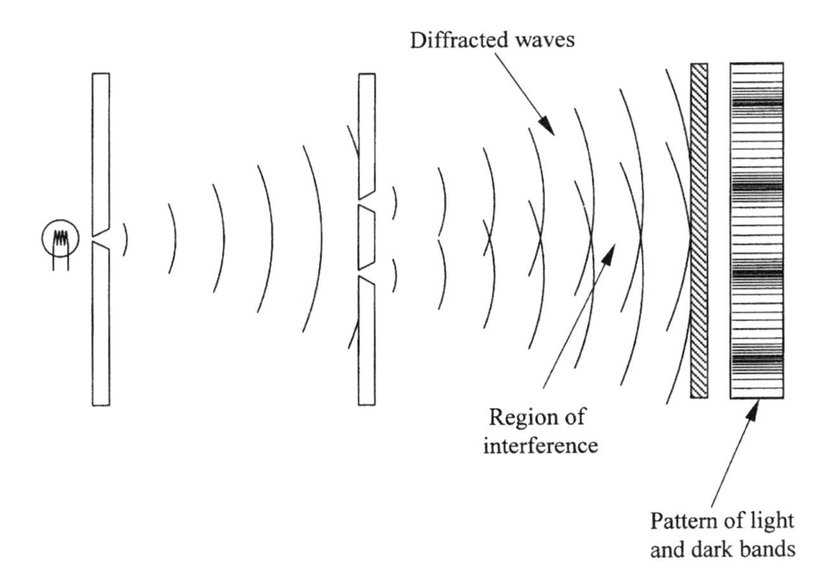

Figure 19.2. Diffraction and interference in the two-slit experiment.

photon passes through *both* slits, forms a superposition state represented by $\Psi_1 + \Psi_2$, and an interference pattern is the result. There is no escape from this weird conclusion. If we close one slit, or otherwise force the photon through one slit, the interference pattern disappears.

How can a single photon pass through two separate slits at the same time? Feynman is not reassuring about finding explanations for the mystery of such interference experiments. "We cannot explain the mystery in the sense of 'explaining' how it works. We will *tell* you how it works. In telling you how it works we will have told you about the basic peculiarities of all quantum mechanics." John Wheeler characterizes a photon in an interference apparatus as a "smoky dragon." It shows its tail, where it originates, and its mouth, where it is detected, but elsewhere there is smoke: "in between we have no right to speak about what is present."

Such quantum weirdness is not restricted to photons. Interference experiments forcing the same conclusions have also been performed with beams of electrons,

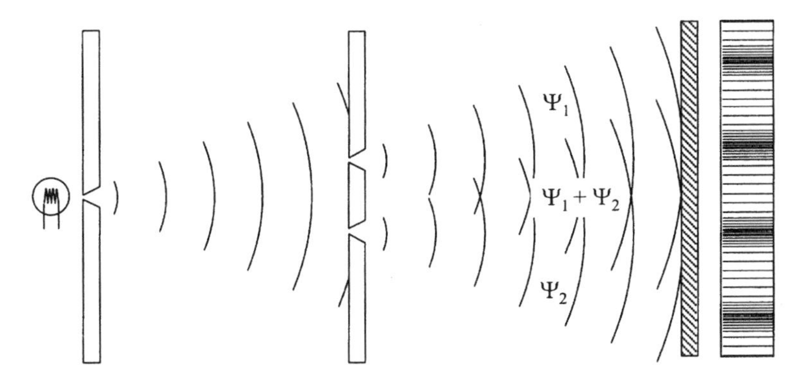

Figure 19.3. Wave functions for the two-slit experiment.

neutrons, and even atoms. All these entities display their wave nature in well-defined interference patterns and are just as smoky as photons as they travel through the apparatus.

Duality and Complementarity

The two-slit interference experiment has another level of meaning. It supplies us with a striking illustration of the wave-particle concept. Waves are demonstrated by the pattern of bright and dark bands. We can even calculate a wavelength λ by simply measuring the distance between bright bands, which is equal to $\frac{D\lambda}{a}$. The experiment viewed another way gives an equally convincing demonstration of particle behavior. When photons arrive at the photographic plate, they are detected in highly localized events: each photon arrival produces a small spot in the photographic emulsion. The interference pattern develops as many of these spots gather into the characteristic bright bands. The individual spots are suggestive of particle-like entities, and taken separately say nothing about wave behavior. But (fortunately) the experiment does not tell us that each photon is simultaneously a wave and particle. We observe the bands of the interference pattern and see waves, *or* we observe spots on the photographic plate and see particles, but never both at the same time.

Note the indispensable role of the observer in this account. We identify waves or particles by making an observation: waves by the interference pattern, and particles by the spots. Bohr insisted that the only route to physical reality allowed by quantum theory is by way of experimental observations. "Quantum mechanics is about only one thing: What can we do with our instruments?" was his credo. The instruments penetrate the smoke and reveal definite entities: waves, or particles.

But, as Feynman reminds us, the instruments do not "explain" anything. The mystery of the two faces of quantum mechanics—wave and particle—is still with us. Bohr responded to the duality problem by simply accepting it at face value and building it into the foundations of quantum theory. This was one feature of Bohr's principle, or philosophy, of "complementarity." The idea applied to the duality question is that, even though experiments find the wave and particle aspects of entities such as photons mutually exclusive, these dual properties are both essential to the physical description. They are, in Bohr's terminology, "complementary" properties of photons (or electrons, neutrons, atoms, and so forth).

For Bohr, this split pattern extended beyond wave-particle phenomena. His reading of Heisenberg's uncertainty principle was that momentum and position are a complementary pair of variables, and so are energy and time. Both variables of a complementary pair are essential to the physics, but they are mutually exclusive in that they cannot be measured simultaneously with certainty.

The True Jacob

Einstein could never accept the absolute necessity of quantum mechanical indeterminacy. "Quantum mechanics is very worthy of regard," he wrote to Max Born. "But an inner voice tells me that this is not the true Jacob. The theory yields much, but it hardly brings us close to the secrets of the Ancient One. In any case, I am convinced that he does not play dice." For many years, in a series of debates with Bohr, he tried all sorts of dodges to outwit Heisenberg's principle. He never quite succeeded, but his final attempt kept Bohr, and Bohr's successors, puzzled for decades.

This attack on quantum-mechanical indeterminacy was launched in a short paper written in 1935 by Einstein with two assistants, Boris Podolsky and Nathan Rosen. The authors argued that quantum mechanics was incomplete because it could not reach certain deterministic elements of physical reality that were, they claimed, experimentally accessible. A physical quantity qualified as an "element of physical reality" if "without in any way disturbing a system we can predict with certainty [its value]."

What Einstein, Podolsky, and Rosen proposed—measuring a value for a quantity with certainty without disturbing the system at hand—was just what quantum mechanical indeterminacy prohibits. But Einstein and his coauthors had one more ingenious scheme to outsmart the indeterminacy. They advocated working with a source of correlated or "entangled" (Schrödinger's term) *pairs* of particles. The members of the pair (usually photons, in the numerous realizations of the experiment) move in opposite directions from the source, and some correlated property, call it P (polarization if photons are involved), is measured at two widely separated locations A and B. The experiment can be designed so that the entanglement guarantees opposite results at points A and B, even if they are miles apart. If $P = +1$ is measured at A (in appropriate units), then the experimenter knows that $P = -1$ at B without making an actual measurement there.

Does this experiment meet the Einstein-Podolsky-Rosen condition for identifying the property P as an element of physical reality? It predicts with certainty the value of P, as required, and if the effect of the measurement at A is "local," so its effect does not reach B, at least not at a speed exceeding that of light, we can design the experiment to give a result at B without in any way disturbing the photon there. Relying on this reasonable "locality assumption" and the remarkable advantages of the particle entanglement, Einstein, Podolsky, and Rosen showed how to confer reality on deterministic physical properties that are not accessible by the methods of quantum mechanics. Quantum mechanics is incomplete, they concluded, because it is blind to this world of "local realism" (not their phrase).

A physics with elements of local reality must have variables to describe those elements, presumably in a deterministic manner. Such variables are hidden to quantum mechanics, but if they exist, as the Einstein-Podolsky-Rosen argument implied, there must be a deeper theory that uses them and banishes the indeterminism of quantum mechanics. Do these "hidden-variable theories" actually exist? The next phase in the prolonged search for the meaning of quantum mechanics focused on this question.

Bell's Theorem

John Bell was an Irish theoretical physicist who was contrary enough to call himself a "quantum engineer." He was based at the mammoth European elementary-particle laboratory on the French border near Geneva, known by the acronym CERN (European Center for Nuclear Research). As an "engineer," he contributed extensively to the theory of beam focusing in large-particle accelerators. For most of his career, he also had a consuming interest in the never-ending debate about the foundations of quantum mechanics. In 1964, when he was thirty-four, Bell published in an obscure journal a short but difficult paper with the title "On the Einstein-Podolsky-Rosen Paradox." The centerpiece of the paper was a theorem formulated as a mathematical inequality that was valid for *any* hidden-variable theory that satisfied the locality assumption, but, as expected, it was not valid for quantum mechanics. Here again was the conflict pointed out by Einstein, Podolsky, and Rosen.

For five years, Bell's paper was mostly ignored. Then suddenly it dawned on experimental physicists that Bell's theorem was more than just another way to reach the Einstein-Podolsky-Rosen conclusions; the inequality could be tested experimentally. "What was at stake in such a test," Jeremy Bernstein writes in a profile of Bell, "was nothing less than the meaning and validity of quantum theory. If Bell's inequality was satisfied, it would mean that all of Einstein's intuitions about the essential incompleteness of the quantum theory had been right all along. If the inequality was violated, it would mean—at least, so many physicists believed—that Bohr and Heisenberg had been right all along and that no return to classical physics was possible."

The experiments were not easy. Suitable methods for producing entangled photons had to be developed. The photons had to be piped to locations separated by many miles, where the correlations could be checked and the time measured between an effect at one location and its result at the other. The first experiments were reported in the early 1970s, and they have been elaborated and refined subsequently to eliminate subtle loopholes. The experimental data are now clearly at odds with Bell's theorem; that means victory for quantum mechanics, and defeat for Einstein's concept of local realism. "The evidence is now overwhelming that Einstein's program to 'complete' the quantum theory with a locally deterministic theory was misguided," writes Bernstein. "Local realism simply does not work."

Bell told Bernstein that he had some regrets: "For me, it is so reasonable to assume that the photons in those experiments carry with them programs, which have been correlated in advance, telling them how to behave [as Einstein's local realism would have allowed]. This is so rational that I think that when Einstein saw that, and the others [for example, Bohr, Heisenberg, Born, Pauli] refused to see it, *he* was the rational man. . . . So for me, it is a pity that Einstein's idea doesn't work. The reasonable thing just doesn't work."

So the locality concept, which Einstein had hoped he could rely on, was not confirmed. And in its place the experiments revealed a "nonlocality" that Einstein would have called "spooky." A measurement on one photon of an entangled pair affects a measurement on the other photon probably *instantaneously*, or at any rate faster than the speed of light. Is this in conflict with Einstein's theory of special relativity, which demands that no signal can be propagated that fast? Not exactly. Quantum-mechanical nonlocality cannot be a vehicle for sending messages because the data in the measurements are entirely random and not under the control of the experimenter; you take what you get and not what you want to put into a message. Thus there is, as Abner Shimony writes, "a peaceful coexistence between quantum mechanics and relativity theory, in spite of quantum mechanical nonlocality."

The Scientist as Humanist

Schrödinger was one of the most eloquent spokesmen of his time for humanism in science. To the conceits that physics was nonexistent before Galileo and that the concepts of quantum physics are new and unique, he responded: "Quantum theory dates 24 centuries further back, to Leucippus and Democritus. They invented the first discontinuity—isolated atoms embedded in empty space. Our notion of the elementary particle has historically descended from their notion of the atom. . . . Physical science in its present form is the direct offspring, the uninterrupted continuation, of ancient science." He feared that theorists were beginning to talk only among themselves: "A theoretical science, unaware that those of its constructs considered relevant and momentous are destined eventually to be framed in concepts and words that have a grip on the educated community and become part and parcel of the general world—a theoretical science, I say, where this is forgotten, and where the initiated continue musing to each other in terms that are, at best, understood by a small group of close fellow travelers—will necessarily be cut off from the rest of cultural mankind; and in the long run it is bound to atrophy and ossify."

Physicists use their theories in the daily tasks of organizing data and planning experiments. Theories "work" if they answer the physicist's special needs, but to the world outside the journals, and worse, to the culture of another era, the preoccupied "musings" of physicists may seem to be written in hieroglyphics. "Would it mean setting too high and proud a goal," Schrödinger wrote, "if we occasionally thought of what will become of our scientific papers 2000 years since? Science will have changed entirely. Will there be anybody to grasp our meaning, as we grasp the meaning of Archimedes?"

Schrödinger believed that physicists and other scientists should venture beyond their specialities. He had some wonderful words of encouragement for practicing and aspiring interdisciplinarians in the preface to his book *What Is Life?* He felt that he should apologize for his lack of expertise in some aspects of his subject:

> A scientist is supposed to have a complete and thorough knowledge, at first hand, of *some* subjects, and, therefore, is expected not to write on any topic of which he is not a master. This is regarded as a matter of noblesse oblige. For the present purpose I beg to renounce the noblesse, if any, and to be freed of the ensuing obligation. My excuse is as follows:
>
> We have inherited from our forefathers the keen longing for unified, all-embracing knowledge. The very name given to the highest institutions of learning reminds us, that from antiquity and throughout many centuries the *universal* aspect has been the only one to be given full credit. But the spread, both in width and depth, of the multifarious branches of knowledge during the last hundred odd years has confronted us with a queer dilemma. We feel clearly that we are only now beginning to acquire reliable material for welding together the sum total of all that is known into a whole; but, on the other hand, it has become next to impossible for a single mind fully to command more than a specialized portion of it.
>
> I can see no other escape from this dilemma (lest our true aim be lost forever) than that some of us should venture to embark on a synthesis of facts and theories, albeit with secondhand and incomplete knowledge of some of them—and at the risk of making fools of ourselves.
>
> So much for my apology.

vii

NUCLEAR PHYSICS

Historical Synopsis

The general theme in this part of the book is again, as it was in part 6, the structure of the atom. In brief, each atom comprises a central positively charged component called the nucleus and surrounding negatively charged shell-like structures containing electrons. The nucleus is very small relative to the rest of the atom ("A fly in a cathedral") but very massive: it contains nearly all the atom's mass. I mentioned the atomic nucleus briefly in chapter 16 as part of the heritage of Bohr and his successors, who gave us the theory of the atom's electronic domain. But the further story of the nucleus, as a separate and fundamental physical entity, remains to be told.

Reset the clock from the 1920s, where we left the development of quantum mechanics in part 6, to the 1890s, which brought the first studies of the most obvious manifestation of the atomic nucleus, radioactivity. The two great pioneers in radioactivity research were Marie Curie, in Paris, and Ernest Rutherford, first in Montreal, then in Manchester, and finally in Cambridge. Marie Curie and her husband Pierre achieved the first separation of a radioactive element, radium. Rutherford identified the three "rays" emitted by radioactive elements and called them α, β, and γ. He and his junior research partner, Frederick Soddy, demonstrated that alchemy (a more polite term is *transmutation*) was involved: one radioactive element could transmute into another. Using α particles (they are actually doubly charged helium ions, not rays) to bombard thin metallic foils, Rutherford, Hans Geiger, and Ernest Marsden made a strong case for the existence of the nucleus in a series of experiments completed in 1913.

Rutherford's model of the nucleus gave it an extremely small, yet finite, size. That raised the further question: What are the structural components of the nucleus? One of them was evidently the proton, the smallest nucleus, that of hydrogen. Another, the neutron, was discovered in 1932 by James Chadwick, Rutherford's second in command at the Cavendish Laboratory in Cambridge. Neutrons add mass, about that of the proton, but no electrical charge, to the nucleus.

Rutherford's favorite experimental tool was the α particle. It gave him the concept of the nucleus, and also the first example of "nuclear chemistry": he found that bombarding nitrogen gas with energetic α particles transmuted nitrogen to oxygen. In the same vein, Chadwick generated neutrons by bombarding beryllium with α particles. In Paris, Irène Joliot-Curie (Marie Curie's daughter) and her husband Frédéric bombarded boron and aluminum with α particles to obtain artificial radioactive elements not found in nature.

Enrico Fermi, in Rome, found another efficient projectile for bombardment experiments: "slow," that is, low-energy, neutrons. Many of the elements beyond oxygen in the periodic table absorb slow neutrons and become radioactive in the process. Neutron capture has a devastating effect on the heaviest element, uranium, causing its nucleus to shatter, or "fission," into two fragments of roughly equal mass. Fermi and his coworkers performed the first neutron bombardment of uranium in 1935, but misunderstood the results. Not until 1938 did Lise Meitner and Otto Frisch, then in Sweden, and Otto Hahn and Fritz Strassmann, in Berlin, introduce the concept of nuclear fission.

One neutron is consumed in a fission event and two or three neutrons are produced. If careful attention is paid to neutron losses and gains, neutrons born in fission events can cause more fissions, thus sustaining a nuclear chain reaction. Each fission releases a formidable amount of energy. If controlled, a nuclear chain reaction is a useful power source. Uncontrolled, it can be used as a bomb capable of flattening cities. In an experiment of unprecedented complexity, Fermi demonstrated in 1942 how to control the uranium chain reaction. During the following three years, a brilliant cast of physicists, engineers, chemists, and mathematicians working in Los Alamos, New Mexico, designed and fabricated the bomb—and horrified themselves when they tested it.

20

Opening Doors

Marie Curie

1867 – 1934

Maria and Marie

Her life was, in a word, heroic. Marie Curie was absolutely unstoppable in any task she undertook, no matter what the obstacles. "First principle: never let one's self be beaten down by persons or by events," she wrote in a letter to a friend when she was twenty-one. At the time, she was enduring life as a governess in a small town in Poland and dreaming of a university education in Paris. Student life at the Sorbonne eventually became a reality, and despite an erratic secondary education in Warsaw, she overcame the deficiencies, placed first in her *licence ès sciences* examination (among 1,825 students, 23 of them women), and second in the *licence ès mathematiques* examination. She chose as a topic for a doctoral thesis an immensely difficult study of the recently discovered phenomena of radioactivity. For that work she received a Nobel Prize, the first woman to do so, and later a *second* Nobel Prize; she was the first scientist, man or woman, to be so honored. She was the first woman to teach at the Sorbonne, and came within two votes of being the first woman elected to the Académie des sciences (the only time in her life she allowed herself to be "beaten down"). During World War I, she designed and directed the operation of a fleet of mobile x-ray radiology units amidst the horrors and chaos of the western front. After the war, she became a superb laboratory director, taking on not only the scientific duties, which she loved, but also the chores of fund-raising and public relations, which she detested.

She was born Maria Sklodowska in Warsaw. Except in the patriotic spirit of its citizens, Poland did not then exist as a nation. At the end of the eighteenth century, the country had been carved into three provinces by Russia, Austria, and Prussia; Warsaw was oppressively ruled by Russia. Maria's father, Wladyslaw, was a professor of physics and mathematics in a government (Russian-controlled) secondary school, but he suffered a series of demotions because of political differences with his Russian superiors. Finally he was forced to run a private boarding school in his home. "I found . . . ready help [in mathematics and physics] from my father, who loved science and had to teach it to himself," Maria wrote later. "He enjoyed any explanation he could give us about Nature and her ways. Unhappily he had no laboratory and could not perform experiments." "Even when we were older," Maria's brother Józef recalled, "we still turned to him with all the questions, as to an encyclopedia."

Bronislawa, Maria's mother, was remembered by her daughter as a woman of "exceptional personality who held . . . in the family remarkable moral authority." Like her husband, she was a teacher, eventually becoming headmistress of a prestigious school for girls, and during the same time bearing five children. Maria, the youngest, was born in 1867. While Maria was still young, the family was crushed by two devastating losses. In 1876, the oldest daughter, Zofia, died of typhus, and in 1878 Bronislawa succumbed to tuberculosis. For Maria, the loss of her mother "was the first great sorrow of my life and threw me into a profound depression. . . . Her influence over me was extraordinary, for in me the natural love of a little girl for her mother was united with a passionate admiration."

All of the Sklodowski siblings did well in school, particularly Maria, who graduated from the gymnasium first in her class. Wladyslaw decided that his daughter needed a change of scene after her graduation. He sent her for a year's visit to the small (nearly impoverished) country estates of her maternal uncles, where she happily entered into the frivolous, gay life of her cousins. As a final fling, Maria and her sister Helena spent a summer on the estate of a wealthy former student of their mother's, where the dancing parties could last for days. Helena recalled in her memoir that once they danced "for three days until we could hardly move." "It is good," Helena wrote, "when a person has had at least one such crazy summer in her life."

Back in Warsaw, life was not so hilarious. One way or another, Maria had to find the ways and means for a university education. The University of Warsaw was not an option because it did not admit women. For a time, she participated in the clandestine "Flying University," which met illegally wherever its organizers could find support. The best route to an advanced education for an ambitious Polish student, the Sorbonne in Paris, seemed out of reach. Wladyslaw's salary was meager, and he had lost what savings he had in an ill-advised scheme concocted by his brother-in-law. With the unlimited optimism and vigor of youth, Maria and her oldest sister, Bronislawa (Bronia), entered into a brave pact that saved the day. Maria would get a job as a governess and help support Bronia while she studied for a medical degree in Paris. Then, with a degree in hand, Bronia could earn enough to turn around and bring her sister to Paris.

Maria spent four years as a governess in other people's homes, and it was a trying experience for the teenager; at times she was in despair. "There have been moments which I will certainly count among the most cruel of my life," she wrote to a friend. But in her isolation she found the time to continue her education in science by herself: "I acquired the habit of independent work, and . . . trying little by little to find my real preferences, I finally turned towards mathematics and physics."

At Maria's first position as a governess, with the Zowarski family in a small town fifty miles north of Warsaw, there were diversions that were not so beneficial. She fell in love with the Zowarski son, Kazimierz. The attraction was mutual and the affair became serious. But a penniless governess, however accomplished, was not what the Zowarskis had in mind for their son, and they brought an abrupt end to the romance.

At last, Bronia wrote from Paris promising deliverance. She had nearly completed work on her medical degree, and was planning marriage to Kazimierz Dluski, who would also soon become a doctor. "And now you, my little Manya: you must make something of your life sometime," wrote Bronia. "If you can get together a few hundred rubles this year you can come to Paris next year and live with us, where you will find board and lodging." Maria hesitated for a year—Kazimierz Zowarski was apparently still on her mind—but in November 1891, she set out on the thousand-mile rail journey to Paris, traveling fourth class, seated on a camp stool, and carrying all the food she needed for the trip. Kazimierz Dluski, now her brother-in-law, met her at the Gare du Nord.

In Paris, Maria Sklodowska began a new life and gave herself a new name, Marie. A few adjustments in the living arrangements proved necessary. Bronia was visiting in Warsaw, and Kazimierz, a gregarious extrovert, hoped that Maria would enjoy long conversations and help him entertain. Maria, now Marie, had other priorities; as Kazimierz wrote to his father-in-law, "Mademoiselle Marie is a very independent young person. . . . She passes nearly all her time at the Sorbonne and we meet only at the evening meals." According to Marie, writing in a letter to her brother Jósef, "my little brother-in-law [disturbs] me endlessly [and is unable] to endure having me do anything but engage in agreeable chatter with him. . . . I had to go to war on him on this subject."

The disagreement was amiable, but Marie had to have her independence. In six months, she moved to a garret in the Latin Quarter. The apartment was one small room, hot in the summer, freezing in the winter, and six flights from the street. She prepared meals with an alcohol lamp, and often could afford no more than bread and a cup of chocolate, with eggs or fruit. Yet she felt no discouragement. "This life, painful from certain points of view," she wrote later in her *Autobiographical Notes*, "had, for all that, a real charm for me. It gave me a very precious sense of liberty and independence. Unknown in Paris, I was lost in the great city, but the feeling of living there alone, taking care of myself without any aid, did not at all depress me. If sometimes I felt lonesome, my usual state of mind was one of calm and great moral satisfaction."

Woman of Genius

Marie intended originally to return to Warsaw to live with her father after completing her *licence* (master's degree) examination, and like both her parents, make a career as a teacher. But those plans were permanently disrupted in 1894, when Pierre Curie came into her life. He was thirty-five when they met, with some solid achievements in theoretical and experimental physics to his credit. He and his brother Jacques had collaborated in the discovery of the "piezoelectric" effect, in which an electric potential is created by the application of a force to the opposite faces of certain crystals, particularly quartz, or conversely, a force is created by applying a potential to the crystal. Working alone, he had completed an experimental study of magnetism, focusing on the effects of temperature changes on magnetic materials. He was an accomplished designer and builder of sensitive electrical instruments, a talent that was to be crucial in the work he later did with Marie on radioactivity.

Marie and Pierre were brought together by a Polish physicist who was an admirer of Pierre's work. Here is Marie's recollection of her first impression of Pierre: "As I entered the room, Pierre Curie was standing in the recess of a French window, opening on the balcony. He seemed to me very young, though he was at the time thirty-five years old. I was struck by the open expression on his face and by the slight suggestion of detachment in his whole attitude. His speech, rather slow and deliberate, his simplicity, and his smile, at once grave and youthful, inspired confidence."

He impressed others as a dreamer, but also as a man who could act. He once wrote in his diary, "It is necessary to make a dream of life, and to make the dream a reality." "He grew up in all freedom, developing his taste for natural science through excursions to the country, where he collected plants and animals for his father," Marie wrote in her biography of Pierre. "These excursions, which he made either alone or with one of the family, helped to wake in him a great love of Nature, a passion which endured to the end of his life."

He always went his own way. We have met other scientific loners in these chapters; Pierre Curie was the most confirmed of the breed. He refused to play the game dictated by the French scientific establishment. He would not accept the customary national honorary awards. He was tardy in completing the ritual of the doctoral thesis, and did not cooperate with friends who wanted to see him elected to the Académie des sciences. During most of his career, he held positions at the École de physique et chemie, recently inaugurated for the education of engineers, and not one of the more prestigious schools.

When Marie and Pierre met, neither was looking for romantic attachments. Marie's memories of the Kazimierz affair were still painful, and Pierre had not recovered from the tragic death of a girl he had loved since childhood. Science had become a priesthood for him, and marriage had not seemed possible. When he was twenty-two, he wrote in his diary, "Women, much more than men, love life for life's sake. Women of genius are rare. . . . [When] we give all our thoughts to some work which removes us from those immediately about us, it is with women that we have to struggle, and the struggle is nearly always an unequal one. For in the name of life and nature they seek to lead us back."

But in Marie Sklodowska, Pierre Curie found the rarity, a "woman of genius," someone with extraordinary talent, and as consecrated as he was to a life in science. He softened his position on love and marriage, and set out to win the slightly overwhelmed Marie. Later he told her that it was the only time in his life that he acted without hesitation.

It was a troubling dilemma for Marie. If she accepted Pierre and a permanent life in France, it meant for her "abandoning my country and my family." She left Paris in doubt in the summer of 1894 and returned to Poland. Pierre wrote letters spinning his hopes and begging her to return in October. "It would be a beautiful thing," he wrote, "a thing I dare not hope, if we could spend our life near each other hypnotized by our dreams: your patriotic dream, our humanitarian dream, and our scientific dream." Marie returned in the fall with her doubts dispelled. Marie and Pierre were married in July 1895 at the town hall in Sceaux, a suburb of Paris where Pierre's parents lived. The wedding party then walked to the Curie family home, where the reception was held. "It was a beautiful day," writes Marie Curie's most recent biographer, Susan Quinn, "and the garden was overflowing with the irises and roses of late July. Marie's father and sister Helena had come from Warsaw. And of course Marie's sister Bronia was there, along with Kazimierz [Dluski], mixing with the more numerous members of the Curie family. It was, Helena remembers, a 'joyous atmosphere.' "

Becquerel Rays

Scientific discoveries are sometimes so contrary and surprising that they can be revealed only by accidents. So it was with the Curies' colleague Henri Becquerel and the discovery of radioactivity. The story begins in 1895 with a report by a quietly efficient German experimental physicist named Wilhelm Röntgen on a new kind of radiation he called "x rays." Röntgen's new rays were something like light: they traveled in straight lines, made shadows, and readily exposed photographic plates. But they had an astounding ability to penetrate almost anything they illuminated, including hands, feet, arms, and legs. The shadows they cast of bones and other parts of the interior anatomy created a popular sensation. Never before had a scientific advance become so rapidly and widely famous.

Röntgen's x rays were generated in an evacuated glass tube supporting a focused electrical discharge, a beam of "cathode rays." The x rays originated where the cathode-ray beam struck the glass wall of the tube. At the same point, the glass also displayed a strong glow, or fluorescence. It occurred to Becquerel, and others, that the fluorescence and the x rays might be generated by the same mechanism. In 1896, that surmise led Becquerel to look for x rays accompanying the known fluorescence of other sources. In particular, he investigated the uranium-containing compound potassium uranyl sulfate, whose fluorescence under the stimulation of sunlight he had studied previously. Sure enough, this uranium salt emitted a penetrating radiation after exposure to sunlight. But then, apparently by a fortunate accident, Becquerel discovered that the uranium salt was a steady source of the penetrating rays even *without* exposure to sunlight: the penetrating rays were an independent phenomenon, not directly connected with the fluorescence. How were the rays generated if not by the fluorescence mechanism, which always required some kind of excitation energy input? The Becquerel rays had no evident energy source; apparently they could even be seen as a denial of the first law of thermodynamics.

This was disturbing news; but it did not get the consideration it deserved, probably because of the preoccupation with Röntgen's more sensational x rays. Two young, ambitious researchers were, however, paying attention: Marie Curie in Paris, just then looking for a suitable doctoral thesis topic, and Ernest Rutherford in Cambridge, beginning a spectacular career we will follow in the next chapter.

Science in a Shed

A detailed experimental study of the Becquerel rays was a perfect choice for the newly formed Sklodowska-Curie research team: they had the means to measure accurately the intensity of the rays from different sources. Becquerel had reported that uranium rays could discharge electrified bodies. If, for example, a uranium salt was placed between the plates of a charged capacitor, a weak electrical current was generated, which slowly discharged the capacitor. One of the Curie brothers' inventions, a sensitive electrometer, was ideal for measuring such currents.

Marie Curie began her work by surveying a list of pure compounds and minerals. She spread each material on one of the plates of a capacitor charged to 100 volts, and then measured the discharge current with the Curie electrometer. As expected, uranium compounds proved to be "active" in this device, and so did those of thorium. Her key observation, which led the way to most of the Curies' further research, was that uranium minerals, in particular pitchblende (mostly uranium oxide), were more active than pure uranium. Because otherwise the activity was proportional to the amount of the element uranium in a sample, this seemed to indicate the presence of small amounts of a yet undetected element with even more activity than uranium. Quickly, surprisingly, the research turned in a direction that has always excited physicists and chemists: the hunt for a new element.

"As we did not know, at the beginning, any of the chemical properties of the unknown substance, but only that it emits rays," Marie wrote later in *Autobiographical Notes*, "it was by these rays that we had to search." Using analytical chemical techniques on pitchblende, the Curies separated fractions containing elements they knew were present and then measured the activity of the fractions separated. This innovative approach soon paid off: fractions that were rich in bismuth were much more active—more "radioactive," as they now put it—than the untreated sample. In an 1898 paper, the Curies laid claim to, and Marie proudly put her stamp on, a new element: "We believe . . . that the substance extracted from pitchblende contains a metal not hitherto distinguished, closely related to bismuth by its analytical properties. If the existence of the new metal is confirmed, we propose to call it *polonium* from the name of the country of origin of one of us." About six months later, the Curies, joined by Gustave Bémont, a colleague at the École de physique et chemie, claimed the discovery of another highly radioactive element. This one was associated chemically with barium, and they named it *radium*.

Both of these claims were plausible, but strictly speaking provisional, until the new elements could be purified and characterized physically and chemically. Isolating the elements proved to be a herculean task. Although they were detectable by their rays, the radioactive elements were present in extremely small amounts, only a fraction of a gram in a ton of raw material. The final product had to be teased out of huge quantities of pitchblende residues in many incremental steps. Marie Curie undertook this awesome task for the isolation of radium. She describes the obstacles: "We were very poorly equipped with facilities. . . . It was necessary to subject large quantities of ore to careful chemical treatment. We had no money, no suitable laboratory, no personal help for our great and difficult undertaking. It was like creating something out of nothing, and if my earlier studying years had once been called by my brother-in-law the heroic period of my life, I can say without exaggeration that the period on which my husband and I now entered was truly the heroic one of our common life."

No suitable laboratory space was available, so

> for lack of anything better, the Director [of the École de physique et chemie] permitted us to use an abandoned shed which had been in service as a dissecting room of the School of Medicine. Its glass roof did not afford complete shelter against rain; the heat was suffocating in summer, and the bitter cold of winter was only a little lessened by the iron stove, except in its immediate vicinity. There was no question of obtaining the needed proper apparatus in common use by chemists. We simply had some pine-wood tables with furnaces and gas burners. We had to use the adjoining yard for those of our chemical

operations that involved producing irritating gases; even then the gas often filled our shed.

It sounds like a nightmare, yet paradoxically it was a time Marie Curie remembered with fondness:

[It] was in this miserable old shed that we passed the best and happiest years of our life, devoting our entire days to our work. . . . Sometimes I had to spend a whole day mixing a boiling mass with a heavy iron rod nearly as large as myself. I would be broken with fatigue at the day's end. Other days, on the contrary, the work would be a most minute and delicate fractional crystallization, in the effort to concentrate the radium. I was then annoyed by the floating dust of iron and coal from which I could not protect my precious products. But I shall never be able to express the joy of the untroubled quietness of this atmosphere of research and the excitement of actual progress with the confident hope of still better results.

The Curie Couple

The radium project was demanding, but during vacations the Curies found the time to travel the countryside, usually on bicycles. They went from inn to inn, and Pierre, an accomplished naturalist, collected and identified plants. In 1897, Marie gave birth to a daughter, Irène, who was often cared for by her grandfather, Pierre's father Eugéne, recently a widower. A second daughter, Eve, was born in 1904. Pierre and Marie were content with an almost nonexistent social life beyond family gatherings, scientific meetings, and a few close friends from the physics community. Included in their circle of close colleagues were Paul Langevin, a former student of Pierre's, known for his early electronic theory of magnetism; Jean Perrin, a physical chemist whose work on the physics of large molecules was mentioned above; and Georges Gouy, whose research seemed to cover the entire territory of physics.

But whether they liked it or not, the world soon caught up with Marie and Pierre Curie when the radium research was published. The radium-enriched samples were radioactive beyond all expectations, and as if to demonstrate their potency they glowed in the dark. Ernest Rutherford visited the Curies on the festive day in June 1903, when Marie successfully defended her doctoral thesis. He joined them at a dinner held in Marie's honor by Paul Langevin. Rutherford recalled: "After a lively evening, we retired about 11 o'clock to the garden, where Professor Curie brought out a tube coated in part with zinc sulphide and containing a large quantity of radium in solution. The luminosity was brilliant in the darkness and it was a splendid finale to an unforgettable day."

Pierre gave a well-received lecture on their work in London before the Royal Society, and a little later Marie and Pierre were awarded the Humphry Davy Medal for the most important discovery of the year in chemistry. His acceptance of the medal reflected a change in attitude for Pierre, who had previously been disdainful of prizes and decorations. Then, in 1903, the Curies shared with Becquerel a Nobel Prize. The Nobel committee reported that "a completely new field of greatest importance and interest has opened for physics research. The credit for these discoveries belongs without doubt in the first place to Henri Becquerel and Mr. and Mrs. Curie. . . . The discovery by Becquerel of the spontaneous radioactivity of uranium . . . inspired diligent research to find more elements with remarkable qualities. The most magnificent, methodical and persistent investigations in this regard were made by Mr. and Mrs. Curie."

This time Pierre reverted partly to his old ways. He did not refuse the prize, but he informed the Swedish Academy that he and his wife could not attend the award ceremonies because "we can't go away at that time of year without greatly upsetting the teaching which is confided to each of us." The obligatory Nobel lecture was finally given by Pierre in the spring of 1905.

The Nobel brought an appointment for Pierre at the Sorbonne, and a deluge of publicity for which the Curies were totally unprepared. Pierre wrote Gouy: "We have been pursued by journalists and photographers from all countries of the world; they have gone even so far as to report the conversation between my daughter and her nurse, and to describe the black-and-white cat that lives with us. . . . Further we have a great many appeals for money. . . . Finally, the collectors of autographs, snobs, society people, and even at times, scientists, have come to see us—in our magnificent and tranquil quarters in the laboratory [the shed]—and every evening there has been voluminous correspondence to send off."

Journalists were fascinated by the "Curie couple" and their "idyll in a physics laboratory." They focused particularly on Marie. She was, as Susan Quinn writes in her recent biography, "a far cry from the conventional wife of a *savant*. And it was this more than anything else, which intrigued the press and public. The idea that a man and a woman could have a loving *and* working relationship was exciting to some, threatening to others."

Deathly Hours

Radioactivity has its dark side. The "rays" emitted by radium and other radioactive elements are extremely energetic; they can destroy living cells, cause deep burns, and damage internal organs. The Curies and their colleagues in radioactivity research were aware of some of these biological effects, but they considered the damage superficial, and seriously underestimated the systemic threats to their health. With hindsight, we can look at the medical histories of the Curies and suspect that they suffered from various forms of what is now called "radiation sickness." During the years immediately following the discovery of radium, both Marie and Pierre were increasingly bothered by fatigue; Marie was anemic and lost weight; Pierre "suffered from attacks of acute pain." Pierre's health problem was first diagnosed as "rheumatism," and then as "a kind of neurasthenia." It was a bone pain, which he felt in his legs and back. A second postponement of his Nobel lecture was prompted by a "violent crisis" in the summer of 1904. In the spring of the following year, he wrote Gouy that with the pains and fatigue "work in the laboratory was barely progressing at all." If biological radiation damage was indeed to blame, we can speculate that Pierre's health might have deteriorated further, making him the first casualty claimed by radiation-induced illness. But for better or worse, Pierre never faced that threat. On April 19, 1906, while attempting to negotiate on foot a jammed intersection in Paris, he was knocked down by a nervous horse, and his head was crushed under the wheels of a wagon. He was in his forty-seventh year.

Marie was overwhelmed by the loss of Pierre. To cope with her grief, she recorded her memories and feelings in diary entries that were addressed to Pierre.

The full diary first became available to researchers in 1990. Susan Quinn makes the diary the centerpiece of her beautiful Curie biography. "This mourning journal, kept sporadically for a year after the fateful day, is an eloquent and profoundly moving document," Quinn writes. "In it, we learn not only of Marie Curie's suffering, but also of some of the pleasures and the tensions in her life with Pierre, and with their two young children. The journal also allows us to know Marie Curie intimately, away from the curious eyes which led her to develop a stiff public persona. The mourning journal gives us the keen emotion under the dignified mask." This dialogue with Pierre is Marie's recollection of the morning of the accident: "You were in a hurry. I was taking care of the children, you left, asking me from below if I was coming to the laboratory. I answered you that I had no idea and I begged you not to torment me. And that is when you left, and the last sentence that I spoke to you was not a sentence of love and tenderness. . . . Nothing has troubled my tranquility more."

In the evening, after she had received the terrible news, Marie sat "for some deathly hours," waiting for the wagon to come bearing Pierre's body. It finally arrived:

> They brought me the objects they found on you. . . . [They are] all that I have left of you, along with some old letters and some papers . . . all I have in exchange for the beloved and tender friend with whom I planned to spend my life. . . . I kissed your face in the wagon, so little changed! [In the house] I kissed you again, and you were still supple and almost warm, and I kissed your dear hand which still flexed. . . . Pierre, my Pierre, there you are calm like a poor wounded one, sleeping with his head wrapped up. And your face is still sweet and serene, it's still you enclosed in a dream from which you cannot emerge.

For herself, she could accept death—but not suicide: "I walk as though hypnotized, without care about anything. I will not kill myself, I don't even have the desire for suicide. But among all those carriages, isn't there one which will make me share the fate of my beloved?"

Scandal and Slander

What Marie needed after the hullabaloo of the Nobel Prize, and during her deep mourning for Pierre, was peace and quiet. What she got was more hullabaloo. First, there was a momentous event at the Sorbonne. Marie was appointed Pierre's successor, becoming the first woman to teach at the Sorbonne. "They have offered that I should take your place, my Pierre. . . . I accepted," she wrote in her mourning journal. "I don't know if it is good or bad. . . . How many times have I said that if I didn't have you I probably wouldn't work anymore? I put all my hope for scientific work in you and here I dare to undertake it without you. You said it was wrong to speak that way that 'it was necessary to continue no matter what' but how many times did you say to yourself that 'if you didn't have me, you might work, but you would be nothing more than a body without a soul.' And how will I find a soul when mine has left with you?" Later in her journal she added, "there are some imbeciles who have actually congratulated me."

Marie's debut lecture attracted a crowd of "men-about-town, artists, reporters, photographers, French and foreign celebrities, many young women from the Polish colony, and also some students," recalled one of the students, a pupil of

Marie's. At exactly the scheduled hour, Marie entered. "She looked very pale to us," the student noted, "her face impassive, her black dress extremely simple; one saw only her luminous, large forehead, crowned by abundant and filmy ashen hair, which she pulled back tight without succeeding in hiding her beauty." Marie's lectures were intended to continue Pierre's course, and that is exactly what they did. With no introduction, she began by repeating exactly Pierre's last sentence in the lecture hall.

In the spring of 1910, Marie startled her friends by discarding her usual black costume and appearing "in a white gown, with a rose at the waist," according to her friend Marguerite Borel. "She sat down, quiet as always, but something signaled her resurrection, just as the springtime succeeding an icy winter announces itself subtly, in the details." The "resurrection" was not a miracle: Marie had fallen in love again. Her friendship of long standing with Paul Langevin had deepened into an intense love affair. "By mid-July of 1910, all the evidence suggests, Marie and Paul were lovers," Quinn writes. "On July 15, they rented an apartment together near the Sorbonne. . . . In their letters to each other, they called it 'our place' (*chez nous*)."

Langevin was married, with four children. His wife Jeanne learned of the affair, and the story became as complicated and lurid as the plot of a romantic novel. Jeanne Langevin accosted Marie on the street; a purloined, incriminating letter was published; there were vicious, slanderous attacks on Marie by the right-wing press; several times Marie was forced to leave Paris and travel incognito; and Langevin fought a duel (with no shots fired).

The slander campaign was whipped up when Marie became a candidate for election to the Académie des sciences. "With her candidacy," Quinn writes, "Marie was to learn what trouble can come to a woman alone, if she is suspected not just of passion, but of ambition." A storm of xenophobic, antifeminist sentiment, often driven by jealousy, descended on her. Her principal rival for election to the Académie won by two votes, but she tried not to be concerned; Pierre's disdain for the Académie was an article of her "religion of memories."

In the midst of this turmoil, and no doubt contributing to it, Marie was awarded a second Nobel Prize. This was the chemistry prize for 1911, given to Marie alone "for services to the advancement of chemistry by the discovery of the elements radium and polonium." The Swedish Academy, represented by Svante Arrhenius, was at first tolerant of the growing scandal in Paris, but had second thoughts later when news of Langevin's duel arrived. Arrhenius now suggested that the award be postponed until Marie's name could be cleared in the impending Langevin divorce trial. To another member of the Swedish Academy, Marie wrote: "In fact the prize has been awarded for the discovery of Radium and Polonium. I believe that there is no connection between my scientific work and the facts of private life. . . . I cannot accept the idea in principle that the appreciation of the value of scientific work should be influenced by libel and slander concerning private life. I am convinced that this opinion is shared by many people. I am very saddened that you are not yourself of this opinion." In December 1911, Marie Curie traveled to Stockholm and collected her prize.

As a final blow in this horrible period of her life, Marie suffered a breakdown in her health, "a severe and complicated kidney ailment," Quinn writes, "undoubtedly exacerbated by the pain of the scandal. She was unable to work."

But she had not forgotten the "first principle" of her youth, not to be "beaten down by persons or by events." By 1913, she was on the road to recovery.

We Will Make Ourselves Useful

To some of Marie Curie's detractors, she was a "foreign woman," even though she had lived in Paris for two decades and had married into a French family. But only the diehards could question her patriotism after her service to France during World War I. When war was declared in August 1914, Marie was in Paris and separated from her two young daughters, vacationing in the north on the Brittany coast. Irène, now a restless teenager, was thrilled when her mother wrote: "You and I, Irène, we will make ourselves useful."

By the time war broke out, civilian doctors were learning the practice of radiology and the use of x-ray apparatus in surgery and diagnosis. But military medical officers had little interest in x-ray methods, and Marie made it her mission to bring them up-to-date. She chose to develop a mobile unit that could carry an x-ray apparatus to the front. She became a radiologist, x-ray technician, ambulance driver, training instructor, fund-raiser, and expert on how to outwit the military bureaucracy. The first mobile unit was a success and, drawing mostly on private contributions, Marie managed to outfit twenty more mobile radiology units, and to install two hundred stationary facilities.

True to her word, Marie made Irène her premier assistant in this hectic work. Irène began by earning a nursing diploma. By September 1916, she was working with other nurses and training a radiology team. Somehow during the war years she completed studies at the Sorbonne with distinction in mathematics, physics, and chemistry. Irène was her mother's daughter.

La Patronne

After the war, Marie turned to the unfinished business of the Radium Institute. The institute was a tribute to Pierre of the kind he would most have appreciated but never had. When the chair was created for him at the Sorbonne, it carried with it the promise of a laboratory for the study of radioactivity. Not much had materialized, however, at the time of his death. At one point when he was offered a medal, he responded with: "I pray you to thank the Minister, and to inform him that I do not in the least feel the need of a decoration, but that I do feel the greatest need for a laboratory."

The radioactivity laboratory was still nonexistent three years after Pierre's death, with Marie installed as his successor at the Sorbonne. Then events turned. Eve Curie tells the story in her biography of her mother, *Madame Curie*:

> [In] 1909, Dr Roux, director of the Pasteur Institute, had the generous and bold idea of building a laboratory for Marie Curie. Thus she would have left the Sorbonne and become a star of the Pasteur Institute.
>
> The heads of the university suddenly pricked up their ears. . . . Let Mme Curie go? Impossible! Cost what it may, she must be retained on the official staff!
>
> An understanding between Dr Roux and Vice-Rector Liard [of the university] put an end to the discussions. At their common expense—400,000 gold francs each—the university and the Pasteur Institute founded the [Radium Institute], which was to comprise two parts: a [radioactivity laboratory], placed under the direction of Marie Curie; and a laboratory for biological research and Curie-therapy, in which studies on the treatment of cancer and the care of the sick

> would be organized. . . . These twin institutions, materially independent, were to work in cooperation for the development of the science of radium.

It was still wartime when the new laboratory was ready in 1915 and Marie moved in. "This was a trying and complicated experience, for which, once more, I had no money nor any help," she recalled in her *Autobiographical Notes*, "so it was only between my journeys that I was able, little by little, to do the transportation of my laboratory equipment, in my radiologic cars."

Marie always appreciated gardens, and she insisted that the courtyard between the two buildings of the institute contain trees and flowers. Undaunted by German shelling, she laid out the plantings herself: "I felt it very necessary for the eyes to have the comfort of fresh leaves in spring and summer time. So I tried to make things pleasant for those who were to work in the new building. We planted a few lime trees and plane trees, as many as there was room for, and did not forget flowerbeds and roses. I well remember the first day of the bombardment of Paris with the big German gun; we had gone, in the early morning, to the flower market, and spent all that day busy with our plantation, while shells fell in the vicinity."

The Radium Institute quickly became a thriving research center. Marie chose the researchers herself, with an unmistakable bias for women and Poles. They called her "la patronne" (the boss). If she thought it was necessary, she could be imperious. One new arrival said that she told him, "You will be my slave for a year, then you will begin work on a thesis under my direction, unless I send you to specialize in a laboratory abroad."

More often she was a sympathetic listener and adviser. One researcher, a young woman who had stalled on her thesis work in another laboratory, recalled her first interview with Marie: "[At first], this woman, pale and thin in a narrow black dress, who scrutinized with her cold, penetrating look, paralyzed me into timidity. . . . But she began to ask me questions with such great simplicity, and her face relaxed into a smile so full of charm, that I allowed myself to go ahead and tell her of my disappointments as a beginning researcher, and she decided to accept me into her Radium Institute."

Once the research project was underway, Marie appeared in the laboratory for a conference every two or three days. "She appeared all of a sudden, noiselessly, always dressed in black, around six in the evening. She sat on a stool and listened attentively to the account of the experiments; she suggested others."

Marie would do anything to advance the cause of the laboratory, even submit to two things she always disliked, travel and publicity. With the energetic assistance of an American journalist, Marie Meloney, usually called "Missy," she earned an impressive gift for the laboratory, and paid for it with an exhausting trip to America and massive publicity.

Like Marie, Missy Meloney was hard to stop. Somehow she managed to break through Marie's formidable defenses and obtain an interview. Later she wrote this account of their first meeting in 1920: "I waited a few minutes in a bare office which might have been furnished from Grand Rapids, Michigan. Then the door opened and I saw a pale, timid little woman in a black cotton dress, with the saddest face I had ever looked upon. . . . Her kind, patient, beautiful face had the detached expression of a scholar." She saw Marie as a "simple woman, working in an inadequate laboratory and living in a simple apartment on the meager pay of a French professor." This was mostly fiction, but it became part of the

Curie legend. Also contributing to the myth was Missy's claim that Marie could cure cancer with her radium. "[Life] is passing and the great Curie getting older, and the world losing, God only knows, what great secret," she wrote in her magazine, the *Delineator*. "And millions are dying of cancer every year!" An editorial in the same issue asserted that "the foremost American scientists say that Madame Curie, provided with a single gram of radium, may advance science to the point where cancer to a very large extent may be eliminated." Such claims were extravagant, to say the least. A by-product of radium was being used with some success in cancer therapy, but no responsible scientist was talking about a cancer cure.

As an enticement to bring Marie to the United States, Missy Meloney organized a huge effort to purchase one gram of radium (the price tag was one hundred thousand dollars), to be presented to Marie by President Harding in a White House ceremony. The fund-raising succeeded, and Marie was eventually persuaded to come. The trip was not at all what Marie wanted or expected. Even in France before her departure, there were celebrations. Quinn describes the festivities in Paris: "[A] French magazine, *Je sais tout*, . . . organized a gala to celebrate 'one of the glories of French science, the discovery of radium.' . . . [The] highest dignitaries in France, including President Aristide Briand, gathered at the Opéra to hear Jean Perrin and others discourse on the accomplishments of Marie Curie and the promise of her discoveries. The great Sarah Bernhardt read an 'Ode to Madame Curie'. . . . The 'foreign woman' of the Langevin scandal was forgotten; Marie Curie was now France's modern Joan of Arc."

In America, the trip was an interminable round of banquets, receptions, academic ritual, and organized sightseeing. Marie emerged from it exhausted and ill, but she got what she came for. In the famous White House ceremony, she accepted the precious gram of radium.

During the 1920s and 1930s, the Radium Institute prospered, but Marie's health slowly declined. In the end, the diagnosis was "pernicious anemia in its extreme form." She died in July 1934. Her coffin was placed over Pierre's in the small cemetery in Sceaux.

In her later years, many nice things, and a few mean things, were said about Marie Curie. Jean Perrin gave this simple appreciation of her finest achievement: "Mme Curie is not only a famous physicist: she is the greatest laboratory director I have ever known."

21

On the Crest of a Wave

Ernest Rutherford 1871–1937

Science as Action

He was large and somewhat clumsy; he had a thundering voice, and piercing eyes that are startling even in old photographs. The conventional role of the intellectual did not appeal to him, so he played it his own way. Once, a distinguished stranger, amazed by his unscholarly accent and appearance, mistook him for an Australian farmer. (His New Zealand origins partly explain the impression.) A nonscientific academic colleague told him he was a "savage—a noble savage I admit—but still a savage!" (This response was not unprovoked. Rutherford had opened the conversation with: "Alexander, all that you have said and all that you have written during the last thirty years—what does it all amount to? Hot air! Hot air!") He was not inclined toward modesty, if undeserved. After moving from McGill University in Montreal, where he did some of his earliest research, to the University of Manchester, which saw the middle period of his career, he reported to his friend and colleague Bertram Boltwood at Yale: "I find the students here regard a full professor as little short of Lord God Almighty. It is quite refreshing after the critical attitude of the Canadian students. It is always a good thing to feel that you are appreciated."

Rutherford's energy and ambition have been described, with only slight exaggeration, as volcanic. In nine years at McGill, his first academic position, he managed to publish some seventy papers, become a fellow of the Royal Society, build a significant research school, and complete the research that later earned him a Nobel Prize. These feats were accomplished with little previous experience (he was twenty-seven when he went to Canada), a handful of students, a meager salary, and the Atlantic Ocean separating him from the scientific centers of Britain and Europe. Frederick Soddy, who assisted in Rutherford's most important work at McGill, sometimes found life under the volcano a bit grim: "Rutherford and his radioactive emanations and active deposits got me before many weeks had elapsed and I abandoned all to follow him. For more than two years, scientific life became hectic to a degree rare in the lifetime of an individual, rare perhaps in the lifetime of an institution." Robert Oppenheimer summarized Rutherford's method in a phrase—"science as action." Oppenheimer was referring to Rutherford's experimental strategies, based on his "strong right arm," the alpha particle, but the description applies as well to Rutherford himself, relentlessly coaxing, driving, and leading his research team forward.

Rutherford was endowed with uncanny scientific intuition. To Charles Ellis, a student during the final period of Rutherford's career at the Cavendish Laboratory in Cambridge, he seemed to have a feeling for the "artistry of nature" so accurate and sensitive that "he almost knew what to expect." In 1920, he predicted the existence of the neutron, twelve years before James Chadwick, second in command at the Cavendish, observed neutrons experimentally. He could rapidly see to the heart of an experimental problem. "You had to talk only about fundamental facts and ideas without going into details in which Rutherford took no interest," says Peter Kapitza, a Russian physicist-engineer who worked at the Cavendish. "He grasped the basic idea of an experiment extremely quickly, in half a word." Rutherford's experimental designs, usually made with the barest minimum of equipment, are legendary among physicists. The wonder of his experimental work, writes Alexander Russell, a student from the Manchester period, was that "it asked of Nature the most pertinent questions." He could "pay attention not so much to what Nature was saying as to what Nature was whispering. In this Rutherford was an artist." The beauty, and at the same time the primitive simplicity, of Rutherford's successful experiments have never been surpassed—"the minimum of fuss with the minimum of error."

Great experiments are no easier to create than great novels, paintings, or symphonies. Kapitza reminds us that Rutherford's and Chadwick's twelve-year search for the neutron was mostly hard, frustrating work, and the final success was not without a bit of luck. Like the artist who cannot bear to reveal, or even fully recognize, the false starts, mistakes, and accidents that finally bring the creation, Rutherford was uncharacteristically reticent about work in progress. "He did not like to speak about his research projects and rather spoke only of what was already performed and [had] yielded results," Kapitza tells us, and adds, "I never heard Rutherford argue about science. Usually he gave his views on the subject very briefly, with the maximum of clarity and very directly. If anybody contradicted him, he listened with interest, but would not answer it, and then the discussion ended." (The contrast is sharp with Bohr, who lived for argument.) C. P. Snow, the physicist turned novelist, also saw this diffidence in Rutherford. Snow's novel *The Search* bothered Rutherford with its realistic portayal of scientists and their methods: " 'What have you been doing to us, young man?' he asked vociferously. . . . He hoped that I was not going to write all my novels about scientists. 'It's a small world, you know,' he said. He meant the world of science. 'Keep off us as much as you can. People are bound to think you are getting at some of us. And I suppose we've all got things that we don't want anyone to see.' "

We can sympathize. After all, do we insist on seeing the pages of corrections, and corrections on corrections, in the manuscript of a distinguished novel? But we can regret that only tidbits remain, like this one told by Mark Oliphant, Rutherford's second in command late in the Cavendish period, to show what it was really like to follow Rutherford in the chase. A call from Rutherford woke Oliphant at three o'clock one morning with what seemed to Oliphant an unreasonable suggestion concerning an experiment in progress. "Reasons! Reasons!" shouted Rutherford. "I feel it in my water!"

Rutherford patrolled the rooms of his laboratory with the relentless regularity of a company commander. At Manchester, his march on the daily inspection rounds was accompanied by a barely recognizable version of "Onward Christian Soldiers" or a dirge, depending on how things were going. A student was likely to find the Rutherford presence before him several times a week with prodding questions such as "Why don't you get a move on?" or "When are you going to get some results?" delivered in a voice so loud it sometimes actuated sensitive counting equipment.

But frightening and stentorian as his comments were, especially to the uninitiated, they had an effect that was subtly instructive and encouraging. Ernest Pollard, a student during the Cavendish period, tells of Rutherford's reactions to an experimental design not without a few "manifest absurdities": "His comments on it, while sharp and sometimes a little cutting, never had the effect of stopping us from trying. At no time did we feel that Rutherford had a contempt for our work, although he might be amused. We might feel that he had watched this sort of thing before and this was a stage we had to go through, but we always had the feeling that he did care, that we were trying the best we could, and he was not going to stop us."

Little in his laboratory escaped Rutherford's attention and interest, and not much escaped his dominance. Most of the physicists who worked with Rutherford found themselves eventually focusing on a problem related directly or indirectly to Rutherford's work. "He was no despot," Russell writes, "not even a benevolent despot." This easy but powerful influence was created partly by Rutherford's greatness as a physicist. His ideas were so numerous and fertile that he could keep an entire laboratory occupied, not just busily, but probing into areas that were new and promising. "Students often began work along lines of their own choosing," says Oliphant, "but rapidly found that the instinct of Rutherford's genius was a surer guide to interesting and important results."

From the Antipodes

Rutherford was born in 1871 near the town of Nelson, on the South Island of New Zealand, the fourth of twelve children. His father, James, supported the large family on a modest income from a variety of activities: flax farming and processing, railroad-tie manufacture, bridge building, and farming. Rutherford's mother, Martha, was a schoolmistress and a woman of exceptionally strong character, who advocated hard work as the "sovereign remedy for many evils of the day." She was the dominant influence in the complicated household of seven sons and five daughters, and the attachment between the mother and her famous son never diminished. Rutherford wrote to her weekly or biweekly until she died at age ninety-two. When he was created baron in 1931, he cabled to her: "Now Lord Rutherford, more your honor than mine."

Education for young Ernest Rutherford came with scholarships and parental sacrifice. He was an excellent student at a good secondary school, Nelson College, and first began to display his extraordinary talent for experimental science at Canterbury College in Christchurch. There he embarked on a study of the effect of electromagnetic waves, recently observed by Heinrich Hertz, on magnetized steel needles. This work quickly led him to the development of a sensitive device for detecting Hertzian waves transmitted over long distances. Invention of the detector made Rutherford, at age twenty-four, a pioneer in the field of research that Guglielmo Marconi would soon exploit to develop wireless telegraphy.

Rutherford's ticket to the world beyond New Zealand was an 1851 Exhibition Scholarship. Proceeds from the Great Exhibition of 1851 in London had been set aside for scholarships to bring deserving students from the Dominions (anywhere in the British Empire) to England for further study in the universities. In 1895, there were two candidates, Rutherford and J. C. Maclaurin, a chemist who had published a paper on the treatment of gold. The examiners did not recognize genius; they awarded the scholarship to Maclaurin. But good luck often found Rutherford at crucial junctures in his career, and it did in this instance: Maclaurin decided to marry and stay in New Zealand, Rutherford was given the award, and in the summer of 1895, with borrowed money for the passage, he was on his way to England.

Rutherford had decided to become a research student at the Cavendish Laboratory. The first Cavendish director had been James Clerk Maxwell. During Maxwell's five-year tenure as Cavendish Professor, he organized a flourishing laboratory for graduate students, the first in England. His successor was John William Strutt, Lord Rayleigh, who expanded the laboratory, partly with his own money, and emphasized accurate and elegant electrical measurements. The third Cavendish Professor, who became Rutherford's mentor, was Joseph John—always called "J. J."—Thomson (no relation to William Thomson).

Thomson was elected Cavendish Professor in 1884 when he was only twenty-eight and not a likely prospect for a chair of experimental physics. He was clumsy with his hands, absentminded, and on occasion had to be restrained from touching delicate instruments. But with a fine intellectual grasp of the inner workings of his experiments, and the services of talented assistants and students, he had, by the time of Rutherford's arrival in 1895, reached the front rank of experimental physicists.

Like his predecessors, Thomson found the Cavendish Professorship less than prestigious in the Cambridge scheme of things. This situation did not improve when the university established a research degree allowing graduates from other universities to receive the Cambridge bachelor of arts degree on the strength of two years' residence and a suitable thesis. The research students were not welcomed into Cambridge society. To many of the dons they were outsiders, even intruders; they were not and never could be Cambridge men. Thomson enthusiastically supported the program, however, and the first of the recruits, Ernest Rutherford, alone confirmed the wisdom of the policy.

Rutherford arrived in London on a gray day in September 1895, and was staggered by the London air. He escaped to Cambridge, where he was warmly received by Thomson. "I went to the Lab and saw Thomson and had a long talk with him," Rutherford reported to Mary Newton, his fiancée back in New Zealand. "He is very pleasant in conversation and is not fossilized at all. As regards appearance he is a medium-sized man, dark and quite youthful still: shaves, very badly, and wears his hair rather long. His face is rather long and thin; has a good head and has a couple of vertical furrows above his nose."

Even before he was recognized as a research student, Rutherford made himself at home in the laboratory, working with such industry and enthusiasm that one of his new colleagues at the Cavendish was moved to write, "We've got a rabbit here from the Antipodes and he's burrowing mighty deep."

First it was the wireless experiments, which were increasingly successful, but to Rutherford scientifically mundane. Then came an invitation from Thomson to collaborate in a study of x-ray effects on gases. For more than a decade, Thomson had been pursuing certain elusive physical entities called "ions," so called (by Michael Faraday) because of their concerted migrations in applied electric fields. Ions are created in a gas—"ionization" occurs—when a high voltage is applied between two metallic plates with a gas between them. For reasons Thomson brought out in later experiments, the applied voltage causes the molecules of the gas to become electrically charged so that some carry negative charges and others positive charges. The charged molecules are the ions.

Thomson's early experiments with gaseous ionization had been frustrating: either the ionization had led to an uncontrollable, complicated spark or glow discharge, or the electrical effect of the ionization was so slight as to be unmeasurable. Then in 1895, the year Rutherford went to Cambridge, Röntgen published his work on x rays. In addition to the more sensational properties of x rays, Thomson took note of their ability to make gases conduct tiny electrical currents with the applied voltages considerably below the spark-producing level. Thomson repeated Röntgen's experiments and found that the small x-ray-induced currents looked much like ionization currents; perhaps this was the long-sought method for creating and studying ions under controlled conditions.

Rutherford willingly left the wireless experiments behind—"I am a little full up of my old subject," he wrote to Mary Newton—and gratefully accepted Thomson's assignment to study the electrical effects of x rays on gases. His experiments confirmed Thomson's ionization conjecture in great detail, showing how the ions were generated, how fast they traveled, and how they could annihilate each other. The ions were so clear to him, Rutherford told his friend and principal biographer, Arthur Eve, he could almost see the "jolly little beggars."

Once he had completed his x-ray work, it was not difficult for Rutherford to map his next experiments. If the electrical effects of x rays were interesting, it was a good bet that similar effects caused by the radiation from uranium and other radioactive elements, recently reported by Becquerel and the Curies, would be an equally rewarding object of study. With no hesitation, he began the line of research that would guide him to his most fundamental discoveries. While Marie Curie was discovering chemical methods for isolating radium, Rutherford was inventing physical techniques for characterizing the radiation accompanying radioactivity. His first important discovery was that uranium radiation had at least two components, one that was easily blocked when layers of aluminum foil were put in its path, and another that was much more penetrating. "For convenience" he called the nonpenetrating component "α rays" and the penetrating one "β rays." This proved to be a nearly general characterization of radioactive emissions; only one other component, called "γ rays," was discovered by Rutherford later.

While he was completing this work, Rutherford was thinking of the future, which did not look encouraging at Cambridge. But there were other prospects, and for an advertisement of his talent he had an enthusiastic testimonial from Thomson: "I have never had a student with more enthusiasm or ability for original research than Mr. Rutherford." Most attractive was a professorship at McGill University in Montreal. In spite of his youth and lack of teaching experience, Rutherford won the McGill competition, and it was a perfect match. The McGill authorities wanted to build their university's research reputation. (As Rutherford wrote later to Mary Newton: "I am expected to do a lot of original work and to form a research school in order to knock the shine out of the Yankees!") For his part, Rutherford got what he needed most, a first-rate physical laboratory, one of the best in the world. The laboratory and the professorship were financed by Sir William MacDonald, a millionaire tobacco merchant who considered smoking a "filthy habit."

Rutherford began his tenure as MacDonald Professor of Physics in September 1898. One of his first research efforts was a collaboration with a young electrical engineer, R. B. Owens, whose task was to study thorium radiation as Rutherford had previously studied uranium radiation. To his amazement, Owens discovered that thorium radiation was affected by air currents. "Something that was neither thorium, nor alphas, nor betas, could be blown about!" as Rutherford's biographer Arthur Eve puts it. Rutherford called the mysterious new component thorium "emanation," and designed experiments to show that it was a radioactive gas. (Eventually it was identified as one of the inert-gas elements and called "radon.") Thorium emanation came wrapped in another mystery: it was a gas but could coat the inside of its containers with a hodgepodge of *solid* radioactive materials. Rutherford first called these radioactive medleys "excited activity," and when he understood them better, "active deposits."

While he was uncovering the complexities of the emanations and active deposits, Rutherford also discovered a third fundamental radiation component generated by radioactive elements. These rays were still more penetrating than β rays, and Rutherford called them γ rays. He suspected that they were similar to x rays, but did not find a way to prove that point until later.

During the summer of 1900, after two years of enormously concentrated work, Rutherford took a vacation. He traveled back to New Zealand to visit his parents and to collect his by now impatient bride (who had endured several postponements of the wedding). The couple returned to Montreal in September 1900, having spent half a year's income ($1,250) on travel expenses and their honeymoon.

Alchemy

When Rutherford returned to his laboratory, one of his first thoughts was that he could not mine the riches of his recent discoveries without the help of a skilled chemist. The radioactive element he had studied, thorium, its emanation, and the constituents of the active deposits clearly differed from each other chemically and needed to be identified and, if possible, isolated. Rutherford found the perfect man for the job next door in the McGill chemistry department. He was Frederick Soddy, a chemistry demonstrator (laboratory instructor) who had recently been graduated from Oxford. Soddy was energetic, ambitious, and broad-minded. He proved to be one of the most gifted of Rutherford's many junior research partners.

Rutherford and Soddy first found the chemical means to take the radioactivity out of thorium. They separated a highly radioactive material, which they called "thorium X," leaving the thorium initially inactive. But the activity and the inactivity did not persist; in a few weeks, the thorium had recovered its original activity, and that of thorium X had decayed. Earlier, Becquerel had found similar behavior in experiments with uranium. By examining this pattern of recovery and decay mathematically, Rutherford and Soddy found evidence for a revolutionary theory. They assumed that thorium X, like thorium, was a chemical element with distinct chemical and physical properties, that it was being formed spontaneously at a very slow rate from thorium in a chemical reaction, and that thorium X was in turn spontaneously converting to another distinctly different element, as yet unidentified. The sequence was

$$\text{Thorium} \rightarrow \text{Thorium X} \rightarrow ?$$

These were chemical reactions from a realm no chemist or physicist had ever visited. One radioactive element or "radioelement" was spontaneously changing into another; on the atomic level, atoms were spontaneously disintegrating to form new atoms. In short, it was transmutation. It was alchemy.

Not surprisingly, in view of its radical nature, the Rutherford-Soddy disintegration theory of radioactivity had its critics, including Pierre Curie and Lord Kelvin (William Thomson when we met him in chapter 7). But the theory proved extraordinarily durable. Rutherford and Soddy applied it to one intermediate radioelement after another, building complete chains of consecutive spontaneous disintegrations. In studies of thorium, radium, and actinium active deposits (all produced by emanations), they discovered many new radioelements. Most of the disintegrations were characterized by emission of one or more of Rutherford's three kinds of rays, α, β, and γ. Each disintegration process was also recognized by the rate of the resulting decay of the activity of the radioelement. It was convenient to express this rate in terms of the time it took for the radioactivity of an element to fall to half its original value. These "half-lives" varied drastically from 3×10^9 years in one case to 1.5 minutes in another. Some radioelements were very stable, hardly radioactive at all, while others were very unstable and highly radioactive.

The amounts of some of the radioelements observed by Rutherford and Soddy were so minuscule they could not be weighed, nor could they be identified chemically, even though, with the extremely sensitive methods used by Rutherford and Soddy, their radioactivity could be accurately measured. Thus until the chemical identities could be established (which came about a decade later), Rutherford and Soddy and others were forced to use arbitrary aliases for the radioelements they observed. For example, Rutherford and Soddy reported that in the radium active deposit they had found radium A, radium B, and so on through radium G. In 1904, Rutherford was invited to deliver the important Bakerian Lecture to the Royal Society in London. (He had been elected a fellow of the Royal Society just the year before.) He used the occasion to summarize all the work he had done in collaboration with Soddy. Figure 21.1 displays his diagram representing the disintegrations he and Soddy had observed during 1902 and 1903. For each disintegration he lists (next to the arrow) the emission, α, β, or γ. ("Pt" in the diagram means "particle"; by this time Rutherford realized that α and β "rays" were actually particles.)

A striking feature of the Rutherford-Soddy scheme is that the radioactive disintegrations are mostly linked in sequential series: for example, disintegration of radium A forms radium B, which in turn disintegrates to form radium C, and so forth, until the series terminates with the production of a nonradioactive element (radium G, not shown in Rutherford's diagram). Later, Rutherford discovered that uranium X and radium are linked, and that the radium series was actually a continuation of the uranium series.

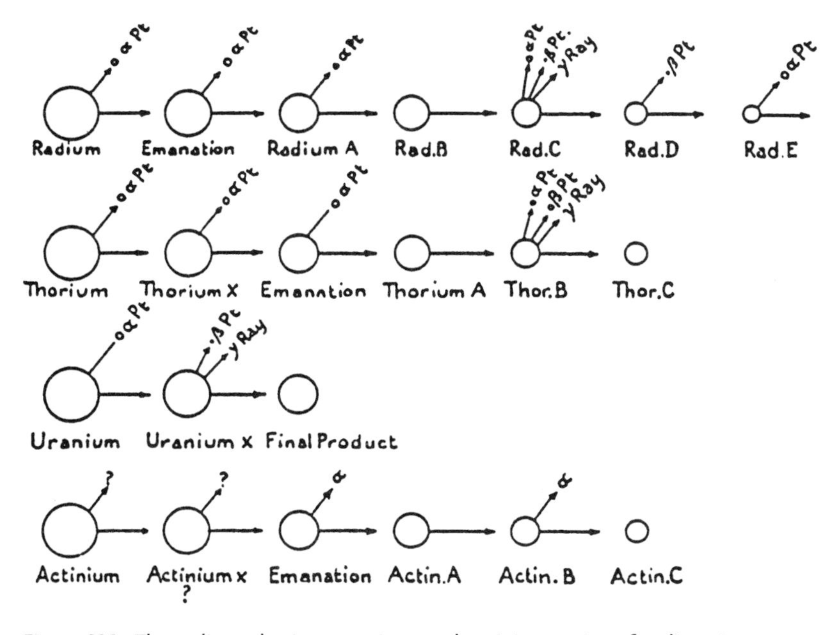

Figure 21.1. The radium, thorium, uranium, and actinium series of radioactive disintegrations, as displayed by Rutherford in his 1904 Bakerian lecture.

Rutherford's work with Soddy on the disintegration theory of radioactivity was his greatest, but not his only, achievement at McGill. At the same time he was beginning an extended study of the physical properties of the ubiquitous α rays. His experimental technique was to direct the rays into a strong magnetic field oriented perpendicularly to the direction of the rays. Newton's laws of motion combined with Maxwell's laws of electromagnetism prescribe that an electrically charged particle in this situation must follow a curved path, with the curvature of the path depending on the charge, mass, and energy of the particle. J. J. Thomson had used this strategy to identify electrons in cathode rays, and Becquerel applied it to β rays, demonstrating that, like cathode rays, β rays are actually streams of high-speed electrons.

Rutherford found that α rays from radium and other sources are also "deviable" in a magnetic field, but much less so than β particles (not rays). Rutherford concluded that α rays are also streams of particles, but with the difference that the α particles (not rays) are considerably more massive than β particles. On the other hand, γ rays are not deviable in a magnetic field and really are rays, similar to x rays. Rutherford illustrated these various aspects of radioactive emissions in a 1904 lecture to the Royal Institution with the sketch in figure 21.2. Notice that the α and β paths are curved in opposite directions. Newton-Maxwell theory interprets this to mean that α and β particles carry electrical charges of different sign, α particles positive and β particles negative.

Once Rutherford and Soddy had sorted out the sequence of the radium series and characterized the accompanying emissions, they could estimate the energy changes driving the chemical transmutations they were observing. What they found would have been hard to believe if the general weirdness of radioactivity had not already become familiar. By their estimate, the energy released during the transmutation of one gram of radium through its entire series to a nonradioactive final product is about two thousand times that produced when one

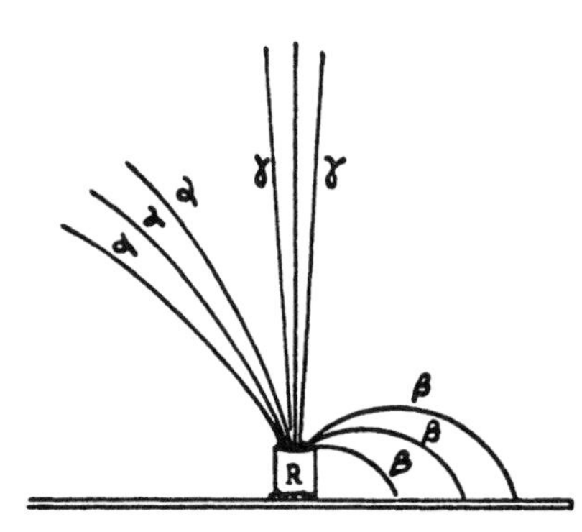

Figure 21.2. Rutherford's sketch of the three kinds of emissions from radium.

gram of water is formed from hydrogen and oxygen, by ordinary standards a very energetic (sometimes explosive) reaction. Rutherford and Soddy concluded in the last paper of their collaboration that "the energy latent in the atom [radioactive or nonradioactive] must be enormous compared with that rendered free in ordinary chemical change." They speculated that if this "atomic energy" were made available by subatomic changes on the Sun, it could account for the maintenance of solar energy.

Alpha Bombardments

Rutherford had practically everything he needed at McGill except proximity to the British and European scientific centers. That disadvantage finally persuaded him to leave Canada when he was offered the Langworthy Chair of Physics at the University of Manchester by Arthur Schuster, who was retiring from the professorship and wanted Rutherford as his successor. Schuster knew that the Manchester physical laboratory, second only to the Cavendish in England, would be an attraction. In the spring of 1907, the Rutherfords (accompanied now by a young daughter, Eileen) arrived in Manchester. The laboratory met Rutherford's expectations. "The laboratory is very good," he wrote to his mother, "although not built so regardless of expense as the laboratory at Montreal." In the same letter, he delighted in telling his mother a story "modesty almost forbids me to relate" about the impression he made on an official visitor: "Baron Kikuchi, Japanese Minister of Education, was here yesterday and was introduced to me by Schuster. Later he said to Schuster, 'I suppose the Rutherford you introduced me to is a son of the celebrated Professor Rutherford'!!!"

The Manchester laboratory included not only an excellent collection of instruments but also a fine young assistant, Hans Geiger, who would make crucial contributions to Rutherford's further studies of α particles. Geiger was German and just what Rutherford needed, a gifted experimentalist with an unlimited capacity for hard work. "Geiger is a good man and worked like a slave," Rutherford wrote to a colleague in 1908 after his first year at Manchester. "I could never have found the time for the drudgery before we got things going in good style. Finally all went well, but the scattering is the devil." As we will see, "the scattering" also harbored an angel.

Rutherford brought with him to Manchester his favorite experimental tool, his "strong right arm," the α particle. He had characterized α particles physically in his experiments at McGill, and guessed that they were charged helium atoms—that is, helium ions. He now wanted to nail down that surmise. With Geiger, he developed an electrical device that registered and counted *individual* α particles by greatly amplifying their ionization effect. (A later incarnation of this instrument came to be called a "Geiger counter.")

At the same time, he took advantage of another method for detecting individual α particles: the observation of the tiny flashes of light produced when α particles strike a screen coated with zinc sulfide. Geiger's counter verified that every arrival of an α particle at the zinc sulfide screen caused one of these "scintillations." Scintillation counting particularly appealed to Rutherford: it was the next thing to seeing an individual α particle.

Rutherford and Geiger used the scintillation technique to count the number of α particles released by one gram of radium in a second. Then with a parallel measurement of the associated electrical charge they obtained an accurate determination of the charge carried by an α particle. The magnitude of the charge was twice that of an electron, and with other confirming evidence, they concluded that α particles are doubly charged helium ions (He^{++}).

To his young recruits at Manchester, Rutherford proclaimed that "all science is either physics or stamp collecting." He knew that one of those "stamp collecting" disciplines had served him well. His distintegration theory of radioactivity was inspired and confirmed by Soddy's excellent technique in the chemical laboratory. In 1908, one year after arriving in Manchester, Rutherford became the Nobel laureate in chemistry. "It was very unexpected," he said, "and I am startled at my metamorphosis into a chemist."

"I was brought up to look at the atom as a nice hard fellow, red or gray in color, according to taste," Rutherford once remarked. The picture of hard, solid atoms was demolished forever by a series of experiments done between 1909 and 1913 by Rutherford, Geiger, and a young undergraduate, Ernest Marsden. They were concerned with the "devil" in experiments with beams of α particles: the tendency for the beam to "scatter"—that is, lose definition—when any material, even a gas, is put in its path. Using the scintillation-counting technique, Geiger had found that when an α-particle beam bombarded a thin metallic foil, most of the particles either passed straight through the foil or were slightly deflected (scattered) by it. But there was a hint that some of the α particles were scattered through larger angles.

Rutherford was always intrigued by improbable experimental results. He decided to have Marsden study the large-angle scattering. As he recalled much later: "I agreed with Geiger that young Marsden, whom he had been training in radioactive methods, ought to begin a research. Why not let him see if any α particles can be scattered through a large angle? I did not believe they would be." But Marsden discovered that, in fact, a few were: about one α particle in eight thousand was deflected through an angle greater than 90° by a platinum foil. For Rutherford this "was quite the most incredible event that has ever happened to me in my life. It was almost as incredible as if you had fired a 15-inch shell at a piece of tissue-paper and it came back and hit you."

Rutherford's α particles were very energetic—they carried the energy they would have had if they were accelerated through millions of volts—and on an atomic scale they were massive. To be turned back occasionally by a metallic foil, they had to encounter something that was charged and equally massive but at the same time small in diameter. Rutherford explained the experimental results with a model that assumed single close encounters between α particles and the intense electrical field of an extremely small "charge center," or "nucleus," as he called it later. He estimated that the diameter of the massive, charged nucleus was 10^{-13} centimeter. Because the size of the atom as a whole, estimated by other means, was 10^{-8} centimeter, Rutherford's model forced the astonishing conclusion that the atom is mostly emptiness, far from the "nice hard fellows" of his earlier acquaintance.

Performing as a theorist for once, Rutherford expressed his model in convenient mathematical language and handed the equations to Geiger and Marsden for a thorough experimental test. Marsden later recalled that "the complete check was a laborious but exciting task. I remember Geiger making a calculation that in the process of the work we counted over a million alpha-particles." The particles were counted by direct observation, one by one, scintillation–by scintillation. The equations and Rutherford's strange model were successful: they passed the test.

But, oddly, Rutherford's colleagues were not persuaded; they were not even interested. "The scientific community . . . was not impressed; this novel theory of the atom was not opposed, but largely ignored," writes one of Rutherford's biographers, Lawrence Badash. Two events brought Rutherford's nuclear atom into the mainstream of physical thought. One was the favorable reception of Bohr's atomic theory, which began with the nucleus concept and then constructed complete atoms by adding orbital electrons. The other was the publication of two papers by a young member of Rutherford's Manchester gathering, Harry Moseley, who showed chemists how to solve some long-standing mysteries concerning their periodic table of the elements without "stamp collecting." I begin the Moseley story with a quick review of the evidence that chemists used in Moseley's time to organize their periodic table.

Chemistry Lessons

In 1869, almost fifty years before Rutherford's atomic model came on the scene, Dmitry Ivanovitch Mendeleev, professor of chemistry at the University of St. Petersburg, called attention to a certain periodicity in the properties of the chemical elements: "Elements placed according to the value of their atomic weights present a clear periodicity of properties." By "atomic weight," a chemist means the mass of an atom on a relative scale that assigns hydrogen, the lightest atom, an atomic weight of about one.

The next eight elements beyond hydrogen listed in order of increasing atomic weight are placed in the top row of the periodic table of figure 21.3, which shows the table as it was known in 1911. (The abbreviations are He = helium, Li = lithium, Be = beryllium, B = boron, C = carbon, N = nitrogen, O = oxygen, F = fluorine.) The tenth element, neon (Ne), illustrates Mendeleev's rule of chemical periodicity: it is a gas, chemically inert, and similar to helium. In the table, it begins a new row and is placed in the same column as its chemical kin, helium. The next element is sodium (Na), with chemical resemblances to the element above it in the table, lithium. This pattern continues across the second row: magnesium (Mg) resembles beryllium, aluminum (Al) resembles boron, and so forth.

In the third and fourth rows, complications arise that won't be elaborated here, except to say that argon (A) and potassium (K) are listed in their legitimate chemical families but not according to Mendeleev's rule: argon is slightly heavier than potassium. There are omissions in this 1911 table, some of them due to elements

0	I	II	III	IV	V	VI	VII	VIII
He 6.94	**Li** 6.94	**Be** 9.1	**B** 11.0	**C** 12.00	**N** 14.0	**O** 16.0	**F** 19.0	
Ne 20.2	**Na** 23.0	**Mg** 24.32	**Al** 27.1	**Si** 28.3	**P** 31.0	**S** 32.07	**Cl** 35.46	
A 39.9	**K** 39.10	**Ca** 40.1	**Sc** 48.1	**Ti** 48.1	**V** 51.0	**Cr** 52.0	**Mn** 54.9	**Fe** 55.8 **Co** 59.0 **Ni** 58.7
	Cu 63.6	**Zn** 65.4	**Ga** 69.9	**Ge** 72.5	**As** 75.0	**Se** 79.2	**Br** 79.92	
Kr 82.9	**Rb** 83.5	**Sr** 87.6	**V** 89.0	**Zr** 90.6	**Nb** 94	**Mn** 96.0	?	**Ru** 101.7 **Rh** 102.9 **Pd** 106.7
	Ag 107.88	**Cd** 112.4	**In** 113	**Sn** 119.0	**Sb** 120.2	**Te** 127.5	**I** 126.9	
X 130	**Cs** 133	**Ba** 137.4	**La** 139	**Ce** 140.25	—	—	—	
	—	—	**Yb** 172.0	—	**Ta** 181.5	**W** 184.0	—	**Os** 191 **Ir** 193.1 **Pt** 195.2
	Au 197.2	**Hg** 200.6	**Tl** 204.0	**Pb** 207.1	**Bi** 208.0	—	—	
	—	**Ra** 226	—	**Th** 232.4	—	**U** 238.5		

Figure 21.3. The periodic table of the elements as it was known in Rutherford's time.

not yet discovered and others because a series of elements simply did not fit. There are many of these misfits between cerium (Ce) and ytterbium (Yb), and they include some of the elements chemists call "rare earths."

In time, chemists would have marshaled the chemical evidence to dispel the confusion in this imperfect 1911 periodic table. But with clues from Rutherford and Bohr, and some x-ray data of his own, Moseley showed them how the table could be quickly put in order with no chemistry at all.

X Rays from Atoms

Harry Moseley went to Manchester after an unimpressive student career at Eton and Oxford. His lack of promise, as judged by academic standards, is reminiscent of that other unsuccessful student, Albert Einstein. (Moseley remarked that at the time of his graduation from Oxford his mind was so "full of cobwebs" he could not think creatively about research.) To Rutherford's discerning eye, however, there was clearly a spark, and he hired Moseley as a demonstrator. Moseley's first project was to count the β particles emitted by radium B and radium C. This work earned him a research fellowship, and he decided to use it to explore the emerging paradox of the wave-particle behavior of x rays. He formed a partnership with Charles Darwin, the resident theorist at Manchester and a grandson of the author of *On the Origin of Species;* together they proposed an x-ray study to Rutherford. The master was dubious because no one at Manchester had experience in the complexities of x-ray research. Moseley solved that problem by going to Leeds and taking lessons from William Bragg, the leading x-ray authority in England.

That satisfied Rutherford, and Moseley and Darwin embarked on a study of x rays as waves. They found x-ray frequencies distributed continuously over a broad range, but missed the discovery of sharp peaks superimposed on the continuous spectrum. The Bragg father-and-son team, William and Lawrence, found these peaks, and Moseley and Darwin followed their lead with a detailed study, making clear that what they were observing was the same thing as a homogeneous x-ray component reported earlier by the Scottish physicist Charles Barkla.

Barkla's radiation was known to be characteristic of the material used as the source of the x rays. Moseley decided to make a systematic investigation of the characteristic x rays generated by a series of elements. His aim at first was to correlate the particular characteristic x rays Barkla labeled "K" with the "atomic number" of the element, the number that designates its position in the periodic table (1 for hydrogen, 2 for helium, 3 for lithium, and so forth).

Working by himself now, Moseley plunged into this project with almost manic intensity. "He was without exception the hardest worker I have ever known," Darwin writes in a reminiscence. Moseley often worked through the night and learned how to find a meal in Manchester at three in the morning. Mental exhaustion was no distraction. "When I told him he ought to be at home in bed," Darwin recalls, "he would answer that when he was feeling well he wanted to be out walking in the country, and that it was only in this condition when he was tired out that he felt inclined for laboratory work." He could never resist making improvements, large or small, in his x-ray equipment: "He was always ready to take the whole apparatus to pieces and set it up again if he could see any possible improvement to be hoped for."

Moseley's initial discovery was an astonishingly simple and precise equation that connected the atomic number Z of the elements calcium through zinc (Z = 20 through 30) with the frequency ν_K of the K characteristic x rays,

$$\nu_K = \left(\frac{3R'}{4}\right)(Z - 1)^2, \tag{1}$$

or

$$\nu_K = R'\left(\frac{1}{1} - \frac{1}{2^2}\right)(Z - 1)^2, \tag{2}$$

with R' identical to the constant in Balmer's equation for the frequencies ν in the hydrogen optical spectrum,

$$\nu = R'\left(\frac{1}{n_1^2} - \frac{1}{n_2^2}\right) \tag{3}$$

in which n_1 and n_2 are integers. Encouraged by this success, Moseley proposed a second empirical equation for the characteristic x rays Barkla had labeled L,

$$\nu_L = R'\left(\frac{1}{2^2} - \frac{1}{3^2}\right)(Z - 7.4)^2, \tag{4}$$

and applied it to many of the elements between zirconium (Z = 40) and gold (Z = 79).

At about the same time as Moseley was pursuing this research, Bohr was also working in the Manchester laboratory and closing in on his impressive theoretical explanation of Balmer's formula (3). For Moseley and Bohr, the obvious resemblances between Moseley's two equations (2) and (4), on the one hand, and the Balmer formula on the other, promised further theoretical developments. They hoped that a theory of atoms containing many electrons would arise from Moseley's equations as Bohr's theory of the hydrogen atom, with its single electron, had grown from Balmer's equation. They never realized that expectation, but two assumptions they introduced in their theoretical efforts—that the atomic number Z for an element counts the number of electrons in each of the element's atoms, and that it also measures in electronic units the balancing positive charge on the nucleus—have lasted and become permanent fixtures in atomic theory.

Moseley observed characteristic x-ray spectra from thirty-nine of the sixty-seven elements between aluminum and gold, and used his equations to determine atomic numbers. Besides verifying the necessity for listing some elements out of order in the atomic weight sequence (e.g., argon and potassium), Moseley's unambiguous evaluation of atomic numbers also showed gaps where there was a number, but no known element that matched it. Four missing elements were indicated, for atomic numbers 43, 61, 72, and 75, and they were eventually found, the last one thirty-four years later. Each element was put "into its right pigeon-hole," as Moseley remarked, even those that had never been seen. The French chemist Georges Urbain, who supplied Moseley with rare-earth samples, wrote to Rutherford of his amazement with Moseley and what he could do: "I was most surprised to find a very young man capable of doing such remarkable work. . . . Moseley's law, for the end as well as for the beginning of the rare earths, has established in a few days the conclusions of my efforts of twenty years of patient work."

Moseley's x-ray work, a distinguished effort if it had required a lifetime, was completed in less than a year. (This included time for a move from Manchester to Oxford and a complete rebuilding of the apparatus with the dubious services of a technician who was a "thorn in the flesh.") Rutherford said that Moseley was "the best of the young people I ever had." He might have been Rutherford's equal.

When England entered the war with Germany in 1914, Moseley quickly volunteered his services. He was commissioned in the Royal Engineers and became a signals officer. In June 1915, his brigade was sent to the Dardanelles. Two months later, a confused action took place in which Moseley's brigade was led, deliberately or mistakenly, by two guides who later disappeared, to a position in front of the British lines. The men slept during the night and awoke to recognize their mistake in the daylight, but by that time the Turks had started an attack. Sometime during the morning Moseley was shot through the head and died instantly.

Not Enough Pigeonholes

Moseley's numbering of the elements added to the atomic weight a fundamentally more important atomic parameter: the atomic number. At about the same time, these two parameters were converging in an entirely different way. It had been assumed for some time that only the heaviest ten or so elements in the periodic table were radioactive. Yet Rutherford, Soddy, and others had claimed discovery of far more radioelements than that, identifying them by their emissions and their half-lives, and giving them noncommittal names such as radium A, thorium X, and ionium. How did the radioelements fit into the periodic table? There were not enough pigeonholes.

By about 1910, it was apparent that, although the proliferating radioelements were physically distinct, they were not all unique chemically. For example, thorium X and radium could not be separated because they had the same chemical properties; likewise, radium D and lead were inseparable, as were ionium and thorium. Soddy resolved the pigeonhole crisis by assuming that inseparable pairs such as these had different atomic weights but the same chemical identity—that is, the same pigeonhole assignment in the periodic table. The genesis of this idea is complicated; it did not originate with Soddy. Two atoms of the same element with different atomic weights are called "isotopes" in Soddy's terminology. In the notation now used, isotopes are designated with the chemical symbol, the magnitude of the nuclear charge (equal to the atomic number) as a subscript, and the approximate atomic weight as a superscript: Rutherford's radium A is now represented ${}_{84}Po^{218}$, polonium with an atomic weight of approximately 218 and an atomic number (or nuclear charge) of 84. If the atomic number is redundant, it is omitted, as in the notation Po^{218}.

Soddy recognized that isotopes are important in the makeup of inactive, as well as radioactive, elements. As it occurs naturally, the element chlorine, for instance, is a mixture of two inactive isotopes, ${}_{17}Cl^{35}$ and ${}_{17}Cl^{37}$. The measured atomic weight for chlorine is 35.47, an average of the atomic weights 35 and 37, with the former dominating.

Soddy's work also helped clarify the evolving concept that the nucleus is the seat of radioactivity. With many examples, he showed how radioelements are displaced in the periodic table when they are transmuted by emission of α and β particles. He suspected that these radioactive atomic number changes were nuclear changes, and Moseley's arguments were in agreement.

Atom Demolition

When war broke out in 1914, most of Rutherford's students took up military service or other wartime duty, and the Manchester laboratory was quickly depopulated. Rutherford himself was busy as a civilian member of a committee assigned the task of developing sonic methods for detecting submarines. But he found time during the war years to complete a course of research that ranked with his best.

This work originated in some intriguing observations recorded by Marsden a few years earlier on the bombardment of hydrogen gas with α particles from a radium C (Bi^{214}) source. Rutherford repeated Marsden's experiment, and concluded that α particles and hydrogen nuclei were knocking into each other like two billiard balls with enough energy to send both nuclei off in recoil motion, ultimately leaving their signatures on a scintillation screen. The hydrogen nuclei—Rutherford was now calling them "protons"—were distinguishable from α particles by their greater ability to penetrate materials put in their paths.

This much was understandable. But then Rutherford did the experiment with nitrogen gas substituted for hydrogen, and *again* he saw proton scintillations. His explanation was that nitrogen nuclei are artificially disintegrated when they are struck by α particles: "It is difficult to avoid the conclusion that the long-range atoms arising from collision of alpha particles with nitrogen are not nitrogen atoms," he wrote in 1919, "but probably atoms of hydrogen, or atoms of mass 2. If this be the case, we must conclude that the nitrogen atom is disintegrated under the intense forces developed in a close collision with a swift α particle, and the hydrogen which is liberated formed a constituent part of the nitrogen nucleus."

After the war, in 1919, Rutherford was again persuaded to leave a place and a job he loved. He wrote to his mother:

> You will have received the news that I have been elected to the Cavendish Chair of Physics held by Sir J. J. Thomson, who is now master of Trinity [College]. It was a difficult question to decide whether to leave Manchester as they have been very good to me, but it is probably best for me to come here, for after all it is the chief physics chair in the country and has turned out most of the physics professors of the last 20 years. . . . It will of course be a wrench pulling up my roots again starting afresh to make new friends, but fortunately I know a good few people there already and will not be a stranger in Trinity College.

One of Rutherford's biographers, Edward Andrade, a Manchester alumnus, writes, "It is generally agreed by those who knew him and have written on the point that the happiest years of his life were spent at Manchester, years that saw the birth of the nuclear atom and the first [artificial] disruption of the nucleus."

To welcome Rutherford at the Cavendish Laboratory, the inhabitants composed a song (à la Gilbert and Sullivan) with this as one of its verses:

What's in an atom,
The innermost substratum?
That's the problem he is working at today.
He lately did discover
How to shoot them down like plover,
And the poor things can't get away.
He uses as munitions
On his hunting expeditions
Alpha particles which out of radium spring.
It's really most surprising,
And it needed some devising,
How to shoot down an atom on the wing.

And the chorus was:

He's the successor
Of his great predecessor,
And their wondrous deeds can never be ignored:
Since they're birds of a feather,
We link them together,
J. J. and Rutherford.

In Cambridge, Rutherford continued aiming α particle at "atoms on the wing." He was joined in this work by James Chadwick, who had moved with him from Manchester and became his closest collaborator, confidant, and second in command at the Cavendish Laboratory. "[Chadwick] knew more intimately than Rutherford just what every person in the Laboratory was doing," writes Mark Oliphant, who eventually succeeded Chadwick as Rutherford's lieutenant. "It was he who trained raw recruits to research in a 'kindergarten' laboratory loft above Rutherford's office. With Rutherford he chose the research tasks of most students and set them on their way. . . . [Rutherford's] commitments outside the laboratory could never have been satisfied without Chadwick's continued, unselfish attention." Chadwick could be cold and impersonal, but he was rarely unreasonable, and he, better than any of Rutherford's other research partners, mastered the art of experimental physics.

Rutherford and Chadwick first demonstrated that α bombardments could disintegrate not only nitrogen nuclei but also the nuclei of other light elements such as boron, fluorine, sodium, aluminum, and phosphorus. But these experiments left an unanswered question. Did the α particles actually penetrate the target nucleus, forming a composite nucleus and a proton, or did they chip protons from the target nucleus and ricochet themselves like bullets glancing off a rock?

As it happened, the equipment for answering this question was at hand in the Cavendish Laboratory, in the work of Charles Thomson Rees Wilson, usually called "C. T. R." Wilson, about Rutherford's age and an old friend, was a student of clouds and cloud formation. His laboratory at first was the summit of Ben Nevis, the highest mountain in Scotland and a dramatic cloud maker. Under J. J. Thomson at the Cavendish, Wilson found that he could do in the laboratory what Ben Nevis did in nature by suddenly expanding and thus cooling air that was saturated with water vapor. He also discovered that his artificial cloud formation was promoted by ionizing the air with x rays. This led, in 1911, to his most important discovery: that the trail of ions left by an α particle in his "cloud chamber" made a cloud track—like a miniature version of an airplane's vapor trail—that clearly marked the particle's trajectory. Like the scintillation screen, Wilson's cloud chamber put the experimenter in touch with individual particles, not only α particles but also β particles, protons, and electrons produced by an x-ray beam.

In the 1920s, Patrick Blackett refined Wilson's cloud chamber and obtained direct evidence for the atom demolitions reported by Rutherford and Chadwick. In photographs of some four hundred thousand α particle tracks, Blackett found eight forked ones with two branches, one for an emitted proton and another for the composite nucleus. No tracks with three branches were found, and thus no ricocheting α particles. It was clear then that the nuclear "reaction" observed by Rutherford in his α bombardment of nitrogen was

$$_2He^4 + {}_7N^{14} \rightarrow {}_1H^1 + {}_8O^{17}.$$

The reaction begins with an α particle ($_2He^4$) and nitrogen ($_7N^{14}$) and produces a proton ($_1H^1$) and a composite nucleus (the oxygen isotope $_8O^{17}$).

Alpha particles were Rutherford's great friends in the laboratory. He favored them because they were readily available in simple radium and polonium sources, and they were powerful. A single radium α particle brings to a nuclear collision the energy it would have if it were accelerated through millions of volts, and that is enough to permit the α particle to disrupt a nucleus by penetrating it. Rutherford surmised that other particles, such as protons, could also be "used as munitions on his hunting expeditions" if they, too, could be obtained at energies of millions of volts. Radioelements are not so generous with protons as they are with α particles, however, so two Cavendish physicist-engineers, John Cockcroft and Ernest Walton, built a machine that produced a beam of protons accelerated through several hundred thousand volts. They bombarded a lithium target with this beam, disrupted lithium nuclei, and produced α particles. The nuclear reaction they induced was

$$_1H^1 + {}_3Li^7 \rightarrow 2\,{}_2He^4.$$

The Cockcroft-Walton machine accelerated protons in a straight tube. At about the same time, Ernest Lawrence at the University of California at Berkeley was developing a circular accelerating machine he called a "cyclotron." Lawrence's cyclotron guided charged particles on spiral paths and accelerated them in incremental steps, twice in each trip around the spiral. This design had the advantage that it required less space and lower voltages across the accelerating gap than the Cockcroft-Walton linear accelerator.

The accelerators at the Cavendish and at Berkeley were the progenitors of a long line of accelerators, both linear and circular, developed by physicists and engineers in many laboratories. Modern practitioners of high-energy physics spend billions of dollars on their accelerating machines and on the descendants of Wilson's cloud chamber, which they use for particle detection. Dozens or even hundreds of scientists, engineers, and technicians are required to mount an experiment with this equipment. Currently the aim is to build accelerating machines that achieve collision energies equivalent to acceleration through hundreds of *billions* of volts.

Rutherford supported Cockcroft and Walton and thus helped inaugurate the era of big accelerators. But big machines and the big money needed to finance them were not his style. The apparatus for a typical Rutherfordian experiment was constructed on a bench top, was operated by one or two research students, and required an annual expenditure of perhaps fifty pounds. He was well aware, however, that bombardment of heavy atoms would not be effective without the high energies only the big accelerators could supply. He approved plans for contruction of a two-million-volt commercial linear accelerator and a cyclotron, the machine he least appreciated, but neither had made important contributions by 1937, the year Rutherford died.

The Elusive Neutron

In 1920, Rutherford gave the prestigious Bakerian Lecture to the Royal Society for a second time. His first Bakerian Lecture, in 1904, had been an account of the spontaneous transmutations accompanying radioactivity. In the second lecture, he spoke of the artificial transmutations he had recently induced with assistance from the admirable α particles. He also included some prognostications, the most important of which introduced his audience to a certain electrically neutral particle: "Under some conditions it may be possible for an electron to combine [with a proton] much more closely [than in the case of the hydrogen atom], forming a kind of neutral doublet. Such an atom would have very novel properties. Its external field would be practically zero, except very close to the nucleus, and in consequence it should be able to move freely through matter. . . . The existence of such atoms seems almost necessary to explain the building of the heavy elements."

In Rutherford's view, the "neutral doublets," which he soon called "neutrons" (borrowing a term that had been used earlier in another connection), joined protons as the fundamental nuclear building blocks. The number of protons in a nucleus determined its positive charge, and the number of protons plus neutrons its atomic weight. The nitrogen isotope $_7N^{14}$, for example, with its nuclear charge of + 7 and atomic weight of 14, contains 7 protons and 7 neutrons. Rutherford's picture of a neutron formed in a close association between an electron and a proton is not supported by modern theory and experimental results, as we will see in chapter 26.

James Chadwick's road to the discovery of the neutron was long and tortuous. Because they carried no electrical charge, neutrons did not leave observable trails of ions as they passed through matter, and left no tracks in Wilson's cloud chamber; to the experimenter they were invisible. As he traveled the meandering road to the neutron, Chadwick took many wrong turns and bumped up against many dead ends. "I did a lot of experiments about which I never said anything," Chadwick told an interviewer.

> Some of them were quite stupid. I suppose I got that habit or impulse or whatever you'd like to call it from Rutherford. He would do some damn silly experiments at times, and we did some together. They were really damn silly. But he never hesitated. At times he would talk in what seemed a rather stupid way.

> He would say things, which, put down on paper, were stupid or would have been stupid. But when one thought about them, you began to see that those words were inadequate to express what was in his mind, but there was something in the back that was worth thinking about. I think the same thing would apply to some of these [neutron] experiments I have said were silly. There was always just the possibility of something turning up, and one shouldn't neglect doing say a few more hours' work or even a few days' work to make quite sure. . . . But I just kept pegging away. I didn't see any other way of building nuclei [i.e., without neutrons].

The final clue that Chadwick needed to make his discovery came from Paris. In 1931, Irène Joliot-Curie, Marie Curie's daughter, and Frédéric Joliot, her husband, described radiation produced by bombardment of a beryllium target with α particles from a polonium source. When they tried to attenuate this radiation with layers of paraffin they got more radiation rather than less—and it consisted of protons. Their explanation for the protons was that they were knocked loose from the hydrogen-containing paraffin by polonium γ rays in an effect discovered in the 1920s by Arthur Compton. They realized that for the γ rays to perform this feat they had to be extremely energetic. "I don't believe it," said Rutherford.

Neither did Chadwick, and he had a better explanation: α particles ($_2He^4$) reacted with beryllium nuclei ($_4Be^9$), forming carbon nuclei ($_6C^{12}$) and neutrons (represented $_0n^1$ because they have an atomic weight of about one, and zero electrical charge),

$$_2He^4 + {_4Be^9} \rightarrow {_6C^{12}} + {_0n^1}.$$

The massive neutrons were much better candidates than the almost massless γ rays for the projectiles that bumped protons out of the paraffin.

Rutherford and Chadwick spent twelve years in pursuit of the neutron. Chadwick made the final discovery after a month of frantic experimentation triggered by the Joliot-Curie paper. "He worked night and day for about three weeks," writes C. P. Snow, who was a Cavendish research student in the 1930s.

> The dialogue passed into Cavendish tradition:
> "Tired, Chadwick?"
> "Not too tired to work."

After he had told the story of his quest to the Cavendish research group, Chadwick asked "to be chloroformed and put to bed for a fortnight."

At Home

Rutherford's origins in New Zealand were unprepossessing, and even with the weight of his later fame and influence he remained a simple man. He never became wealthy. The homes he made with his wife Mary were unpretentious. Newnham Cottage, their rented home in Cambridge, was a "comfortable, tasteless, academic home, lacking in grace or inspiration, run by three or four servants in the manner of the times, with a wife whose main interest was in her garden, for a husband whose main interest was in the laboratory," writes David Wilson, Rutherford's most recent biographer.

Mary Rutherford (Lady Rutherford after 1913) "was a blunt, down to earth woman, round of face and dumpy of figure, but quick of movement," Mark Oliphant, Chadwick's successor as Rutherford's assistant director at the Cavendish, tells us. Oliphant continues with a sketch of the Rutherfords' domestic life:

> The Rutherfords occupied separate bedrooms, both at Newnham Cottage and when at their country cottages. There were not overt acts of affection between them. Yet they were devoted to one another. Lady Rutherford understood little or nothing of her husband's work, but she was proud of the honours that came to him and reacted violently to any criticism. She treated him in many ways as she would a child, still attempting to correct his faults when eating, for instance. I never heard him reply impatiently to her, as would most men when treated in this way.

Lady Rutherford's contribution to conversation at the breakfast table might include "Ern, you're dribbling," or "Ern, you've dropped marmalade down your jacket."

The Rutherford's only child, Eileen, married Ralph Fowler, the chief theorist at the Cavendish, and the Fowlers had four children. Eileen died, tragically, shortly after the birth of the fourth child. Rutherford loved his grandchildren and could enter into their world. His biographers include a photograph taken at the seashore, of the grandfather lending a hand in the construction projects of a young granddaughter. The grandmother, on the other hand, was more inclined to give the children lessons in manners; they called her "Lady Rutherford" or "Lady R."

Elements of Success

Rutherford was as straightforward and unpretentious as a physicist as he was elsewhere in life, and that no doubt was one of the secrets of his success. "I was always a believer in simplicity, being a simple man myself," he said. If a principle of physics could not be explained to a barmaid, he insisted, the problem was with the principle, not the barmaid.

For Rutherford, simplicity meant concrete, visualizable concepts, with minimal mathematics and elementary apparatus. "One [Rutherford] experiment after the other is so directly conceived, so clean and so convincing as to produce a feeling almost of awe, and they come in such profusion that one marvels that one man could do so much," Chadwick wrote in 1937, shortly after Rutherford's death. "He had the most astonishing insight into physical processes, and in a few remarks he would illuminate a whole subject. . . . To work with him was a continual joy and wonder. He seemed to know the answer before the experiment was made, and was ready to push with irresistible urge to the next." Others may "play games with their symbols," Rutherford said, "But we at the Cavendish turn out the real solid facts of Nature."

Rutherford had a powerful voice. When a friend heard that a Rutherford speech would be carried across the Atlantic in a radio broadcast, he asked, "Why use radio?" Rutherfordian black moods and eruptions could terrify students and assistants, but the storm would soon pass and was likely to be followed by an apology. Remarkably, though, beneath the irritability and impatience was an exquisite tact. A longtime friend once said, "Rutherford never made an enemy and never lost a friend." That is an exaggeration; a man of Rutherford's fame and influence inevitably had some enemies. But it would be hard to find even one of his countless friends who defected. He shunned scientific controversy and avoided political and religious arguments. When contentious issues did occasionally arise, he found ways to settle them amicably for all involved. For his gift of tact and kindness, Rutherford got in return the priceless services of dozens of talented, hardworking students and associates. They loved him.

Like his peers in our pantheon—Newton, Faraday, Maxwell, Gibbs, Einstein, and Bohr—Rutherford could concentrate on a difficult, frustrating problem for long periods without losing acuity or enthusiasm; it seemed that he never got stale. A story told by Harold Robinson, a Manchester alumnus, shows Rutherford's pure pleasure in the business of the laboratory, even under the worst circumstances. Robinson found himself, not by choice, in the laboratory with Rutherford, wasting a fine Saturday afternoon

> in an obviously rather hopeless effort to purify, with a few dregs of liquid air, a very little sample of radon [emanation] with which we hoped to work. The attempt ended with a momentary lapse on Rutherford's part, which resulted in the admission of a much larger volume of air than we had previously succeeded in extracting—a slip which brought the characteristic remark, "Well, it's a good job *I* did that and not you." I am afraid I felt that the afternoon might have been better spent, but Rutherford's final comment, as he sucked contentedly at his pipe while we cleared up the mess, was: "Robinson, you know I *am* sorry for the poor fellows that haven't labs to work in!"

To close this list of Rutherford's secrets of success I add one more: luck. Just when he began his career, radioactivity was discovered, inviting research that exactly suited his style. Would he have done as well if he had been born, say, thirty years later? Perhaps not, but we should not underestimate Rutherford's astonishing ability to make his own luck. After all, when he began his radioactivity research there were "other sprinters in this road of investigation," as he told his mother, but their *combined* effort was less than his. Arthur Eve once said to him, "You are a lucky man, Rutherford, always on the crest of a wave!" And Rutherford replied "Well! I made the wave didn't I?" Then he added "At least to some extent."

22

Physics and Friendships

Lise Meitner

Vienna

So far, we have met twenty-two of the great physicists. Have you wondered how entertaining it would be to spend a few hours with one of them in casual conversation? Might Newton be too neurotic to engage in a satisfying conversation? Might Einstein be too detached? Heisenberg too formal? Rutherford too loud? Faraday too busy? Maxwell too ironic? Boltzmann too distracted? Schrödinger too self-centered? About our next subject, Lise Meitner, you would have no such reservations. An evening spent with her would be pleasant and stimulating. She was good company.

Lise Meitner was born in Vienna in 1878 into a middle-class, liberal, Jewish family, the third child of eight. Her father was a lawyer and a man of diverse interests. He and his wife Hedwig "made their home a gathering place for interesting people—legislators, writers, chess players, lawyers," writes Ruth Sime, Meitner's principal biographer. "The children stayed up and listened. Years later when Meitner was asked about her childhood, she remembered most of all 'the unusual goodness of my parents, and the extraordinarily stimulating intellectual atmosphere in which my brothers and sisters and I grew up.' "

The Meitner children had talent and they were rewarded. Lise's older sister Auguste (Gusti) was a musical prodigy; she became a composer and a concert pianist. Lise, too, loved music, but lacked the temperament of a performer. From as early an age as eight, she had a well-developed interest in mathematics and physics, and aimed for a university education. But in nineteenth-century Austria, a girl's public school education lasted to age fourteen, far short of the preparation needed for university entrance. Lise Meitner, like Marie Curie, was not stopped by deficient secondary education. With the help of a tutor and incessant hard work, she passed the Matura, the university entrance examination. The family joke was that Lise would fail the Matura if she did not have a book in her hand every minute of the day.

At the University of Vienna, Meitner had the extraordinary good fortune to attend the last full cycle of lectures on theoretical physics given by Ludwig Boltzmann, who was pleased to have women attend his courses. "Boltzmann gave her the vision of physics as a battle for the ultimate truth, a vision she never lost," writes Otto Frisch, Meitner's nephew, who later collaborated with Meitner in her most important research. Boltzmann was the best physics teacher in the world at the time. He told his students, in the first lecture of the cycle,

> Forgive me if I have not accomplished much today with respect to all these things, involved theorems, very highly refined concepts, and complicated proofs. . . . I think that much will become clearer later in the course of the work. Today I only wanted to offer you something quite modest, admittedly for me all I have, myself, my entire way of thinking and feeling. Likewise I shall have to ask a number of things of you during the course of the lectures: strict attention, iron diligence, untiring will. But forgive me, if before we go on I ask for something that is most important to me: your confidence, your sympathy, your love, in a word the greatest thing you are able to give, yourself.

Meitner was enthralled. She gave everything Boltzmann demanded, and in return acquired a superb background in the theoretical physics of the time. Her careful notes on Boltzmann's lectures attracted Paul Ehrenfest, another Boltzmann student, who would later become an outstanding theorist. The two studied together, and Meitner profited as much from Ehrenfest's imaginative teaching as from that of Boltzmann. "[He] was an excellent and stimulating teacher," Meitner wrote later. "I am sure that working with him was a great help in my scientific development." Meitner was still shy and naïve, however, and the charming, more worldly Ehrenfest sometimes put her off. "I must confess," she wrote, "that sometimes I was disturbed by his inclination to put questions about altogether personal things."

With Boltzmann and Ehrenfest, Meitner studied the role of the theorist. To take some of the theory she had learned into the laboratory, she chose a doctoral thesis topic involving an experimental test of one of Maxwell's equations. She took her oral examinations in 1905 and passed summa cum laude; she was the second woman to earn a doctorate at the University of Vienna.

In the fall of 1906, physicists everywhere were devastated by the news that Boltzmann, in deep depression, had committed suicide. Boltzmann's act was difficult for Meitner to understand; she could recognize it only as "mental instability." But it brought her closer to a career in physics. As Ruth Sime writes, "Boltzmann's death strengthened her determination to remain in physics, so that the spark he kindled in her would remain alive."

Suddenly, unexpectedly, Meitner found the research path she would follow for the rest of her career. Stefan Meyer, a pioneer in radioactivity research, took over Boltzmann's institute, and he invited Meitner to study the behavior of α and β radiation passing through metals. She focused on the scattering phenomenon, which Rutherford would later find so "devilish" and then so profitable in the pursuit of the atomic nucleus.

For Meitner, physics was always as much a human endeavor as a technical one. She chose her mentors and colleagues for their human qualities and worked with them as close friends. In 1907, after a year of successful research with Stefan Meyer, she decided to go to Berlin to work with the man she admired most in the physics community, Max Planck.

Berlin Conquests

In Berlin, Meitner experienced the warmth of Planck's friendship and the hostility of German attitudes toward women in universities. Throughout the nineteenth century, women were tolerated in German university classes only as unmatriculated auditors; Meitner was obliged to ask Planck for permission to attend his classes. He was kind and sympathetic, but skeptical. Meitner describes their meeting in her reminiscences, *Looking Back*: "He received me very kindly and soon afterwards invited me to his home. The first time I visited him there he said to me, 'But you are a Doctor already! What more do you want?' When I replied that I would like to gain some real understanding of physics, he just said a few friendly words and did not pursue the matter further. Naturally, I concluded that he could have no very high opinion of women students, and possibly that was true enough at the time."

Planck may have had reservations about women as professionals, but he was delighted to welcome Meitner into his household. Musical evenings were the favorite form of entertainment. Planck on the piano, the famous concert violinist Josef Joachim, and (later) Einstein on the violin were often the performers. Meitner did not play but she deeply appreciated the music. Among Meitner's close friends of her own age were Planck's identical twin daughters, Emma and Grete.

Also in attendance at Planck's musical evenings, and lending a good tenor singing voice, was Otto Hahn, a young radiochemist who had trained with Rutherford in Montreal. Hahn was gregarious and informal, and without much ado he proposed that Meitner join him in radioactivity research. Meitner, who still felt shy and insecure in the bustling Berlin atmosphere, quickly recognized Hahn as a friend and a valuable colleague. "Hahn was of the same age as myself and very informal in manner," Meitner recalls in *Looking Back*, "and I had the feeling that I would have no hesitation in asking him all I needed to know. Moreover, he had a very good reputation in radioactivity, so I was convinced he could teach me a great deal."

In this casual way, a unique scientific collaboration began. Hahn was a chemist, expert in the chemical separation techniques practiced in radiochemistry, and Meitner was a physicist who was rapidly developing as both a theorist and an experimentalist. Together they could meet the interdisciplinary demands of radioactivity research. They worked in the same institute from 1907 to 1938, when Meitner was forced out of Germany by Nazi racial laws. During this time they worked together not only as colleagues, but also as close friends. Hahn was dapper and handsome, and Meitner was petite and lovely, but there was never a romantic attachment between them. The Victorian proprieties had to be observed: they did not eat together or go out for walks together, and for sixteen years they addressed each other as Herr Hahn and Fräulein Meitner. It was not a partnership like that of the Curies.

Hahn had an appointment as an assistant in the University of Berlin Chemistry Institute, which was directed by the renowned organic chemist, Emil Fischer. Meitner, however, had no professional standing; worse, women were not allowed in Fischer's institute. (One of Fischer's fears was that women's hairstyles were a fire hazard.) But Hahn was persuasive, and Fischer agreed that Meitner could work in a basement room, a former carpenter's shop, with an outside entrance. Meitner used the room without setting foot elsewhere in the institute; she endured this and other indignities stoically. The only toilet accessible to her was in a restaurant down the street. For some of the institute's assistants she became an invisible woman. They greeted Meitner and Hahn together with, "Good day, *Herr Hahn*!"

Outside the institute, Meitner formed many lasting friendships in the physics community. At the physics colloquia held by Heinrich Rubens, she met James Franck and Max von Laue. Franck was an experimentalist who, with Gustav Hertz, had performed a Nobel Prize–winning experiment that demonstrated the reality of quantization of atoms. Meitner and Franck knew when they first met that they spoke the same language. Laue was best known for his pioneering work on x-ray diffraction. Laue and Meitner supported each other during the dark days of the Nazi regime. Many years later, Laue expressed his gratitude to Meitner: "Did you realize how deeply your words affected us? . . . Your goodness, your consideration had their effect. The notion of humanity acquired substance. For this I am grateful to you. . . . I was saved from things for which I would never have forgiven myself."

Gradually, incrementally, the barriers against women in Prussian universities were lowered. By 1909, university education for women was officially sanctioned and Meitner was given access to the laboratories of the chemistry institute (and a ladies' room was installed). But she was still unpaid and living frugally on an allowance from her parents. In spite of the primitive working conditions, Meitner and Hahn published three major articles in 1908 and six in 1909. Their main focus was on "β-emitters," those radioelements that give off β particles when they decay. The perplexing theory of "β-decay" would be a continuing theme for Meitner for more than a decade.

The essential operation of radiochemistry is the separation of one element from another. This can be done by chemical means. For example, a mixture can be treated chemically so compounds of certain elements precipitate and others remain in solution. Chemical methods are versatile but rarely "clean"—that is, capable of producing a perfectly pure product. Hahn and Meitner developed a method that was more efficient. They discovered that a "daughter" atom formed in a radioactive disintegration might be so energetic that it was driven away from the solid surface where it was formed, and could be collected in pure form on another surface. It was like a kernel of popcorn leaping off a hot plate and landing elsewhere.

At last, in 1912, Meitner had an opportunity to move out of the carpenter's shop, and to take the first steps on the academic ladder. Institutes for chemistry and physical chemistry were opened in Dahlem, a Berlin suburb, under the sponsorship of Kaiser Wilhelm. Hahn was appointed a "scientific associate" and given the responsibility for a radioactivity section within the Kaiser-Wilhelm Institute of Chemistry. Meitner joined him as an unpaid "guest physicist."

At about the same time, Planck appointed Meitner as his assistant. It was menial work; she graded student papers. But she loved Planck, and it was her first paid academic position. A few years later, Fischer, who had lost his paranoia about women in laboratories, saw to it that Meitner was given the same title as Hahn, scientific associate, but with a considerably lower salary than the one Hahn received. The radioactivity section was now the Hahn-Meitner Laboratorium. Salary discrepancies aside, Meitner now knew that she had arrived. "I love physics with all my heart," she wrote to a friend. "I can hardly imagine it not being a part of my life. It is a kind of personal love, as one has for a person to whom one is grateful for many things. And I, who tend to suffer from a guilty conscience, am a physicist without the slightest guilty conscience."

In their new quarters, Meitner and Hahn began an arduous hunt for the long-lived radioelement they believed to be the precursor, the "mother substance," of actinium. That work, done mostly by Meitner during the war years while Hahn was on active duty in the army, extended to four years. Their conferences were mainly by mail; the final paper was published in 1918. Stefan Meyer, always a fan of Meitner's, had some suggestions for naming the new element: "lisonium" or "lisottonium." The official name, less charming, was protactinium.

Even before the protactinium success, Meitner could see her career blossoming further. She received an attractive offer from Prague, a junior academic position with good prospects for advancement. Planck took note of this development, and saw to it that Fischer was informed. Fischer doubled her salary to three thousand marks, and Meitner gratefully remained in Berlin.

Her progress in the academic world continued. In 1917, she was given her own physics section, and an increase in salary to four thousand marks (essentially equivalent to Hahn's salary of five thousand marks, which included a marriage allowance), and the Laboratorium Hahn-Meitner was divided into the Laboratorium Hahn and the Laboratorium Meitner. Two years later Meitner had a new title, professor in the institute, and probably became the first woman in Germany to have the title of professor. She had little appreciation for the title, but "enjoyed the real pleasure my friends took in it."

The β-Decay Problem

From the beginning of their partnership, Meitner and Hahn had been interested in radioelements that disintegrated with production of β particles. There was a deep mystery about β decay as opposed to α decay that Meitner was determined to unravel. Alpha particles produced by a given radioelement always appeared with about the same energy. Beta particles, on the other hand, were emitted with energies covering a broad continuous range, from practically zero to a certain maximum value. Where did these electrons come from? Meitner believed that they were partly of "secondary" origin—that they were emitted as "primary" electrons from the nucleus and then in secondary processes lost energy as x rays in the strong electric field of the nucleus.

One of Rutherford's associates at the Cavendish Laboratory, Charles Ellis, disagreed. He was convinced that the secondary effects proposed by Meitner were too small to account for the observed continuous spectrum of β particles. In a letter written in 1925, Ellis summarized their points of agreement and disagreement: "We both agree that once the [β particles] are outside the parent atom they are already inhomogeneous in velocity [covering a continuous range of energies]. We both agree that a quantized nucleus ought to give [β particles] of a definite [energy]. Whereas you think various subsidiary effects are sufficiently large to produce the observed inhomogeneity, I think they are much too small."

Ellis and his student William Wooster decided the issue by performing an experiment that strongly supported their point of view. Meitner repeated their experiment and wrote to Ellis: "We have verified your results completely. It seems to me now that there can be absolutely no doubt that you were completely correct in assuming β radiations are primarily inhomogeneous [covering a broad spectrum]. But I do not understand this result at all."

The lengthy Meitner-Ellis debate was concluded, but the fundamental nature of β decay was as shrouded in mystery as ever. The problem was this. Those on both sides of the debate believed that the total energy of the β decay process was constant. Some of that energy went to the β particle and some to a new nucleus. But if the β energy was small, as it could be in the spectrum advocated by Ellis, and now by Meitner, the two energies did not add up to the necessary total. What happened to the rest of the energy? Theorists were in crisis. For a while Bohr was willing to abandon the principle of conservation of energy on an atomic scale.

That drastic measure proved to be unnecessary: Wolfgang Pauli had a different idea. Unconventional as ever, he outlined his theory in an open letter written in 1930 and addressed to Lise Meitner and Hans Geiger, and to those attending a conference in Tübingen. He proposed a new particle he called a "neutron," but this was not the neutron, companion to the proton in the nucleus, observed by Chadwick two years later.

> Dear Radioactive Ladies and Gentlemen [Pauli wrote]. As the bearer of these lines, for whom I ask your gracious attention, will explain to you in more detail, I have, faced with . . . the continuous β-spectrum, stumbled upon a desperate remedy. Namely the possibility that in the nucleus there could exist electrically neutral particles which I will call neutrons, which have a spin of one-half and obey the exclusion principle and in addition also differ from light quanta in that they do not travel at the speed of light. The mass of the neutron must be the same order of magnitude as the mass of the electron and in any case not larger than 0.01 proton mass. The continuous beta spectrum would then be understandable assuming that in β-decay a neutron is emitted along with the electron in such a way that the sum of the energies of neutron and electron is constant. . . .
>
> At the moment I don't trust myself enough to publish anything about this idea and turn confidently to you, dear radioactives, with the question of how one might experimentally prove such a neutron, if its penetrating ability is similar [to] or about 10 times that of γ-radiation. I admit that my remedy may at first seem only slightly probable, because if neutrons do exist they should have been observed long ago! But nothing ventured, nothing gained, and the gravity of the situation with the continuous spectrum is illustrated by a statement of my respected predecessor in this office, Herr Debye, who told me recently in Brussels: "Oh, it is best not to think about it all, like the new taxes!" Thus one should discuss every means of salvation. Therefore, dear radioactives, test and decide! Unfortunately I cannot appear in Tübingen in person, since I am indispensable here due to a ball which will take place the night of December 6 and 7 in Zürich. With many greetings to you all, your most humble and obedient servant, W. Pauli.

Pauli's proposal was indeed a "desperate remedy," only slightly less so than Bohr's willingness to abandon the principle of conservation of energy for elementary particles. Pauli's "neutron" had little mass (his estimate turned out to be generous by orders of magnitude) and no electrical charge. Pauli asked his colleagues to "test and decide," but how could they? No experimental equipment of the time could detect such a particle, directly or indirectly. It was enough to raise the ghosts of nineteenth-century positivists who could not abide anything that could not be directly observed. But desperation breeds confidence in weird theories. In 1934, Enrico Fermi proposed a more complete theory of β decay in which Chadwick's neutrons (observed in 1932) were primary inhabitants of the nucleus. In β decay they were transformed into an electron (a β particle), a proton, and one of Pauli's "neutrons," which Fermi now called "neutrinos." Fermi's theory tied up the loose ends of β decay, and some nuclear statistical matters as well; the theory was quickly accepted. Neutrinos joined electrons, protons, neutrons, and the newly discovered "positrons" (positively charged electrons) as one more kind of elementary particle, even though they were not detected experimentally for another twenty-two years. The mass of Fermi's neutrino is still an open question.

Berlin Nightmares

On January 30, 1930, Adolf Hitler was sworn in as chancellor of the German Reich. If there were any doubts about his intentions before the installation, few remained thereafter. By March the Reichstag was dissolved, new elections were scheduled, and Hitler unleashed his private militia to stifle the opposition. Meitner, like most scientists, could hardly believe what was happening, and hoped that sanity would prevail. "The political situation is rather strange," she wrote to Hahn, who was visiting in the United States, "but I very much hope it will take a calmer, more sensible turn." Two weeks later she wrote, "Everything and everyone is influenced by political upheavals."

Soon the racial policies of Hitler's Nazi Party were implemented. A nationwide boycott of Jewish businesses was called for April, and a general campaign began to remove Jews from professions of all kinds, in government, medicine, law, education, and the arts. Bernhard Rust, the Prussian minister of education, had hoped to dramatize the Jewish boycott by firing Germany's, and indeed the world's, most famous Jew, Albert Einstein, from the Prussian Academy of Sciences. But before the Nazi functionary could act against him, Einstein announced from abroad that he would not return to a Germany that was without "civil liberty, tolerance, and equality of all citizens before the law," and that was ruled by a "raw and rabid mob of the Nazi militia."

Thus began the exodus of some of Germany's finest scientists and intellectuals. James Franck, director of the Second Physics Institute in Göttingen, was one of the first to resign. He was followed by Max Born, director of the Institute for Theoretical Physics in Göttingen, and Richard Courant, a prominent mathematician. David Hilbert, the greatest of the Göttingen mathematicians, was asked by the new minister of education if the institutes had suffered from "the departure of the Jews and their friends." "Suffered? No they didn't suffer, Herr Minister," replied Hilbert. "They just don't exist anymore!"

In this atmosphere of crumbling moral standards, Lise Meitner's response was dangerously equivocal. She listened to Planck, who according to Born, "trusted that violence and oppression would subside in time and everything [would] return to normal. He did not see that an irreversible process was going on." Planck, Meitner's most trusted mentor, advised her to stay, and so did Hahn. As Meitner wrote later to a friend, she sensed the increasing danger, but was "only too willing to let myself be persuaded by Planck and Hahn." Her plight was agonizing.

Ruth Sime writes: "Emigration was hard: the world was gripped by depression and positions were scarce. Lise could not bring herself to leap into the unknown, to relive her earlier days in Berlin, to be a frightened outsider again, a stranger in a foreign land. She clung to her physics section: 'I built it from its very first little stone; it was, so to speak, my life's work, and it seemed so terribly hard to separate myself from it.' "

She passed up a grant for a year's stay at Bohr's Copenhagen institute, and rejected the possibility of a position at Swarthmore College in the United States. (Swarthmore could not meet her needs for laboratory space, equipment, and staff.)

"Lise Meitner would not leave until she lost everything and was driven out," writes Sime. First, she was dismissed from the University of Berlin, and not allowed to attend meetings or colloquia there. Planck, Laue, and Hahn were all anti-Nazis, and for a time they could protect Meitner, using her Austrian citizenship as a shield. Then in 1938 came the Austrian *Anschluss* (annexation) and she no longer had even that thin defense. Hahn was constantly under attack from ambitious pro-Nazis in his institute, and he began to lose his nerve. Pressured by his superiors, he asked Meitner not to come to the institute anymore. "He has, in essence, thrown me out," she wrote in her diary. There was a reprieve, but it was not to be trusted. Later in her diary, she wrote: "Promises are of no use, they are not kept. Possibilities narrowing." She knew she had to get out of Germany, but now she was trapped without a valid passport and restricted by a recent edict forbidding technical and academic personnel to leave Germany.

Meanwhile, Meitner's many friends were making heroic efforts to get her out. In Holland, two physicists, Dirk Coster and Adriaan Fokker, were trying to raise enough money for at least a year's stipend at a Dutch university. Bohr searched in vain for a grant to support her work in Copenhagen. Paul Scherrer, a physical chemist in Zürich, repeatedly wired her to come for a "conference," but that route was closed because the Swiss would not accept her without a valid passport. During these increasingly frantic rescue efforts, lines of communication were kept open by Peter Debye, director of the Kaiser Wilhelm Institute for Physical Chemistry in Berlin, and protected by his Dutch citizenship. Finally, Bohr produced one more possibility: Manne Siegbahn, an experimental physicist, might have space for Meitner in his new Stockholm institute.

Urgency turned to desperation. Debye wrote to Coster a coded message with a clear meaning: "If you come to Berlin may I ask you to stay with us, and (providing of course that the circumstances are still favorable) if you were to come rather soon—as if you received an SOS—that would give my wife and me even greater pleasure." Coster went to Berlin, and with elaborate care not to arouse suspicion, Meitner packed a few things. Accompanied by Coster, she escaped to the Dutch border, where Coster had made some discreet arrangements with the border guards. When Hahn said goodbye to Meitner in Berlin he gave her an inherited diamond ring: "I wanted her to be provided for in an emergency."

Meitner was at last safe, but still not permanently situated. Coster and Fokker were still seeking money for a stipend. Then the Swedish offer, on again and off again for months, at last became firm, and Meitner decided to take it. This was a wise choice, as it turned out, because she would have been vulnerable again when Germany later invaded Holland. There were a few more anxious moments when Meitner flew to Copenhagen and then to Stockholm—bad weather might have brought the plane down in Germany. On August 1, 1938, she arrived in Stockholm, a world safe from the Nazi menace but with its own depressing problems. "One dare not look backward," she wrote to Coster, "one cannot look forward."

Isolation in Stockholm

Meitner's benefactor in Stockholm, Manne Siegbahn, was a man with an agenda that left little encouragement for Lise Meitner. He was an experimentalist who had won a 1924 Nobel Prize for his work on x-ray spectroscopy. In 1937, he began a program of nuclear research with the construction of a cyclotron. Siegbahn and Meitner came from different generations. As Sime writes, "Siegbahn may well have regarded her as old-fashioned. Eight years older than he, she had come to nuclear physics much earlier and made important discoveries with simple equipment. He had always tied his experiments to the advancement of his instruments: she had looked for problems where theory and experiment progress together. She assumed he would be glad to have her; he may have thought that she would be content with laboratory space and nothing more."

Meitner was paid the salary of a junior assistant by the Swedish Academy. Her bank account was frozen in Berlin, and she had no prospects of receiving her pension money. She was living on borrowed money in a small hotel room. In letters to Hahn, she told him of her plight and pleaded with him to find some way to release her possessions and bank account in Berlin. For his part, Hahn complained that he was under attack in his institute by ambitious, ruthless Nazi underlings. They were like a married couple forcibly separated under the worst circumstances.

Feeding Meitner's discontent was her position—or lack of it—in Siegbahn's institute. "The Siegbahn institute is unimaginably empty," she wrote Hahn, "a very fine building, in which a cyclotron and a large x-ray spectroscopic instrument are being prepared, but with hardly a *thought* for experimental work. There are no pumps, no rheostats, no capacitors, no ammeters—nothing to do experiments with, and in the entire large building four young physicists and a very hierarchical work organization." "And in that organization," writes Sime, "Meitner seemed to have no place. Neither asked to join Siegbahn's group nor given the resources to form her own, she had laboratory space but no collaborators, equipment, or technical support, not even her own set of keys to workshops and laboratories." As it was during her first days in Fischer's chemistry institute in Berlin, she was again an invisible woman. Yet in those dismal circumstances, Lise Meitner, in collaboration with her nephew, Otto Frisch, made one of the most important discoveries in twentieth-century physics, certainly ranking with Rutherford's discovery of the nucleus.

Nuclear Fission, or the Transuranes That Weren't

Theoretical physicists are fundamentally conservative (in their professional activities, if not in their politics). They develop their theories along previously traveled intellectual routes if at all possible. Only when they are persuaded by indisputable evidence to the contrary do they depart from the traveled path and head into the unknown, and then with trepidation. Remember Pauli's hesitant proposal of the role of his "neutron" in β decay ("At the moment I don't trust myself enough to publish anything about this idea"). The story of the Meitner-Frisch discovery teaches the same lesson.

In 1935 Enrico Fermi, who was experimenting in Rome with neutron bombardment of uranium, observed some new radioelements. The conservative assumption was that they were formed when uranium absorbed a neutron, becoming both heavier and β active. Emission of the β particle advanced the atomic number beyond uranium's 92, into the realm of artificial "transuranic" elements that do not occur in nature. Hahn and Meitner soon took up the study of these "transuranes."

Also active in the pursuit of the transuranes were Irène Joliot-Curie and Pavel Savitch in Paris. Just before Meitner made her perilous escape from Berlin, she discussed with Hahn and Fritz Strassman, a young analytical chemist, the strange Joliot-Curie-Savitch finding that one of the radioelements resulting from neutron bombardment of uranium behaved chemically like lanthanum, whose atomic weight is almost half that of uranium. A radical interpretation of this result would have been that neutron bombardment caused the uranium nucleus to split into two smaller nuclei, each with an atomic weight of about half that of uranium.

No one anticipated this, but Hahn and Strassmann repeated the Joliot-Curie-Savitch experiment and made their own astonishing discovery: among the products of the neutron-uranium bombardment were radioelements that behaved like radium, except that they had much shorter half-lives than radium. Careful analytical work by Strassmann showed that they were isotopes of barium, another element with almost half the atomic weight of uranium. Here was more evidence that uranium was splitting in the neutron bombardment.

Hahn was still incredulous. "We know ourselves that [uranium] can't actually burst apart into [barium]," he wrote to Meitner late in 1938. "If there is anything you could propose that you could publish, then it would still in a way be work by the three of us." (Hahn could no longer publish with his Jewish colleague.) A few days later he wrote: "How beautiful and exciting it would be just now if we could have worked together as before. We cannot suppress our results, even if perhaps they are physically absurd. You see, you will do a good deed if you can find a way out of this." Meitner wrote that she could not then see "a way out," but her experience as a theorist told her that strange concepts occasionally succeed. "[In] nuclear physics we have experienced so many surprises, that one cannot unconditionally say: it is impossible," she wrote to Hahn. When Hahn published his and Strassmann's results he hedged: "As chemists the experiments we have briefly described force us to substitute for the [heavy] elements formerly identified as radium, actinium, thorium the [much lighter] elements barium, lanthanum and cerium, but as 'nuclear chemists' close to physics we cannot yet take this leap which is contrary to all experience of nuclear physics." This was Hahn's opinion. Strassmann later recalled that he was more willing to "take the leap" and propose that neutron bombardment could split the uranium nucleus.

At that time, just before Christmas 1938, Otto Frisch, Meitner's favorite nephew (he was the gifted son of Lise's older sister Gusti), went to Sweden to spend the holidays with his favorite aunt in the town of Kungälv, on the Swedish east coast. He came from Copenhagen, where he was working in Bohr's institute. "When I came out of my hotel room after the first night in Kungälv," he writes in his autobiography, *What Little I Remember*, "I found Lise Meitner studying a letter from Hahn and obviously worried about it." Frisch wanted to tell her about his work in Copenhagen, "but she wouldn't listen; I had to read that letter." In the letter, Hahn reported his and Strassman's finding that barium resulted in the neutron irradiation of uranium, and asked Meitner to solve the mystery.

Hahn was an accomplished radiochemist, and Meitner did not think he was wrong about the barium. That seemed to force the conclusion that the uranium nucleus was indeed splitting. But how could it? Bohr and George Gamow, a young Russian theorist who was a frequent and entertaining visitor at Bohr's institute, had suggested earlier that a nucleus was like a liquid drop. One could imagine that the drop might elongate, become constricted near the center, and finally divide into two drops. Something like the surface tension of an ordinary liquid drop would oppose such a division, but each uranium nuclear fragment would carry a large positive charge, and the repulsion between the charges would strongly assist the division process.

On a walk through the snowy Swedish woods, with Frisch on skis and Meitner "making good her claim that she could walk just as fast without," the two physicists began to glimpse the makings of a theory. "At that point," writes Frisch, "we both sat down on a tree trunk and started to calculate on scraps of paper. The charge of the uranium nucleus, we found, was indeed large enough to overcome the effect of the surface tension almost completely; so the uranium nucleus might indeed resemble a very wobbly, unstable drop, ready to divide itself at the slightest provocation, such as the impact of a single neutron."

Meitner and Frisch could now visualize the uranium-splitting process, but they had to cope with another problem. The two positively charged fragments would be driven apart by their mutual repulsion with an immense energy, about 200 MeV—that is, 200 million electron volts, the energy acquired by an electron when it is accelerated through 200 million volts. That was about ten times any energy previously observed in a nuclear process. Where could it come from? Meitner remembered the formulas needed to calculate the masses of two typical fragments formed in the splitting process. Taking the difference between the uranium mass and the total mass of the fragments, and converting the difference to energy via Einstein's $E = mc^2$ equation, they could fully account for the 200 MeV likely to accompany the splitting of the uranium. And so, on a snowy December day beside a ski trail, Meitner and Frisch sketched a theory that accounted for the splitting of the uranium nucleus under neutron bombardment.

Frisch took the news back to Bohr in Copenhagen, who understood and accepted the theory immediately. "I had hardly begun to tell him," writes Frisch, "when he struck his forehead with his hand and exclaimed: 'Oh, what fools we all have been! Oh, but this is wonderful! This is just as it may be! Have you and Lise Meitner written a paper about it?' Not yet, I said, but we would at once and Bohr promised not to talk about it before the paper was out." Hours after this conversation Bohr sailed to the United States for a series of lectures.

Meitner and Frisch composed their historic paper by way of several lengthy long-distance telephone calls. They decided to appropriate the term "fission," which was used by biologists when speaking of the dividing of a living cell. The Meitner-Frisch process became "nuclear fission."

A skeptical colleague of Frisch's, George Placzek, challenged Frisch to test his theory by designing an experiment that detected the highly energetic "fission fragments" produced when the uranium nucleus splits. "Oddly enough that thought hadn't occurred to me," Frisch writes, "but now I quickly set to work, and the experiment (which was really very easy) was done in two days and a short note about it was sent to *Nature* together with the other note I had composed with Lise Meitner."

En route to America, Bohr discussed the Meitner-Frisch process with his associate Léon Rosenfeld, and he became even more convinced of its importance and validity. When they landed in New York, however, Bohr neglected to tell Rosenfeld that the news was to be kept quiet until Meitner and Frisch could publish and be guaranteed their priority. While Bohr stayed in New York for a few days, Rosenfeld went on to Princeton, attended a seminar, and told an astonished audience all about nuclear fission. The news created a sensation. Experimentalists rushed to their laboratories to repeat Frisch's experiment. Many succeeded, but to Bohr's great relief, Meitner and Frisch did not lose priority for their theory, nor did Frisch for his detection of fission fragments.

With the acceptance of uranium nuclear fission, the original crop of transuranic elements died, except for two that lived on; these would have been Meitner's discovery if she could have commanded an intense neutron source. Edwin McMillan and Emilio Segrè had used the Berkeley cyclotron for neutron bombardment of uranium and discovered a radioelement with β activity and a half-life of 2.3 days, which Segrè identified as a fission fragment. Meitner did not believe this interpretation because the 2.3-day activity remained with the uranium: a fission fragment would have had ample energy to recoil from the thin samples of uranium used by McMillan and Segrè. To Meitner, it was clear that the 2.3-day activity was a true transurane, element 93, and that its β decay led to another, element 94.

To prove her point, Meitner had to repeat the McMillan-Segrè experiment. For months she waited in vain for access to Siegbahn's cyclotron. Finally, in April 1940, she traveled to Copenhagen to use the cyclotron in Bohr's institute. The day after her arrival Germany invaded Denmark, the Danes surrendered, and Meitner's plans were again frustrated. Seven weeks later, McMillan and Philip Abelson identified element 93 as Meitner had anticipated, and they called it neptunium. "That was terribly difficult for her to accept," writes Ruth Sime, "more so as McMillan and Abelson's neptunium, a beta emitter, was the precursor to yet another transuranic, element 94 [eventually called plutonium]. Of the many heartaches Meitner suffered after leaving Berlin, her failure to find element 93 grieved her most. It would remain a 'crêve coeur,' as she put it, the rest of her life."

Celebrity, Deserved and Undeserved

The further story of nuclear fission is a complex tale not only of physics, but also of national and international politics, bureaucracy, military control, chemistry, and engineering. One result was the atomic bombs dropped on Hiroshima and Nagasaki in August 1945, which will be mentioned in the next chapter. I note here two more features of nuclear fission that emerged in the first months of 1939. Bohr and John Wheeler, a Princeton theorist, showed that fission resulting from bombardment of natural uranium with "slow" (low-energy) neutrons was mainly due to the rare isotope U^{235} and not to the much more abundant isotope U^{238}. In addition, experiments in Paris and at Columbia University showed that each uranium fission not only *consumed* a neutron but also *released* two or three more neutrons. This raised the exciting possibility that neutrons produced in a fission could induce one or more further fissions, and those fissions could produce still more neutrons, and so on. Such rapid neutron multiplication might sustain a nuclear chain reaction that released energy at a fantastic rate, especially if the process were uncontrolled.

By this time, German armies were overrunning Europe, and the military possibilities of a uranium fission bomb were obvious to all nuclear physicists. A German effort, involving Werner Heisenberg and Otto Hahn among others, was ultimately an embarassing failure. After a slow start, a massive effort at several sites in the United States produced the two devastating bombs dropped on Japan.

Lise Meitner was absolutely opposed to nuclear weapons. She was invited to join a group of British and refugee physicists and engineers who were assigned to the rapidly growing laboratory at Los Alamos, New Mexico, where the bombs were being designed. She flatly refused, and was the only nuclear physicist of note on the Allied side to do so. Service in Los Alamos would have meant escape from stagnation in Stockholm, and an opportunity to work again with her friends. But no enticement would change her mind. "*I will have nothing to do with a bomb!*" she declared.

When the horrific news came from Hiroshima and Nagasaki, Meitner was in the uncomfortable position of being the only nuclear physicist who was not locked up somewhere and inaccessible to the press. The German physicists were by that time interned in England. American, British, and refugee nuclear experts were behind fences at Los Alamos and other nuclear facilities in the United States, Britain, and Canada. Meitner was besieged by reporters. She became the "fleeing Jewess," who stole the secret of the atomic bomb from Hitler's scientists and handed it over to her British friends. A respected *New York Times* science reporter told of Meitner wiring the secret to Otto Frisch in Copenhagen, who then passed it on to his "father-in-law" Niels Bohr. Some reporters became fond of calling Meitner the "Jewish mother of the bomb."

The bomb celebrity was unwanted and acutely embarrassing for Meitner, while the celebrity she deserved, and had every reason to expect, was denied to her. In 1944, the Nobel Prize for chemistry was awarded to Otto Hahn *alone* for the discovery of nuclear fission. Meitner had no objections to Hahn's award, but she and many of her friends could not understand why her contributions and Frisch's were ignored. As she explained to a friend, "Surely Hahn fully deserved the Nobel Prize in Chemistry. There is really no doubt about it. But I believe that Frisch and I contributed something not insignificant to the clarification of the process of uranium fission—how it originates and that it produces so much energy, and that was something very remote from Hahn."

Hahn's (and Strassmann's) radiochemical experiments were essential to the discovery, but so also were the physical concepts established by Meitner and Frisch. With hindsight, we can see that a discovery as important as nuclear fission deserved two awards, the chemistry prize to Hahn (and Strassmann), and the physics prize to Meitner and Frisch. The inscrutable Nobel committee did not see it that way. Some detected the hand of Meitner's nemesis, Siegbahn. "[In] Sweden, Lise's friends were furious," writes Sime. "They viewed her exclusion as neither omission nor oversight but deliberate personal rejection, the work of Manne Siegbahn."

If Lise Meitner was denied the first prize, her work was certainly not unappreciated. In 1946, she traveled to the United States for the first time, where she was swept away by a round of receptions, meetings, awards, lectures, and honorary degrees, and a flood of congratulatory letters. It was like a second coming of Marie Curie. Hollywood had a script for her called *The Beginning of the End.* It was "nonsense from the first to the last," she wrote to Frisch. "It is based on the stupid newspaper story that I left Germany with the bomb in my purse, that Himmler's people came to Dahlem to inform me of my dismissal and more along the same lines." She refused to cooperate. "I would rather walk naked down Broadway," she said.

Last Days

Soon after the war, in 1947, Meitner retired from the Siegbahn institute and began work in a small laboratory created for her by the Swedish Atomic Energy Commission at the Royal Institute of Technology. Later she moved to the laboratory of the Royal Academy for Engineering Sciences, for research connected with an experimental nuclear reactor. Finally, in 1960, after twenty years in Sweden, she retired to Cambridge, England, to be near Otto Frisch and his family. She continued an active life of traveling, lecturing, and attending concerts.

After Meitner left Berlin in 1938, the Meitner-Hahn partnership was dissolved, but their friendship continued, sometimes leaving a residue of pain and bitterness for Meitner. This was especially true during the Nobel season of 1946 when she had the unenviable duty of entertaining and celebrating Hahn, the new Nobel laureate, in Stockholm. To Meitner, it seemed that Hahn presented the discovery of uranium fission as a one-man show. She wrote to a friend, "I found it quite painful that in his interviews [Hahn] did not say one word about me, to say nothing of our thirty years of work together. His motivation is somewhat complicated. He is convinced that the Germans are being treated unjustly, the more so in that he simply suppresses the past. Therefore while he was here [in Stockholm] his only thoughts were to speak for Germany. As for me, I am part of the suppressed past."

But these memories eventually faded and Meitner's friendship with Hahn, with all its trials (suffered more by Meitner than Hahn) remained. Max Perutz, a molecular biologist who knew Meitner in Cambridge, reports that according to Otto Frisch's widow, Meitner "never voiced anything but deep affection for Hahn."

Lise Meitner died a few days before her ninetieth birthday. Otto Hahn had died several months earlier. She was buried in an English country churchyard. The inscription on her headstone, prepared by Frisch, is

> Lise Meitner: a physicist who never lost her humanity.

She got what she richly deserved: a superb career, a long life, many honors, and countless enduring friendships.

23

Complete Physicist

Enrico Fermi

1901–1954

Prodigy

As a rule, scientists display their talents either as theorists or as experimentalists, but not both. Einstein, Maxwell, and Gibbs, for example, were great as theorists but not creative as experimentalists, while Faraday and Rutherford, great as experimentalists, were limited as theorists. Only Newton, in our company of physicists seen so far, displayed great talent as both an experimentalist and a theorist (and also as a mathematician). The subject of this chapter, Enrico Fermi, is another exception to the rule that physics is a bipartisan community. Fermi was, as his biographer and colleague, Emilio Segrè, remarks, "from the first a complete physicist for whom theory and experiment possessed equal weight."

He began as a theorist in 1926 by showing how to count the quantum states of atoms according to Pauli's exclusion principle. In the 1930s, he built a complete theory of β decay beginning with another Pauli idea, that β particles always appear in company with tiny particles that carry no electrical charge and almost no mass. This work was a pioneering effort in what is now known as quantum field theory. Fermi could have continued in this direction and become a dedicated theorist. Instead, he chose to become an experimentalist armed with the technique of neutron bombardment. These efforts were also pioneering, and they led him finally to one of the landmark achievements of modern experimental physics: control of a nuclear chain reaction.

Enrico Fermi was born in Rome in 1901, the youngest of three children. His mother, Ida, was a schoolteacher; his father, Alberto, a railroad administrative employee. The Fermi family had few luxuries. Their apartment "had no heating of any kind," reports Fermi's wife Laura in her charming biography and reminiscence, *Atoms in the Family*. While he was studying, Enrico was obliged to sit on his hands to keep them warm, and somehow contrived to "turn the pages of his book with the tip of his tongue, rather than pull his hands out of their snug warming place."

The dominant influence in the Fermi household was Ida. "It was [Ida's] thorough and intelligent devotion that kept them together," writes Laura Fermi. She had rules and they were enforced: "Her devotion was mixed with an overstressed sense of duty and an inflexible integrity, which the children inherited, although they occasionally resented it. Into her affection she brought a certain rigidity that made her expect from others as much as she would give. Her children were to work hard to maintain the high moral and intellectual standards that she had set for them and exacted of them."

Enrico and his older brother, Giulio, were constant companions and partners in endless boyhood projects. The partnership was tragically broken when Giulio died in what was to have been minor throat surgery. The family was devastated: Ida had an emotional breakdown, and Enrico was left alone without his best friend. To escape the melancholy around him, he began an intense, personal study of mathematics and physics.

At first these studies were haphazard, guided mainly by readings in whatever books he could find at a book market. But the teenaged Enrico had a guardian angel. He was Adolfo Amidei, a colleague of Alberto Fermi's with an engineering background and a generous spirit. Amidei was impressed by Enrico's questions about geometry and he lent the boy a book on projective geometry. When Enrico returned the book two months later, he had, to Amidei's astonishment, mastered the proofs and completed all the practice exercises, some of them strenuous. Amidei, himself, had not done so much. He looked at Enrico's proofs, concluding that "the boy, during the little free time that was left to him after he had fulfilled all the requirements of the high school studies, had learned projective geometry perfectly and quickly solved many advanced problems without encountering any difficulties. I became convinced that Enrico was truly a prodigy, at least with respect to geometry."

Amidei continued to engage his young friend over a period of four years with books on trigonometry, algebra, calculus, and theoretical mechanics. Even as a teenager, Fermi could consume and retain the contents of a book on mathematics or physics. "I had already ascertained that when he read a book, even once, he knew it perfectly and didn't forget it," Amidei recalled. When he had finished a book, he did not keep it for reference. "As a matter of fact," Fermi told Amidei, "after a few years I'll see the contents in it even more clearly than now, and if I need a formula I'll know how to derive it easily enough." This ability to hold in his memory or "know how to derive easily enough" any passage of physics he needed was an essential ingredient of Fermi's talent. It was like the musician's ability to memorize a musical score. "[In] his later years," Segrè tells us, "Fermi mentally rehearsed chapters of physics as a director rehearses a symphony. He would do this on long cross-country drives or similar occasions."

Amidei had the wisdom to see that his protégé was being held in check by the atmosphere of grief at home, and he convinced Ida and Adolfo that their son should attend the University of Pisa as a fellow at the affiliated Scuola Normale Superiore. The Scuola Normale (Normal School) had once been a school for teachers but had evolved into an institution that accommodated forty of the brightest students in Italy. Even in this elite environment, Fermi stood out. His admission essay on the topic "Characteristics of Sound" was an advanced mathematical analysis, including statements of the differential equations for the propagation of sound and their solutions. The examiner was so astounded by this performance that he insisted on meeting the applicant (not the usual procedure) to tell him that in a long academic career he had never seen such a student essay. The director of the physics laboratory, an amiable and gifted man, but no longer at the forefront of his profession, casually accepted Fermi's superiority, and would often say to him, "teach me something."

Fermi left Pisa in 1922, returned to Rome, and met another guardian angel, Orso Mario Corbino, director of the physics laboratory at the University of Rome. Corbino, like Amidei, quickly recognized Fermi's genius and became his friend and patron. Corbino's dream was to restore Italian physics to its former eminence. In Fermi, he saw another Galileo—and he was right. At the time, Italian physicists and mathematicians were making important contributions to the theory of general relativity, but were not even teaching in the other new fields that were then flourishing elsewhere in Europe, particularly quantum theory. On Corbino's advice, Fermi traveled to the outside world of science on fellowships, first to Max Born's institute at Göttingen, where he did not feel at home in the Heisenberg-Pauli club. His next stop was Leiden; he had been invited by Paul Ehrenfest, professor of theoretical physics and successor to the great classical physicist Hendrik Lorentz. Ehrenfest's acquaintance with contemporary physics was encyclopedic (as Fermi's later would be). There was a role, he said, for Fermi in the revolutionary developments to come.

Back in Italy, Fermi took an appointment at the University of Florence and lived in Arcetri, where Galileo spent his last days. There he displayed another facet of his talent, an outstanding teaching ability. "A serious reason for his wanting a professorial appointment was his love of teaching, apparent in all his activities from the time of his boyhood," Segrè writes. At the same time, he was reading, and as always efficiently assimilating, the current physics journals. "He thought deeply about what he read and was often inspired to add something new," Segrè observes. "This habit, which lasted until the time of his neutron work, helps to explain the vastness and universality of his knowledge." In one of these excursions into the contemporary literature of physics, he made his first major contribution by "adding something new" to Pauli's exclusion principle.

Quantum Statistics

Fermi reached Pauli's principle in a roundabout way. He aimed to make an entropy calculation for an ideal gas of atoms using Boltzmann's statistical entropy equation,

$$S = k \ln W,$$

with careful attention to the rules of quantum mechanics. One of those rules is that atoms can exist only in certain discrete states and no others. Another is that like atoms in an enclosure cannot be labeled and distinguished from each other. This is because the wave function representing an atom has a long enough reach that it overlaps wave functions for other atoms in the enclosure. (Wave functions for electrons were discussed in chapter 19. With a suitably constructed Schrödinger equation, wave functions for atoms, or any other physical entity, can be defined.) Fermi's model thus departed from Boltzmann's, which was based on the assumption that like atoms (or molecules) of a gas are distinguishable from each other.

To succeed in his entropy calculation, Fermi had to include one more departure from Boltzmann. Taking his cue from Pauli, he added the rule that each quantum state can accommodate one and only one atom. Even at low temperatures, the atoms must all be found in different quantum states. With his rejection of Boltzmann's rule of distinguishability, and his adaptation of Pauli's rule, Fermi got the entropy calculation he wanted. He published his new statistical model in 1926.

Fermi was not the first to find uses for statistical models modified to meet the demands of quantum mechanics. Two years before Fermi's paper was written, the Indian physicist Satyendranath Bose proposed a model based on Einstein's concept that light and other forms of radiation behave like an ideal gas of particles, later called photons. Bose found that he could reconcile Einstein's theory of radiation with Planck's by deriving Planck's radiation law with a statistical model that accepted the indistinguishability of photons, and also that each quantum state could accommodate any number of photons, not just one as in Fermi's model.

Bose was beginning his career as a theorist when he found this connection between the otherwise unreconciled theories of Planck and Einstein. He sent his manuscript to Einstein, who was impressed, translated the paper into German, and had it published in the *Zeitschrift für Physik*. Einstein added the note: "In my opinion Bose's derivation of the Planck formula signifies an important advance."

As Fermi was pursuing his statistical model, Paul Dirac was independently exploring the same territory from a broader point of view. He emphasized the difference between the Bose-Einstein model for photons and a model he proposed for electrons in atoms based, like Fermi's theory, on the requirements of Pauli's principle. Fermi's paper preceded Dirac's, but Dirac failed to mention it in his own paper, even though, as he later admitted, he had seen the Fermi work but failed to appreciate its importance. This brought an objection from Fermi. "Since I suppose that you have not seen my paper," Fermi wrote to Dirac, "I beg to attract your attention to it."

The Fermi-Dirac model was limited to atoms and electrons, and the Bose-Einstein model to photons, but the two models have proved to be far more encompassing. Contemporary particle physicists assume that *all* particles—not only electrons and photons, but protons, neutrons, neutrinos, and many other particles—fit one model or the other. Dirac atoned for his sin of omission by proposing that all particles following Fermi's (and Dirac's own) scheme be called "fermions." Similarly, he introduced the term "boson" for particles obeying the Bose-Einstein model.

Physics Reawakens in Rome

In the fall of 1926, Fermi went back to Rome. Largely through the efforts of Fermi's patron, Orso Corbino, a chair of theoretical physics had been established at the University of Rome, and Fermi easily won the competition for the new post. At age twenty-five, he had, Segrè writes, "practically attained the zenith of a university career in Italy."

Corbino expected Fermi to bring modern physics to Italy. As Segrè remarks, "a new generation had to take over, and Fermi was to be its leader." Fermi's first step to make himself and his subject known was to give popular lectures and write textbooks. The writing was done during summer vacations in his favorite mountain country, the Dolomites of northern Italy. There, according to Segrè, he sometimes worked "lying on his stomach in a mountain meadow, armed with an adequate supply of pencils and bound blank notebooks, [writing] page after page, without a book for consultation, without an erasure (there are no erasers on Italian pencils) or a word crossed out."

A year after Fermi's arrival, Corbino brought another protégé to Rome, the young experimentalist Franco Rasetti. He was "an elongated man with thin hair, a determined chin, and a steady gaze that went through people," Laura Fermi tells us. Rasetti and Fermi had been classmates at the Scuola Normale in Pisa and confederates in mischief-making, ranging from "fights with pails of water on the roofs of Pisa to protect young damsels' honor, which had never been in danger," says Laura Fermi, to "make-believe duels for reasons that were unknown both to challengers and to challenged." One escapade, a stink-bomb in a classroom, nearly brought permanent expulsion from the university. According to Laura Fermi, Rasetti was the ringleader in these merry pranks: "I do not believe Fermi would have given himself so thoroughly to this kind of life if he had not been dragged into it and held fast by . . . Franco Rasetti."

Fermi and Rasetti, with two more recruits, Edoardo Amaldi, a former engineering student, and Emilio Segrè, Fermi's first graduate student, formed the core of Corbino's School of Rome. Corbino called them "his boys." They were young, talented, intensely devoted to their work, and convinced that great discoveries would come their way, as indeed they did. In a casual way, Fermi was their leader. In theoretical matters he was infallible, so they called him the "pope." Otto Frisch, who knew Fermi later, remarked that he had "never met anyone who in such a relaxed and unpretentious way could be so completely dominant."

Fermi's style as a theorist was always pragmatic and as simple as possible. He aimed for the concrete and avoided the abstract. Hans Bethe, another colleague of Fermi's in later work, contrasts Fermi's style with another, mainly German, tradition:

> My greatest impression of Fermi's method in theoretical physics was its simplicity. He was able to analyse into its essentials every problem, however complicated it seemed to be. He stripped it of mathematical complications and of unnecessary formalism. In this way, often in half an hour or less, he could solve the essential physical problem involved. Of course there was not yet a mathematically complete solution, but when you left Fermi after one of these discussions, it was clear how the mathematical solution should proceed.
>
> This method was particularly impressive to me because I had come from the school of Sommerfeld in Munich who proceeded in all his work by complete mathematical solution. Having grown up in Sommerfeld's school, I thought that the method to follow was to set up the differential equation for the problem (usually the Schrödinger equation), to use your mathematical skill in finding a solution as accurate and elegant as possible, and then to discuss this solution. In the discussion, you would find out the qualitative features of the solution, and hence understand the physics of the problem. Sommerfeld's way was a good one where the fundamental physics was already understood, but was extremely laborious. It would take several months before you knew the answer to the question.
>
> It was extremely impressive to see that Fermi did not need all this labor. The physics became clear by an analysis of the essentials, and a few order-of-magnitude estimates. His approach was pragmatic. . . .
>
> Fermi was a good mathematician. Whenever it was required, he was able to do elaborate mathematics; however, he first wanted to make sure that this was

worth doing. He was a master at achieving results with a minimum of effort and mathematical apparatus.

On a hot day in July 1928, Enrico Fermi married Laura Capon. They had met four years earlier on an outing of young people to the countryside south of Rome. Laura was not impressed by a "short-legged young man in a black suit and a black felt hat, with rounded shoulders and neck craned forward," but he took charge and organized a soccer game, and Laura did as she was told when he assigned her to goalkeeping. Two years later they met again, this time on a mountain-climbing excursion. Fermi, whom Laura remembered as "the queer guy who made me play soccer," was again in command. He mapped out twelve-mile conditioning hikes, and accepted no excuses. "It was always thus," Laura tells us. "Fermi would propose, and the others would follow, relinquishing their wills to him."

By the fall of 1926, the soccer captain and hiking companion had become "Professor Fermi" at the University of Rome, but did not wear the "overwhelming halo of importance and solemnity" expected of a full professor. "[The] young physicist who could inspire respect in his older colleagues showed a remarkable ability to put himself on the level of the young," writes Laura, "and I found I could still talk to him without restraint. Often on Sundays I joined him and his group for a hike in the country or a stroll in Villa Borghese, the main park of Rome. Our companionship did not break up." Then, on the hot July day in 1928, Laura became a partner in the Fermi enterprise. The story of the marriage is a happy one, and Laura Fermi has told it with style and candor in *Atoms in the Family*.

Beta Decay, Continued

Theorists of the late 1920s and early 1930s were mystified and frustrated by the behavior of β particles found in the emissions of radioactive elements. Beta particles were clearly emitted by radioactive nuclei with energies covering a broad range. Where did they come from? The earliest theories simply assumed that electrons inhabited nuclei in company with protons, and that they occasionally escaped as β particles. Heisenberg's uncertainty principle put an end to this concept by showing that if an electron were confined to a nucleus, its position uncertainty (Δx in the Heisenberg inequality) would be very small, requiring that the momentum uncertainty (Δp_x) be so large that the nuclear electron could not be stable.

The further mystery of β particles was their energy spectrum: they could have any energy in a continuous range from zero up to some usually large maximum value. This feature prompted Wolfgang Pauli to address the letter to his colleagues ("Dear Radioactive Ladies and Gentlemen") in which he proposed without much conviction that each β particle appeared in tandem with another particle that bordered on the nonexistent: it had no electrical charge and little or no mass.

Bizarre as it was, Fermi accepted Pauli's phantom particle and named it the "neutrino" (little neutron). He also accepted the concept, recently introduced by Heisenberg, that the two principal building blocks of the nucleus are the proton and the neutron, and thus banished electrons from the nuclear habitat. To account for the appearance of electrons as β particles, he constructed a theory of an interaction—now known as the "weak interaction"—that takes place in the field manifested by a neutron and produces a proton, an electron, and a neutrino. In symbols, this is

$$n \rightarrow p^{+} + e^{-} + \nu,$$

with n, p^{+}, e^{-}, and ν representing the neutron, proton, electron, and neutrino. (The neutrino in this weak interaction is actually an antineutrino; antiparticles and antimatter will be explored in the next chapter.)

This is a process of neutron decay, much like the decay of a radioactive nucleus. Any neutron, inside or outside the nucleus, can decay in this fashion. It transforms the neutron into a proton (which remains in the nucleus if that is the site of the decay) and *creates* an electron-neutrino pair (which appears outside a nucleus). In the mathematical construction of his theory, Fermi represented the field responsible for the interaction with a mathematical entity known in quantum mechanics as a "Hamiltonian function." Here is his summary of the theory: "Electrons (or neutrinos) can be created or disappear. . . . The Hamiltonian function of the system consisting of heavy and light particles must be chosen such that to every transition from neutron to proton there is associated a creation of an electron and a neutrino. To the inverse process, the change of a proton into a neutron, the disappearance of an electron and a neutrino should be associated."

Fermi submitted a note on his theory to the British journal *Nature* in December 1933, and to his everlasting annoyance, the paper was rejected "because it contained speculations too remote from reality to be of interest to the readers." A longer version appeared, however, in two installments in *Zeitschrift für Physik*, and it is now accepted as Fermi's most important theoretical paper.

Neutron Work

Fermi had no more to say about the theory of β decay, although others were happy to build on his foundations. The complete physicist next turned to experimental work. The Joliot-Curies had reported from Paris that new radioactive isotopes of nitrogen and phosphorus could be created by bombarding boron and aluminum, respectively, with energetic α particles. Among other things, this was a remarkable lesson in electrostatics: even though the nuclei and the α particles were positively charged and thus inclined to repel each other (the rules of electrostatics are that like charges repel and unlike charges attract), they could overcome the electrostatic barrier and merge to produce a radioactive nucleus. What if neutrons, carrying *no* charge, were used as the bombarding particles? Fermi expected that they should be even more efficient in nuclear processes. He was right, but unprepared for some surprises.

Fermi began this project by himself; Rasetti, the group's experimentalist, was on an extended vacation in Morocco. With the help of Professor Giulio Trabacchi, another occupant of the university's physics building, Fermi assembled a neutron source. As director of the physics laboratory of the Bureau of Public Health, Trabacchi was well equipped. He was also well organized, and always seemed to have the materials and equipment needed by the physicists: they called him the "Divine Providence." This time he bestowed his greatest gift, a steady supply of radon extracted from his one-gram store of radium. Fermi constructed a neutron source, as Chadwick had, by bombarding beryllium with α particles emitted by the radon. His experimental plan was simple: he would bombard different elements with neutrons and look for induced radioactivity, as the Joliot-Curies had done. Laura Fermi tells the story of these first neutron experiments:

> Being a man of method, [Fermi] did not start by bombarding substances at random, but proceeded in order, starting with the lightest element, hydrogen, and following the periodic table of elements. Hydrogen gave no results: when he bombarded water with neutrons nothing happened. He tried lithium next, but again without luck. He went on to beryllium, then to boron, to carbon, to nitrogen. None were activated. Enrico wavered, discouraged, and was on the point of giving up his researches. He would try one more element. That oxygen would not become radioactive he knew already, for his first bombardment had been on water. So he irradiated fluorine. Hurrah! He was rewarded. Fluorine was strongly activated, and so were other elements that came after fluorine in the periodic table.

Fermi had opened a promising new line of research and he was quick to exploit it. Amaldi and Segrè joined the project, Rasetti was summoned from Morocco, and a radiochemist, Oscar D'Agostino, who had trained at the Curie Institute in Paris, was recruited.

The group soon found that as bombardment projectiles neutrons had some peculiarities. First, there was the mystery of the wooden tables. Amaldi tells about it: "[There] were certain wooden tables near the spectroscope in a dark room which had miraculous properties, since silver irradiated on those tables gained much more activity than when it was irradiated on a marble table in the same room."

To pursue this anomaly, Fermi decided to filter the bombarding neutrons with a lead wedge, and then, for reasons he did not quite understand, he changed his mind. In an interview, Fermi later tried to explain his ambivalence:

> I will tell you how I came to make the discovery which I suppose is the most important one I have made. We were working very hard on the neutron-induced radioactivity and the results we were obtaining made no sense. One day, as I came to the laboratory, it occurred to me that I should examine the effect of placing a piece of lead before the incident neutrons [between the neutron source and the target]. Instead of my usual custom, I took great pains to have the piece of lead precisely machined. I was clearly dissatisfied with something; I tried every excuse to postpone putting the piece of lead in its place. When finally, with some reluctance, I was going to put it in its place, I said to myself: "No, I do not want this piece of lead here; what I want is a piece of paraffin." It was just like that with no advanced warning, no conscious prior reasoning. I immediately took some old piece of paraffin and placed it where the lead was to have been.

Here is the legendary Fermi intuition in action. The lead insert would have produced a result of no particular interest. The paraffin insert, as someone remarked later, was "black magic": when it was in place, the neutron-induced radioactivity was dramatically increased.

Fermi the theorist came forward, and in a few hours he proposed an explanation for the paraffin effect. Neutrons coming from the source were born "fast"—that is, very energetic. But as they passed through the paraffin they collided billiard-ball fashion with hydrogen nuclei (paraffin is a hydrocarbon), and in each such collision lost an appreciable fraction of their energy. This was a slowing down or "moderating" effect, which converted the fast neutrons from the source into "slow neutrons." As they traveled past silver (and other) nuclei, slow neutrons had more time to be taken in by the heavy nuclei and cause activation. Wood contains hydrogen, marble does not, and that, Fermi assumed, explained the strange business of the wooden and marble tables. Hans Bethe quipped that the efficacy of slow neutrons "might never have been discovered if Italy were not rich in marble."

With slow neutrons in their arsenal of bombardment projectiles, Fermi and his group went through the list of elements again, looking for new effects. Finally, at the end of the periodic table, they came to uranium—and confusion. The activities they observed in slow-neutron bombardment of uranium had half-lives of 15 seconds, 13 minutes, and 100 minutes. They found that the new activities could not be caused by elements between lead and uranium in the periodic table, and thus surmised that the observed activities came from a uranium isotope and from elements 93 and 94, new transuranic elements beyond uranium, arising in successive β-decay events. If isotopes of elements 93 and 94 were present, Fermi and his colleagues could claim to have manufactured two artificial elements that are unstable and not found in nature. "The simplest interpretation consistent with the known facts," they wrote in 1935, "is to assume that the 15-second, 13-minute, and 100-minute activities are chain products [successive products of β decay], probably with atomic numbers 92, 93 and 94 respectively and atomic weight 239."

Alas, the simplest interpretation in physics is not always the complete or correct one. The full story of neutron bombardment of uranium is vastly more complicated than Fermi and his coworkers imagined. Meitner, Frisch, Hahn, and Strassmann eventually concluded in 1938 that capture of a neutron, fast or slow, can shatter a uranium nucleus by causing it to fission into two fragments of roughly equal mass. The activities observed by Fermi were evidently due to a few of these fission fragments, and not to transuranic isotopes. No doubt elements 93 and 94 were also produced in the neutron bombardment, but Fermi and his coworkers did not observe and identify them.

The uranium confusion extended beyond the laboratory. Fermi's patron, Orso Corbino, always ready to advertise the accomplishments of his protégé, gave an important speech before the ancient Academy of Lynxes in which he assured his audience that element 93, at least, was a sure thing: "From the progress of these investigations, which I have followed day by day, I feel I can conclude that production of this element has already been definitely ascertained." In a further irony, Fermi received a Nobel Prize in 1938 (just before Meitner, Hahn, Frisch, and Strassmann straightened out the matter) partly for his discovery of "new radioactive elements."

Fermi may or may not have been comforted by a congratulatory letter he received from Rutherford in 1934, soon after completion of the neutron work:

> Dear Fermi,
>
> I have to thank you for your kindness in sending me an account of your recent experiments in causing temporary radioactivity in a number of elements by means of neutrons. Your results are of great interest. . . .

> I congratulate you on your successful escape from the sphere of theoretical physics! You seem to have struck a good line to start with. You may be interested to hear that Professor Dirac [England's most prominent theoretician during the 1920s and 1930s] also is doing some experiments. This seems to be a good augury for the future of theoretical physics!
>
> Congratulations and best wishes. Yours sincerely, Rutherford

The American Branch

At about the same time as Fermi was beginning to make his reputation in the early 1920s, a Fascist dictatorship under Benito Mussolini was rising in Italy. Fermi, like most of his associates, was apolitical, but he was also ambitious. In 1929, he accepted an appointment to Mussolini's Academia d'Italia, where "he found himself automatically among Fascist bigwigs," as Segrè remarks. But honorary societies with obligatory titles and regalia were not Fermi's natural habitat. He was invited to official ceremonies but avoided them whenever possible. On one occasion, the wedding of the crown prince, Fermi chose to spend the day in his laboratory. "To get to the laboratory," Segrè relates,

> he had to cross a street on the procession route that had been closed to traffic and was guarded by lines of soldiers. Fermi, driving his shabby little car in his usual clothes instead of the brilliant uniform of the academy, nevertheless had the invitation card in his pocket, and when stopped by soldiers, he showed it to an officer. "I am the chauffeur of His Excellency Fermi," he said. "I have to fetch him for the wedding. Could you please let me cross the soldiers' lines?" Whereupon he was led through the lines and spent the rest of the day in the laboratory.

Until 1937, Fermi's political stance was an acceptable and profitable coexistence with the Mussolini regime, neither anti-Fascist nor pro-Fascist. Then the political and ideological climate took a sharp turn for the worse when Italy joined forces with the German Third Reich under Hitler. From the beginning, the partnership was dominated by Hitler. By 1938, Hitler had imposed models of the Nazi racial laws on Italian society, resulting in dismissals and harassment of Jews. Laura Fermi was Jewish, and when the first anti-Semitic laws were passed in September 1938, the Fermis decided to leave Italy as soon as possible.

They had a fortunate escape route. The news came prematurely that Fermi would be the 1938 Nobel laureate in physics. They went on a shopping spree to spend their Italian lire, traveled to Stockholm for the prize, and then, without warning anyone but close friends, sailed directly to the United States, where Fermi had accepted a position at Columbia University.

To Laura Fermi, the end of their journey on a cold January day brought conflicting emotions. "Soon the New York skyline appeared in the gray sky [she writes], dim at first, then sharply jagged, and the Statue of Liberty moved toward us, a cold, huge woman of metal, who had no message yet to give me." Laura turned to her husband and he responded, "as a smile lit his face tanned by the sea: 'we have founded the American branch of the Fermi family.' " But they had been torn too abruptly from their homeland. "This is no American family," Laura thought to herself. "Not yet."

Manhattan Engineer District

While the Fermi family was crossing the Atlantic, Lise Meitner and Otto Frisch discovered the idea of nuclear fission. Frisch passed news of the discovery to Niels Bohr, who promptly brought it to America in January 1939. Fermi and other neutron experts were fascinated. It was quickly established that capture of slow neutrons by the rare isotope U^{235} was mostly responsible for the fission, and that each fission produced more neutrons than it consumed.

Neutron multiplication raised the sobering possibility that a carefully constructed uranium assembly could sustain a chain of fission reactions, with the release of energy at an immensely high rate. The chain reaction might be controlled, permitting its use as the energy source in a power plant, or uncontrolled in a nuclear bomb capable of unprecedented destruction. While physicists discussed these developments, World War II broke out in Europe, and to some refugee scientists who had recently come from Europe, the danger of a nuclear weapon in the hands of German scientists was urgent and frightening. In the summer of 1939, a trio of Hungarian physicist-refugees, Leo Szilard, Eugene Wigner, and Edward Teller, took matters in their own hands. They drafted a letter to President Roosevelt and persuaded Einstein to sign it. The letter warned that "the element uranium may be turned into a new and important source of energy in the immediate future. . . . This new phenomenon would also lead to the construction of bombs, and it is conceivable—though much less certain—that extremely powerful bombs of a new type may thus be constructed."

The United States was still in an isolationist mood, and the warning brought an unimpressive response: a small appropriation of funds and the appointment of a sluggish Advisory Committee on Uranium. In about a year, however, with German armies advancing in Europe, war preparations in the United States became a reality. The bureaucratic organization of war-related scientific efforts began, with the establishment first of the National Defense Research Council (NDRC), with the Uranium Committee as a subcommittee, and then of the larger and more inclusive Office of Scientific Research and Development (OSRD). The OSRD was directed by Vannevar Bush, a physicist and engineer who was plainspoken and a shrewd administrator; the NDRC was led by James Bryant Conant, an organic chemist and former president of Harvard. Nuclear efforts were still mainly theoretical, but a bomb project was taking shape. It was clear that for a bomb to be prepared, the fissionable but rare isotope U^{235} had to be separated from the abundant isotope U^{238}. That could be no less than a mammoth task, because large quantities would be needed and the two isotopes had identical chemical properties and only slightly different physical properties. Another route to a bomb had also been discovered via element 94—now identified and named "plutonium" by Glenn Seaborg and his coworkers in Berkeley. Like U^{235}, the plutonium isotope Pu^{239} was fissionable, and it could be prepared as a by-product of a controlled uranium chain reaction.

The Japanese bombing of Pearl Harbor in late 1941 brought another escalation of the uranium project. Over the next two years, physics gave way to engineering on a grand scale. Construction was started on a vast gaseous diffusion plant in Oak Ridge, Tennessee, for separating U^{235} from U^{238}. An electromagnetic process for isolating Pu^{239} was developed in Berkeley. The uranium chain reaction was safely harnessed, first on a small scale by Fermi in Chicago, and then on a vastly greater scale near Richland, Washington, for production of Pu^{239}. As these efforts advanced, a stellar group of physicists, chemists, engineers, and mathematicians gathered at a lonely site in New Mexico called Los Alamos, to design, build, and test the bomb.

By late 1942, this gigantic effort was beyond anything the OSRD could handle with the urgency that was anticipated, and the entire project was put in the hands of the military, specifically the Army Corps of Engineers. The man in charge was one of the army's chief expediters, Brigadier General Leslie Groves. His previous assignment had been supervision of the construction of the Pentagon, and he was at first unimpressed by the budget for his new project. For no good reason except to supply a code name, Groves gave the entire bomb effort the name "Manhattan Engineer District," soon shortened to the "Manhattan Project."

Groves could hardly have differed more from his new colleagues, who came mostly from the academic world. He knew next to nothing about physics, had little tact, and sometimes got his way with bullying, but he knew how to handle formidable construction projects. After meeting Groves, Bush wrote in a memo, "I fear we are in the soup," but quickly changed his mind when he saw Groves in action. Groves's junior officer, Lieutenant Colonel Kenneth Nichols, had this to say about working with the talented general:

> [He was] the biggest sonovabitch I've ever met in my life, but also one of the most capable individuals. He had an ego second to none, he had tireless energy—he was a big man, a heavy man but he never seemed to tire. He had absolute confidence in his decisions and he was absolutely ruthless in how he approached a problem to get it done. But that was the beauty of working for him—that you never had to worry about the decisions being made or what they meant. In fact I've often thought that if I were to have to do my part all over again, I would select Groves as boss. I hated his guts and so did everybody else but we had our form of understanding.

The Pile in the Squash Court

Soon after he arrived in New York in 1939, Fermi took on the challenge of designing an experiment that would sustain the uranium chain reaction. From the materials point of view, he needed two things, fissionable uranium—that is, the scarce isotope U^{235}—and a "moderator" capable of slowing fast neutrons born in fission events to slow neutrons capable of causing more fissions. The great plants that would separate U^{235} from the abundant U^{238} were still several years in the future, so Fermi had no choice but to use natural uranium containing only 0.7 percent of U^{235}. That meant the chain-reaction device, however it was constructed, would require many tons of uranium and moderator.

For the moderator, Fermi chose graphite, whose carbon atoms were light enough to slow fast neutrons efficiently. Graphite also had the structural integrity needed for the assembly of the room-sized experiment. Fermi adopted a design for the assembly composed of graphite bricks supporting a lattice of shaped lumps of uranium.

Beyond these structural features, designing the "pile," as Fermi called it, was a matter of neutron budgeting. To sustain the chain reaction, more neutrons had to be produced in fission events than were lost in other processes. Some neutrons escaped through the surfaces of the graphite structure, and others were captured by uranium or other nuclei in nonfission events. Losses of the latter kind are minimal in graphite; its carbon nuclei have little appetite for neutrons. Water, a more obvious moderator, does not have this advantage.

Fermi adopted a performance factor called "the reproduction factor," represented by k, which calculated the average number of secondary neutrons produced by a single original neutron. If we think of neutrons as being born in "generations" of fission events, k neutrons are produced by the original neutron in the first generation, k^2 in the second generation, k^3 in the third, and so forth. If $k < 1$, these numbers get smaller with each generation and the chain dies out; if $k > 1$, the numbers get larger and the chain diverges, eventually going out of control; and if $k = 1$, the chain proceeds at a steady rate, with the production and loss of neutrons balanced.

At first, Fermi and his group gathered crucial data on uranium and graphite in "subcritical assemblies," those for which $k < 1$. These piles were not large enough to sustain the nuclear chain reaction, but were already outgrowing space available on the Columbia campus. Similar work was being conducted at the University of Chicago at a site code-named the "Metallurgical Laboratory." Arthur Compton, who directed all of the Manhattan Project nuclear research, decided in early 1942 to consolidate the pile research in Chicago.

Another move was not good news for the members of the Fermi family, who had recently settled, permanently, they thought, in the suburban town of Leonia, New Jersey. They had become "the happy owners of a house on the Palisades, with a large lawn, a small pond, and a lot of dampness in the basement," writes Laura Fermi. She was glad to have her children "where the dirt on [their] knees would not be gray, as in New York, but an honest brown." And she was beginning to understand the American suburbanite's mania for perfect lawns (crabgrass was the enemy). For a few months, Fermi divided his time between Chicago and New York, but in June 1942, the family bowed to the bureaucracy, left its Palisades lawn behind, and moved to a rented house in Chicago near the university.

Fermi and his forces now aimed for a full-scale "critical assembly" with a reproduction factor k larger than one (very slightly larger). It is an impressive measure of Fermi's self-confidence, and Arthur Compton's confidence in Fermi, that Compton approved Fermi's plan to build the critical pile on the University of Chicago campus in a doubles squash court under the West Stands of the university stadium. Compton made the decision quickly, which was not his usual habit, without consulting the university president, Robert Hutchins. Compton reasoned "that he should not ask a lawyer to judge a matter of nuclear physics," writes Richard Rhodes, the best among many chroniclers of the Manhattan Project. "The word *meltdown* had not entered the nuclear engineer's vocabulary—Fermi was only inventing that specialty—but that was what Compton was risking, a small Chernobyl in the midst of a crowded city. Except that Fermi, as he knew, was a formidably competent engineer."

What reassured Compton was Fermi's meticulous plan for control of the pile. The primary control device was a set of "control rods," cadmium sheets nailed to wooden strips, which could be inserted or withdrawn from the pile. Cadmium behaves like an efficient neutron sponge, so with the control rods fully inserted the neutron population was low enough to keep the pile subcritical. Slow withdrawal of the control rods in a pile large enough to become critical would increase the reproduction factor k finally to values larger than one. Fermi was also gratefully aware of a gift of nature. Not all of the neutrons generated in a pile are "prompt"—that is, born immediately in fission events. A small fraction, called "delayed neutrons," appears a few seconds later. Fermi predicted that the fortunate effect of the delayed neutrons would be to slow the rate of increase in the neutron population to allow the operators of the pile to respond to any signs of danger.

Laura Fermi gives us a dramatic account of the events on an icy day in early December 1942, when CP-1 (Chicago Pile Number One), containing 6 tons of uranium, 40 tons of uranium oxide, and 385 tons of graphite, was safely brought to criticality:

> Only six weeks had passed from the laying of the first graphite brick, and it was December 2.
>
> Herbert Anderson [one of Fermi's collaborators in the design of the pile] was sleepy and grouchy. He had been up until two in the morning to give the pile its finishing touches. Had he pulled a control rod during the night, he could have operated the pile and have been the first man to achieve a chain reaction, at least in a material, mechanical sense. He had a moral duty not to pull that rod, despite the strong temptation. It would not be fair to Fermi. Fermi was the leader. He directed research and worked out theories. His were the basic ideas. His were the privilege and the responsibility of conducting and controlling the chain reaction. . . .
>
> There is no record of what were the feelings of three young men who crouched on top of the pile. . . . They were called the "suicide squad." It was a joke, but perhaps they were asking themselves whether the joke held some truth. They were like firemen alerted to the possibility of a fire, ready to extinguish it. If something unexpected were to happen, if the pile should get out of control, they would "extinguish" it by flooding it with cadmium solution.
>
> [An audience of about twenty] climbed onto the balcony at the north end of the squash court; all, except the three boys perched on top of the pile and except a young physicist, George Weil, who stood alone on the floor by a cadmium [control] rod he was to pull out of the pile when so instructed.
>
> And so the show began.

Fermi explained the purpose of the control rod, and instructed Weil to withdraw it, leaving thirteen feet inserted in the pile. The counters measuring neutron intensity responded by clicking faster, and the trace of the pen on a chart recorder, also measuring neutrons, climbed and then leveled off. The chain reaction, not yet self-sustaining, ceased generating neutrons. All morning Fermi continued the experiment in this way, instructing Weil to withdraw the control rod in six-inch increments, and each time the observers watched the recorder pen climb and level off in rounded steps. At 11:30 A.M., Fermi, "a man of habits," as Laura Fermi remarks, announced that it was time for lunch, "although nobody else had given signs of being hungry."

At two o'clock in the afternoon, Fermi and his audience, now doubled, returned to the squash court. With a calculation and an extrapolation, Fermi could see that the pile was nearly critical. He told Weil to withdraw the control rod twelve more inches. "This is going to do it," Fermi told Compton. "Now it will become self-sustaining. The trace [on the recorder] will climb and continue to climb, it will not level off."

The moment had arrived. This is what followed, as Herbert Anderson recalled:

> At first you could hear the sound of the neutron counter, clickety-clack, clickety-clack. Then the clicks came more and more rapidly, and after a while they began to merge into a roar. The counter couldn't follow any more [and it was turned off]. . . . [Everyone] watched in the sudden silence the mounting deflection of the recorder's pen. It was an awesome silence. Everyone realized the significance [of the recorder trace]. . . . Again and again, the scale of the recorder had to be changed to accommodate the neutron intensity which was increasing more and more rapidly. Suddenly Fermi raised his hand. "The pile has gone critical," he announced. No one present had any doubt about it.

"Fermi allowed himself a grin," writes Rhodes. "He would tell the technical council the next day that the pile achieved a k of 1.0006. Its neutron intensity was then doubling every two minutes [a leisurely rate, thanks to the delayed neutrons]. Left uncontrolled for an hour and a half, that rate of increase would have carried it to a million kilowatts. Long before so extreme a runaway it would have killed anyone left in the room and melted down."

Fermi calmly ordered the pile shut down after 4.5 minutes of operation, bringing it to a power of ½ watt, hardly enough to light the bulb of a flashlight. "When do we become scared?" Leona Woods, the only woman in the Chicago group, whispered to Fermi.

Compton's telephoned report to Conant on Fermi's success was in code: "The Italian Navigator has reached the New World," Compton said. "And how did he find the natives?" Conant asked. "Very friendly," Compton responded.

The sequel to Fermi's Chicago Pile (or "nuclear reactor," the generic term for any controlled chain-reacting system) was designed by Eugene Wigner, another physicist turned engineer. It was built near Richland, Washington, on the Columbia River, and operated at 250,000 kilowatts, enough power to light a small city. Its purpose was plutonium production.

To the Mesa

"Enrico thought we would be in Chicago for the duration and then we would go back to Leonia," writes Laura Fermi. "He was an optimist." The definition of "duration" going around the Manhattan Project later was the time "it would take for all the physicists on the East Coast to reach the West Coast and for all the physicists on the West Coast to reach the East Coast." Another move was in store for the Fermi family in the summer of 1944, to "Site Y," many miles west of Chicago.

Their new home was on a remote New Mexico mesa high above the Los Alamos Canyon forty-five miles northwest of Santa Fe. This site—which soon came to be called Los Alamos—had been chosen by an odd couple: General Groves, stout, blunt, pragmatic, and opinionated, and Robert Oppenheimer, gaunt, subtle, erudite, and opinionated. The two men had met in the fall of 1942 in Berkeley, where Oppenheimer was a part-time member of the physics department. Groves was in the beginning stages of organizing the Manhattan Project. Oppenheimer advised him to establish "a central laboratory devoted wholly to [bomb design and fabrication], where theoretical ideas and experimental findings could affect each other, where the waste and frustration and error of the many compartmentalized experimental studies could be eliminated, where we could begin to come to grips with chemical, metallurgical, engineering, and ordnance problems that so far had received no consideration."

Groves was convinced, not only of the need for a central laboratory, but also that Oppenheimer was the exceptional leader such a laboratory demanded. "He's a genius," Groves said in a postwar interview. "A real genius. While Lawrence [father of the cyclotron] is very bright he's not a genius, just a good hard worker. Why, Oppenheimer knows about everything. He can talk to you about anything you bring up. Well, not exactly. I guess there are a few things he doesn't know about. He doesn't know anything about sports."

Dividing his time between Caltech in Pasadena and the University of California in Berkeley, Oppenheimer had developed a fine reputation as a teacher and a theorist; he was among the first to make a convincing case for the existence of gravitational black holes. Hans Bethe, who would go to Los Alamos as the leader of the theoretical division, was impressed by Oppenheimer's "exquisite taste" in his role as a research mentor: "He always knew what were the important problems, as shown by his choice of subjects. He truly lived with those problems, struggling for a solution, and he communicated his concern to his group. . . . He was interested in everything, and in one afternoon [he] might discuss quantum electrodynamics, cosmic rays, electron pair production and nuclear physics."

Groves and Oppenheimer selected the secluded mesa in New Mexico for security reasons, and also because Oppenheimer loved the mountain country of northern New Mexico. His family owned a ranch in the Sangre de Cristo Mountains to the northeast of Los Alamos.

Physicists, mathematicians, chemists, engineers, and military personnel from all over the country moved (east or west) to Los Alamos. The city had no official existence. It was not on the map, its inhabitants could not vote, and to outsiders it was Site Y or P.O. Box 1663. Housing construction could never keep up with the influx of new arrivals. Vegetation on the high mesa (at an altitude of seventy-two hundred feet) was flattened by trucks and construction equipment. The result was perpetual mud, from rainstorms in the summer and from melting snow in the winter. Beyond the chaos of the town was the beauty of the mountains: trout streams, ski trails, forest, and peaks to climb.

Like their Los Alamos neighbors, the Fermis managed to make themselves at home in this extraordinary environment. They were assigned to apartment D in building T-186, Laura Fermi reports,

> one of a dozen identical four-apartment houses down a street that started near the water tower on the summit of town, sloped leisurely toward the virgin country, and faded away into it. . . . The apartment was small but adequate and comfortable. In its three bedrooms were army cots on which their previous occupants, boys in the armed forces, had carved their names and ranks. Sheets and blankets were stamped USED in big black letters that shocked us greatly until we realized that they stood for United States Engineer Department. Everything provided by the project was either USED or GI, even light bulbs and floor mops. But through the three contiguous windows of our living-room I could see the round green tops of the Jémez hills slanting down against the sky, as in a three-panel picture by an old master. There were no man-made marks on the hills, and I could call them mine.

Fermi became director of the Los Alamos F Division (F for Fermi) with an omnibus mission. "The general responsibility of the F Division was to investigate problems that did not fit into the work of other divisions," writes Segrè. "Fermi was a sort of oracle to whom any physicist could appeal and more often than not come away with substantial help. There was no limit to the variety of problems that were brought to him."

For Fermi, bomb research and development was a necessary evil and he approached it as a distasteful duty. But this was not the attitude he found in Los Alamos. There was a pervasive enthusiasm that he did not at first understand. "After he had sat in on one of his first conferences here," Oppenheimer remembered, "he turned to me and said, 'I believe your people actually *want* to make a bomb.' " It was the spirit of Los Alamos: the bomb was their obsession. And by the time a plutonium bomb had been designed and was ready to be assembled for the test Oppenheimer called "Trinity," Fermi, too, was under the spell, according to Segrè:

> To my knowledge there are no written accounts of Fermi's contribution to the testing problems, nor would it be easy to reconstruct them in detail. This, however, was one of those occasions in which Fermi's dominion over all physics, one of his most startling characteristics, came into its own. The problems involved in the Trinity test ranged from hydrodynamics to nuclear physics, from optics to thermodynamics, from geophysics to nuclear chemistry. Often they were interrelated, and to solve one it was necessary to understand all the others. Even though the purpose was grim and terrifying, it was one of the greatest physics experiments of all time. Fermi completely immersed himself in the task. At the time of the test he was one of the very few persons (or perhaps the only one) who understood all the technical ramifications of the activities at Alamogordo [the site of the test in southern New Mexico].

Half Empty

An essential element of Fermi's genius was his intellectual restlessness. Even when he had opened promising new fields of research, he was content to have done the pioneering work and to leave to others further exploration of the new territory. There were no sequels to his major papers on quantum statistics and β decay. Once the Chicago Pile was a reality, nuclear reactors were no longer a major research interest. While he was still in Chicago, he demonstrated the usefulness of neutron beams for the study of the solid state, and other physicists followed his lead. A research field that had settled into maturity was not his cup of tea. Always the adventurer, the "Italian Navigator" sought new fields, and he never failed to find them.

He left Los Alamos and bomb physics in late 1945 and returned to Chicago. The next year the University of Chicago inaugurated its Institute for Nuclear Studies, and Fermi accepted an influential position in the new institute with no administrative duties. His experimental tool now was a new cyclotron, constructed just across the street from the West Stands of the university stadium, where the Chicago Pile had made its debut. The cyclotron was his "new toy," Laura Fermi writes. "He played with the cyclotron at all hours of day and evening during [the] summer of 1951. He allowed the cyclotron to upset his routine." His theoretical tool, replacing the neutron, was the meson, the particle then believed to mediate between nucleons (protons and neutrons) and hold them together in the nucleus. He used the cyclotron in experimental studies of interactions between mesons and nucleons.

Fermi was active again as a research supervisor and teacher, just as he had been about a decade earlier in Rome. His Chicago school became a center for research in nuclear and high-energy physics. He was a fixture at conferences and traveled frequently to other research centers. Discussions with younger physicists were particularly valuable, both for them and for him.

When the cyclotron began operation in 1951, Fermi was still in his prime as a theorist and experimentalist. He complained that his memory was not what it used to be, but he knew how to handle that with his "artificial memory"—scrupulously organized notes and reprints. He should have lived another twenty or thirty years and done as much for particle physics and high-energy physics as he had already done for nuclear physics, but that did not happen. In 1954, his health suddenly declined. The diagnosis was an incurable stomach cancer, and he died a few months later, at age fifty-three.

Fermi told Segrè in 1945, at the end of the war, that he had then completed about one-third of his life's work. By that reckoning, when he died nine years later, Enrico Fermi had given us no more than half of what he had to offer.

viii

PARTICLE PHYSICS

Historical Synopsis

For about a century now, physicists have been occupied with fracturing atoms and sorting out the subatomic particles that are produced. The first experiments of this kind, leading to the discovery of the negatively charged electron, were done by J. J. Thomson in 1897. Rutherford's discovery of the atomic nucleus, and studies of radioactive elements by Rutherford, the Curies, Meitner, Fermi, and many others, penetrated further the mysteries of the subatomic realm. A wealth of energetic subatomic particles coming from outer space—so-called cosmic rays—has been observed in cloud chambers and other detectors. And experimentalists have learned how to create their own high-energy beams of subatomic particles in enormous accelerating machines. The beams are aimed at a target or another beam, and the particles produced are observed in highly sophisticated detectors. These experiments have produced so many different kinds of particles that at first theorists hardly knew what to do with them.

This part of the book introduces the three physicists who were most prominent in building the theories that brought order to the jungle of data from the domain of the subatomic. The first is Paul Dirac, who formulated one of the great theories of mathematical physics, ranking with Maxwell's theory of electromagnetism. Dirac's theory describes electrons moving at speeds high enough to demand the restrictions of Einstein's theory of special relativity. The theory shows that electrons have a kind of spin motion, and that they behave like tiny magnets, with north and south poles. Dirac also extracted from his theory the completely unexpected prediction that the electron has a positively charged counterpart, later called the "positron" after it was observed by Carl Anderson in a cloud chamber. The electron and the positron, which differ in other respects as well, are "antiparticles." *All* other particles, from neutrons to neutrinos, also have their anti partners.

The second particle theorist is Richard Feynman, who came from the quantum generation following that of Dirac. Feynman's theory responded to the experimental discovery by Willis Lamb and Polykarp Kusch that Dirac's theory was slightly in error. The

Feynman method demands lengthy calculations, but with the difficulty comes phenomenal accuracy. To ease the pain of the calculations, Feynman invented a visual approach that represents each calculational step with an ingenious diagram.

The third particle theorist is Murray Gell-Mann, whose theories probe not only the atomic nucleus but the particles within the nucleus and all the nuclear debris produced when nuclei are blasted apart in accelerators. Gell-Mann identified the ultimate units of matter as the fractionally charged, forever-confined particles he called "quarks." Baryons (for example, protons and neutrons) contain three quarks and mesons (for example, pions) two. Quarks come in six "flavors," as particle physicists whimsically put it, and in three "colors." Quark color, like electrical charge, generates a field, and the quanta of this field, called "gluons" and analogous to the photons of the electromagnetic field, carry the strong force that holds quarks together and keeps them confined.

A word on the term "particle." In modern usage, the term refers to *subatomic* particles of matter, and also to particles (or quanta) found in fields. Examples of the former are electrons, protons, and neutrons. The photon of the electromagnetic field is an example of a field particle. Electrons, photons, quarks, neutrinos, and gluons are all *elementary* particles: they are not made up of smaller particles. Neutrons and protons do not qualify as elementary particles: they have structural components, quarks and gluons.

24

$$i\gamma \cdot \partial \psi = m\psi$$

Paul Dirac 1902–1984

Isolation

The story of Paul Dirac's life reads like a dark psychological novel. During his childhood, adolescence, and early adulthood in Bristol, England, he was dominated by a misanthropic father. Charles Dirac had little use for social contacts, and he imposed his bleak outlook on his family. He taught French at the University of Bristol, and brought the French lessons into the home by forcing Paul to converse in French at the dinner table, while the rest of the family—Paul's mother, Florence; his older brother, Reginald; and his younger sister, Beatrice—ate in the kitchen. The roots of this domestic disaster were deep: Charles Dirac himself suffered an unhappy childhood in Switzerland and had run away from home at age twenty. Paul did not reach that extremity, but he had no love for his father; when he became a Nobel laureate in 1933, he did not invite his father to attend the ceremony. When Charles Dirac died in 1936, Paul wrote to his wife, "I feel much freer now."

The French lessons at the dinner table left young Paul with limited verbal skills. He was, in a word, silent. "Since I found I couldn't express myself in French," he wrote later, "it was better for me to stay silent than to talk in English. So I became very silent at that time—that started very early." Many are the anecdotes from his later life about his unabashed silence, and his economy with words when he did say something. A colleague at Cambridge who had known him for years said, "I still find it very difficult to talk with Dirac. If I need his advice I try to formulate my question as briefly as possible." The response would come as if from the witness stand: "He looks for five minutes at the ceiling, five minutes at the windows, and then says 'Yes' or 'No.' And he is always right." He responded factually to direct questions, and the five-word answer might take five days to comprehend. He told Bohr, who was as voluble as Dirac was silent, that when he was young he learned that he should not start a sentence unless he knew how to finish it: not a recipe for spontaneous conversation.

In a negative way, parental dominance helped steer Dirac to his destiny. "He was not able to revolt against his father's influence," writes Helge Kragh, Dirac's biographer, "and compensated for the lack of emotional and social life by concentrating on mathematics and physics with a religious fervor." His high school teachers recognized and encouraged his talent. The University of Bristol, where his father taught, was the natural choice for his further education. But his university courses were not those prescribed for aspiring physicists. Charles Dirac forced both of his sons to study engineering. Reginald wanted to be a doctor, did poorly in the engineering program, and was eventually driven to suicide by severe depression. Paul, more passive and less concerned about the future, did well in his engineering courses and learned a valuable lesson from them. He came to appreciate that the laws of physics may not be suitably expressed in the language of pure mathematics: approximate, intuitive mathematical statements may sometimes better suit the purpose.

But practical work in engineering and technology was not for Dirac. After graduation he was unable to get a job, and he remained at the University of Bristol for two more years, studying mathematics.

Research Student

Cambridge was Dirac's salvation. In 1923, at age twenty-one, he went to the university as a research student, the same course Rutherford had taken about two decades earlier. Dirac's first choice was to pick a research topic in the theory of relativity under Ebenezer Cunningham. But relativity was popular (Einstein had recently published his general theory of relativity), Cunningham was worried that his many research students would "run away" from him, and Dirac was assigned to Ralph Fowler instead.

It was, as Kragh writes, "undoubtedly a happy choice." Fowler was Rutherford's son-in-law and perhaps the only link between theorists and experimentalists at the Cavendish Laboratory. Crucially for Dirac, he was "the main exponent of modern theoretical physics at Cambridge and the only one with a firm grip on the most recent developments in quantum theory as it was evolving in Germany and Denmark." Fowler was a hard man to track down for consultations, but that did not bother Dirac, who worked alone and needed no day-to-day guidance. Under Fowler's somewhat remote tutelage, he entered the worlds of atomic theory and statistical physics and found what had already impressed him in relativity: elegant mathematical pictures of nature. As he recalled later: "Fowler introduced me to quite a new field of interest, namely the atom of Rutherford, Bohr, and Sommerfeld. Previously I had heard nothing about the Bohr theory, it was quite an eyeopener to me. I was very much surprised to see that one could not make use of the equations of classical electrodynamics in the atom. The atoms were always considered as very hypothetical things by me, and here were people actually dealing with the structure of the atom."

Dirac soon learned how to nurture and express his own scientific creativity. After about a year at Cambridge, he advanced from student to published scientist. In two more years, he made the first of his major breakthroughs in quantum mechanics. He rarely worked in collaboration with another physicist; only a handful of his more than 250 publications were written with a coauthor. He hesitated even to discuss his theories with colleagues before they were published. He was not being secretive; like Einstein and Gibbs, he had full confidence in his ability as a theorist, and simply had no need to seek the approval of others. Visitors from Göttingen and Copenhagen, where collegiality was a necessity for progress, were puzzled by Dirac's working habits.

Throughout his life, Dirac recovered from the fatigue of concentrated intellectual activity by taking a break, usually a strenuous one. In Cambridge, he did his work every day, "except on Sundays when I relaxed and, if the weather was fine, I took a long solitary walk out in the country. The intention was to have a rest from the intense studies of the week, and perhaps to try and get a new outlook with which to approach the problem the following Monday. But the intention of these walks was mainly to relax, and I had just the problems maybe floating about in the back of my mind without consciously bringing them up. That was the kind of life I was leading."

Later he took his breaks by traveling far and wide (often alone), across continents and three times around the world. He saw more of Russia than most Russians do, and more of America than most Americans. He often sought the mountains and was willing to perform at a conference if the trip included a trek in mountain territory new to him.

Three Brands of Quantum Mechanics

When Dirac emerged from his studies under Fowler, and began to scrutinize the rapidly developing world of quantum mechanics, he found what appeared to be two different methods. On the one hand, there was the matrix mechanics espoused by the Göttingen school (Heisenberg, Born, and Jordan), and on the other hand, the wave mechanics of Schrödinger in Zürich. The matrix method works with tables of numbers (matrices) and follows certain rules of algebra involving addition, subtraction, multiplication, inversion, transformation, and so forth. Wave mechanics is rooted in calculus; its master equation, named after Schrödinger, is an energy equation formulated as a differential equation.

To Dirac, this was mathematically unacceptable: quantum mechanics did not need two voices. Matrix mechanics and wave mechanics treated the same problems and provided the same answers. They had to be representations of a single, more elegant mathematical language. Dirac first focused on Heisenberg's strange multiplication rule that xy does not equal yx for the matrices of his mechanics, that is, $xy - yx$ does not equal zero. The "commutator" $xy - yx$, which Dirac represented with a bracket symbol,

$$[x,y] = xy - yx,$$

has a certain formal resemblance to a mathematical entity called a "Poisson bracket" used by nineteenth-century theorists to put Newton's mechanics in a particularly concise and general form. In his first major paper, published in 1925, Dirac presented a striking correspondence between classical mechanics written with Poisson brackets and quantum mechanics expressed with his own bracket notation. Others were traveling this theoretical path, but none with Dirac's mastery. Max Born was astonished when he saw the paper. "The name Dirac was completely unknown to me," he recalled later. "The author appeared to be a youngster, yet everything was perfect in its way and admirable." Heisenberg's admiration came more grudgingly. In a letter to Pauli he wrote, "An Englishman working with Fowler, Dirac, has independently re-done the mathematics for my work. Born and Jordan will probably be a bit depressed about that, but at any rate they did it first and now we really know that the theory is correct."

The "youngster" soon had much more to say. In 1926, he took quantum mechanics to the higher mathematical plane that he knew existed beyond matrix mechanics and wave mechanics. This was his "transformation theory." It showed, in broad terms, how to transform from one version of quantum mechanics to another, and demonstrated that whether one chose matrix mechanics or wave mechanics was simply a matter of taste or expedience.

In his transformation theory, Dirac revealed the logical essence of quantum mechanics, and he was proud of it. "This work [transformation theory] gave me more pleasure in carrying it through than any of the other papers which I have written on quantum mechanics either before or after," he wrote later.

Out of the Sky

Dirac's habit of working in isolation left his colleagues, even those at Cambridge, wondering what would come next. Nevill Mott, as close as anyone to Dirac, remarked that "all Dirac's discoveries just sort of fell on me and there they were. I never heard him talk about them. . . . They just came out of the sky." In 1928, from out of the sky, came what most commentators rate as Dirac's greatest contribution to physics: his relativistic theory of the electron.

Schrödinger had supplied a preliminary theory of electron behavior with his differential equation, but the equation had two serious faults: as an energy equation, it did not follow the dictates of Einstein's special theory of relativity, and it took no notice of spin motion, which by 1925 had become an accepted electron attribute, as important as the electron's mass and charge. It was suspected that somehow relativity and spin were connected, but no one had even come close to uncovering the connection.

We can glimpse the energy problem by looking closer at Schrödinger's equation. For a free electron (outside the confines of the atom), it calculates the electron's kinetic energy in terms of its momentum. In Newton's mechanics, the kinetic energy E of a particle is calculated from the particle's mass m and speed v with

$$E = \frac{mv^2}{2}.$$

The momentum p of the particle is the product of the mass and speed,

$$p = mv,$$

so

$$E = \frac{p^2}{2m}. \tag{1}$$

Schrödinger supplied rules that translated this classical equation into a quantum mechanical equation that describes free electrons. The form of Schrödinger's equation for this case is

$$\hat{E}\psi = \frac{\hat{p}^2\psi}{2m}, \tag{2}$$

in which ψ is another fundamental ingredient of quantum mechanics, the wave function; $\hat{p}^2\psi$ and $\hat{E}\psi$ denote certain derivatives which need not be specified in detail here. Equation (2) is a differential equation; when it is solved, it supplies information on both the energy and the wave function of the free electron.

Special relativity theory does not accept the energy equation (1), or anything remotely like it. The relativistic energy equation for a free electron is

$$E^2 = p^2c^2 + m^2c^4, \tag{3}$$

with c equal to the speed of light. Note that the energy E is squared in the relativistic equation (3), but not in the nonrelativistic equation (1). This seemingly innocent mathematical feature had far-reaching consequences. It eventually opened the door to a new realm of physics.

Several of Dirac's contemporaries investigated the differential equation derived directly from the energy equation (3), in the same way Schrödinger had derived his successful equation (2) from the classical equation (1). Such an equation has the form

$$\hat{E}^2\psi = (\hat{p}^2c^2 + m^2c^4)\psi, \tag{4}$$

another differential equation.

Dirac saw a subtle difficulty in this equation: it was not sanctioned by his transformation theory, and that for him was a fatal flaw. "The transformation theory had become my darling," he wrote later. "I was not interested in considering any theory which would not fit my darling. . . . I just couldn't face giving up the transformation theory." He found harmony with his transformation theory by simply taking the square root of equation (3), thus eliminating the E^2 on the left,

$$E = \pm\sqrt{p^2c^2 + m^2c^4}. \tag{5}$$

This is actually two equations, as indicated by the $\pm$ notation,

$$E = +\sqrt{p^2c^2 + m^2c^4} \text{ and } E = -\sqrt{p^2c^2 + m^2c^4}. \tag{6}$$

Either of these equations squared gives equation (3) (the square of a negative number yields a positive number).

Dirac's task, as he saw it, was to invent a differential equation in the image of the relativistic equation (5), following as much as possible the rules that had served Schrödinger so well. It was a formidable undertaking, requiring much mathematical ingenuity, "playing with equations and seeing what they give," as Dirac put it. After about two months of this, Dirac had what he wanted. In its most elegant form, Dirac's relativistic electron equation (which appears as the title to this chapter) is

$$i\gamma \cdot \partial\psi = m\psi. \tag{7}$$

It is appropriately carved on the Dirac memorial in Westminster Abbey.

Like some other grand equations of theoretical physics (for instance, Einstein's gravitational field equation), Dirac's electron equation (7) is not so simple as it looks. All of the differential aspects of the equation are compressed into the symbol ∂; the factor γ represents four 4×4 arrays of numbers (matrices); and ψ, no longer the simple wave function of Schrödinger's theory, now has four components. The symbol i is a ubiquitous mathematical symbol representing the constant $\sqrt{-1}$. To make calculations easier, the units for equation (7) are chosen so the speed of light c, an inconveniently large number in ordinary units, and Planck's constant h, an inconveniently small number, both have a value of one.

Equation (7) describes a free electron, unaffected by an external field. When the equation is elaborated so it includes the influence of an applied electromagnetic field, a truly astonishing thing happens: the equation reveals, with no ad hoc prompting from the theorist, that electrons have spin motion. Here was elegant proof of the prevailing suspicion that relativity and spin were connected.

If some of Dirac's contemporaries failed to see the importance of his transformation theory, few had reservations about his relativistic electron theory. "[It] was regarded as a miracle," Léon Rosenfeld, one of Bohr's associates, recalled. "The general feeling was that Dirac had more than he deserved! Doing physics in that way was not done! . . . It [the Dirac equation] was immediately seen as *the* solution. It was regarded as an absolute wonder." The miracle it performed was the *deduction* of electron spin motion. Previous theories had done no better than to graft the spin concept onto the Schrödinger equation more or less as an afterthought.

The ± Difficulty

But there was a price to be paid. Although Dirac's equation answered long-standing questions about electron spin, it raised a profoundly puzzling further question begged by the ± in the energy equation (5). The equation tells us that both positive and negative energies are allowed, but before Dirac's work, negative energies had had no place in relativity theory. In some cases, physicists avoid problems like this by simply discarding mathematically valid, but physically meaningless, quantities. That was not a way out of what became to be known as the "± difficulty," however, because Dirac's theory permitted electrons to make transitions from positive to negative energy states and vice versa. Like it or not, the negative energy states had to remain in the picture. There was a continuum of positive energy states beginning at m_0c^2 (for $p = 0$ in equation [5] with the + sign in effect); m_0 is the electron mass when the electron is at rest. Mirroring these is a continuum of negative energy states beginning at $-m_0c^2$ (also for $p = 0$). See figure 24.1.

Dirac proposed a radical solution to the ± problem. He supposed that all the negative energy states were occupied by a "sea" of electrons according to the Pauli principle, one and only one electron to each state. If, in addition, some electrons also occupied positive energy states, they could not make transitions to the already completely occupied negative energy states, and would thus behave normally. Dirac supposed further that if energy were available, say in the form of a γ-ray photon, an electron could be promoted from a negative energy state to a positive one, where it became an observable positive-energy electron, leaving behind in the sea of negative-energy electrons a vacancy, a "hole," which

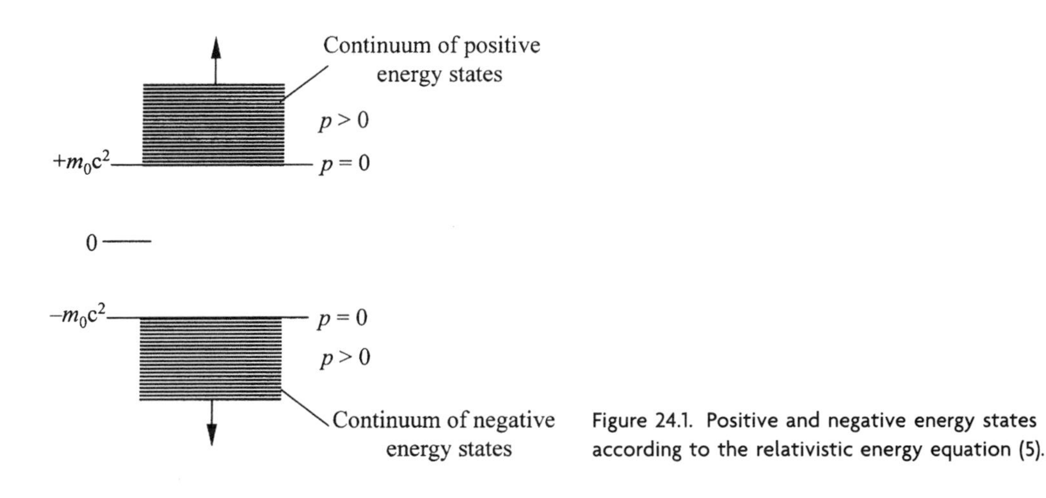

Figure 24.1. Positive and negative energy states according to the relativistic energy equation (5).

gave rise to observable physical effects (see fig. 24.2). What kind of physical effects? Because a one-electron hole was an absence of negative charge, it had an effective positive charge. That charge was $+e$ if positive-energy electrons had the charge $-e$.

The charge on the proton is also $+e$, so Dirac proposed first that holes in the sea of negative-energy electrons were observable and that they were protons. It was an appealing idea. If acceptable, it meant that Dirac's theory explained both of the elementary particles known at the time, the electron and the proton. But it did not survive the critics. Robert Oppenheimer pointed out that an ordinary piece of matter would, according to Dirac's theory, annihilate itself in 10^{-10} second as electrons fell into proton holes, the downward transition being accompanied by the emission of γ-ray photons (the reverse of the process shown in fig. 24.2). Pauli was prompted to announce his "second principle," to the effect that any theory should first be tested by applying it to the theorist who invented it. Pauli's test would have been spectacularly negative when applied to Dirac and his theory: in 10^{-10} second theorist Dirac would have disappeared in a burst of γ rays.

But Dirac had the last laugh. In 1931, he changed his mind and identified the holes as a new kind of elementary particle, an electron with a positive charge, which he called an "antielectron." While he was at it, he speculated that "antiprotons," protons with negative charges, also existed. That same year

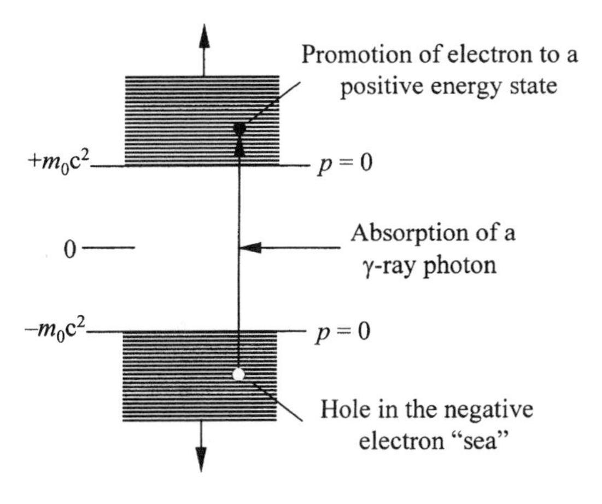

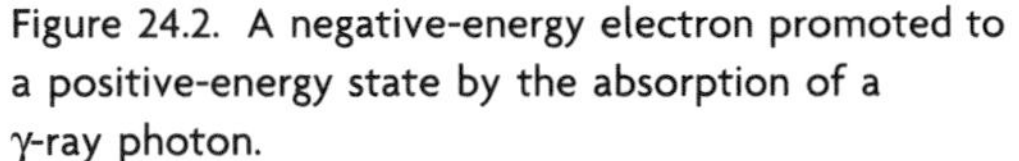
Figure 24.2. A negative-energy electron promoted to a positive-energy state by the absorption of a γ-ray photon.

Carl Anderson, a research fellow at the California Institute of Technology (Caltech), published strong experimental evidence for the existence of Dirac's antielectrons. Anderson was studying the energies of secondary electrons produced by cosmic radiation reaching Earth from outer space. His apparatus was a Wilson cloud chamber. He expected that his cloud chamber photographs would show secondary electrons of the kind described by Arthur Compton. He soon found that he was wrong, however; most of the secondary electrons were *positively* charged.

Anderson naturally assumed that the positive particles were protons, but this assumption was not supported by the cloud chamber tracks, which showed properties belonging to particles much less massive than protons. In 1933, Anderson reported the discovery of positive electrons, with the same physical properties as Dirac's antielectrons. Anderson called them "positrons."

According to Dirac's theory, positrons form in the process shown in figure 24.2, when negative-energy electrons absorb energy from one of the components of the cosmic radiation, perhaps γ rays, and are promoted to positive energy states. Because the hole- or positron-creating events are rare, the density of holes is sparse, and the loss of electrons by falling into holes is no longer the disaster it was with Dirac's identification of the holes as protons. But each *individual* electron-hole—that is, electron-positron—encounter is a disaster for the electron and positron involved, because both particles lose their identities in the process and two γ-ray photons appear in their place (see fig. 24.3).

Although Anderson was familiar with Dirac's theory, he writes that "the discovery of the positron was wholly accidental." Dirac's theory, which could have guided any "sagacious" experimentalist to "discover the positron in a single afternoon . . . [in] any well-equipped laboratory . . . played no part whatsoever in the discovery of the positron." Like Planck's quantized blackbody resonators, Einstein's light quanta (photons), Bohr's stationary states, and de Broglie's matter waves, Dirac's sea of negative energy states and the holes it contains were at first too "unphysical" for most physicists to accept them with confidence.

Anderson's positron discovery was the first in a long series of antiparticle discoveries, including (in 1955) the antiproton and antineutron. Physicists are now convinced that all particles, elementary and otherwise, have their antiparticle counterparts. And it is conceivable that our universe, or another, contains entire antiworlds built of antimatter.

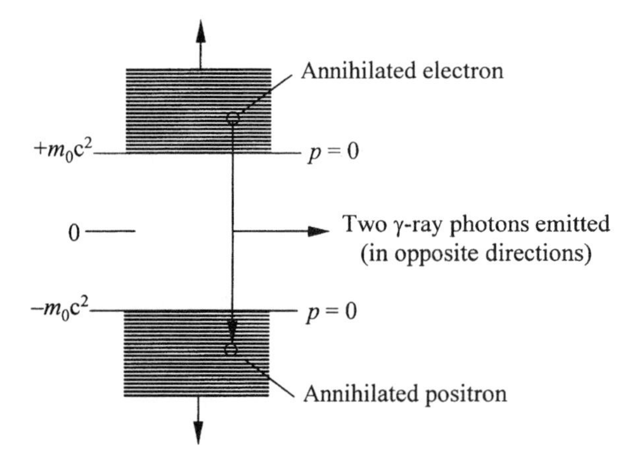

Figure 24.3. Annihilation of a positron and an electron resulting in formation of γ-rays.

Beauty in the Equations

Theoretical physicists build their theories as much with a kind of inspired artistry as with the strict procedures of logic. A great theory can no more be put together only with straightforward reasoning than a great symphony can be written with just the textbook principles of composition. The composer must have a finely tuned ear, the theorist a finely tuned insight. Both the theorist and the composer search for, and hope to achieve, a kind of beauty in their work. "A physical law must have mathematical beauty," Dirac wrote in 1956.

This was Dirac's credo. "[It] summarizes the philosophy of science that dominated Dirac's thinking from the mid-1930s on," writes Helge Kragh. "No other modern physicist has been so preoccupied with the concept of beauty as was Dirac. Again and again in his publications we find terms like beauty, beautiful, or pretty, and ugly and ugliness."

What is mathematical beauty? "Mathematical beauty is a quality which cannot be defined, any more than beauty in art can be defined," Dirac wrote, "but which people who study mathematics usually have no difficulty appreciating." To the theoretical physicist, Dirac says: Let mathematics be your guide, at least initially. "[First] play with the pretty mathematics for its own sake, then see whether this leads to new physics," he advised. Look for the common ground: "One may describe the situation by saying that the mathematician plays a game in which he himself invents the rules while the physicist plays a game in which the rules are provided by Nature, but as time goes on it becomes increasingly evident that the rules which the mathematician finds interesting are the same as those which Nature has chosen."

"The creative principle is mathematics," Einstein said. And Dirac agreed: "One could perhaps [say] . . . that God is a mathematician of very high order, and he used very advanced mathematics in constructing the universe."

Dirac emphasized that mathematical beauty and simplicity are not synonymous. The simplest theory may be the best, but not necessarily. "It often happens that the requirements of simplicity and beauty are the same," he wrote, "but where they clash the latter must take precedence." Newton's theory of gravitation is simpler than Einstein's, but the Einstein theory is deeper and more admired for its beauty. "Many outstanding physicists have shared Dirac's belief in Einstein's theory of gravitation as a theory that was created without empirical reasoning," writes Kragh, "and that has to be true because of its aesthetic merits."

But paradoxically, Dirac did not believe that the theoretical physicist should follow the road of the *pure* mathematician. Late in his life he wrote that "the pure mathematician who wants to set up all his work with absolute accuracy is not likely to get very far in physics." Even theories relying on approximations could have mathematical beauty. One of Dirac's most famous inventions, called the "δ function," was a mathematical orphan when Dirac started using it. Long after he had casually built the δ function into his quantum mechanics, mathematicians demonstrated its formal legitimacy.

Dirac felt that at any given stage in the development of physics, theorists might not have all the mathematical equipment they need to proceed. He doubted that there ever would be an all-encompassing "Theory of Everything." "One should separate the difficulties in physics one from another as far as possible," he wrote, "and then dispose of them one by one." He had in mind not only the "difficulties in physics," but also those in mathematics.

Not surprisingly, Dirac had his critics from the ranks of the mathematicians. An American mathematician, Garrett Birkhoff, wrote in a letter: "Contrary to my expectations, I have found that while Dirac's method of representation of physical systems is formally convenient, it does not embody any mathematical principles which are not thoroughly familiar. . . . Dirac permits himself a number of mathematical liberties. . . . He impresses me as being at least comparatively deficient in appreciation of quantitative principles, logical consistency and completeness, and possibilities of systematic exposition and extension of a central theory."

Like Einstein

Dirac did his most creative work during the years from 1925 to 1933. This was his "heroic period," as Abraham Pais puts it, "during which he emerged as one of the principal figures in twentieth-century science and changed the face of physics." In 1932, he was appointed to the Lucasian Chair at Cambridge, once held by Newton, and as C. P. Snow writes, brought back the elements of Newton's style: "candor," "rationality," "strong and prevailing aesthetic sense," "lucidity," and "austerity, that is, a dislike for unnecessary frills, indeed frills of any kind." A Nobel Prize in physics (shared with Schrödinger) came the next year, and Dirac, fearful of the publicity the prize would bring, considered refusing it. But Rutherford persuaded him to change his mind, pointing out that the refusal would generate even more publicity.

The new attention brought Dirac more into the social world. The man who had been described by a journalist as "the genius who fears all women" married in 1937. His wife was Margit Wigner, the sister of Eugene Wigner, another well-known theoretical physicist. Margit had two children from a former marriage, and the couple had two children of their own. "That history repeats itself is only too true in the Dirac family," Margit writes in a reminiscence. "Paul, although not a domineering father, kept himself too aloof from his children." In this and other ways, it was "a very old-fashioned Victorian marriage."

Having constructed a theory of free electrons in the 1930s, Dirac took the next step and embarked on a lengthy attempt to build a theory of quantum electrodynamics (called "QED" in the physicist's vernacular)—that is, the theory of an electron interacting with another electron or other elementary particle. Sadly, in this endeavor his creativity deserted him. His sense of mathematical beauty was not the heuristic guide to success it had been in his earlier work. The theory that took the stage in the late 1940s and finally solved the problems of electrodynamics had some idiosyncrasies that Dirac could not accept. The new theory, developed by Richard Feynman, Julian Schwinger, Sin-Itiro Tomonaga, and Freeman Dyson, was spectacularly successful as a device for calculating properties of the electron and treating all kinds of elementary-particle interactions. But the equations of the theory contain hidden infinite quantities, which mar the theory's mathematical form: mathematicians cannot even do arithmetic with infinities. An essential mathematical maneuver of the theory, called "renormalization," is to include measured parameters that absorb the infinities. Calculations of both the mass and the charge of the electron are handled this way. As a result, the measured electron mass m and charge e enter the equations, and the offending infinities are swept out of sight, if not out of mind.

For Dirac, this was mathematical ugliness. He spent decades trying to restore beauty to the equations of electrodynamics and never found what he was looking for. In 1979, at age seventy-seven, he gave this bleak assessment: "I really spent my life trying to find better equations for quantum electrodynamics, and so far without success, but I continue to work on it." The prevailing theory of electrodynamics, "with such inelegant infinities, which [have] to be subtracted [cannot] possibly be correct," he insisted. He became isolated from his colleagues, who were using the renormalized theory with unprecedented accuracy.

Dirac shared the fate of Einstein, who spent many years in a fruitless search for a unified theory of gravitation and electromagnetism. Pais reminds us of this parallel, and others, between the two great theorists: "In some, but only some, ways [Dirac] reminds me of Einstein: one of the century's great contributors, always going his own way, not making a school, compelled by the need for beauty and simplicity in physical theory, in his later years more addicted to mathematics than was good for his physics, continuing his activities in pure research until close to his death."

Bohr's tribute to Dirac was the best: "Of all physicists, Dirac has the purest soul."

25

What Do *You* Care?

Richard Feynman

1918–1988

Curious Character

Great scientists are geniuses, and geniuses have a tendency to lead eccentric lives. We have seen ample evidence for this in earlier chapters. Our subject for this chapter, Richard Feynman, was one of the greatest scientific geniuses, and at the same time one of the most eccentric in this company of eccentrics.

He spent most of his career in the academic community, briefly at Cornell University, and then for the rest of his life at Caltech. But he went well out of his way to find antidotes for too much of the academic life. One escape was to Las Vegas, where he sought out gamblers, con artists, and beautiful women. He did not gamble or drink by the time he reached Caltech, but he was addicted to attractive women, and they were often willing. He was tall, handsome, a skilled dancer and drummer, and equipped with a never-ending fund of entertaining stories featuring himself as a picaresque hero. Feynman dictated some of these monologues to his drumming partner, Ralph Leighton, and they were published in two volumes subtitled *Adventures of a Curious Character;* both became best-sellers.

One of Feynman's haunts in Pasadena was Giannoni's topless bar. The bar was, as his remarkably understanding wife Gweneth said, Feynman's "club." He always ordered 7-Up; then, after watching the dancers for a while, he would turn to physics or prepare a lecture. Gweneth also had no lasting objections when Feynman took up drawing from nude female models (one of them formerly the subject of a centerfold feature) in his home studio. More on the wise and tolerant Gweneth, and her role in a happy marriage, later.

Feynman's first wife, Arline, was at least partly responsible for his fascination with the unconventional. They were married when both of them knew that she was slowly dying of tuberculosis of the lymphatic system. Friends, parents, and doctors opposed the marriage, but it was, as Feynman said, "a love like no other I know of." While Feynman was a graduate student at Princeton University, Arline stayed in a nearby hospital, and later, when Feynman joined the war effort at Los Alamos, she lived until her death in 1945 in an Albuquerque sanitarium. He called her "Putsy," and she laughed at his embarrassment when she sent a box of pencils inscribed with "Richard darling, I love you! Putsy." She reproached him with, "What do *you* care what other people think?" and elaborated with poetry on a postcard:

> If you don't like the things I do
> My friend, I say, Pecans to you!
> ....
> If deep inside sound notions brew
> And from without you take your cue
> My sorry friend, Pecans to you!

Arline's admonition was always with him. As Freeman Dyson, one of Feynman's associates at a crucial point in his career, wrote in an appreciation written shortly after Feynman's death in 1988, "[Arline's] spirit stayed with him all his life and helped make him what he was."

Feynman was always an adventurer, as a physicist as well as in his other life. He could not approach a research problem without reconstructing the entire subject in his own way, "turning it around," as he said. On his blackboard at Caltech was the message: "What I cannot create I cannot understand." In his Ph.D. thesis, he composed an entirely fresh approach to quantum mechanics, unlike anything Heisenberg, Schrödinger, or Dirac had done, then applied the method with spectacular success to the theory of interactions of electrons and photons.

He had no fear of doubt and uncertainty. "I don't feel frightened by not knowing things, by being lost in a mysterious universe without any purpose," he wrote. "It doesn't frighten me." Doubt was his motivation; it led to discovery, and "the pleasure of finding things out." He didn't care that solving one mystery usually led to another. "With more knowledge comes a deeper, more wonderful mystery, luring one to penetrate deeper still. Never concerned that the answer may prove disappointing, with pleasure and confidence we turn over each new stone to find unimagined strangeness leading to more wonderful questions—certainly a grand adventure!"

It was a "religious experience" for Feynman "to contemplate the universe beyond man, to think what it means without man—as it was for the greatest part of its long history, and as it is in the great majority of the places." This is the grand objective view of the physicist. "When [it] is attained, and the mystery and majesty of matter are appreciated, to then turn the objective eye back on man viewed as matter, to see life as part of the universal mystery of greatest depth is to sense an experience which is rarely described." Poets do not write about it, he noted, so he wrote a poem himself, which ended with the lines

> Stands at the sea,
> wonders at wondering: I
> a universe of atoms
> an atom in the universe.

In Far Rockaway

"Some facts about my timing: I was born in 1918 in a small town called Far Rockaway, right on the outskirts of New York, near the sea," begins the first of the monologue books, *Surely You're Joking, Mr. Feynman.* "Charmed lives were led by the children of Far Rockaway, a village that amounted to a few hundred acres of frame houses and brick apartment blocks on a spit of beach floating off Long Island's south shore," writes Feynman's biographer James Gleick. "The neighborhood had been agglomerated into the political entity of New York City as one of the more than sixty towns and neighborhoods that merged as the borough of Queens in 1898."

The citizens of Far Rockaway worked hard and maintained a respectable middle-class existence. Richard's father, Melville, struggled through the Depression era with a series of entrepreneurial endeavors that never quite met his expectations: sale of a car wax called Whiz, a real estate business, and a chain of dry-cleaning stores. Finally he took a job as a sales manager for a large company that dealt in uniforms.

Melville would have been a scientist if he had had the means. Just before Richard was born he said, "If it's a boy, he'll be a scientist." (Note that Feynman's sister, Joan, also became a physicist.) Melville had no formal training in science but he had common sense in depth. "In a way, he never knew the *facts* very well. But he knew *truths*," Feynman told another biographer, Jagdish Mehra, and continued, "I find him now, when I look back, a very remarkable man. . . . [He] had a complete understanding of what not to pay attention to, the difference between naming things and the facts, the fact that if you looked into things you always found exciting things. He had a complete understanding of the deeper flavor of science, which he communicated to me."

Melville was Richard's first teacher. His principal medium was the *Encyclopaedia Britannica*, which he read to his son, explaining as they went along. "Everything we read had to be translated into understandable things," Feynman told Mehra, "and this also became a characteristic of my own and my work all the time. I [still] find great difficulty in understanding most things that other people are doing or how they do them. I always try to translate back, back, back, into some way that I can understand."

Richard's mother, Lucille, came from a higher economic stratum than Melville. Her father was a successful designer of women's hats, and the family had moved up from a tenement on the Lower East Side to a townhouse uptown on Ninety-second Street near Park Avenue. Moving from Park Avenue to Far Rockaway may have been trying for Lucille, but she did not complain. Her defense was a sense of humor, which shaped her son's personality as much as Melville's science lessons. Feynman writes, in another one of his monologue books, *What Do You Care?* "Although my mother didn't know anything about science, she had a great influence on me as well. In particular, she had a wonderful sense of humor, and I learned from her that the highest forms of understanding we can achieve are laughter and human compassion." One of Lucille's lines later in her life, when her son was famous and celebrated as the "smartest man in the world": "If that's the world's smartest man, God help us."

Richard's main interests in high school were mathematics, science, and girls, perhaps not in that order. He taught himself the rudiments of calculus from the first book on the subject added to the school library, *Calculus for the Practical Man*. In his senior year in high school, he joined Abram Bader's honors physics course. Bader had studied statistical mechanics and quantum mechanics under Isidor Rabi at Columbia University, but had dropped out of the Ph.D. program when his money ran out. He was a well-informed and sympathetic teacher, and quickly recognized Richard's talent. "Feynman was *sui generis*," Bader told Mehra. "In only one day he stood out as the top student in a class of top students." The top student was bored, however, and Bader gave him Frederick Woods's *Advanced Calculus* to study. In a month, Richard had read the book, and answered questions Bader had noted in the margins when he had studied the book.

Bader also taught Richard some impressive physics lessons. He emphasized the importance of energy functions. There is, he pointed out, the sum of the function T for kinetic energy and the function V for potential energy. This function, called the "Hamiltonian" (for William Rowan Hamilton, the Irish mathematician and physicist, who based an elaboration of Newton's dynamics on this function), is represented with H,

$$H = T + V. \tag{1}$$

Equally important is the function L, equal to the *difference* between T and V,

$$L = T - V, \tag{2}$$

and called the "Lagrangian" (for Joseph Lagrange, the eighteenth-century French physicist and mathematician, who based *his* elaboration of Newton's dynamics on L).

The Hamiltonian function appears in differential equations, which have to be solved (that is, integrated) to define the motion of an object, which is sometimes a difficult task. The Lagrangian function leads to the same result by an entirely different mathematical route. It is integrated with respect to time over any possible path followed by a moving object, to define another function called the "action," represented with S,

$$S = \int_{\text{any path}} L dt. \tag{3}$$

The distinctive thing about the action, called the principle of least action, is that action has its lowest value, a minimum, for the actual path followed by a moving object. Thus the action, constrained so it gives a minimum value, defines that path. For Richard this was a revelation, the awakening of his physical intuition. As he told Mehra, "Instead of differential equations, it tells the property of the whole path. And this fascinated me. That was the greatest thing ever. The rest of my life I have played with action, one way or another, in all my work. I loved it always."

As for the girls in Richard's adolescent life, there was only one he took seriously. She was the popular and artistic Arline Greenbaum. (Curiously, Feynman, or at any rate his editors, spell her name "Arlene" in the monologue books.) Richard and Arline met when they were both thirteen. "She was a very lovely and pleasant girl, rather sweet. She had deep dimples, and everybody liked her," reports one of Richard's former rivals.

Richard graduated from Far Rockaway High School in triumph. He captured most of the honors, and the romance with Arline was forever. University was next. His rejection by Columbia, after a steep application fee, can probably be blamed on a quota for Jewish students. He was accepted at the Massachusetts Institute of Technology (MIT), and arrived in Cambridge in the fall of 1935, in the company of prospective fraternity brothers.

Wild about Problems

At first, Feynman wandered intellectually at MIT. Majoring in mathematics seemed attractive until he asked the mathematics department head what he could do with a mathematics degree besides teach mathematics. "Well," the department head answered, "you can become an actuary, calculating insurance rates for an insurance company," and added that students who asked such questions usually did not have the makings of mathematicians. Next came electrical engineering, but he found that he had no taste for that either. Almost by elimination, Feynman finally came to physics.

It was a wise choice; MIT had a strong physics department. The department head was John Slater, a Bohr collaborator and a prolific textbook author. One of his texts was used in the advanced course, "Introduction to Theoretical Physics," required of all seniors and graduate students. It was taught by Julius Stratton, who would become president of MIT, and Philip Morse, an expert on atomic and molecular calculations. Feynman and an equally precocious friend, Ted Welton, decided with some trepidation to take the course in their sophomore year. They need not have worried; both had reached the senior level in physics. On their own, they had read and thoroughly discussed introductory texts in quantum mechanics and relativity. In a reminiscence of those days, Welton recalls Stratton's class with Feynman in attendance: "Stratton, who was certainly an admirable lecturer, would occasionally skimp on his preparation with the usual consequence that he would come to an embarrassed halt, with a little red creeping into his complexion. With only a moment's hesitation he would ask, 'Mr. Feynman, how did you handle this problem,' and Dick would diffidently proceed to the blackboard and give the solution, always correctly and frequently ingeniously."

At the time (the mid-1930s), quantum mechanics was still a new subject, and in the standard courses it was taught with little depth. Morse offered to teach Feynman and Welton, together with a third student, the *real* workings of quantum mechanics in a special course that met once a week. They began with Dirac's abstruse *Principles of Quantum Mechanics*, and then Morse suggested a research problem involving some detailed calculations of atomic properties beginning with hydrogen, or hydrogenic (which Feynman pronounced "hygienic"), wave functions. Computers and electronic calculators were still in the distant future. "Dick and I set to work with a will," writes Welton, "first learning how to use the 'chug-chug-ding-chug-chug-ding' [mechanical] calculators of those prewar days." This was the other side of quantum mechanics, the beautiful equations in use. The calculations were tedious but at the same time exciting. "Morse brought us to calculate really interesting things," Feynman remarked. "He made an effort with us. He came across somehow; he knew quantum mechanics."

At about this time, Feynman had advanced enough in his comprehension of relativity and quantum mechanics to propose a valid equation for relativistic quantum mechanics. He did not know that he was considering the problem that Dirac had solved in 1928. Feynman arrived at a precursor to Dirac's equation, now known as the Klein-Gordon equation. The equation did not work for electrons, however, as Feynman and Welton learned when they tried to apply it to the electron in the hydrogen atom. "That was the end of it," Feynman concluded. But not actually: perhaps for the first time he had enjoyed what he called "the kick in the discovery."

In his work on the senior thesis at MIT, Feynman made an important discovery. Slater asked him to explain the remarkable fact that quartz expands much less than most materials when it is heated. "My mind was wild about the problem," Feynman told Mehra. "I began to think, how am I going to calculate the expansion? The way to do that would be to imagine that the crystal is fixed in space, fixed in size, and ask what forces and stresses are generated to hold it." To begin his analysis, Feynman stated and proved a theorem that has since been adopted as a standard tool by physical chemists for molecular calculations. The theorem asserts that in general, in quartz or elsewhere, the nucleus of an atom feels the electrostatic forces of the surrounding nuclei and electrons. The electrons are pictured as three-dimensional cloudlike smears, as demanded by quantum mechanics. A shortened version of the thesis was published in the major physics journal, *Physical Review*.

Now it was time for graduate school, and Feynman expected to stay at MIT. He went to Slater anticipating approval, but did not get it. The conversation went like this:

> SLATER: Why do you want to go to MIT?
> FEYNMAN: Because it is the best school in the country for science and engineering.
> SLATER: Do you think so?
> FEYNMAN: Yes.
> SLATER: That's why you have to go to another school.

Forced to look elsewhere, Feynman settled on Princeton. Slater and Morse enthusiastically supported his application for admission, but had some explaining to do. Feynman's scores on the Graduate Record Examination were the best the graduate admission committee had seen in mathematics and physics, and about the worst in history and English. And there was another problem, which the Princeton head of the physics department, H. D. Smyth, explained to Morse without bothering to be subtle: "One question always arises, particularly with men interested in theoretical physics. Is Feynman Jewish? We have no definite rule against Jews but have to keep their proportion in our department reasonably small because of the difficulty of placing them." Slater and Morse continued their campaign on Feynman's behalf, and overcame the admission committee's objections, real and imagined.

One of Feynman's first experiences at Princeton led to the curious title of the first of his monologue books. The day he arrived on the Princeton campus in the fall of 1939 he attended the obligatory Sunday tea given by the dean of the graduate school. "I didn't even know what a 'tea' was, or why!" Feynman writes. "I had no social abilities whatsoever. I had no experience with this sort of thing." At the door, the dean somehow recognized Feynman and welcomed him by name. Then he faced the room:

> It's all very formal and I'm thinking about where to sit down and should I sit next to this girl, or not, and how I should behave, when I hear a voice behind me.

"Would you like cream or lemon in your tea, Mr. Feynman?" It's [the dean's wife], pouring tea.

"I'll have both, thank you," I say, still looking for where I'm going to sit, when suddenly I hear "Surely you're *joking*, Mr. Feynman."

Joking? Joking? What the hell did I just say? Then I realized what I had done. So that was my first experience with this tea business.

Princeton offered Feynman a research assistantship, and he was told that his mentor would be Eugene Wigner, an Hungarian quantum theorist who belonged to the generation of Dirac, Fermi, Heisenberg, and Pauli. "When I got there," Feynman told Mehra, "it turned out that they had shifted it around and I worked with John Archibald Wheeler, which was just fine." Indeed it was. Wheeler was young, just six years older than Feynman, and a remarkable mixture of the conservative and the daring. "He dressed like a businessman," writes James Gleick, "his tie tightly knotted and his white cuffs starched, and fastidiously pulled out a pocket watch when he began a session with a student (conveying the message: the professor will spare just so much time . . .)." Another member of the Princeton department, Robert Wilson, gave this impression of his colleague: "Somewhere among those polite facades there was a tiger loose; a reckless buccaneer . . . who had the courage to look at any crazy problem." As we have seen, Wheeler had collaborated with Bohr in important studies of nuclear fission.

In his first session with Wheeler, Feynman noted the pocket-watch gesture, and responded at the next meeting by pulling out his own dollar watch and deliberately putting it down next to Wheeler's. They both laughed extravagantly, and got down to the business of physics.

One of Wheeler's "crazy problems" concerned a difficulty of long standing in electron theory, the so-called "self-energy" of the electron. Electrons are charged, and the charge generates an electromagnetic field, which in turn interacts with the electron. The energy of this self-interaction can be calculated, in classical or quantum theory, if the electron has a finite size. But the constraints added by relativity theory do not permit electrons to be anything but points; that is, they must have a radius of zero. That seemingly innocent fact had frustrated theorists for decades. When they attempted to calculate the self-energy for a point-sized electron, they got a result of infinity, and that, mathematically speaking, was an absurdity. Infinities are not valid numbers; they cannot be reliably added, multiplied, or divided. The infinity difficulty arises because the theory assumes that electronic interactions are mediated by electromagnetic fields. Wheeler proposed to radically modify the theory by discarding the field concept, making it possible to avoid the offending concept of an electron interacting with itself. This was a return to the "action-at-a-distance" principle that had been suppressed by the advent of field theory.

Feynman had thought about the infinities that plague field theories while he was at MIT, and, always willing to challenge authority, even that of Faraday, Maxwell, and Einstein, he was happy to join Wheeler in his pursuit of an action-at-a-distance theory. They succeeded in developing a valid classical theory, but never managed to find a corresponding quantum theory. Feynman did, however, take an important step in the right direction. He found a way to rewrite quantum mechanics so that it did not rely on a differential equation such as the Schrödinger equation: differential equations were not the right mathematical language for Wheeler's program. Instead, Feynman calculated the probability for an event to occur, let's say the passage of an electron from one point in space and time to another, by summing contributions from every conceivable path that connected the two events. Each path had the same weight or amplitude, but a different "phase," in the summation. Paths reinforced each other in the summation to the extent that they were in phase. As Feynman anticipated from hints in Dirac's writings, the factor that determined the phase of a path turned out to be his old friend the action S, and he was able to prove, entirely from the point of view of quantum mechanics, the principle of least action.

Wheeler was so impressed by his student's achievement that he mentioned it to Einstein:

> Feynman has found a beautiful picture to understand the probability amplitude for a dynamical system to go from one specified configuration at one time to one at another specified configuration at a later time. He treats on a footing of absolute equality every conceivable history that leads from the initial state to the final one, no matter how crazy the motion in between. The contributions of these histories differ not at all in amplitude, only in phase. And the phase is nothing but the classical action integral. This prescription reproduces all of standard quantum theory. How could one ever want a simpler way to see what quantum theory is all about!

It was enough to make anyone a believer in quantum theory, Wheeler said, maybe even its most famous critic, Einstein. "I still cannot believe that God plays dice," Einstein answered. "But maybe. I have earned the right to make mistakes."

While Feynman and Wheeler were probing the foundations of physics, Feynman's other life was in turmoil. While he was still at MIT, Arline had developed health problems, an unexplained growth on the side of her neck and an accompanying fever. After two fumbled diagnoses, a biopsy confirmed tuberculosis of the lymphatic system, and Richard and Arline had to face the probability that she would not recover. They decided, nevertheless, to marry, in spite of harsh opposition from Lucille and Melville. "Your marriage at this time, seems a selfish thing to do, just to please one person," Lucille wrote. "I was surprised to learn such a marriage is not unlawful. It ought to be."

Finally, during Richard's third year at Princeton, he picked up Arline at her home on Long Island and the couple eloped. They were married by a justice of the peace on Staten Island in June 1942, with a bookkeeper and an accountant as witnesses. The ferry ride from Brooklyn to Staten Island was their "romantic boat ride." Richard arranged for Arline to stay at a charity hospital near Princeton. "[Love] is so good & powerful. . . . I know we both have a future ahead of us with a world of happiness—now & forever," Arline wrote in a letter from the hospital.

A Second Dirac, But More Human

In the fall of 1939, while Feynman was adjusting to Princeton tea parties, Wheeler's "crazy problems," and Arline's uncertain future, World War II was beginning in Europe. Earlier in the year, Bohr brought the news of nuclear fission to American physicists, and with Wheeler, fashioned a theory of uranium fission. In the summer of 1939, Einstein received two Hungarian refugees, Leo Szilard and Eugene Wigner, and signed their letter to President Roosevelt warning of the possibility and unimaginable dangers of nuclear weapons. The response was at first bureaucratic and slow, but in 1941, with German armies advancing in Europe, the Japanese bombing of Pearl Harbor, and the knowledge that uranium fission had been discovered in Germany, the uranium project was absorbing money and personnel at an unprecedented rate. In late 1942, the entire effort was taken over by the Army Corps of Engineers, with General Leslie Groves the chief administrator.

After some initial hesitation about war work, Feynman joined an effort at Princeton directed by Robert Wilson. The aim was to develop a device called an "isotron," which would separate the rare isotope U^{235} from the dominant U^{238}. Wilson's method lost out in the competition with other separation schemes, and the Princeton contingent impatiently waited for another assignment. It came in 1943, when the bomb laboratory at Los Alamos, New Mexico, under the direction of Robert Oppenheimer, opened its gates. Oppenheimer had traveled the country to recruit his staff of physicists, chemists, engineers, and mathematicians, sometimes poaching from other branches of the Manhattan Project.

Feynman was among the first to report to Oppenheimer, in the spring of 1943. Richard and Arline traveled to Santa Fe by train, in a private compartment to make Arline as comfortable as possible. Oppenheimer had found a sanitarium in Albuquerque for Arline. Richard hitchhiked or drove a borrowed car to Albuquerque every weekend to be with her.

Feynman's talent was quickly appreciated at Los Alamos, particularly by the nuclear physicist, Hans Bethe, who had come from Cornell to lead the theoretical division. Bethe was a German with a broad background: he had studied in Munich, Cambridge, and Rome; in 1933, he had joined the Nazi-inspired exodus from Europe to the United States. He was a large man with a homely face who was generous, tactful, pragmatic, and far from the rigid, authoritarian German prototype. In *Surely You're Joking*, Feynman describes his first meetings with Bethe at Los Alamos:

> All the big shots except for Hans Bethe happened to be away at the time, and what Bethe needed was someone to talk to, to push his ideas against. Well, he comes in to this little squirt in an office and starts to argue, explaining his idea. I say, "no, no you're crazy. It'll go like this." And he says, "Just a moment," and explains how *he's* not crazy, *I'm* crazy. And we keep on going like this. You see when I hear about physics, I just think about physics, and I don't know who I'm talking to, so I say dopey things like "no, no, you're wrong," or "you're crazy." But it turned out that's exactly what he needed.

Bethe made Feynman a group leader, the youngest at Los Alamos. (Feynman was then twenty-five). His group was assigned computational problems. These were the prehistoric days when "computers" were people working on mechanical calculators. In the fall of 1943, the computer age arrived at Los Alamos in the form of an IBM plug-programmable, punch-card machine, which Feynman and his crew wired and assembled before the astonished IBM technician arrived to do the job.

The Los Alamos confinement affected all the inhabitants, but none more than Feynman. He and Arline jousted with the censors. He got drunk one night and marched the streets singing and beating on pots and pans. This incident led to his first resolution to stop drinking. (He wrote Arline that he was getting "moraller and moraller.") He found holes in the fence that surrounded the base, and pondered the morality of cheating the Coca-Cola machine. His specialty was picking locks and opening safes. With a combination of patience, sensitive fingers, and applied psychology, he became a skilled safecracker. His victims found stern notes criticizing their security lapses.

Feynman's fidgets were not so arbitrary as they seemed. He and his colleagues were working at a furious pace under constant stress on the bomb project, and added to that, Feynman had to watch the slow but sure decline in Arline's health. It all came to an end in 1945. Arline died in the spring. In one of his monologues Feynman tells the story of his last, wild trip to Arline's bedside. The story includes Klaus Fuchs's car, three flat tires en route to Albuquerque, and a spooky clock that stopped at the time of Arline's death. There was no "dramatic collapse . . . her breathing gradually became less and less, until there was no more breath—but just before that there was a very small one." When he got back to Los Alamos (with still another flat tire on the return trip), he responded to solemn looks from his colleagues with, "She's dead. And how's the program going?" He did not feel the full pain of the loss until about a month later: "I was walking past a department store in Oak Ridge [Tennessee, where he had been sent on a special assignment] and noticed a pretty dress in the window. I thought, 'Arlene would like that,' and then it hit me."

On July 16, 1945, the bomb project came to its spectacular and sobering climax with the Trinity test of a plutonium bomb. Feynman returned from a leave of absence in Far Rockaway just in time to witness the event. With the dubious argument that "bright light can never hurt your eyes," he did not wear the dark glasses he had been issued, and watched the explosion through a truck windshield to block ultraviolet radiation, "so I could *see* the damn thing. . . . I'm probably the only guy who saw it with the human eye."

Except perhaps in the security division, Feynman's performance at Los Alamos brought accolades. Oppenheimer, who wanted to recruit him for the Berkeley physics department, wrote to the department chairman, Raymond Birge: "He is by all odds the most brilliant young physicist here, and everyone knows this. He is a man of so thoroughly engaging a character and personality, extremely clear, extremely normal in all respects, and an excellent teacher with a warm feeling for physics in all its aspects. He has the best possible relations both with the theoretical people of whom he is one, and with the experimental people with whom he works in very close harmony."

Bethe, not to be outdone, wrote to *his* department head, R. C. Gibbs, at Cornell: "We have here an exceedingly brilliant young theoretical physicist, Richard Feynman. He is in the opinion of all the wise men here as good as Schwinger [soon to be Feynman's competitor in the construction of theories of quantum electrodynamics], but at the same time quite an extrovert and, therefore much more useful to any department such as this. I wonder whether it would not be possible to secure this man for our department before he gets other offers, which he undoubtedly will."

Bethe and Cornell got their man in the fall of 1944 by offering Feynman an assistant professorship and a leave of absence until his work at Los Alamos was completed. "I got offers from other places, but I just did not consider them because I wanted to be with Hans Bethe," Feynman told Mehra. "I like him very much, and I never regretted that decision. I just decided to go to Cornell."

Feynman was the first of the group leaders to leave Los Alamos. He arrived on the Cornell campus in the town of Ithaca, located in central New York, in late October 1945. Not surprisingly, after the wrenching events earlier in the year, he found it difficult to settle into the academic life; for the first time in his life, he was worried and frustrated. But he was in no danger of a breakdown; as Bethe said, "Feynman depressed is just a little more cheerful than any other person when he is exuberant." The female students were a distraction. He went to parties masquerading as an undergraduate (he was twenty-seven, and looked younger), and found "girls" at the cafeteria who needed help with their physics homework. He was, as always, a storyteller, and his reputation sometimes made it difficult to tell the truth, as in this exchange:

> GIRL: Are you a student, or a graduate student?
> FEYNMAN: No, I'm a professor.
> GIRL: Oh? A professor of what?
> FEYNMAN: Theoretical physics.
> GIRL: I suppose you worked on the atomic bomb.
> FEYNMAN: Yes, I was at Los Alamos during the war.
> GIRL: You're a damn liar!

Feynman's mental block lifted suddenly in a kind of epiphany. He was eating in the cafeteria when "some guy, fooling around, throws a plate in the air." The plate was spinning and at the same time wobbling. With nothing else to do, Feynman wrote the plate's equations of motion and derived an equation that related the two kinds of motion. The derivation was insignificant; the dynamics of spin and wobble had been known for many years. But for Feynman translating the complex motion of the flying plate into differential equations was pure fun.

He went to Bethe: "Hey Hans! I noticed something interesting." But Bethe, who sometimes could not fathom Feynman's moods, was mystified: "Feynman, that's interesting," he said, "But what's the importance of it? Why are you doing it?" "Hah!" Feynman responded. "There's no importance whatsoever. I'm just doing it for the fun of it."

Like a writer or artist recovering from a block in creativity, Feynman had found his medium again. In a short time, he was thinking about electrons and the plague of infinities brought on by the electron self-energy calculation. "It was effortless. It was [now] easy to play with these things. It was like uncorking a bottle: Everything flowed effortlessly. I almost tried to resist! There was no importance to what I was doing, but ultimately there was. The diagrams and the whole business that I got the Nobel Prize for came from that piddling around with a wobbling plate."

Three Conferences

In the summer of 1947, twenty-five physicists gathered at the Ram's Head Inn on Shelter Island, located between the two clawlike prongs of eastern Long Island. Attendance was by invitation only, and the elite of the physics community responded, including Oppenheimer, Bethe, Feynman, Wheeler, Willis Lamb, Isidor Rabi, and Julian Schwinger. It was an American version of the Solvay conferences.

Julian Schwinger was a contemporary of Feynman's, and like Feynman he was working hard on a theory of electron-photon interactions, the endeavor known as quantum electrodynamics or QED. His family background—middle class, Jewish, and centered in New York—was also similar to Feynman's background. But Schwinger's life growing up in well-to-do Manhattan neighborhoods was a world away from Feynman's in Far Rockaway on the city's distant outskirts. Schwinger's father was a gifted designer of women's clothing. As a teenager, Schwinger was shy, precocious, and intensely focused on physics and mathematics. By the time he entered the City College of New York (CCNY), hardly anything else mattered. He stayed away from classes, and spent his time in the library reading advanced physics; Dirac's papers were a major influence.

Schwinger's career at CCNY was faltering when his older brother Harold introduced him to Lloyd Motz, a graduate student at Columbia University. Motz found Schwinger "very, very shy, introverted, kind and musical." His grasp of mathematics and physics was "so far above anybody else—there was no way to compare him to anyone." Motz thought his mentor at Columbia, Isodor Rabi, might also be impressed. He was: Schwinger straightened Rabi out on a point raised in an Einstein paper on the interpretation of quantum mechanics. "[At] one point there was a bit of an impasse," Rabi recalled later, "and this kid spoke up and used the completeness theorem to settle an argument. . . . I was startled. What's this, What's this? So then I wanted to talk to him." Rabi, who was the physics department chairman and an influential, skilled campus politician, got Schwinger admitted to Columbia in spite of his mediocre record at CCNY. (A favorable opinion by Bethe on a paper written by Schwinger helped.) "Everything changed for Julian after that," Rabi continues. "He actually became a member of Phi Beta Kappa. A reformed character."

No doubt Rabi was responsible for the reformation. The Columbia physics department under his leadership was a major research center. "All [in the department] were deeply committed to physics," writes the science historian Silvan S. Schweber. "All were spurred on by the brilliance, wit, charisma—and at times the wrath—of Rabi." One who was there, Morton Hammermesh, recalled that "lots of people worked at Columbia who had absolutely nothing to do with the place. You knew someone and so you came to the party."

Schwinger completed his graduate research at Columbia before he graduated. While still an undergraduate, he published a paper on neutron scattering, and that work, extended, became his Ph.D. thesis. He shed some of his shyness and developed an accomplished lecturing style. When Rabi was away, Schwinger substituted for him in the quantum mechanics course.

Willis Lamb was another one of Rabi's stars. Lamb was a Berkeley graduate in chemistry. At Berkeley, that meant physical chemistry, so he had the background in mathematics and physics to enter the graduate program in physics, where the attraction was Oppenheimer's group of talented theorists. Lamb stayed long enough with Oppenheimer to finish his Ph.D. thesis, and in 1938 accepted an invitation from Rabi to go to Columbia.

During the war, Lamb worked on the theory of the "magnetron" devices that generate microwaves for radar signals. Like Fermi, he was skilled not only as a theorist but also as an experimentalist. He learned the demanding fabrication and vacuum techniques required to construct and operate magnetrons from another Rabi protégé, Polykarp Kusch. Lamb continued in this experimental vein after the war with an emphasis on the peculiarities of metastable atomic hydrogen. His major discovery, sensational news at the Shelter Island conference, was that two hydrogen states, labeled $2S_{1/2}$ and $2P_{1/2}$ by spectroscopists, had slightly different energies. The difference, which came to be known as the "Lamb Shift," was minuscule, but of major importance to theorists because the prevailing theory, based on Dirac's equation, incorrectly calculated exactly the same energy for the two states: the Dirac theory was vulnerable. As Schwinger put it: "The facts were incredible: to be told that the sacred Dirac theory was breaking down all over the place!"

Freeman Dyson, an Englishman who joined Bethe at Cornell as a graduate student, recalled later that Lamb's experiments launched the "wave of progress" that carried theorists to one of their grandest achievements, a new theory of quantum electrodynamics. In a congratulatory letter to Lamb on his sixty-fifth birthday, Dyson wrote: "Those years, when the Lamb Shift was the central theme of physics, were golden years for all the physicists of my generation. You were the first to see that the tiny shift, so elusive and hard to measure, would clarify in a fundamental way our thinking about particles and fields."

That clarification began to emerge in 1948, at a meeting of the American Physical Society in New York, and later at another invitational conference, this time in Pocono Manor, Pennsylvania, located about midway between Scranton and the Delaware Gap. Most of the Shelter Island participants were back, and they were joined by Bohr, Dirac, and Wigner. Now it was Schwinger's turn to take the stage. He had developed an elaborate theory of quantum electrodynamics that accurately calculated the Lamb Shift. An indispensable feature of Schwinger's theory, which he borrowed from some of his predecessors, was to sweep all the infinite terms out of sight into factors that contained the measured electronic charge and mass. This "renormalization" procedure left the theory with dubious mathematical qualifications: the infinities were still there, but they were no longer an obstacle to calculations.

Schwinger's theory was complicated. He presented it in an all-day marathon session. Few, other than Oppenheimer, followed him all the way to the end. It was, as Dyson comments, "built on orthodox principles and was a masterpiece of mathematical technique."

Feynman also performed at the Pocono conference, but not so impressively. He, too, had a working theory of quantum electrodynamics that calculated the Lamb Shift, but its basis was intuitive and pictorial, and that was not what the distinguished Pocono conferees were accustomed to. Dyson explains: "The reason Dick's physics was so hard for ordinary people to grasp was that he did not use equations. . . . Dick just wrote down the solutions out of his head without ever writing down the equations. He had a physical picture of the way things happen, and the picture gave him the solutions directly with a minimum of calculation. It was no wonder that people who had spent their lives solving equations were baffled by him. Their minds were analytical; his was pictorial."

Feynman knew that he was right but he lacked the mathematical means to convince others. As he said to Mehra, "[The] problem for me was that all my thinking was physical. I did things by cut and dry methods, which I had myself invented. I didn't have a mathematical scheme to talk about. Actually I had discovered *one* mathematical expression, from which all my diagrams, rules, and results would come out. The only way I knew that one of my formulas worked out was when I got the right result from it."

The man who brought the seemingly disparate theories of Feynman and Schwinger together, and unified them for use by ordinary physicists, was the Englishman Freeman Dyson. Like Schwinger, Dyson was precocious; even as a child he was fascinated by mathematics and science. His parents "cared deeply about intellectual matters—and with Freeman they got exactly the child they wanted," writes Schweber. At age eight, he was sent to a prestigious boarding school, Twyford College. The school was only three miles from Freeman's home, but his parents did not come for visits, nor did he go home except on holidays. It was, he remembered later, a "strange and forbidding environment."

But his academic performance was outstanding, and when he was twelve he took the scholarship examination for Winchester College, "the intellectual summit of the English public [i.e., private] school system," notes Schweber. Dyson not only won the scholarship but placed first in the competition. His father could not contain his pride: "It was a bigger event in our family than getting a Nobel Prize," he said. Twyford declared a holiday.

At Winchester, Dyson began his mathematical studies in earnest, and won many prizes. He also showed a remarkable aptitude for languages, and found time to become a competent violinist and the school's champion in the steeplechase. A friend who was Dyson's contemporary at Winchester recalled "a very bright, slightly built boy with the same piercing eyes and infectious, slightly sardonic laugh that he still has."

From Winchester, Dyson naturally went to Cambridge, where he entered Trinity College on a scholarship. Most of the Cambridge mathematicians and physicists had left for war service (it was 1941), but Dirac was still on hand, and so were the mathematicians G. H. Hardy and Abram Besicovitch. Dyson had read Dirac's *Principles of Quantum Mechanics*, "without any understanding," at Winchester, and hoped to learn more from Dirac's lectures. He was frustrated to find that the lectures came directly from the book, almost word for word. Hardy was deeply depressed at the time, remote, and "not encouraging." Besicovitch, on the other hand, proved to be a fine teacher and a friend.

For excitement, Dyson practiced a perilous variation on rock climbing called "night climbing." Instead of roped climbing on faces and ridges in the mountains, its practitioners climbed free on the chimneys, drainpipes, and window sills of the university buildings. The sport was nocturnal so they could avoid getting caught. Dyson and a friend, Peter Sankey, "did" most of the buildings described in the guide to the sport, *Night Climbers of Cambridge*. "It was beautiful to go out at night, to be up there on top of the building . . . [and] listen to the bells chime," Dyson told Schweber.

Dyson's Cambridge career was interrupted by two years' service as a civilian scientist with the Royal Bomber Command. His task was to analyze the strategies of the intense and dangerous bombing raids then directed at German cities, and suggest improvements. His advice was often no more than common sense, but it was not what the commander in chief wanted to hear. Dyson concluded that "the Bomber Command might have been invented by some sociologist to exhibit as clearly as possible the evil aspects of science and technology."

Back at Cambridge after the war, theoretical physics looked more attractive to Dyson than mathematics, for the same reason others were deserting the physics community. "Theoretical physics is in such a mess, I have decided to switch to pure mathematics," a friend said. Dyson's response was: "That's curious. I have decided to switch to theoretical physics for precisely the same reason!" He knew what he was doing. Physics was in ferment, especially in America, and there were opportunities for major contributions. One of Dyson's mentors at Cambridge, G. I. Taylor, a Los Alamos alumnus, advised him to go to Cornell and work with Bethe. In a recommendation letter to Bethe, Taylor wrote that Dyson was, in his opinion, "the best mathematician in England."

Dyson arrived at Cornell in the fall of 1947, and enrolled as a regular graduate student (he had not yet earned his doctorate, and never would). At first he expressed some doubts about Bethe in a letter to his parents: "Bethe himself is an odd figure, very large and clumsy and with an exceptionally muddy old pair of shoes. He gives the impression of being very friendly, but rather a caricature of a professor; however, he was second in command at Los Alamos, so he must be a first-rate organizer as well."

A month later Dyson saw much more than Bethe's shoes and mannerisms. He wrote to his family that he "was bowled over by Bethe's complete generosity and unselfishness." He was also impressed by others in the Cornell department, particularly Feynman, who was soon a close friend.

Dyson joined Bethe's group in the year of the Shelter Island conference, followed the next year by the Pocono gathering, where Schwinger presented his theory in an exhausting display of mathematical virtuosity, and Feynman tried to explain his theory without enough mathematics to suit his audience. Dyson did not attend either conference—he was a mere graduate student. But from notes prepared by Wheeler at the Pocono meeting he got the gist of Schwinger's and Feynman's arguments. He planned to learn more by attending a University of Michigan summer symposium in Ann Arbor, where Schwinger was to lecture on his theory. Two weeks before the seminar Feynman mentioned that he was about to drive to Albuquerque in pursuit of a girlfriend, and he casually asked Dyson if he would like to come along. Dyson seized the opportunity; he would see more of the country, and with Feynman's more or less undivided attention for a few days, he would have a chance to probe more deeply the mathematical and physical meaning of his friend's theory.

The Albuquerque trip was a typical Feynman adventure including a night in a brothel (the hotels were full), the company of two persevering, hitchhiking Indians, and a speeding fine as they drove into Albuquerque (reduced from fifty to ten dollars after Feynman and the justice of the peace got acquainted). Dyson then traveled back east to Ann Arbor on a Greyhound bus, his favorite mode of travel in America.

Schwinger's lectures were "excellent," and Dyson explored the details with him in conversations. But the Schwinger theory, although coherent mathematically, seemed to Dyson to be "unbelievably complicated." It could not "be the right way to do it." As the theory came across in Schwinger's lectures, it was "something which needed such skills that nobody besides Schwinger could do it. If you listened to the lectures you couldn't see the motivation; it was all hidden in this wonderful mathematical apparatus."

From Ann Arbor, Dyson took another scenic Greyhound bus trip, this time to San Francisco and Berkeley. "On the return trip," Dyson writes,

> [as] we were droning across Nebraska on the third day, something suddenly happened. For two weeks I had not thought about physics, and now it came bursting into my consciousness like an explosion. Feynman's pictures and Schwinger's equations began sorting themselves out in my head with a clarity they never had before. For the first time I was able to put them all together. For an hour or two I arranged and rearranged the pieces. Then I knew they all fitted. I had no pencil or paper, but everything was so clear I did not need to write it

> down. Feynman and Schwinger were just looking at the same set of ideas from two different sides. Putting their methods together, you would have a theory of quantum electrodynamics that combined the mathematical precision of Schwinger with the practical flexibility of Feynman. Finally there would be a straightforward theory of the middle ground.

Dyson presented his "theory of the middle ground" at a meeting of the American Physical Society at the end of January 1949. It was quickly successful, and Dyson became a celebrity. "Well, Doc, you're in," was Feynman's remark.

As it happened, there was a fourth principal player in the saga of quantum electrodynamics, in addition to Feynman, Schwinger, and Dyson. He was Sin-Itiro Tomonaga, born in Tokyo in 1906, the son of a philosophy professor. As a child, he was not well coordinated and often was sick, so he turned to homemade experiments for amusement. The family moved to Kyoto in 1913, where he performed well at a prestigious high school. Another superior student in the school, and a close friend of Tomonaga's, was Hideki Yukawa, who would become famous for his meson theory of nuclear forces.

In 1923, Tomonaga and Yukawa entered Kyoto University. During their final year, they concentrated on quantum mechanics by reading the papers of Heisenberg, Dirac, Jordan, Schrödinger, and Pauli. Heisenberg and Dirac they saw in person when the two quantum theorists lectured in Tokyo in 1929. Jobs were scarce when Tomonaga graduated—Japan had entered the depression years—but he had the good fortune to attract the attention of Yoshio Nishina, the Bohr of Japanese physics. Between 1921 and 1928, Nishina had studied at the major European centers where physicists gathered: Cambridge, Göttingen, Copenhagen, and Hamburg. When he had returned to Japan he had been appointed director of research at the Institute for Physical and Chemical Research (Rikagen Kenkyusho, "Riken" for short) in Tokyo. Nishina invited Tomonaga to join his laboratory at Riken as a tenured assistant. It was a golden opportunity. As Tomonaga recalled later, "The Nishina Laboratory in those days was full of freshness. All the members were young; even our great chief Nishina was still in his early forties. We all got together every day, an eager group of people discussing various matters, not only physics but also plans for beer parties, excursions and so on."

Through the 1930s, the theoretical group at Riken, including Nishina and Tomonaga, closely followed developments in quantum electrodynamics. They brought Dirac to a Japanese audience by submitting to the "heavy labor" of translating Dirac's book from English to Japanese. In 1937, Tomonaga traveled to Leipzig to work for two years with Heisenberg. The projects he attempted in Leipzig were not promising, and in his impatience he became depressed. "Ill-starred work indeed!" he wrote in his diary. "Recently I have felt very sad without any reason. . . . Why isn't nature clearer and more directly comprehensible?"

Encouraged, and protected by Nishina, Tomonaga continued his theoretical work through the early 1940s as war broke out in the Pacific and the fighting approached Japan. He was focusing now on quantum electrodynamics, attempting to go beyond Dirac and cope with the electron self-energy problem and the curse of infinities it seemed to imply. In 1943, he published a series of papers in a Japanese journal, *Progress in Theoretical Physics*, that belong in any anthology of great scientific literature. As Dyson tells us, the papers "set out simply and lucidly, without any mathematical elaboration, the central idea of Julian Schwinger's theory."

This was in 1943, five years before Schwinger published his theory and four years before Lamb announced his crucial experimental results. By that time, Japan was isolated from the Western world and the papers went unnoticed. Recognition finally came after the war. With his home in rubble and little food available, Tomonaga found a task "that required no thinking," the translation of his wartime papers into English. Translation became more than busywork when he read about Lamb's experiments in the science column of *Newsweek*. He could now see the full importance of his earlier work, and how to extend it to include the renormalization program. In 1948, he sent a summary of his wartime and recent research to Oppenhemier, who responded in a telegram urging him to "write a summary account of present state and views for prompt publication in *Physical Review*."

The third of the postwar invitational conferences took place in April 1949 at Oldstone-on-the-Hudson in Peekskill, New York. The favorite topics for discussion were Feynman's method, and Dyson's synthesis of Feynman's, Schwinger's—and now Tomonaga's—points of view.

A postscript to this story: Feynman, Schwinger, and Tomonaga shared the Nobel Prize in physics in 1965. Dyson was not included, although many felt that he deserved the honor. He seems to have been a victim of the rule that no more than three can share a Nobel Prize.

QED and *QED*

The theory of quantum electrodynamics is "the jewel of physics—our proudest possession," Feynman writes in his fine book for the lay reader, *QED*. Few would disagree. The theory has been applied to a world of physical effects with complete success. Some of these phenomena permit measurements of astonishing accuracy, and calculations dictated by Feynman's approach to QED are in agreement. For example, the strength of the magnetic field carried by an electron, the electron's "magnetic moment," has been measured as 1.00115965221 (in certain units), with an uncertainty of about 4 in the last digit. (This number is predicted to be exactly 1 in Dirac's theory.) The theory calculates 1.00115965246, with an uncertainty of about 20 in the last digits, for the electron's magnetic moment. If you could measure the distance from New York to Los Angeles with this accuracy, you would have it accurate to within the thickness of a human hair.

QED concerns two kinds of elementary particles, electrons and photons, and a multitude of ways they can interact. As usual in quantum mechanics, the theory is limited to statistical calculations. An "amplitude" is calculated for a certain event from which the probability for the event can be determined. There is no way to dig deeper than the probabilities. "There are no 'wheels and gears' beneath this analysis of Nature," Feynman writes. "If you want to understand Her, this is what you have to take."

The theory does its impressive work by recognizing just three kinds of actions among electrons and photons: Feynman describes them in *QED* as (1) "a photon goes from point to point"; (2) "an electron goes from point to point"; and (3) "an electron emits or absorbs a photon." Each action has a certain amplitude. If the two points are A and B, we can follow Feynman in *QED* and use $P(A \text{ to } B)$ to represent the amplitude for the first action, and $E(A \text{ to } B)$ for the second. The amplitude for the third is simply a number whose value is about -0.1 (in certain units), which also equals the electric charge on the electron. Feynman uses j in *QED* to represent this number.

Thus in a QED calculation Feynman-style for a process, let's say two electrons starting at points 1 and 2 in space and time, and ending at points 3 and 4, as shown in the simple "Feynman diagram" on the left in figure 25.1, we need to know the two amplitudes E(1 to 3) and E(2 to 4).

In a detailed calculation, we would consult a set of rules, called "Feynman Rules," to find the mathematical form of the two E functions. No photon is involved, so no *P*s or *j*s enter the calculation. The amplitude for the process proceeding this way has the form E(1 to 3) × E(2 to 4). This is only one way the electrons can get from points 1 and 2 to points 3 and 4. The diagram on the right in figure 25.1 shows another way whose amplitude calculation has the different form E(1 to 4) × E(2 to 3). Combining (adding or subtracting) the two amplitudes gives a first approximation for the event.

Because the two electrons carry electrical charge, they interact with each other through a force carried by a photon. That possibility is not recognized in the diagrams of figure 25.1. The Feynman diagrams shown in figure 25.2 represent two ways our event can be realized with the interaction mediated by a single photon. An electron first travels from point 1 to point 5. Then it emits a photon, represented by a wavy line, and continues on from point 5 to point 3. A second electron begins at point 2, absorbs the photon at point 6, and continues from point 6 to point 4. Now the calculation must include the amplitudes E(1 to 5), E(5 to 3), E(2 to 6), and E(6 to 4) for the electrons, P(5 to 6) for the photon, and j twice, once for the emission of the photon and again for the absorption. The form of the calculation, read directly from the diagram, is

$$E(1 \text{ to } 5) \times j \times E(5 \text{ to } 3) \times P(5 \text{ to } 6) \times E(2 \text{ to } 6) \times \mathrm{j} \times E(6 \text{ to } 4).$$

For a better approximation to the amplitude for our event we add this result to the first two. Notice that j occurs twice in the calculation, as the product $j \times j = j^2$. Because j is small in magnitude (about equal to 0.1), and the product j^2 is still smaller (about equal to 0.01), this term compared to the first one makes a minor

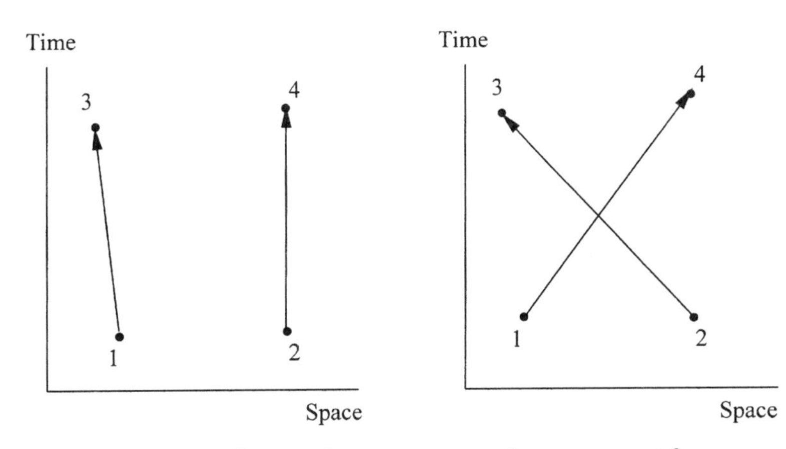

Figure 25.1. Feynman diagrams for two ways two electrons can get from points 1 and 2 in space and time to points 3 and 4.

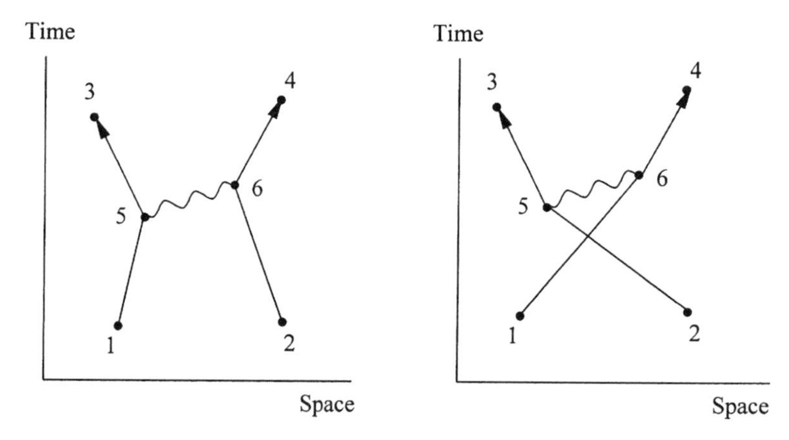

Figure 25.2. Feynman diagrams for two more ways two electrons can get from points 1 and 2 in space and time to points 3 and 4. These paths involve interaction between the electrons mediated by a photon.

contribution to the total amplitude for the event. The Feynman diagram on the right in figure 25.2 shows another way the event can occur with a photon mediating the interaction. The photon that appears in these diagrams as a wavy line represents an electromagnetic interaction between the two electrons, but it has only a transient existence. It is not an ordinary photon that can travel great distances. It exists only for the very short time the uncertainty principle allows it to. In the parlance of particle physics, it is a "virtual photon."

The Feynman calculation is an open-ended affair. We can go on drawing Feynman diagrams and formulating the corresponding calculations ad infinitum. Figure 25.3 depicts one more possibility for the two-electron event we have been considering. Here two photons are involved and both electrons emit and absorb a photon. The calculation now includes four j factors, that is, the product j^4, which is equal to about 0.0001, and its contribution to the total amplitude for the event is even smaller compared to those already considered. Fortunately for those who pursue calculations of this kind, the pattern we are seeing is a general one. As the Feynman diagrams proliferate and get more complicated, the calculations they dictate get less important and more expendable.

This account has so far omitted a major aspect of electron behavior, spin, and

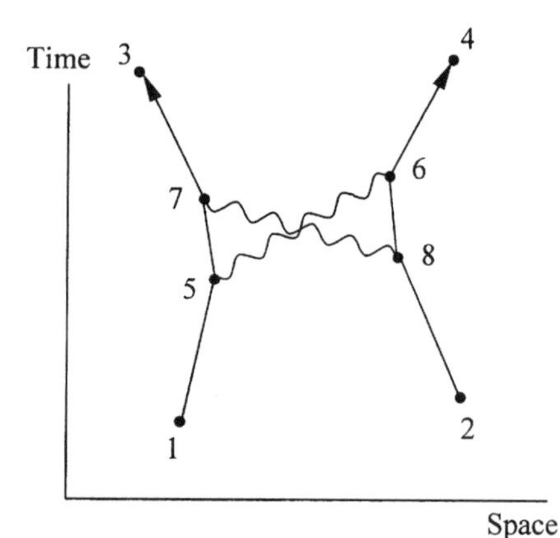

Figure 25.3. Feynman diagram for still another way two electrons can get from points 1 and 2 to points 3 and 4. Each electron emits and absorbs a photon.

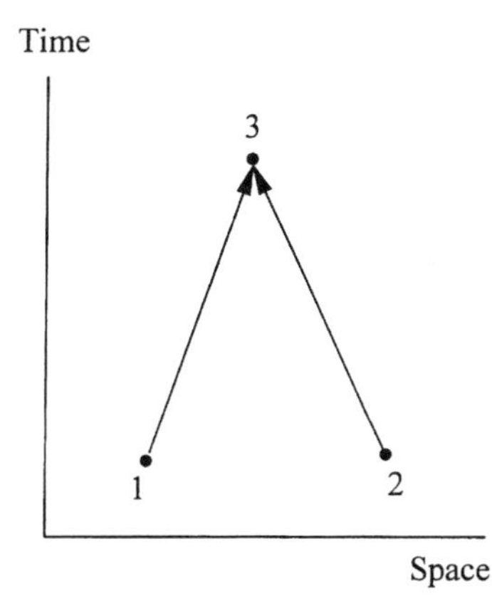

Figure 25.4. Feynman diagram for two electrons attempting to occupy the same point in space and time. The amplitude for this process is zero if the two electrons have the same spin state, nonzero if they have different spin states.

a powerful principle: two electrons in the same spin state cannot occupy the same point in space and time. Thus the amplitude corresponding to the diagram in figure 25.4, showing two electrons with the same space-time destination, must be equal to zero if the two electrons have the same spin state. This is another way to put the Pauli exclusion principle, which we introduced earlier in the language of quantum numbers. The general idea is that two electrons with the same physical description refuse to come together; in fact, they deliberately avoid each other.

Feynman's diagrams, aided by the exclusion principle, tell much of the story of electrons and photons, and the electromagnetic force that governs their behavior. As Feynman puts it in *QED*, the theory offers "a terrific cornucopia of variety and excitement that comes from the exclusion principle and the repetition again and again of the three very simple actions *P*(A to B), *E*(A to B), and *j*."

The Girl in the Polka-Dot Bikini

Feynman began to have negative thoughts about Cornell and Ithaca on a snowy winter day in 1950. His car skidded in the slush, he got out to struggle with tire chains, and thought of better climates. "I remember that *that* was the *moment* when I decided that *this* is *insane*," says Feynman in one of his monologue books. "There must be a part of the world that doesn't have this problem."

There was, and Feynman found it when he was invited to visit Caltech in Pasadena, California, by Robert Bacher, formerly of Cornell and a friend of Bethe's, but not above some poaching. "He was very smart when I visited," writes Feynman in *Surely You're Joking*. "He knew me inside out, so he said, 'Feynman, I have this extra car, which I'm gonna lend you. Now here's how you go to Hollywood and the Sunset Strip. Enjoy yourself.' So I drove his car every night to the Sunset Strip—to the nightclubs and the bars and the action. It was the kind of stuff I liked from Las Vegas—pretty girls, big operators, and so on. So Bacher knew how to get me interested in Caltech."

Another part of the world that beat Ithaca in the weather department was Brazil. Feynman had visited Rio de Janeiro for six weeks in 1949, and hoped to return. Bacher said he could work that out. Caltech offered him a professorship *beginning* with a sabbatical leave for 1951–52 in Brazil. With some regrets—he still had the highest regard for Hans Bethe—he accepted.

In Brazil, Feynman stayed at the Miramar Palace Hotel in Copacabana. He lectured (in Portuguese) at the Center for Physical Research in Rio during the mornings and went to the beach in the afternoons. He found friendly women on the beach and at the bars, and began to drink too much. One afternoon when he had an urge for a solitary drink he realized that he was slipping into alcoholism. "[That] strong feeling that I didn't understand frightened me," he writes in *Surely You're Joking*. "You see, I get such fun out of *thinking* that I don't want to destroy this most pleasant machine that makes life such a great kick." After that, he gave up alcohol, but saw no reason why he could not *pretend* to be drunk at parties.

Near the end of his stay in Brazil, Feynman took a friend ("a very lovely girl with braids") to a museum. In the Egyptian section, he lectured to her about burial practices, and suddenly remembered that he had learned these esoteric bits of Egyptology from Mary Louise Bell, a former girlfriend in Ithaca, now living near Pasadena. He had a severe attack of nostalgia and proposed to her in a letter. She accepted and they were married in June 1952, when Feynman returned from Brazil. The marriage was a mistake, however, a case of mismatched personalities; it ended in divorce four years later.

In his third attempt at a lasting marriage, Feynman was exceptionally lucky. He met Gweneth Howarth at a beach on Lake Geneva in Switzerland; she was wearing a polka-dot bikini. Feynman was attending a conference in Geneva. Gweneth was fleeing from the provincialism of her hometown in Yorkshire. She planned to work her way around the world, earning her expenses as she went; her first stop was Geneva, where she was working as an au pair. On the beach, Feynman started a conversation. He told her about California; she told him about her odyssey and her boyfriends. And then, without much hesitation, Feynman offered her a job as his housekeeper. This was, to say the least, an unusual job offer; she would have to think about it. Feynman went back to his home in Altadena determined to bring Gweneth to America. Eventually he did, but the employer-employee relationship did not last long: they were married in the fall of 1960.

Thanks in large part to Gweneth's tolerance and equanimity, the marriage was a success. They had two children, and family life suited them both. In an interview given in 1977, Gweneth told about her side of the marriage:

> I'm typically happy with what I do, and don't feel I have to compete. I don't feel [I'm a] shadow; I'm perfectly happy—not being a servant to him—we get along very well. I know he's happy because he says it. When he comes home at night he says, "Oh, it's nice to come home." Like on a rainy winter night when we have a big fire in the fireplace and the curtains are drawn and good smells coming from the kitchen. I don't do it just for him—I do it for the family, and I like it—I like to feel comfortable. This is where my satisfaction lies and I don't have to feel important. I do things that [Richard] doesn't, and I do them well.

The V-A Way

In his book *The Character of Physical Law*, Feynman writes, "Nature uses only the longest threads to weave her patterns, so that each small piece of her fabric reveals the organization of the entire tapestry." Many of these long threads are based on symmetry principles. For the physicist, symmetry is not exactly a matter of aesthetics. Hermann Weyl, one of the first to express quantum mechanics in the mathematical language of symmetry called "group theory," had this definition of symmetry (as paraphrased by Feynman): "A thing is symmetrical if there is something you can do to it, so that after you have finished doing it, it looks the same as it did before." Of great interest in theoretical physics are the symmetries of the physical laws. One of these is based on translation from one point to another: if we move an experiment, and all the objects with which it interacts, from one location to another, the experiment and the physical laws that govern it do not change. Similarly, translation in time leaves an experiment and its physical laws unchanged, and so does rotation in space. Until the late 1950s, a fourth fundamental symmetry rule was included with these three: an experiment reflected through a point does not change.

One of the triumphs of quantum mechanics is that it links with each of these symmetries a fundamental conservation law. Translational symmetry in space guarantees conservation of momentum, meaning that if two particles A and B interact to form C and D,

$$A + B \rightarrow C + D,$$

the total momentum of A and B must be equal to the total momentum of C and D. Translational symmetry in time guarantees conservation of energy in the same way, and rotational symmetry conservation of angular momentum. (Angular momentum is always used in quantum mechanics to describe rotational motion, like electron spin motion or electron orbital motion around the nucleus of an atom.) When physicists still believed in the inviolability of reflection symmetry, they proclaimed its related conservation law, conservation of parity. (Unlike energy and momentum, parity is a property that can be appreciated only in the context of quantum mechanics. It measures the mathematical effect of reflection on a wave function.)

Actually, reflection symmetry and parity conservation do hold in many instances, but in the middle 1950s two young Chinese theorists, Tsung-Dao Lee and Chen Ning (or Frank) Yang, who were based at Columbia University and the Brookhaven National Laboratory, respectively, began to have some doubts. Where the weak interaction (responsible for beta decay) is involved, they could find no convincing evidence that parity is, in fact, conserved. They published this heretical view in 1956. The next year Chien-Shiung Wu, a colleague of Lee's at Columbia, published the results of a cleverly designed experiment that demonstrated the nonconservation of parity in the β-decay of Co^{60}.

The news brought by Wu and her compatriots, Lee and Yang, left theorists in consternation. Just before the experimental results were known, Pauli had said that he was willing to bet a very large sum on the side of parity conservation. Feynman was not placing any bets. As always, he was intrigued and excited by ideas that others found crazy; even before the experimental confirmation was reported, he was in hot pursuit of a new theory of the weak interaction that permitted reflection symmetry to be broken and the violation of parity conservation.

For his starting point, Feynman was indebted to Pauli. Drawing on the demands of special relativity, Pauli had proved that interactions of any kind between particles could be described in terms of just five kinds of mathematical entities taken two at a time in the calculation. The five entities are called scalars, vectors, tensors, axial vectors, and pseudoscalars, abbreviated S, V, T, A, and P. (Scalar and vectorial quantities were introduced in chapter 12. A tensor quantity is a mathematical elaboration of a vector: vectors express change in one direction, tensors in two directions. Axial vectors and pseudoscalars differ from vectors and scalars in their reflection symmetry.)

The job for the theorist following Pauli's prescription was to find which of these, taken in pairs, was correct. In quantum electrodynamics, the combination is V-V. Before Lee and Yang's conjecture and Wu's experiments, theorists used the S-T combination for theories of β decay and other instances of the weak interaction. Feynman, and at the same time several others, realized that if reflection symmetry was broken, V-A was the correct recipe. When he came to this conclusion, Feynman felt once again the "kick in the discovery." He said to Mehra,

> As I thought about it, as I beheld it in my mind's eye, the goddamn thing was sparkling, it was shining brightly! As I looked at it, I felt that it was the first time, and the only time, in my scientific career that I knew a law of nature that no one else knew. Now, it wasn't as beautiful a law as Dirac's or Maxwell's, but my equation for beta decay was a bit like that. . . . This discovery was completely new, although, of course, I learned later that others had thought of it about the same time or a little bit before, but that did not make any difference. At the time I was doing it, I felt the thrill of a new discovery!

Among the others who had thought about the V-A pattern for weak interactions was Murray Gell-Mann, whose office was just down the hall from Feynman's at Caltech. The two theorists were competitors and sometimes friends. Each recognized in the other a valuable partner for debating ideas and theories; probably there were none better at the time. Their debate about V-A theory was refereed by Robert Bacher, the chairman of the physics department, who persuaded them to write a joint paper.

Partons and Quarks

By the middle 1960s, Feynman had contributed in a major way to the theory of three of the four fundamental kinds of forces recognized by physicists, involving electromagnetic interactions (in his QED theory), weak interactions (in his V-A theory), and gravitational interactions (another interest of Feynman's, which won't be explored here). After the hoopla of the Nobel Prize in 1965, Feynman turned his attention to the fourth of the fundamental forces, the one that mediates "strong interactions" among and within the constituent particles of nuclei, protons and neutrons. It had been clear for many years that protons and neutrons have structure, unlike electrons and neutrinos, which are treated as structureless, point-sized objects. Feynman's model for hadron structure ("hadron" is the generic term for protons, neutrons, and other particles held together by strong interactions) was designed to fit the experiments then being done at the Stanford Linear Accelerator Center in which extremely high-energy electrons bombarded protons.

Taking his usual visual approach, Feynman asked himself what he would see if he were a high-speed electron approaching and interacting with a hadron. His basic assumption was that hadrons contain hard, point-sized, charged particles called "partons," which float almost freely within the hadron's confines. Relativistic effects would be dramatic: from the point of view of the electrons, hadrons would be flattened like pancakes, and time would be slowed so that the partons would seem almost static. Most of the electrons would pass through the hadron pancake with no interaction, but a few would collide billiard-ball fashion with partons. It was a simple model (with analogies to the picture used by Rutherford to account for the scattering of α particles), and it became popular among theorists and experimentalists at the Stanford accelerator.

Murray Gell-Mann also had a theory of hadron structure that included confined point-sized particles. Gell-Mann called his particles "quarks," from a line in James Joyce's *Finnegans Wake*: "Three quarks for Muster Mark!" The quark model was more explicit than Feynman's parton model: the quarks carried electrical charges that were fractions of the charge on the proton, and three of them brought together by the strong interaction made a proton.

The quark model had its critics, including Feynman, and for a while Gell-Mann himself was skeptical about the reality of his invention. The fractional charge was particularly hard for theorists to swallow; the charge on the proton had always been considered indivisible. It didn't help either that there was, and still is, no evidence for the existence of *free* quarks outside the confines of the hadron.

Feynman remained skeptical about quarks until 1970, when he completed a study with two students of a large collection of particle data. He was finally convinced and became a "quarkerian." "A quark picture may ultimately pervade the entire field of hadron physics," he wrote in the paper reporting the survey. The data displayed the "mysteriously good fit of a peculiar model." Gell-Mann was annoyed that Feynman took so long to come to terms with the quarks. He could not resist calling Feynman's version of the confined particles "put-ons."

Nature Cannot Be Fooled

Feynman rarely played the part of the conscientious, distinguished scientist by accepting honorary degrees, invitations to deliver lectures, and appointments to committees. But when he did accept a public responsibility the consequences could be dramatic. His service on the commission that investigated the *Challenger* disaster brought him more public attention than anything else he did.

On January 8, 1986, the space shuttle *Challenger* left its Florida launch pad in unusually cold weather. Seventy-three seconds into the flight the shuttle exploded, killing all seven astronauts on board. A few days after the accident William Graham, head of the National Aeronautics and Space Administration (NASA), called Feynman and asked him to join the commission that would investigate the accident. Feynman had reservations about the scientific importance of the shuttle program and further doubts about going to Washington, where the commission was to meet. "I have a principle of not going anywhere near Washington or having anything to do with government," he says in *What Do* You *Care?* He had another reason for choosing to stay home, which he does not mention. Since 1978, he had been fighting a battle against a rare form of abdominal cancer, and in 1986, just before the NASA summons, he had found that he was also suffering from a rare form of bone marrow cancer.

Feynman's friends told him he should go to Washington, and Gweneth said, "If you don't do it, there will be twelve people, all in group, going around from place to place together. But if you join the commission, there will be eleven people—all in a group going around from place to place together—while the twelfth one runs around all over the place, checking all kinds of unusual things. There probably won't be anything, but if there is, you'll find it. There isn't anyone else who can do that like you can." "Being very immodest I believed her," Feynman continues. He told Gweneth, "I'm gonna commit suicide for six months"—and accepted the appointment.

It was his penultimate adventure. First, he rounded up some friends of a friend at the Jet Propulsion Laboratory in Pasadena, who quickly briefed him on shuttle engine design. Then he took an overnight flight to Washington to be in time for the first meeting. The chairman of the commission was William Rogers, secretary of state in the Nixon administration, and the commission also included Neil Armstrong, the first man on the moon; Sally Ride, the first American woman in space; Chuck Yeager, formerly a test pilot; and Major General Donald Kutyna, who had represented the shuttle program in the defense department.

Kutyna proved to be Feynman's only ally on the commission. Two of the members revealed their allegiances soon after they were appointed. Armstrong said the investigation was unnecessary, and Rogers said, "We are not going to conduct this investigation in a manner which would be unfairly critical of NASA, because we think—I certainly think—NASA has done an excellent job, and I think the American people do."

Some of the members of the commission had technical backgrounds and expertise in shuttle operation. Their detailed questions brought few satisfying answers from the NASA administrators who were appearing before the commission. "We'll get that information to you later," was the usual response.

Eventually, it became clear that good candidates for the cause of the accident were the immense rubber O-rings intended as seals between sections of the solid fuel rockets. They were about a quarter of an inch in diameter, and thirty-seven feet in circumference. Ordinary O-rings seal a static gap, "but in the case of the shuttle," Feynman explains, "the gap *expands* as the pressure builds up in the rocket. And to maintain the seal, the rubber has to expand *fast* enough to close the gap—and during the launch the gap opens in a fraction of a second. Thus the resilience of the rubber became a very essential part of the design."

The O-rings were further implicated when an engineer from the Thiokol Company, manufacturer of the O-rings, reported that Thiokol engineers had "come to the conclusion that low temperatures had something to do with the seals problem, and they were very, very worried about it," writes Feynman. "On the night before the launch, during the flight readiness review, [Thiokol engineers] told NASA the shuttle shouldn't fly if the temperature was below 53 degrees—the previous lowest temperature—and on that morning it was 29." With some pressure and twisted reasoning from NASA, Thiokol reversed itself, and the launch proceeded.

Feynman was getting frustrated. He tried to pursue the problem of the resilience of the O-ring rubber at low temperatures, and got evasive answers from the project manager for solid rockets, Lawrence Mulloy. He finally decided to make his point as he would in the lecture hall, with a demonstration. The next day, when a public meeting was scheduled, was an opportune time. Early that morning he found a hardware store and bought a pair of pliers and a small C-clamp. With some coaching from General Kutyna, he did his experiments when the television cameras were pointed in the right direction. The demonstration was dramatic because it was so extremely simple. Here Feynman explains it to the witness, the commission, and the cameras: "Dr Feynman: This is a comment for Mr. Mulloy. I took this stuff [the O-ring rubber] that I got out of your seal and I put it in ice water, and I discovered that when you put some pressure on it for a while [with the C-clamp] and undo it it doesn't stretch back. It stays the same dimension. In other words, for a few seconds at least and more seconds than that, there is no resilience in this particular material when it is at 32 degrees. I believe that has some significance for our problem."

Feynman spent months conducting his own investigation, traveling to the space centers in Florida, Alabama, and Texas, and to the headquarters of contractors. He found a pattern of "exaggeration at the top being inconsistent with the reality at the bottom," so "communication got slowed up and ultimately jammed." He wrote his own report, which was relegated by the commission to an appendix (an "inflamed appendix," he called it). He concluded with the comment: "For a successful technology, reality must take precedence over public relations, for nature cannot be fooled."

Nothing Is Mere

Feynman was judged by his peers to be a great teacher. As in his other endeavors, however, he did it his own way. At Caltech, he never attended physics division board meetings, avoided committee assignments, never sought research grants, tormented seminar speakers with relentless pointed questions, and accepted few Ph.D. students. "Most of us were afraid of him," a Caltech colleague reports. "At faculty meetings, if you said something with which he disagreed he would put you in your place with a sharp tongue. He didn't suffer fools at all. My impression is that no one on the faculty got close to him." On the other hand, his door was always open to students. And for many years, he conducted an informal course called Physics X, which took up topics chosen by the students. Anyone could attend, except faculty members.

For Feynman, the lecture hall was a theater and teaching a performance. There had to be drama, surprise, comedy, and eloquence. Faraday would have appreciated his lectures. In the early 1960s, he was persuaded by one of his colleagues, Matthew Sands, to take on the formidable task of teaching introductory physics. "Look, Richard, you have spent [many] years trying to understand physics," Sands said. "Now here is your chance to distill it down to the essence at the level of the freshmen." Feynman thought about it for several days, and then asked Sands, "Do you know if there ever has been a great physicist who lectured on freshman physics?" Sands said he didn't think so. "I'll do it!" concluded Feynman.

It was no casual effort. Feynman worked on the lectures full time. Sands and Robert Leighton, a Caltech colleague and father of Feynman's drumming partner, Ralph Leighton, produced the written three-volume version of the lectures. The combined effort succeeded admirably; the books are still popular almost forty years later.

Here is Feynman telling his audience about the electrical force and the ultimate electrical charge carried by electrons and protons:

> [All] matter is a mixture of positive protons and negative electrons which are attracting and repelling with this great force. So perfect is the balance, however,

> that when you stand near someone else you don't feel any force at all. If there were even a little imbalance you would know it. If you were standing at arm's length from someone and each of you had *one percent* more electrons than protons the repelling force would be incredible. How great? Enough to lift the Empire State Building? No! To lift Mount Everest? No! The repulsion would be enough to lift a "weight" equal to the entire earth!

He wonders about the qualitative, as well as the quantitative, meaning of the equations of physics: "Our equations for the Sun, for example, as a ball of hydrogen gas, describe a Sun complete without sunspots, without the rice-grain structure of the surface, without the prominences, without coronas. Yet, all these are really in the equations, we just haven't found the way to get them out."

He answers the critics who complain about the narrowness and unimaginative objectivity of science:

> Poets say science takes away from the beauty of the stars—mere globs of gas atoms. Nothing is "mere." I too can see the stars on a desert night, and feel them. But do I see less or more? The vastness of the heavens stretches my imagination—stuck on this little carousel my little eye can catch million-year-old light. A vast pattern—of which I am a part—perhaps my stuff was belched from some forgotten star, as one is belching here. Or see them [the stars] with the greater eye of Palomar, rushing all apart, from a common starting point when they were perhaps all together. What is the pattern, or the meaning, or the *why*? It does not do harm for the mystery to know a little about it. For far more marvelous is the truth than any artists of the past imagined! Why do poets of the present not speak of it? What men are poets who can speak of Jupiter if he were like a man, but if he is an immense spinning sphere of methane and ammonia must be silent?

Presence

For students, lecture audiences, colleagues, friends, adversaries, and detractors, the Feynman presence was above all an indestructible vitality. He missed nothing and was curious about everything. He delighted in "being something I'm not." He became an accomplished bongo drummer. He learned some biology and hired himself out to the Caltech biology department as a teaching assistant during a sabbatical leave. He worked for several summers as an ordinary staffer with a fledgling computer company. He was one of the first to see the possibilities of the technology of very small machines and manipulations, the field now known as nanotechnology. He acquired respectable techniques in the arts of drawing and safecracking. He remained creative longer than any other great physicist, with the exceptions of Gibbs and Chandrasekhar.

Even in the final decade of his life, as cancers were slowly killing him, he was still going strong. In a picture taken a few weeks before he died, he still appears to be having fun, and probably he was. But his last great adventure proved to be a disappointment. At the end, he said to Gweneth, "I'd hate to die twice. It's so boring."

26

Telling the Tale of the Quarks

Murray Gell-Mann 1929–

Prodigy Story

It is said that a student once asked Enrico Fermi about the name of a fundamental particle, and that Fermi responded, "Young man, if I could remember the names of these particles, I would have been a botanist." At that time (the mid-1950s), only a dozen particles were known, but their taxonomy was already a tangled problem. Ten years later the list was approaching one hundred and getting longer as powerful new accelerators came on line and increasingly sensitive devices for particle detection were developed. The "botanists" among particle physicists were in despair.

The physicist who dominated the effort to bring order to this seemingly chaotic jungle of particles was Murray Gell-Mann, a man with a deep faith that beneath it all there were simple patterns dictated by symmetry principles. One of his tools was a mathematical device that quantum physicists had recognized since the early work of Bohr and Pauli: the quantum number. Some of the most mysterious of the elementary particles could be classified with a new quantum number appropriately called "strangeness." In Gell-Mann's scheme, the strangeness quantum number became half of a more complicated system of particle groupings that led straight to the heart of the structural symmetries of protons and neutrons and their exotic relatives. Symmetry was the key because it could tell its story with no recourse to the still unknown details of the appropriate quantum dynamics. By proceeding with his eyes on symmetry theory, Gell-Mann gave birth to the now ubiquitous quark concept. That, in a nutshell, is one of the stories told in this chapter. Another is the story of Murray Gell-Mann himself.

He was born in 1929, when his family lived on Fourteenth Street in lower Manhattan. Arthur Gell-Mann, Murray's father, had gone to New York from Vienna as Isidore Gellmann. He soon adopted the name Arthur, and shortly after his marriage, added the distinctive hyphen to his last name. He was a distinguished-looking man with intellectual aspirations that never were realized. In Vienna, he had started philosophical and mathematical studies, but his parents, who had emigrated to the Lower East Side in New York, needed his help and support, and he too emigrated.

Arthur had a good, if not brilliant, mind. His outstanding facility with languages led to the founding of the Arthur Gell-Mann School for language instruction. Unfortunately, Arthur's nearly perfect mastery of English was not matched by his ability as a teacher. As George Johnson, Murray Gell-Mann's biographer, writes, "He [Arthur] developed a pedantic, overbearing style that would later drive young Murray [and, presumably, Arthur's pupils] up the wall." He acquainted his students, all immigrants, with an overwhelming list of grammatical rules and devices, all supplied with proper nomenclature. "Here was a man who mastered a language, flowery rhetoric and all," writes Johnson, "a man in love with learning and with the sound of his own voice. It's hard to imagine who he thought his audience was—other than himself."

The language school did not survive the years of the Great Depression, and by 1932 the family was in dire straits. In desperation, Arthur took a job as a guard in a bank. He withdrew from his family and from the rest of the world, and exercised his intellect on the abstractions of Einstein's relativity theory. Murray's mother, Pauline, also withdrew by simply denying her troubles. "She was obsessively cheerful even when there was nothing to be cheerful about, rarely complaining, losing herself in a dream world . . . the beginning of mental illness," Johnson writes. The preoccupied parents could offer little guidance to their talented son, so Murray turned to his brother Ben, who was about ten years older and also precocious. Together the boys educated themselves in the great New York museums, the Museum of Natural History on the West Side, and the Metropolitan Museum of Art across Central Park on the East Side. The park itself was a living museum, where Murray began to learn the great lesson that nature is endlessly diverse.

To become a great physicist, one does not have to be a prodigy, but it helps. Think of Thomson, Maxwell, Rutherford, Heisenberg, Dirac, and Pauli. Murray Gell-Mann belongs in this company. At age three, he could multiply large numbers in his head. A few years later he knew enough Latin and Roman history to correct his elders (in an appealing way). When he was seven, he won a spelling bee in competition with twelve-year-olds. His teachers hardly knew what to do with him, but a kind and perceptive piano teacher, Florence Freint, did. She became a close friend and took him to see the headmaster of a private school on the Upper West Side of Manhattan, Columbia Grammar School. Freint got the appointment by insisting that the boy had to be seen to be believed. If the headmaster had any doubts, they were soon dispelled. He found that Murray's intellectual ability (at age eight) was equal to that of most college students. Murray entered Columbia Grammar with a full scholarship, and for once Arthur and Pauline responded to their remarkable son's needs. The family moved to an apartment on West Ninety-third Street, on the same block as the school.

At Columbia Grammar, Murray was still ahead of his teachers and classmates, and at the same time the youngest in his class by several years. Later he disparaged the school, claiming that he learned absolutely nothing there, but it opened the door to an Ivy League education at Yale, once again on a full scholarship. Yale brought Murray into a WASP milieu he had not experienced in New York. He was one of a quota of Jewish students (exactly 10 percent), and the freshman dean made it clear that no allowances would be made: "This is a Christian school run on a Christian calendar, and we want you to be sure to know that we expect you to be in class on Jewish holidays." When some students protested, the Yale president reversed the dean's edict but not the sentiment behind it.

Murray's first choice of a major at Yale was archaeology. That was foolishness, Arthur told him. As an archaeologist he would starve. He should emulate the great German archaeologist Heinrich Schliemann: make a fortune first and *then* organize archaeological expeditions. And where would he find the obligatory fortune? "In engineering," said Arthur. Murray quickly vetoed that option; any structure he designed would collapse, he insisted. Finally father and son met on a compromise, which was probably doubtful to both of them at first: physics.

Yale can take at least partial credit for confirming Murray Gell-Mann as a physicist. The Yale faculty member who most impressed Murray as a thinker and teacher was Henry Margenau, a quantum theorist from the generation of Dirac, Heisenberg, and Pauli. Margenau was also a science philosopher, and he taught an uncommon course that probed the foundations of physics. From Margenau's teaching and writing, Murray first began to appreciate the intellectual architecture of relativity and quantum theory.

Although Murray was reaching advanced levels of both mathematics and physics, the course work was still easy for him, and studying was hardly necessary. To the dismay of his toiling classmates, he was always ready to go out for a beer or argue politics, usually from the left.

In his senior year at Yale, Murray manifested symptoms of a malady that would never be cured: writer's block. He could not finish—he could hardly even start—his senior thesis. He did not know the formalities of researching and writing a paper, and would not ask Margenau, his thesis adviser, for help. Worse than that, he could not face a blank page "with the image of his father over his shoulder," writes Johnson. "Nothing would ever be good enough." Perhaps because of the missing thesis, or a deciding vote cast against him by Margenau, Murray was not accepted at Yale for graduate work, or at any Ivy League university except Harvard, which offered no scholarship.

There was a place for him, however, at MIT, as an assistant to Victor Weisskopf, one of the leaders in the physics community. "Viki" Weisskopf had started his career by studying with almost all the founders of quantum mechanics: with Schrödinger in Berlin, Born in Göttingen, Heisenberg in Leipzig, Bohr in Copenhagen, Pauli in Zürich, and Dirac in Cambridge. During the war, he had joined Oppenheimer's illustrious company of theorists at Los Alamos. Gell-Mann was unaware of these credentials—he had not even heard of Weisskopf—and he was unenthusiastic about MIT. "How could I go to that grubby place?" he wanted to know. But the alternatives were limited: "A little reflection convinced me that I could try MIT and then commit suicide later if I wanted to, but not the other way around."

MIT was worth trying. From Weisskopf, Gell-Mann began to learn about physics as it was actually practiced, about "prizing agreement with the evidence above mathematical sophistication," striving, "if at all possible, for simplicity," and "avoiding cant and pomposity." He had ambitious plans for his thesis research: he would find a topic that matched the importance of one of Einstein's great 1905 papers. In the end, Weisskopf gave him a problem that was hardly Einsteinian, but still important as a unification of the two reigning models of nuclear behavior. This theoretical task should have been straightforward and comparatively brief for a physicist of Gell-Mann's talent, but once again writer's block was an impediment. A deadline went by, and in the meantime Gell-Mann got an appointment from Oppenheimer at the Princeton Institute for Advanced Study that was contingent on completion of the dissertation. With that as a prod, Gell-Mann finally faced the writing task, and early in 1951 he finished it in a few days.

At the Institute for Advanced Study, Gell-Mann shared an office with Francis Low, who was struggling with the residue of problems remaining in quantum electrodynamics. Both men had recently received doctorates, but Gell-Mann was twenty-one, while Low, whose career had been delayed by war service, was thirty. "They put this child in my office," Low said later in an interview. But he soon found that the "child" had uncommonly mature insights into the principles of physics. Low learned, as others would later, that Gell-Mann's gift was an ability to look beyond the immediate details of a problem, and with the "eye of analysis," as Feynman called it, reveal the underlying patterns.

Chicago and Strangeness

Gell-Mann's stay at the Princeton institute was productive but temporary. Oppenheimer was impressed with his work but could not offer a faculty position. A friend from MIT, Marvin Goldberger, who was now an assistant professor at the University of Chicago and would eventually become president of the California Institute of Technology (Caltech), promoted his cause at Enrico Fermi's Institute of Nuclear Studies. The only position available was a lowly instructorship, but Fermi's institute was the place to be in the early 1950s. For a former New Yorker, Chicago had its limitations, and the weather could be abominable, but the disadvantages were countered by Fermi's presence. At the time, Fermi was the greatest practicing physicist, a notch above Oppenheimer, Rabi, Weisskopf, and even Bethe.

Fermi was a pragmatist who avoided excessive mathematics. Pauli called him a "quantum engineer," but the direct approach had brought Fermi triumphs in both experimental and theoretical physics. Fermi's style, with its similarities to Weisskopf's, appealed to Gell-Mann. It was just what was needed to reap a rich harvest from the data then emerging from the Chicago cyclotron and other particle accelerators in Berkeley and at the Brookhaven National Laboratory on Long Island.

The unidentified particles that announced themselves most distinctively in the particle detectors left a V-shaped track. At first, they were called "V-particles," and as the mystery of their origin deepened, they were dubbed "strange." In the mid-1950s, Gell-Mann opened the door to the world of strange particles with two tools that had been valuable to quantum physicists for many years: the concepts of the quantum number and the conservation law. Reduced to its essentials, Gell-Mann's scheme assigns a quantum number called "strangeness" to each strange particle, and adopts the principle that this quantum number does not change—it is conserved—in strong interactions, although it may change if weak interactions come into play. (Remember that strong interactions take place among neutrons and protons in atomic nuclei, and also *within* neutrons and protons among their constituent particles, for which Gell-Mann later supplied the name "quarks.")

Particle physicists use the label "hadron" for all particles such as protons and neutrons that are subject to strong interactions. They further classify hadrons as "baryons," with comparatively large masses, and "mesons," with intermediate masses. Another category of particles, called "leptons," includes the lightest ones: electrons and neutrinos.

Gell-Mann's rules of strangeness are illustrated by the particle reaction induced when a high-energy beam of mesons called "pions" enters a chamber filled with liquid hydrogen. The pions react with the protons present as hydrogen nuclei, and two strange particles are produced:

$$\mathrm{p} + \pi^- \rightarrow \Lambda^0 + \mathrm{K}^0, \tag{1}$$

in which p is a proton, π^- is a pion carrying a negative electrical charge (it has two siblings, π^0 and π^+, which are, respectively, neutral and positively charged), and Λ^0 and K^0 are both neutral strange particles, the former a baryon and the latter a meson. When the pressure is suddenly dropped in a liquid-hydrogen chamber, *charged* particles in transit through the chamber reveal their presence by leaving behind a trail of bubbles. It is said that this detection device, called a "bubble chamber," was suggested to its inventor, Donald Glaser, as he meditated on the rising bubbles in a bottle of beer. The bubble chamber was a successor to the Wilson cloud chamber.

The tracks in a bubble chamber shown in figure 26.1 tell the story of the particle reaction described above and represent Gell-Mann's strangeness scheme in action. An interpretation of the photograph, which omits all extraneous tracks, is supplied in figure. 26.2. Notice first the curvature of many of the tracks, caused by a strong magnetic field applied perpendicularly to the bubble chamber. (Maxwell's equations tell us that any electrical current follows a curved path in a magnetic field.) That curvature can be accurately measured and it reveals both the charge and the mass of the particle responsible for the track.

Figure 26.1. Photograph of tracks induced in a hydrogen-filled bubble chamber by a beam of high-energy, negatively charged pions. This photograph is reproduced by permission from the Lawrence Berkeley National Laboratory.

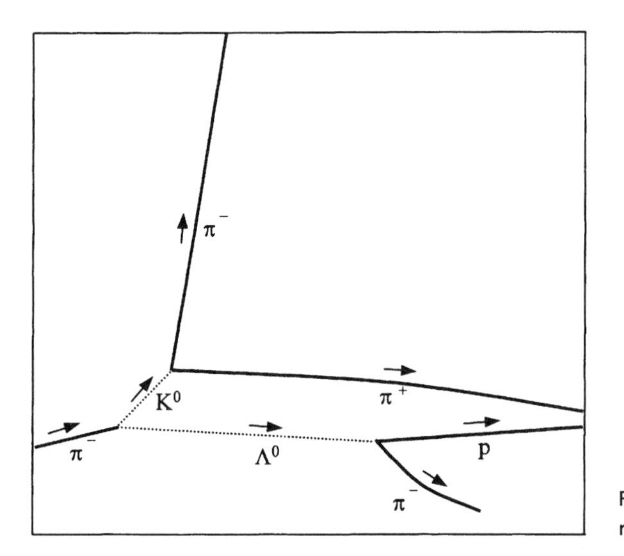

Figure 26.2. Interpretation of fig. 26.1. Only the relevant tracks are shown.

A highly energetic pion enters the bubble chamber on the left side of the picture in figure 26.2, encounters a proton (a hydrogen nucleus), and reaction (1) takes place, producing the Λ^0 and K^0 particles. They are both electrically neutral, and therefore leave no tracks in the bubble chamber. But they subsequently decay (break apart) into charged pions and protons according to

$$K^0 \rightarrow \pi^- + \pi^+ \tag{2}$$

and

$$\Lambda^0 \rightarrow p + \pi^- \tag{3}$$

Gell-Mann's rules assign values of the strangeness quantum number S as follows: $S = 0$ for nonstrange particles (p, π^+, and π^- in the example), $S = +1$ for K^0, and $S = -1$ for Λ^0. Thus the total strangeness quantum number for the left side of reaction (1) is $0 + 0 = 0$, and for the right side the total is the same: $-1 + 1 = 0$. The strangeness is therefore conserved in the reaction, as required for a strong interaction by Gell-Mann's scheme.

On the other hand, reactions (2) and (3) do entail a change in the strangeness quantum number, from $+1$ to 0 in reaction (2), and from -1 to 0 in reaction (3). Such changes in strangeness rule out strong interactions (and electromagnetic interactions), and indicate weak interactions. The length of a track in a bubble chamber, actually seen or inferred, is a measure of the "lifetime" of the particle before it "dies" in a subsequent reaction. The paths followed by the K^0 and Λ^0 particles in figure 26.2 indicate comparatively long lifetimes of particles that decay through weak interactions. The times are not really long—typically 10^{-8} second—but they are longer by many orders of magnitude than they would be if the particles decayed through strong interactions. There would be no direct evidence for K^0 and Λ^0 in the bubble chamber if their decay could follow the strong route.

Spin and Isospin

The particles of matter are known by their mass, electrical charge, mode of motion, and assorted other properties such as parity. The motion of a particle can take it from one place to another, as seen in the paths traced in the bubble chamber, and also give it a kind of spin. The term "spin" is a crude name for a property that actually exists only in the quantum realm. Feynman suggests that we should emphasize the abstract nature of particle spin by calling it "quantspin" rather than just spin. It is *nothing* like the spin of a golf ball or baseball. For one thing, electrons, neutrinos, and quarks have spin motion even though the theory does not allow them to have measurable size: they are points.

Another peculiarity is that these particles, like all of the elementary particles of matter, have just two spin modes or states. One says in the parlance of quantum mechanics that electrons, neutrinos, and quarks have spin ½ and that their two spin states have the quantum numbers −½ and +½. It is sufficient for our purposes to interpret these two spin modes as simply clockwise and counterclockwise.

All particles with spin ½ are called "fermions," for their statistical behavior, first mentioned by Fermi and a little later by Dirac. The statistical rule, which was also clarified by Pauli, is that two fermions cannot be found in the same quantum state. This profoundly important rule dictates the electronic shell structures of atoms and the electronic bonding between atoms in molecules. It guided Gell-Mann and others to some of the fundamental features of quark theory. Particles that are not elementary can also have two spin states, or more. For example, a particle with spin 3/2 has four spin states whose quantum numbers are −3/2, −½, +½, and +3/2. Notice that the recipe here is that only quantum numbers separated by one unit are allowed. This particle, and any other with half-integer spin (e.g., 7/2, 9/2, etc.), is also classified as a fermion.

Photons are elementary particles, and they too have spin. Their behavior indicates a spin of 1, and allows three spin states with quantum numbers −1, 0, and +1. In direct contrast to fermions, any number of them can inhabit the same quantum state, a pattern that was discovered by Satyendranath Bose and elaborated by Einstein. Photons and all other particles with integer spin (1, 2, 3, etc.), elementary or otherwise, are called "bosons."

Early in the history of particle physics (1932), Heisenberg took the concept of spin one step further into abstraction. He assumed that, as a model for nuclear structure, the constituent particles in nuclei are neutrons and protons, and that they are affected primarily by strong nuclear forces and much more weakly by electrical forces (among the positively charged protons). Noting this relative indifference to electrical charge, that the neutron and proton have nearly the same mass, and that the neutron can convert into the proton and vice versa, he constructed a theory based on the concept that the neutron and the proton are simply different states of a single entity called the "nucleon." The two states of the nucleon reminded Heisenberg of the two spin states of fermions with spin ½, and he introduced the "isospin" concept: the nucleon has an *isospin* of ½ (analogous to the *spin* ½ of an electron), and has two *isospin* states with quantum numbers −½ and +½, which are observed as the neutron and the proton (analogous to the electron's two *spin* states with the same quantum numbers). Heisenberg's motivation was strictly mathematical: he did not imagine any kind of real spin motion. But abstract as it is, the isospin concept, extended and combined with Gell-Mann's strangeness rules, displays just what theorists want to know: the underlying symmetries of the nucleon and its hadron relatives.

We will need a few more items of nomenclature. The isospin quantum number is designated by I, and the separate isospin states by I_3. (The subscript 3 pertains to the convention adopted for expressing isospin states; mathematical details on that convention are not important for our discussion). Thus for the nucleon $I = 1/2$, and for its "doublet" of isospin states—the neutron and the proton—$I_3 = -1/2$ and $+1/2$. The pion (a meson) has isospin $I = 1$ and a "triplet" of isospin states designated $I_3 = -1$, 0, and $+1$. Later we will meet the Δ hadron. It has isospin $I = 3/2$, and a "quartet" of isospin states for which $I_3 = -3/2, -1/2, +1/2$, and $+3/2$. The Λ strange particle mentioned earlier stands alone; it has zero isospin, $I = 0$, and a "singlet" isospin state with $I_3 = 0$. These are all examples of hadron "multiplets." From the examples, we can see that if a hadron has isospin I it belongs to an isospin multiplet with $2I + 1$ members.

More Symmetry Lessons

The deepest and most reliable principles of physics have their origins in nature's symmetries. Symmetries imply conservation laws, and vice versa. The diagnosis of symmetry also provides the theorist with important clues concerning structure. If, for example, we can identify the symmetry of a molecule using experimental techniques (for example, spectroscopy), we can probably deduce the molecule's shape and the arrangement of its atoms. Both strategies are completely independent of the applicable dynamical theories. Dynamical theories may come and go, but any symmetry that has a secure experimental foundation will remain a fixture.

Gell-Mann was one of the first theorists, and the most successful one, to follow the symmetry route to a solution of the formidable problem of hadronic structure. He knew, as others had before him, that the isospin multiplets are signatures, or "representations," as mathematicians call them, of a kind of symmetry designated formally as the "symmetry group" SU(2). The "SU" part of the notation stands for "special unitary," which guarantees that certain demands made by quantum mechanics are met. The 2 in parentheses reflects that the simplest of the isospin multiplets is the twofold doublet comprising the neutron and the proton. Gell-Mann and his fellow theorists looked at the emerging evidence for SU(2) symmetry among the hadrons then being discovered in profusion, and wondered if the data would reveal higher symmetries. Any new symmetry patterns could perhaps be mined for clues concerning the structures of hadrons. It was a race, first to uncover the hidden symmetries, and then to deduce the hadronic structures.

Caltech and the Eightfold Way

While he was pondering the data of hadron spectroscopy, Gell-Mann was uprooting himself from an increasingly unsatisfactory lifestyle. Through a mutual friend, Gwen Groves (the daughter of General Leslie Groves, director of the Manhattan Project), he met a young Englishwoman, Margaret Dow, in Princeton during his second appointment at the Institute for Advanced Study. She was an assistant to an archaeologist at the institute, and came from a family with a background similar to Gell-Mann's; she had experienced hard times when her father failed in business. Gell-Mann shared her interest in archaeology (he probably would have been an archaeologist if his father had not warned him off), and she sympathized with his love of ornithology. They even joined forces on a puffin-seeking expedition to a remote island off the west coast of Scotland. A single bird obliged, and the puffin became their talisman.

Gell-Mann proposed to Margaret Dow in November 1954, she accepted, and they were married in the spring of 1955. In the meantime, Fermi had died, and Gell-Mann, again at the Princeton Institute for Advanced Study, decided not to go back to Chicago. Other opportunities were coming his way. Julian Schwinger backed an appointment for Gell-Mann at Harvard. Feynman invited him to Caltech, and that was his choice. At first, Margaret was not enthusiastic about Pasadena and the rest of southern California, but the marriage was stronger than the annoyances of life in California. Margaret adapted, and for Gell-Mann it was a priceless gift. "She . . . changed his life," Johnson writes. "He began to realize how rarely he used to think about anyone but himself. Before Margaret, he would say, he was like 'a malfunctioning computing machine' or 'an atom that is bounced around' by sentient forces. Now he felt like a person."

By the early 1960s, Gell-Mann was beginning to solve the mysteries of the hadron data. One of his breakthroughs in the search for higher symmetries was the discovery of a revealing grouping of isospin multiplets. He prepared a simple map of the known baryons with spin (not isospin) ½ and a certain parity. The coordinates of the plot were the strangeness S and the isospin state I_3, as shown in figure 26.3. The baryons plotted are the nucleons and the three particles labeled Λ, Σ, Ξ. The plot places the nucleon doublet (n and p for neutron and proton) in the bottom row. Then comes the Λ^0 singlet and the Σ triplet comprising Σ^-, Σ^0, and Σ^+. (Λ^0 and Σ^0 are both supposed to be located at the point $S = -1$, $I_3 = 0$.) At the top of the diagram is the Ξ doublet consisting of Ξ^- and Ξ^0. The map in figure 26.3 groups eight baryons, and it is called an "octet." In a sense, it is a multiplet of isospin multiplets, but not exactly, because of the mass differences. Gell-Mann prepared a similar map of an octet of mesons.

Beyond the octets, Gell-Mann could glimpse a larger grouping of ten baryons—a "decuplet." The decuplet construction is shown in figure 26.4. It contains a quartet of Δs with strangeness $S = 0$, a triplet of Σ^*s (not the same as the Σs in the baryon octet, hence the asterisk) with $S = -1$, a doublet of Ξ^*s (different from the Ξs of the octet) with $S = -2$, and at the top of the diagram with $S = -3$ the particle Ω^-, a singlet. When Gell-Mann began to envision this decuplet in the early 1960s, not all of the ten baryons had been discovered; only four of

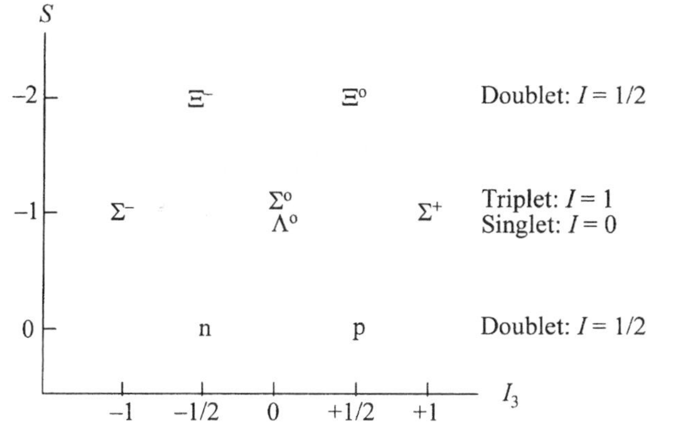

Figure 26.3. An octet of baryons plotted according to their isospin I_3 and strangeness S. Each symbol represents a point on the plot. For example, "n" locates a point at $I_3 = -\frac{1}{2}$ and $S = 0$.

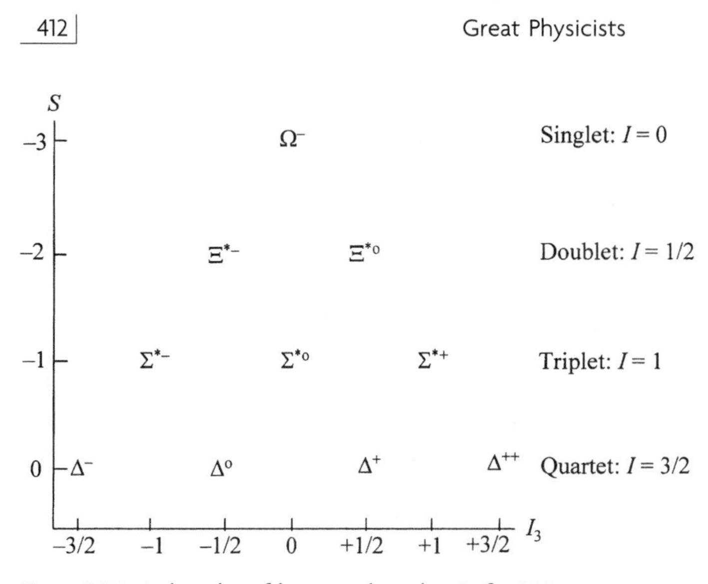

Figure 26.4. A decuplet of baryons plotted as in fig. 26.3.

them (the Δs) had. In spite of these "holes" in the decuplet diagram, Gell-Mann published his findings in 1961, deliberately emphasizing the octets, and dubbing his scheme the "Eightfold Way."

Multiplets of the octet and decuplet kind are vital in Gell-Mann's scheme because they are representations of a symmetry known formally as the symmetry group SU(3), a higher symmetry than the SU(2) of isospin theory. If confirmed by further hadron data, SU(3) symmetry could be not only the key to hadron taxonomy, but beyond that a major clue in the search for hadron structure.

Gell-Mann's approach to the organization of hadron data is something like the periodic scheme proposed in the late 1860s by Dmitry Ivanovich Mendeleev for listing the chemical elements. Mendeleev's table, like Gell-Mann's tables, had gaps where elements should have appeared, but did not, and they inspired some remarkable discoveries of new elements.

Gell-Mann's decuplet plot prompted similar searches for missing baryons. All were duly discovered, except for the Ω^- particle located at the peak of the decuplet mountain, with the strangeness $S = -3$. It was particularly inaccessible because of its extraordinarily large mass, meaning that a high energy had to be supplied in an accelerator to create the particle. A heroic effort was mounted in 1964 by Nicholas Samios and Robert Palmer at the Brookhaven National Laboratory to catch an Ω^-. More than ninety-seven thousand images of bubble-chamber tracks were photographed, consuming more than a million feet of film. But when the experimenters got the right picture, it was a beauty. It revealed not only production of the elusive Ω^- but the cascade of subsequent events that brought the system to stable particles. From the tracks of the final particles, the mass of the Ω^- could be inferred. In energy units, the mass was 1682 ± 12 MeV (MeV stands for a million electron volts). Gell-Mann had predicted 1685 MeV. The case for SU(3) symmetry was clinched.

Quarks in Three Flavors

While experimenters were chasing down the Ω^- particle and its cousins, Gell-Mann was confidently exploring the meaning of SU(3) hadron symmetry. The central mystery was the 3 in SU(3). What, if anything, did it mean physically? Mathematically speaking, there is a threefold representation of SU(3) symmetry, but Gell-Mann could find no direct physical counterpart for it in the hadron data. Isospin SU(2) symmetry had its twofold nucleon (neutron + proton) doublet, but Gell-Mann searched the data in vain for an analogous SU(3) triplet.

A more subtle possibility was that threefold symmetry was built into the *structure* of hadrons. Gell-Mann began to entertain the idea that neutrons and protons, and all the other baryons, are constructed from three elementary particles, which come in three types, or "flavors," as Gell-Mann put it. He first called his elementary particles "kworks"; then he noticed the line in James Joyce's *Finnegans Wake*, "Three quarks for Muster Mark!" and converted the name to the now-famous "quark." (Particle physicists are addicted to shamelessly whimsical nomenclature. Gell-Mann claims that the term quark was "just a gag . . . a reaction against pretentious language.")

The model had some problems. The most stubborn was the matter of electrical charge. Presumably, the quarks carried charge, but how could the total charge from *three* of them add up to the proton's unit charge, or the neutron's zero charge? The only way out was to endow the quarks with something that had never been observed in nature: *fractions* of the proton's unit charge. And to be taken seriously, any such model would have to explain why the fractionally charged quarks were so tightly locked up in matter that they never showed themselves in the laboratory.

Another peculiarity of Gell-Mann's model was that one of the three quarks had to be assigned a mass substantially different from those of the other two. This damaged the aesthetics of the model: it made the threefold symmetry imperfect. The symmetry is "broken," as particle physicists say, but not so badly that the hallmarks of SU(3) symmetry, octets and decuplets, cannot clearly be seen in Gell-Mann's octet and decuplet plots.

Gell-Mann called his three quarks "up," "down," and "strange," or (better) u, d, and s. The u and d quarks have about the same mass, and s is heavier by roughly 40 percent. The u quark has the fractional electrical charge $+ 2/3$ (in units of the proton charge), the d quark $- 1/3$, and the s quark also $- 1/3$. (Different quark charges spoil the threefold symmetry only slightly because, within the confines of a hadron, electrical interactions are almost irrelevant compared to the strong interaction.)

Gell-Mann's theory builds a proton with two u quarks and one d quark, or uud for short. The total charge is $2/3 + 2/3 - 1/3 = 1$, as required. The neutron has the quark composition udd, providing the necessary zero charge: $2/3 - 1/3 - 1/3 = 0$. As its name implies, the s quark is found only in strange particles, and it has a strangeness of -1, while the u and d quarks have zero strangeness. The members of the strange triplet Σ^-, Σ^0, and Σ^+ have the quark contents dds, uds, and uus, respectively, and the charges $- 1/3 - 1/3 - 1/3 = -1$, $2/3 - 1/3 - 1/3 = 0$, and $2/3 + 2/3 - 1/3 = 1$. The famous Ω^- particle contains three s quarks, each contributing a strangeness of -1, and the charge is $- 1/3 - 1/3 - 1/3 = - 1$.

Mesons also have quark structures in Gell-Mann's theory, but they are fundamentally different from baryon structures. They always contain two quarks, one a normal quark and the other an antiquark. A quark and its anti version are opposite in their electrical charges and also in their strangeness. Thus, for example, the d quark and the anti-d quark, written $\overline{d}$, have the charges $- 1/3$ and

+ 1/3, respectively. The s quark and its anti partner $\bar{s}$ have the charges − 1/3 and + 1/3, and the strangeness assignments − 1 and + 1.

An example is the meson triplet K^-, K^0, and K^+ whose quark structures are $s\bar{u}$, $d\bar{s}$, and $u\bar{s}$, respectively, with the charges $-\frac{1}{3} - \frac{2}{3} = -1$, $-\frac{1}{3} + \frac{1}{3} = 0$, and $\frac{2}{3} + \frac{1}{3} = 1$. The corresponding strangenesses are − 1, + 1, and + 1. It is a general rule in particle physics that particles and antiparticles do not get along in close proximity; the result is mutual destruction. The rule is illustrated by the observation that mesons, which bring a quark and an antiquark together, are always unstable.

Gell-Mann's quark model was hardly an instant success. "Quarks went over like a lead balloon," Gell-Mann recalled later. The quark's fractional charges and their everlasting confinement in hadrons were particularly hard for theorists to accept. Included among the skeptics was Gell-Mann himself. "Even I thought the idea of [unobservable] fractionally charged particles was crank," he recalled. Because the quarks were "permanently stuck inside" (a colleague, Sheldon Glashow, quipped, "You can't even pull one out with a quarkscrew"), Gell-Mann hedged by saying that they had a mathematical existence but were not real. Convinced that a quark paper would be rejected by the reviewers and editors of *Physical Review*, Gell-Mann submitted it to the more liberal journal *Physics Letters*, published at CERN, the European accelerator center near Geneva. The paper, titled "A Schematic Model of Baryons and Mesons," appeared in 1964, just one day after the Ω^- discovery was reported.

Quarks in Three Colors

Gell-Mann believes that physicists should be rated by subtracting, from the number of correct ideas they have published, twice the number of wrong ideas. When he published his first quark paper, Gell-Mann still had reason to worry that the theory might be a detraction. There was a clear-cut conflict with a statistical principle that theorists had relied on for more than forty years, Pauli's exclusion principle. Quarks, like electrons, are particles with spin ½. That means they qualify as fermions and must obey the exclusion stricture that two or more of them in the same quantum state cannot be found in close proximity. But the Ω^- particle, for example, contains three s quarks in close confinement, and at times apparently in the same quantum state.

Nothing is more distasteful to theorists than having to abandon a principle that has served them well for decades; few were willing to discard the Pauli principle. Some would not have mourned the loss of quark theory, but there was a good alternative. The theory could be elaborated by equipping the quarks with an additional quantum property, which was different for the three quarks that gather in the Ω^- particle. A theory of this kind was first suggested by Moo-Young Han and Yoichiro Nambu. Their idea was that quarks have a property called "color" (more whimsy) analogous to electrical charge. There are two kinds of electrical charge, negative and positive; Han and Nambu postulated three kinds of a different and independent charge possessed by quarks. This idea was taken up by other theorists, including Gell-Mann, who designated the three kinds of color charge red, white, and blue. Other physicists (perhaps mindful that white is not a pure color) changed the color designations to the primary red, green, and blue. Thus in the curious language of quark theory one says that the three quark flavors u, d, and s (and more, as we will see later) come in three colors, red, green, and blue.

When Gell-Mann received what he called the "Swedish Prize" in 1969, quark theory was still tenuous enough that the Nobel citation said nothing about it: Gell-Mann was honored "for his contribution and discoveries concerning the classification of elementary particles and their interactions." The speaker who introduced Gell-Mann on the occasion of his Nobel lecture saw only the "great heuristic value" of the quark concept. Gell-Mann was not willing to go much further. He could say only that the "quark is just a notion so far. It is a useful notion, but actual quarks may not exist at all." The question of quark existence or nonexistence was "immaterial."

Quarks in Captivity

At about the same time Gell-Mann was accepting his Nobel award, quark theory was receiving some impressive experimental support. Experiments in the late 1960s at the Stanford Linear Accelerator Center, in which protons were bombarded with beams of high-energy electrons, revealed hard, pointlike, charged particles within the protons. Feynman called them "partons," and noted that within their proton confinement they seemed to move freely.

For Gell-Mann, Feynman's partons were "put-ons" and should have been called quarks. "The whole idea of saying that they weren't quarks and antiquarks but some new thing called 'put-ons' seemed to me an insult to the whole idea we had developed," Gell-Mann said in an interview with Robert Crease and Charles Mann, authors of *The Second Creation*, a chronicle of particle physics. Gell-Mann went on to say that an important point was made by the Stanford theorist James Bjorken, which was explained "in a somewhat more—ah, what should I say?—popular manner by Feynman. And that was that deep in the interior of the nucleon the quarks were always free."

Odd beasts, these quark/partons. They are unable to escape from their hadronic confinement, and yet within their cages they do not seem to influence one another. It was one thing to say that this is the way things are in the domains of the hadrons, and quite another to construct a dynamical theory that explained the behavior. Color was the key, Gell-Mann told Crease and Mann: "We gradually saw that that variable [color] was going to do *everything* for us. It fixed the statistics, and it could do that without involving us in crazy new particles. Then we also realized that it could fix the dynamics."

The dynamical "fix" was a new quantum field with the quark's color charge as its source; it was analogous to the electromagnetic field, which has electrical charge as its source. The quanta of the new field, analogous to the photons of the electromagnetic field, carry the strong force that binds the quarks together, and came to be called "gluons." There are eight types of gluons (an octet; SU[3] symmetry is again at work), and they also have the color attribute.

One of the finishing touches to the color dynamics, or "quantum chromodynamics" (QCD), as Gell-Mann later called it, came in 1973 when three theoreticians, David Politzer, Frank Wilczek, and David Gross, showed that the field does what is necessary: it keeps free quarks captive in their hadronic cages. To do so it must generate an uncommon type of force, one that increased with increasing distance. Most familiar forces operate the other way: they diminish with increasing distance. Examples are the gravitational and electrical forces, which have a decreasing $1/r^2$ dependence, where r measures the distance between two objects influenced by the force. The force manifested in quantum chromodynamics is like that of a rubber band. When two objects held together by the rubber band separate, the band becomes taut, stretches, and pulls the objects back together. When the two objects are close to each other, the band hangs loose and the objects are free and unrestricted.

The confinement feature, which theorists call "asymptotic freedom" for reasons that are not important here, gave physicists hope for a full understanding of the strong force. As Crease and Mann put it, "Asymptotic freedom came like the opening of a curtain onto a previously hidden stage, revealing the strong interaction in its full dimension. Many physicists had pictured elements of the scene—Gell-Mann came closest to encompassing its entirety—but none till then had fully grasped the flawless elegance of nature's conception."

Quark Generations

Three quark flavors were not enough. As the energy of accelerator beams climbed higher and higher, short-lived particles began to appear in the increasingly sophisticated detectors, particles that could not be explained by building from the three known quarks u, d, and s. Tracks of the first of these particles appeared unexpectedly and simultaneously in two laboratories using completely different techniques. A team lead by Samuel Ting at the Brookhaven National Laboratory found a particle they labeled J, and a collaboration at the Stanford Linear Accelerator Center under Burton Richter reported a particle with the same mass, which they named ψ. Both claims were valid, and neither had precedence, so the particle was christened J/ψ. It was heavy, more than three times the mass of the proton, and for a particle of such large mass (and energy) surprisingly long-lived. A model based on quantum chromodynamics and on the existence of a fourth quark, which acquired the name "charm," or c, gave a good account of the remarkable J/ψ. J/ψ is a meson and its quark structure is assumed to be $c\bar{c}$ ($\bar{c}$ is an anticharm quark). The success of quantum chromodynamics in the J/ψ episode did a lot to bring the quark concept into mainstream physics.

With the discovery of a fourth quark, a pattern of quark families or generations began to emerge. The two lightest quarks, u and d, belong to the first generation, and the distinctly heavier quarks, s and c, to a second generation. In the late 1970s, the pattern was extended with the discovery of Y (upsilon) particles (mesons), which demanded a fifth, still heavier quark belonging to a third generation, and called, for no particularly good reason, the "bottom" or b quark.

Because there was a fifth quark, there evidently had to be a sixth to complete the third quark generation. Two immense projects were mounted (as friendly checks on each other) in the middle 1990s at the Fermilab National Accelerator in Batavia, Illinois, to gather evidence for the existence of the bottom quark's sibling called (of course) the "top" or t quark. The two collaborations required the services of about a thousand physicists, and an army of technicians. Out of some trillion events in one of the detectors, twelve were deemed to have produced a top-antitop pair. This was big science with a vengeance; Rutherford would have been appalled.

Three generations of quarks are enough to give theorists the building blocks they need to construct models of all the known mesons and baryons. The third category of particles, leptons, have no structure (they are points), according to current quantum field theory; quark theory has nothing to say about them. Yet they, too, seem to belong to three families or generations.

The first lepton generation comprises the electron e, both negative and positive (also known as the positron), and the neutrino v. Because there are more kinds of neutrinos to come, I will represent this one with v_e, to stress that it is associated with the electron. The muon μ, for a long time an embarrassment to particle physicists ("The muon. Who ordered that?" complained Isidor Rabi), was finally recognized as a heavy counterpart to the electron. It is electrically charged, either positively or negatively, and its mass is about two hundred times that of the electron. The electron-muon parallel demands a special muon neutrino, v_μ. An immense experiment at Brookhaven designed by physicists from Columbia University in the early 1960s gave evidence for the existence of this second-generation neutrino.

In the 1970s, experimenters moved on to the third generation of leptons. The electron analogue, named τ, was discovered unexpectedly at the Stanford Linear Accelerator Center. Like e and μ, it can carry negative or positive charge, and its mass is about thirty-five hundred times that of the electron. Once the τ lepton was discovered, the existence of its neutrino v_τ was also accepted, and has recently been observed. Here is a tidy summary of the three generations of leptons and quarks:

First generation:
quarks, d and u
leptons, e and v_e.
Second generation:
quarks, s and c
leptons, μ and v_μ.
Third generation:
quarks, b and t
leptons, τ and v_τ.

Remember that each of these particles has an anti version.

Is there a *fourth* generation of quarks and leptons? The answer is no, as shown by two complex and clever experiments carried out in the late 1980s at the Stanford Linear Accelerator Center and at the European accelerator center, CERN. The experiments demonstrated indirectly the nonexistence of a fourth kind of neutrino. If there is no fourth-generation neutrino, theorists concluded, there are no fourth generation quarks and electron counterparts. The list is closed.

Competition

The theoretical path that Gell-Mann followed to his major discoveries in the 1950s and 1960s was a crowded one. This story would be incomplete without mentioning some of those who traveled with Gell-Mann as competitors in close races to the same destinations.

Gell-Mann's idea that strange particles deserved a new quantum number was shared by the Japanese theorist Kazuhiko Nishijima. The codiscoverer of the symmetry principles behind the Eightfold Way was Yuval Ne'eman, a colonel in the Israeli army and an amateur physicist. Ne'eman studied physics at the Imperial

College in London while serving as a military attaché there. His mentor was Abdus Salam, renowned for his work on the field theory of the weak interaction, who was at first unenthusiastic about Ne'eman's claims for SU(3) symmetry, but changed his mind when he learned that Gell-Mann was also following the SU(3) route.

The man who shared with Gell-Mann the discovery of the quark concept was George Zweig, who had been one of Gell-Mann's research students. His theory was developed during a visit to CERN, with no influence from Gell-Mann. Zweig called his subhadronic particles "aces." Baryons, containing three aces, were "treys," and mesons, with two aces, were "deuces." Zweig constructed his theory in detail, but with the proviso that his aces might be no more than "rather elaborate mnemonic devices." Nevertheless, he could also see an "outside chance that the model is a closer approximation to nature than we may think, and that fractionally charged aces abound within us."

Zweig was willing to bet on another outside chance. He was expected to submit his paper to the CERN journal, *Physics Letters.* Instead he chose to send it to the prestigious *Physical Review,* where it was emphatically rejected. Gell-Mann, older and wiser, anticipated a negative reception at the *Physical Review* to such bizarre entities as unobservable, fractionally charged elementary particles, and he published his first quark paper in *Physics Letters.* Zweig's theory went unpublished except in a CERN report, but it and its author acquired a certain reputation. When Zweig sought an appointment at a major university, the head of the department pronounced him a "charlatan."

Quarks and Jaguars

Murray Gell-Mann is not an easy man to understand. He has many sides, some good, some bad, and some baffling. All who have known him agree that he has a genius for theoretical physics and a flair for promoting his ideas with memorable terminology. He has broad, often esoteric, interests. His biographer lists classical history, archaeology, linguistics, wildlife ecology, ornithology, numismatics, and French and Chinese cuisine. For years he proposed interdisciplinary efforts at Caltech without much effect. In the mid-1980s, he joined with several like-minded colleagues in planning an institute to be located in Santa Fe, which would be devoted to "complexity studies." For Gell-Mann this meant exploring the paths from the simple to the complex. The Santa Fe Institute became a reality in 1987, with George Conway, formerly the research director at Los Alamos, as president, and Gell-Mann as head of the institute's science board.

More than any of the other physicists in this book, Gell-Mann has taken on high-level public service. In the 1960s, he joined an advisory group attached to the Institute of Defense Analysis. The members called themselves "Jason," and they considered such matters of defense policy as antiballistic missile systems and the detection of nuclear explosions. It was an elite group, including Hans Bethe, Edward Teller, John Wheeler, Eugene Wigner, and Freeman Dyson. Gell-Mann also served on Richard Nixon's science advisory board, and would have been a science advisor to Robert Kennedy had Kennedy not been assassinated. Gell-Mann's many other prestigious associations outside physics have included appointments to the boards of the Smithsonian Institute and the MacArthur Foundation.

Contrasting with this enviable record of committee work, presumably without serious disagreements with other committee members, is Gell-Mann's habit of handing out put-downs to selected colleagues, usually those in competition with him. Gell-Mann's favorite target was Richard Feynman, who responded in kind. Their offices at Caltech were close to each other, and they shared the services of a secretary. At first, their relationship was collaborative and congenial. Both had a need for a human sounding board to test new ideas. "When we were together discussing physics," Gell-Mann writes, "we would exchange ideas and silly jokes in between bouts of mathematical calculation—we struck sparks off each other, and it was exhilarating." They called these grand discussions "twisting the tail of the cosmos."

But Gell-Mann and Feynman had personalities that were anti versions of each other. When they got together there were bound to be, if not gamma rays, at least fireworks. Feynman was Gell-Mann's match in the art of the put-down. After one exchange in which Gell-Mann displayed linguistic knowledge not shared by Feynman, the last word was Feynman's: "Murray, in a hundred years nobody will know whether your name is hyphenated or not." Gell-Mann liked to call Feynman's monologue books "Dick's joke books." In a collection of memorial essays on Feynman, *Most of the Good Stuff*, Gell-Mann noted that he was not impressed by a "well-known aspect of Richard's style. He surrounded himself with a cloud of myth, and he spent a great deal of time and energy generating anecdotes about himself. Sometimes it did not require a great deal of effort."

Gell-Mann has had two happy marriages. In the late 1970s, his first wife, Margaret, developed symptoms of colon cancer, and before the cancer was diagnosed and a course of treatment could be started, the cancer had spread to the liver. She died in 1981, after painful and futile attempts to suppress the liver cancer. It was a devastating loss for Gell-Mann. Margaret had been one of his mainstays.

At about the same time, Gell-Mann was finding his role as a father agonizing. His daughter, Lisa, who had been exemplary and dutiful as a child, plunged passionately into left-wing politics. She was persuaded by a young man who came from a wealthy New York background to join a revolutionary group called the Central Organization of United States Marxist-Leninists. Their hero was Joseph Stalin, and their political ideal was the dictatorship of Enver Hoxha in Albania. Gell-Mann saw Lisa only intermittently, and his attempts to redirect her interests failed. A few years later his relationship with his son, Nick, was deteriorating. These estrangements were to nag Gell-Mann for about a decade.

For a scientist, Gell-Mann has led a lavish lifestyle. At one point, he maintained homes in Santa Fe, Aspen, and Pasadena. The Santa Fe house "was like a museum," writes Johnson, "with Murray's collection of indigenous American pottery, African art, rare books, and ancient weapons—an Eskimo harpoon, a North African mace, a blowgun complete with poison darts, a Sumatran dagger, a Chinese beheading sword." The Pasadena house was "dark and imposing, with an anonymous, almost institutional façade," Johnson writes. "It struck some visitors as more of a museum than a house, a place to keep Murray's growing collection of antiquities." Gell-Mann also has a taste for luxurious automobiles; he spent part of his Nobel Prize money on a Jaguar sedan.

Gell-Mann is articulate, an accomplished lecturer, and a charming conversationalist. Yet for some reason, perhaps extending back to his father's stern influence, he has always been afflicted by a case of writer's block. He could not finish his senior thesis at Yale. His doctoral thesis at MIT was tardy by about six months. He gave his Nobel lecture but procrastinated over delivering a written version of the lecture for publication in the distinguished volume *Le Prix Nobel*.

After six months of cabled reminders from the editor of the volume, and apologies from Gell-Mann for the delay, he did not meet the deadline.

He could do no better with other requests to contribute papers. One editor of a Festschrift got this response: "Unfortunately when I make promises to contribute articles to such publications, I almost never keep the promises. Therefore, I have learned not to make them in the first place, and I must decline with regret your invitation." Even letters of recommendation were difficult. "Unfortunately, I have great difficulty writing letters and requests for written recommendations are likely to be ignored," he warned one former student. "Please feel free to have people call me about you."

Gell-Mann's capacity for procrastination when faced with a writing task reached epic proportions when he decided to write a book for a popular audience. He had in mind the successes of Feynman and Stephen Hawking, both of whom had published best-sellers. A flamboyant literary agent, John Brockman, who was also thinking of the Hawking and Feynman fortunes, eagerly promoted the project. He hired a well-known science writer to prepare a proposal in collaboration with Gell-Mann. Bantam Books, publishers of Hawking's *Brief History of Time*, offered $550,000 for the rights to the book in the United States and Canada, with Gell-Mann receiving 25 percent of this as an advance. Brockman also received promises for hefty advances from foreign publishers. Brockman's deals were reputed to be the most lucrative ever made for a popular science book. The proposal went out in late 1990, and the manuscript was to be completed by June 1992. The book had the title *The Quark and the Jaguar*. It was to take the reader from the simple to the complex patterns of nature. The quark exemplified nature reduced to its simplest level, and Gell-Mann's beautiful metaphor for complexity in nature was the jaguar.

It was not a good time for Gell-Mann to focus on a writing task that was sure to be demanding. In the summer of 1991, he and Marcia Southwick became engaged, and the wedding was set for June 1992, on a collision course with Gell-Mann's deadline for delivery of his manuscript. Southwick is a poet who had met Gell-Mann in Aspen, where she sometimes teaches at the Aspen Writer's Conference.

The wedding was a success, but the book project was not. Assistance from two editors and coaching by Southwick were not enough to salvage it. Bantam rejected the manuscript, and Gell-Mann had to return his advance. A smaller publisher, W. H. Freeman, then stepped in and took the project for about one-tenth of the original Bantam offer. With more editorial assistance, including a review of the manuscript by the novelist Cormac McCarthy, *The Quark and the Jaguar* was finally published in 1994. "I have never worked so hard on anything in my life," Gell-Mann notes in his preface. The book has repaid the publisher—and, one hopes, the author.

Gell-Mann still lives in Santa Fe. As ever, he enjoys the high mountains, the desert, and the canyons of New Mexico. And he still travels widely to witness the beauty and diversity of nature. (His list of bird species sighted now stands at the amazing figure of more than four thousand.) His second marriage is a success, and he is reconciled with his son and daughter, Nick and Lisa. Intellectually, he is doing what elder statesmen of science rarely do: he is stretching to the utmost the limits of his scientific studies. He once told a lecture audience: "For me two things are inseparable, the love of the beauty of nature and the desire to explore further the symmetry and subtlety of nature's laws."

ix

ASTRONOMY, ASTROPHYSICS, AND COSMOLOGY

Historical Synopsis

Our story ends where it began, with physicists scrutinizing the night skies for clues about the universe we inhabit. In the first two chapters, we saw Galileo confirming with his telescope the Copernican message that Earth and other planets orbit the Sun, and Newton building his universal gravitation theory to calculate the planetary orbits and the motion of all the other heavenly bodies. In this part of the book, we see some of the great strides taken in the twentieth century by physicists in their efforts to map our universe, define its dynamics, and write its history. The main characters in the modern story are an astronomer, Edwin Hubble; an astrophysicist, Subrahmanyan Chandrasekhar; and a cosmologist, Stephen Hawking.

Edwin Hubble was the first to identify galaxies beyond our own. He used the greatest telescopes of his time to estimate distances in this extragalactic realm. Then he made a careful study of the colors of distant galaxies and found that the greater the distance to a galaxy, the more its color shifted toward the red. He proposed a simple linear relationship between distance and this "redshift." Because the redshift of a galaxy can be interpreted to mean that the galaxy is moving away from us, Hubble's data suggested that the universe as a whole is expanding. Hubble was at first cautious about adopting this interpretation, but others were more easily convinced, and the first steps were taken in the development of the now dominant "big bang" theory of the origin and history of the universe.

Subrahmanyan Chandrasekhar (or "Chandra," as he was known) was a man who excelled in probing the complexities of stellar physics. In his long career as an astrophysicist, he studied stellar structure, dynamics, and evolution. One of his last efforts was an investigation of the mathematical theory of "black holes," massive objects that have been completely and violently crunched by extreme gravitational collapse. His principal theoretical tool in black-hole research was the theory of gravity embodied in Einstein's theory of general relativity.

The last of our company of physicists, Stephen Hawking, was also fascinated by the weirdness of black holes. He and a colleague, Roger Penrose, focused on the unwelcome infinities that general relativity told them were harbored at the centers of black holes. They looked on these "singularities" with suspicion, particularly when they found that a universe governed by general relativity and the big-bang scenario had to begin with an exposed singularity. One escape from the unfortunate singularities is to fashion a theory of gravity that combines general relativity with quantum theory. That unification has yet to be discovered, but Hawking and other cosmologists are, as always, optimistic.

27

Beyond the Galaxy
Edwin Hubble

Missourian

Edwin Hubble was a man with grand aspirations, and to a remarkable degree he attained his goals. As a pioneer in the observation of realms lying beyond our galaxy, he became the preeminent astronomer of his time, and indeed of the twentieth century. His astronomical observations gave us the first glimpse of our modern cosmology based on a universe whose space is expanding. He married into a wealthy southern California family, and counted among his friends many from the California intellectual elite.

But the successes came with a price. As he rose through the social and economic strata, Hubble reinvented himself, sometimes with dubious credentials. There was a discontinuity between the one Hubble, with an ordinary midwestern background, and the other, a wealthy Anglophile who mingled with the Hollywood greats. Along the way, Hubble partly disowned his family members, not allowing any of them to meet his wife or her family. Hubble's youngest sister, Betsy, said in an interview, "I always wondered if Edwin didn't feel guilty about not having done more [for his family]. But great men have to go their own way. There is bound to be some trampling. We never minded." Some of his colleagues considered him "arrogant and self-serving," and blocked one of the prizes he coveted most, directorship of the great Mount Palomar Observatory.

Hubble was handsome almost to a fault. He was tall, athletic, usually equipped with a pipe, and as an admiring neighbor put it, "very, very masculine." Anita Loos, a writer best known for her novel *Gentlemen Prefer Blondes* and the Broadway production of the same title, would not have hired him to play the part of a famous astronomer: "You can't have him look like a blooming Clark Gable!"

He was born in 1889, far from the scenes of his triumphs, in Marshfield, Missouri. Edwin was the third of seven surviving children, three boys and four girls. Their father, John, was trained in the legal profession, but preferred the insurance business and all the traveling that went with it. His wife, Virginia Lee ("Jennie") James, seems to have tolerated John's many absences by living close to her parents in Marshfield. Edwin Hubble's biographer Gale Christianson provides this grim sketch of John: "John Hubble was the product of a puritanical upbringing and strict education. A stern, hard-bitten blend of moral high-mindedness and relentless ambition, he became a demanding taskmaster who 'ruled the roost' in no uncertain terms. . . . Following [his father's] example, he renounced alcohol and rarely, if ever, cursed. The only vice inherited from his father was the love of tobacco. John smoked both a pipe and large cigars."

Jennie, on the other hand, was always accessible to her children. Her daughters remembered her as "a lovely lady in every way," writes Christianson. Like her husband, she was deeply religious, but the "belief in salvation held a more prominent place in her thinking than the threat of damnation."

Edwin seems to have been closer to his grandfathers than to his father. Grandfather Hubble told some fine stories, many of them based on real family history. Grandfather James, who was distantly related to Jesse James, found the materials to build a telescope, which Edwin was permitted to use far into the night on his eighth birthday.

By the time Edwin was ten, John had made a reputation as an insurance underwriter, and had settled into a good management job in Chicago. The family moved to the suburban community of Wheaton, where it remained for almost a decade. At the Wheaton Central School, which included all the grades, Edwin was a good student but no prodigy. Marks in the deportment column of the report card were not so admirable, but some stern conferences with John were the remedy for that. Taller than his classmates, Edwin used his size to excel in sports. He was practically a one-man track team; he was also a star in football at the tackle position, and in basketball at center. At the graduation ceremonies, the school superintendent sized up Edwin's promise and presumption: "Edwin Hubble, I have watched you for four years and I have never seen you study for more than ten minutes." Then, after a dramatic pause, he smiled and said, "Here is a scholarship to the University of Chicago."

As he entered the university in 1906, Edwin had two goals in mind: he would study astronomy and he would earn a Rhodes scholarship. He had apparently been dreaming of astronomy ever since the night with Grandfather James's telescope. The recently established Rhodes awards were granted to American students for postgraduate study at one of the Oxford colleges. Only one of these goals had parental approval. John would be proud if his son succeeded in the Rhodes competition, but a career in astronomy was out of the question; Edwin belonged in the legal profession.

John's objections notwithstanding, astronomy at the University of Chicago in the early 1900s was an excellent choice. The university was a new institution, but already it had a high-ranking physics department. Albert Michelson, who made the sophisticated measurements that confirmed the constancy of the speed of light, was there. So was Robert Millikan, who would soon accurately measure the electronic charge by following the motion of minute oil drops in applied electric fields. Both men received Nobel Prizes, Michelson in 1907, during Edwin's sophomore year, and Millikan in 1923. Also affiliated with the Chicago physics department, and no doubt a factor in Edwin's thinking, was the Yerkes Observatory, located in Williams Bay, Wisconsin. This facility had been founded in the late 1890s by George Ellery Hale, who would later be a major influence in Edwin's life. The Yerkes establishment housed one of the great telescopes of the time and also supported a large physical laboratory.

Edwin's sister Betsy recalled that during her brother's career at the University of Chicago he had "only one thought in his mind, and he wasn't going to let anyone else bother that." The thought was astronomy, but John remained an obstacle, as another sister, Helen, remembered: "Papa wouldn't have let him go through school if he was going to be a thing as outlandish as that." So the aspiring astronomer followed two tracks: he took the scientific courses he needed for graduate study in astronomy, and at the same time fulfilled the prerequisites for admission to law school.

The Hubble sisters may not have noticed, but Edwin had not lost sight of his other major objective: a Rhodes scholarship. Wiser than the other contestants, he anticipated the classical component of the qualifying examinations. "Study has been 'my middle name,' " he wrote Grandfather Hubble. "This summer I have taken nothing but Latin in preparing, as I am, for the Rhodes Scholarship Exams." The Rhodes examiners expected more than academic achievement. They evaluated the applicant's leadership potential, and looked with favor on athletic performance. Edwin had not been the star in sports at the university that he had been in high school, but he performed competently, often as a reserve, on a championship basketball team.

Edwin was the best in the Rhodes arena, however. He was voted the 1910 Rhodes scholarship for Illinois, entitling him to three years' study at an Oxford college in a field of his choice, with an annual stipend of fifteen hundred dollars. When a campus reporter asked him how he would direct his studies at Oxford, he had a prudent answer: "Although I have diverted most of my attention while at Chicago to the sciences, especially physics, I expect to take up law and international law at Oxford. Most excellent courses in both these subjects are offered in the English institution."

Rhodes Scholar

At Queen's College, Oxford (Edmund Halley's college), Edwin became an instant and lasting Anglophile. His letters were soon full of Oxford vernacular, and he tailored his accent, but not always with consistency. Some of his fellow Rhodes scholars were amused. One of them recalled: "We laughed at his effort to acquire an extreme English pronunciation while the rest of us tried to keep the pronunciation we brought from home. We always claimed that he could not be consistent, so that he might take a bäth in a băthtub."

Always intrigued by a new sport, Edwin took up rowing, and endured the traditional freshman apprenticeship. Still well proportioned, and now heavier, he was a good prospect for the Queen's rowing team, but a dislocated ankle at the wrong time put an end to his rowing career.

True to his word, Edwin pursued the law curriculum. He had hoped that his Chicago studies would be adequate for an exemption from the Law Preliminary Examination. The Warden of Queen's (the dean) was not impressed by all the physics courses, however, and refused Edwin's request. That meant three months of concentrated, disciplined preparation. As always when confronted with a challenge, he rose to the occasion. He earned his pass in the examination, but it was a trying experience, and it taught him a lesson he would not forget. He told Jennie about it in a letter (with attention to British spellings): "Labour which is labour and nothing else becomes an aversion. . . . Work, to be pleasant, must be toward some great end; an end so great that dreams of it, anticipation of it overcomes all aversion to labour. So until one has an end which he identifies with his whole life, work is hardly satisfactory." "Some great end": Jennie could guess that this would be the key to her son's motivation.

The Missourian was becoming a man of the world. He traveled to Europe when he could, and was drawn to Germany. German efficiency and military power impressed him, and he feared that if a European war broke out things would not go well for the English. He found a new athletic activity in Germany: affairs of honor settled by saber dueling. He horrified his parents with a sensational account of this "sport": "One must not move the head a hair's breadth—scarcely an eyelash during the whole fight—[lest] his whole cheek be laid open, his ear sliced off or his nose divided." The blood did not bother him in the duels he watched, but sometimes he thought the scars were misplaced.

John and Jennie did not hear about Edwin's friendship with Herbert Turner, Savilian Professor of Astronomy and director of the observatory at Oxford. Turner and his wife, Daisy, invited Edwin to their lovely home several times. On one occasion, he reverted to his Missouri manners, and charmed his hosts. "It's mighty good of you, ma'am, to have had me to lunch like this," he said to Daisy as he departed. After a dinner party including Edwin's remarkable presence, another guest commented to Daisy, "You said you had asked a Queen's undergraduate to dinner, but you never said he was an Adonis."

While Edwin was finding a new identity in England and on the Continent, his father's health was slipping. John was suffering from a disease of the kidney called Bright's disease, or nephritis. By the summer of 1912, the prognosis was total kidney failure in a few months. The news reached Edwin that fall, and he asked his father's permission to return home. John responded with the order that Edwin stay in England and finish his task. John Hubble died in mid-January 1913.

For Edwin, John's death was, as Gale Christianson puts it, "a deliverance as well as a blow." Edwin had been living under John's strict edicts, and his reaction now was to feel the lifting of that constraint. It was a time to think freely about "some great end," the destiny Jennie had heard about earlier. He fulfilled his obligations at Oxford, and departed, leaving behind this evaluation from the Warden of Queen's: "Considerable ability. Manly. Did quite well here. I didn't care [very] much for his manner—but he was better than his manner. Will get A."

The Major

To the astonishment of his younger sisters, Edwin arrived home—the family was now living in Louisville, Kentucky—wearing knickers, a wristwatch (an affectation for midwesterners at the time), and a ring on his little finger, sporting an Oxford cape and a cane, and communicating with a sometimes incomprehensible British accent. John had left his family with limited means. Edwin and his older brother Henry made matters worse by persuading their mother to invest in a business that ultimately failed. In later years, Hubble told friends that he had passed the Kentucky bar examination and briefly practiced law in Louisville. When asked about this later in an interview, one of Hubble's sisters, Helen, was incredulous: "Where did that information come from?" she asked. "He did not practice law."

Hubble did do a successful stint as a high school teacher and coach in the town of New Albany, Indiana, across the Ohio River from Louisville. His subjects were science, mathematics, and Spanish. Hubble's students were fascinated by his exotic mannerisms, and even more impressed by the performance of his basketball team in an undefeated season and a respectable third-place finish in the Indiana state championship tournament. They fondly dedicated their 1914 yearbook to him.

But neither teaching nor a law practice (if any) could meet Hubble's still fervent aspirations: no "great end" was yet in view. So he pinned his hopes once again on astronomy, and wrote to his former astronomy professor, Forest Ray Moulton, about opportunities for graduate study. Moulton recommended him to Edwin Frost, director of the University of Chicago's Yerkes Observatory, seventy-five miles north of Chicago. "Personally he is a man of the finest type," Moulton wrote. "Physically he is a splendid specimen. In his work here, altogether, and especially in science, he showed exceptional ability." Frost was losing talented astronomers to the new Mt. Wilson Observatory, near Pasadena in southern California, and he welcomed the new applicant.

Early in 1915 Hubble was granted "seeing time" on one of the Yerkes telescopes and embarked on his forty-year explorations of the night sky. He concentrated first (and, as it turned out, for most of his career) on photographic studies of the dim objects he called "faint nebulae." In modern usage, a nebula is a cloud of gas in space, usually originating in an expiring star. The objects observed by Hubble are what we now call "galaxies"; they are vastly distant, more or less independent, stellar systems containing tens or hundreds of billions of stars. Hubble never changed his terminology, partly because the term "galaxy" was favored by his archenemy, Harlow Shapley.

While Hubble was gathering data for his dissertation at the Yerkes Observatory, he was becoming increasingly aware of the rival Mt. Wilson Observatory in California. Under the directorship of George Ellery Hale (who had previously been the director at Yerkes), Mt. Wilson was rapidly becoming the world's leading observatory. It would soon have an immense reflector telescope with a hundred-inch mirror, almost twice the size of its nearest competitor at Yerkes. More astronomers would be needed to staff the new facility, and Hale offered Hubble a position, contingent on completion of the dissertation.

It was a perfect opportunity; in a normal world, Hubble would have accepted the offer and hastened to Pasadena. But the year was 1916, and normality was not the rule. General war had broken out, spreading devastation across Europe. Many of Hubble's Oxford friends were in the front lines, and some had been killed; the United States would soon enter the war. Hubble persuaded Hale to hold the Mt. Wilson position, hastily completed his dissertation, took his final oral examination (passing magna cum laude), and reported for reserve officer training in May 1917.

Hubble and his fellow trainees had only the sketchiest understanding of the trench warfare that was dragging on in Europe and slaughtering thousands of soldiers. Hubble wrote to Frost in high excitement (wayward spelling included): "This military game seems to be a nitch in which I fit. I was the fourth man to be made student captain . . . and am now an instructor in everything from bayonet work to signaling. This next Sunday I am chosen to represent the Company in a delegation to visit some model trenches." He was commissioned an infantry captain and given command of a battalion.

In September 1918, Major Hubble (recently promoted) and his command boarded a troopship and sailed for Europe. After a rough crossing, Hubble reached France in October and witnessed the final stages of the war. He told his wife later about combat duty during the German retreat. He was, he said, injured and knocked unconscious by an exploding shell. When he awoke in a field hospital, he dressed and departed without a word to anyone. The story may well be true: anything is possible in the chaos of war. But it is not corroborated by the record included in Hubble's discharge papers. On the line following "battles, engagements, skirmishes" the entry is "none."

After the armistice, Hubble lingered in England through the following summer. With astronomy on his mind again, he rented an apartment in Cambridge, where he met the esteemed astrophysicist Arthur Eddington and a wealthy astronomer, H. F. Newall, who proposed Hubble for membership in the Royal Astronomical Society. At a dinner honoring a visiting Mt. Wilson delegation, the American astronomers were amazed to see the thirty-year-old Hubble seated between two of Britain's most distinguished scientists: the physicist Arthur Schuster, and Frank Dyson, the Astronomer Royal.

Meanwhile, Hale was getting impatient. He wanted his young astronomer in Pasadena, not Cambridge. "Please come as soon as possible," he wrote, "as we expect to get the 100-inch telescope into commission very soon, and there should be abundant work by the time you arrive." Hubble's annual salary would be fifteen hundred dollars and he could expect promotion "as rapidly as your work and the funds at our disposal will warrant." Hubble was soon on his way. En route to Pasadena he stopped for one day in Chicago to see his mother and sisters (who had traveled from their current home in Madison, Wisconsin), and paid a courtesy call at the Lick Observatory near San Jose, Calfornia. Still in uniform and playing the military role, he introduced himself as Major Hubble. The Lick astronomers were duly impressed; they called him then and ever after "the Major."

On the Mountain

In September 1919, Hubble joined the select company of astronomers on the staff of the Mt. Wilson Observatory. The offices of the observatory were located in Pasadena and the telescopes with supporting facilities on nearby 5,714-foot Mt. Wilson in the San Gabriel Mountains. Hubble's first "runs" on the mountain used the ten-inch and sixty-inch telescopes. By Christmas Eve 1919, he was exposing his first photographic plates with the new hundred-inch Hooker telescope.

Life for the astronomers on the mountain seemed monastic. Access via the steep, narrow, switchback road was an adventure. Because a telescope cannot be warmer than its surroundings, the astronomer's vigil through a night of observing in the telescope's dome was not only lonely but sometimes bitterly cold. But there were amenities. The lodge where the astronomers stayed, aptly called "the Monastery," had comfortable, if spartan, rooms. There was a well-stocked library, and a sitting room with a fireplace and comfortable rocking chairs facing windows that looked out on a spectacular view of the valley. Hale had established a dinner ritual. The astronomers, each wearing a coat and tie, were seated in order of the importance of their assignments for the night's observing. At the head of the table sat the astronomer scheduled on the hundred-inch telescope and his night assistant, then those scheduled for the sixty-inch, and so forth.

Milton Humason, a man with little formal education, frequently served as Hubble's assistant and later became a respected astronomer in his own right. In a reminiscence, written after Hubble's death, he recalled the young astronomer in action:

> He was photographing at the 60-inch, standing while he did his guiding. His tall, vigorous figure, pipe in mouth, was clearly outlined against the sky. A brisk wind whipped his military trench coat around his body and occasionally blew sparks from his pipe into the darkness of the dome. "Seeing" that night was extremely poor on our Mount Wilson scale, but when Hubble came back from developing his plate in the dark room he was jubilant. "If this is a sample of poor seeing conditions," he said, "I shall always be able to get usable photographs from the Mount Wilson instruments." He was sure of himself—of what he wanted to do, and how to do it.

Hubble directed his observations, as he had before, at the faint "nebulae." What he wanted to do was measure the distances from Earth to these objects. And he knew how to do it, by taking advantage of the work of two of his predecessors, Henrietta Leavitt, a research assistant at the Harvard College Observatory, and Harlow Shapley, for a few years one of Hubble's colleagues at Mt. Wilson, and later director of the Harvard Observatory

In the early 1900s, Leavitt had made a series of studies of "Cepheid variable stars," whose brightness changes in predictable cycles. Her major discovery was that the period of a Cepheid's changing brightness was accurately related to the star's average intrinsic brightness: the brighter the star, the longer the period of its change from maximum to minimum brightness. When a Cepheid is observed, its apparent brightness, the image recorded by the telescope, depends on the intrinsic brightness, and also in a simple way on the distance to the star.

Shapley had developed these relationships into a cosmic yardstick. His method was to observe a Cepheid's apparent brightness, use Leavitt's results to calculate the star's intrinsic brightness, and then from the observed apparent brightness, calculate the distance to the star. In the late 1910s, Shapley applied his yardstick to the stars of the Milky Way galaxy and found the equatorial diameter of the galaxy to be about three hundred thousand light-years. (A light-year is a distance measure popular with astronomers. It is the distance traveled by a beam of light through empty space in one year, and is equivalent to 5.88 trillion miles.) This was more than ten times the previous estimate. Shapley's observations also placed the Sun at a distance of about sixty thousand light-years from the center of the galaxy, another drastic revision of previous results, which had located the Sun near the center of the galaxy.

Shapley's observations reached to the edge of the galaxy. In his first great discovery, Hubble looked beyond the galaxy. During the fall of 1923, he directed his attention to a large spiral nebula in the Andromeda constellation, called M31 by astronomers. He resolved stars in M31, and some of them were the useful Cepheid variables. Invoking the Leavitt-Shapley methodology, he calculated a distance of about a million light-years to the stars and to the nebula, certainly outside our galaxy.

Hubble's 1924 paper, "Cepheids in Spiral Nebulae," provided the first solid evidence for a speculation that some astronomers were already entertaining: that the most distant objects resolved by telescopes were "island universes," that is, independent galaxies, some as large as, or larger than, our own Milky Way stellar system. Hubble had focused on one of these external objects. He called it an "extragalactic spiral nebula," but it was nothing less than a complete galaxy similar to (and slightly larger than) our own.

Rants

Hubble's paper brought accolades from the astronomy community. Princeton's esteemed Henry Norris Russell pronounced Hubble's discovery "a beautiful piece of work" and saw to it that the paper earned a major award from the American Association for the Advancement of Science. But to Hubble's annoyance, there were dissenting voices. In particular, there was the work of the Dutch astronomer, Adriaan van Maanen, a Mt. Wilson colleague. Van Maanen had studied spiral galaxies (nebulae to Hubble) for years and claimed that his observations showed a kind of galactic rotational motion. When he combined his observed rates of rotation with Hubble's calculated distances, the calculation implied that some of the stars in the outer reaches of spiral galaxies had to be moving at incredible speeds, faster than the speed of light. His conclusion was that Hubble's measurements were wrong: the nebulae Hubble observed were inside, not outside, our galaxy. Never gracious under criticism, Hubble treated van Maanen with contempt, and van Maanen returned the favor. Shapley, a good friend of van Maanen's, took his friend's side in the controversy, and similarly earned Hubble's disdain.

Hubble's feud with van Maanen dragged on. Van Maanen would not retreat, and Hubble stubbornly searched for the error in van Maanen's analysis. Walter Adams, director of the Mt. Wilson Observatory, tried in vain to persuade Hubble and van Maanen to settle their differences like scientists and gentlemen. He admitted that it was "one of the most difficult problems with which the observatory has had to deal. . . . Due to the attitudes and temperaments of both men there was no cooperation in the matter but much feeling developed." The story is told that one evening on the mountain Hubble defied etiquette and displaced van Maanen, the hundred-inch observer that night, from his rightful place at the head of the dinner table. Finally, Hubble made a strong case that van Maanen was the victim of subtle systematic errors. Adams proposed that Hubble and van Maanen publish a joint paper that would put the controversy, now about a decade old, to rest. Van Maanen consented, Hubble refused, and Adams was exasperated. In a report to the president of the Carnegie Institution, a major source of Mt. Wilson funding, he wrote, "I do not feel that Hubble's attitude in this matter was in any way justified."

Hubble and Shapley remained implacable enemies for the rest of their lives. It was a battle of egos. Both were intensely ambitious and unforgiving rivals. Shapley had this snub for Hubble: "He was a Rhodes Scholar, and never lived it down." And Hubble's colleagues listened to tirades against Shapley. The Princeton astronomer Martin Schwarzschild said in an interview, "Hubble was the worst. I have suffered under a couple of sermons from him, ranting in the most unreasonable way against Shapley." And then he added, "Shapley was no angel either."

Redshifts

One aspect of Hubble's genius was his ability to take the incomplete work of others and carry it further with his own extensive and carefully planned observations. In one of his rare compliments for Hubble, Shapley said (more than a decade after Hubble's death), "Hubble, by the way, was an excellent observer, better than I. He was patient." We have seen Hubble's masterful use of the Leavitt-Shapley method for measuring cosmic distances. His greatest achievement, probably the most important by an astronomer in the twentieth century, was inspired by a series of observations made in the 1910s by a largely self-taught astronomer named Vesto Slipher at the Lowell Observatory in Flagstaff, Arizona.

Slipher worked with a large refracting telescope equipped with a spectroscope, a device that analyzes light by spreading it into its rainbow components. He discovered that characteristic spectral lines in light received from galaxies were slightly shifted in wavelength from what is observed for the same lines in an earthly laboratory. When he observed the M31 galaxy, for example, the shift was toward short wavelengths, a "blueshift" because blue is on the short-wavelength end of the visible spectrum. Most of the galaxies Slipher observed exhibited "redshifts," however, toward longer wavelengths. The terms "red" and "blue" do not denote actual colors, only the direction of the shift to longer or shorter wavelengths. Figure 27.1 illustrates the effect.

Slipher interpreted these wavelength shifts by referring to an effect first described in 1842 by Johann Doppler. The Doppler effect has several important manifestations, all connected with changes in wave behavior when the source and the observer of the waves move with respect to each other. In modern applications, a redshift of light represented by z is defined by

$$z = \frac{\lambda_o - \lambda_s}{\lambda_s}, \tag{1}$$

in which λ_s is the wavelength of the light as it is emitted by the source, and λ_o is the longer wavelength of the light seen by the observer. The actual change in wavelength is $\lambda_o - \lambda_s$, and z calculates the fractional change. For redshifts, the wavelength λ_o is greater than λ_s, and z in equation (1) is positive. Equation (1) also applies to blueshifts, which reverse the situation: λ_o is less than λ_s, and z is negative.

If the relative speed v between observer and light source is not large, the Doppler connection between the redshift (or blueshift) z and the relative speed v is

$$z = \frac{v}{c}, \tag{2}$$

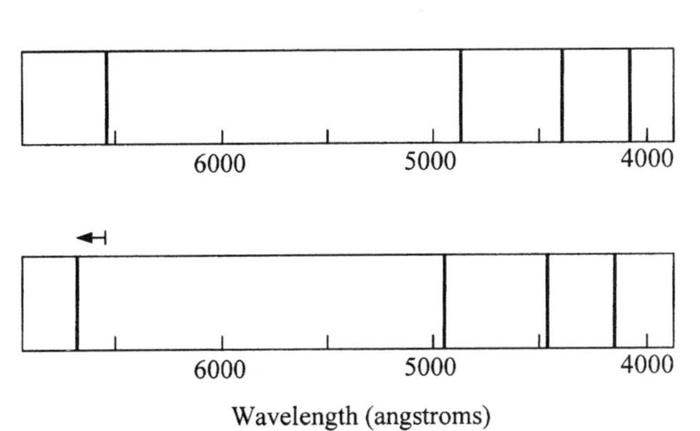

Figure 27.1. Wavelengths from the spectrum of light emitted by hydrogen. Above: as they appear normally. Below: the same wavelengths redshifted by about 2%, with $z = 0.02$. An angstrom is a very small measure of length, equal to 10^{-8} centimeter. Adapted with permission from John Hawley and Katherine Holcomb, *Foundations of Modern Cosmology* (New York: Oxford University Press, 1998), 263.

where c is the speed of light. A positive value of v calculated from this equation for a redshift suggests that the source of the light, say a galaxy, is moving away from the observing astronomer on Earth. Data reported by Slipher in 1914 and thereafter implied speeds of galactic recession from Earth as high as an astonishing 1,100 kilometers per second. Blueshifts—that is, negative values of z—are uncommon for galaxies; when they occur, they calculate negative relative speeds, meaning that the galaxy is approaching Earth.

In the late 1920s, Hubble got into the redshift business. Using Slipher's data at first, he plotted Doppler speeds of recession calculated with equation (2) against his own distance measurements. The plot showed that galactic recession speeds increase with distance: the more distant the galaxy, the faster it seems to move away from Earth. From inconclusive data published in 1929, Hubble guessed that there was a linear connection between the calculated speed of recession v of a galaxy and the distance l from Earth to the galaxy. The mathematical statement of this conjecture, as it is now written, is

$$v = Hl, \tag{3}$$

in which H is a constant, now known as "Hubble's constant." Written in terms of the redshift z, the equation is

$$z = \frac{H}{c}l. \tag{4}$$

By 1931, Hubble had better data and more confidence in the linear relation expressed by equation (3). But he was cautious about accepting the reality of the recession speeds. He preferred to call them "apparent" speeds, and consider the linear law to be an "empirical relation between observed data." Others saw the redshifts and the related recession speeds as evidence for a cosmology based on an expanding universe—the idea that the universe is continually expanding at a rate that increases in proportion to distance.

The expansion carries the galaxies with it, and the result is a (possibly never-ending) increase in distances among galaxies. Figure 27.2 gives an impression of the expansion, as it would appear if it occurred in a two-dimensional curved space rather than our own three-dimensional space. Note that all points move

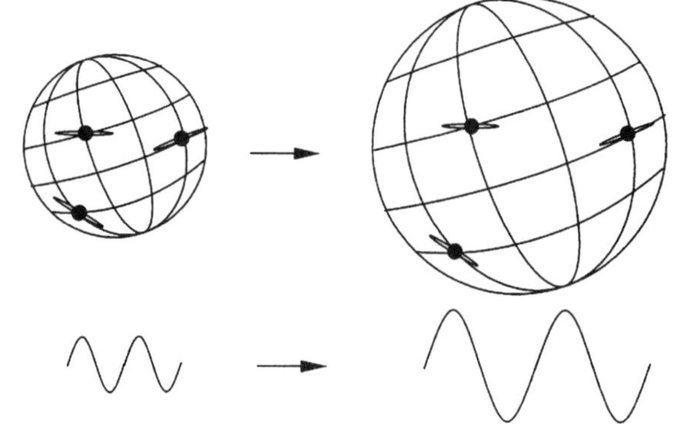

Figure 27.2. Above: an impression of an expanding universe as it might be seen in two-dimensional curved space. The two-dimensional expansion is like the inflating of a balloon with the galaxies pasted on its surface. Below: the corresponding increase in the wavelength of light. Adapted with permission from John Hawley and Katherine Holcombe, *Foundations of Modern Cosmology* (New York: Oxford University Press, 1998), 282.

away from one another, and that there is no center of expansion. The wavy lines in the bottom part of the figure represent schematically part of a light wave and its stretching (redshifting) in spacetime with the expanding universe.

If it is appropriate to say that the expansion is driven by a force, that force is weak compared to the gravitational force within the confines of the solar system or a galaxy. So, as suggested by the figure, the galaxies themselves retain their gravity-controlled shapes and do not necessarily expand.

Equation (2) has a shortcoming. The recession speed v, like other speeds, cannot exceed the speed of light c, and that seems to tell us that the redshift z cannot be greater than one. Yet redshifts approaching eight are now being reported. The trouble is that equation (2) does not recognize the requirements of special relativity. If the Doppler equation is made relativistic, it becomes

$$1 + z = \sqrt{\frac{1 + v/c}{1 - v/c}} \tag{5}$$

for a universe with no curvature. This equation allows z to increase to infinity. (As v approaches c, $1 - v/c$ in equation [5] becomes very small, and division by the small number gives a very large value for z.)

Astronomers look outward in space and backward in time. Because light travels at a finite speed, it takes a finite time, possibly a very long time, to travel from a star or galaxy to Earth. If the distance to a galaxy is eight billion light-years (a distance covered by the largest modern telescopes), the image of the galaxy recorded by the astronomer shows the galaxy not as it *is* now, but as it *was* eight billion years ago. And during the time a light ray was making its eight-billion-light-year journey, the universe was expanding. The observed redshift gives a simple measure of the extent of that expansion. Rearrange equation (1) to

$$z = \frac{\lambda_o}{\lambda_s} - 1,$$

or

$$z + 1 = \frac{\lambda_o}{\lambda_s}. \tag{6}$$

Suppose we find that the light mentioned has the redshift $z = 1$. Then the last equation calculates $\frac{\lambda_o}{\lambda_s} = 2$: the wavelength of the light doubled while it traveled from the galaxy to Earth, reflecting a twofold expansion of the universe during that time. The galaxy is finally zipping away from us on Earth at 180,000 kilometers per second.

Hubble and his talented assistant, Milton Humason, made many redshift observations following Hubble's initial analysis of Slipher's data. They extended the linear equation (3) to a distance of more than one hundred million light-years and gave it a permanent and prominent place on the list of nature's most fundamental physical laws.

Patrician

In February 1924, Edwin Hubble and Grace Burke Leib were married in a simple and private wedding (no members of Hubble's family attended). She was the daughter of a wealthy and influential Los Angeles banker, John Burke. She had been married before, twelve years earlier, to Earl Leib, who also came from a wealthy and socially prominent family. Leib was a geologist who specialized in assaying coal deposits. In June 1921, he was overcome by gas and killed while descending into a coal mine. A year later Hubble and Grace Leib were seeing each other regularly.

On their honeymoon, the couple traveled first to England and enjoyed a grand reception at Oxford and Cambridge. It was 1924, and Hubble had just announced his measurement of the distance to the M31 galaxy, giving astronomers their first clear view of the great realm beyond our galaxy. At Cambridge, the Hubbles stayed at the grand estate of H. F. Newall and lunched with Arthur Eddington, Britain's best-known astrophysicist. In London, Hubble was the guest of honor at a dinner held by the Royal Astronomical Society. He lectured on his cosmic distance measurements and his scheme for the classification of galaxies according to their shapes. From England, they traveled to Paris, Switzerland, and Italy. In Florence, they saw rooms in the Palazzo Vecchio that they promised themselves would be models for their dream house.

As he entered Grace's world of money and privilege, Hubble seems to have decided that his already secure and promising future as an astronomer was not impressive enough, and he deemed it necessary to tell his story with some major and minor inventions. His war record, career as a lawyer, athletic ability, and exploits in the north woods and in Europe all became greater than they really were. Gale Christianson writes, "With his family two thousand miles away, Hubble completed the long process of reinventing himself. His stories of heroism in the Wisconsin woods, of boxing the greats, of saving young women from drowning, of practicing law, and of leading frightened men into battle were dutifully recorded by Grace [in her journal] seemingly without the slightest question as to their validity. It was an impressive display which helped win her over, and her parents as well."

The Hollywood stars became almost as attractive to Hubble as the nebulae. The Hubbles counted among their friends, acquaintances, and neighbors many of the Hollywood elite, as well as others from the California intelligentsia. Hubble was appointed to a seat on the board of trustees of the Huntington Library and Art Gallery, which housed a vast collection of rare books and eighteenth-century paintings, and 250 acres of finely manicured gardens. Many other honors had previously come Hubble's way. When a *Los Angeles Times* journalist reported on the Huntington trusteeship, he doubted that Hubble could expect further honors because there were no more to give, "with the possible exception of a minor medal or two."

Perhaps because he was caught in the web of his inventions, Hubble felt that he had to separate himself from his family and his real past. His sister Betsy told Christianson that Hubble gave his mother, Jennie, little financial support and did not visit her during the last decade of her life. Grace never met Jennie or any of Hubble's brothers or sisters. "Whenever one of them came west, as Lucy had after her late marriage," Christianson writes, "Edwin arranged a meeting at his office or elsewhere in Pasadena, never offering to take his visitor home. The [Hubble sisters] thought that this intense desire for privacy was Grace's wish as much as her brother's."

In Hubble's absence, his younger brother Bill, a dairy farmer, became the head of the family. "Bill is really our 'unsung hero,' " one of the sisters, Helen, wrote in a letter. "Betsy and I felt that Bill is really indirectly responsible for Edwin's accomplishments. Bill gave [up] his *dreams* to do the mundane things of necessity."

The Constant

Hubble's most important legacy is embodied in the law expressed by equation (3), called "Hubble's law." The hallmark of the equation is the factor H, Hubble's constant. It is fundamental in cosmology because it expresses the rate of expansion of the universe. We can see that specifically by writing an equation for the expansion as expressed by the changing distance l between two galaxies, separated by, say, ten million light-years to make gravitational effects negligible,

$$l = l_0 R, \tag{7}$$

in which l_0 is the present separation between the galaxies, and R plays the role of a cosmic "scaling factor." At the present time, $R = 1$ and $l = l_0$, as assumed. At some time in the past, when the expansion was half what it is now, $R = 1/2$ and $l = l_0/2$. At some time in the future, when the expansion has doubled, we will have $R = 2$ and $l = 2l_0$.

The rate of the expansion, that is, its speed, is calculated as the time derivative $\frac{dR}{dt}$. Hubble's constant H is directly related to this rate, and takes us to the heart of the expansion problem. The connection is

$$H = \frac{1}{R}\frac{dR}{dt}. \tag{8}$$

(Proof of this equation is not difficult. Take time derivatives on both sides of equation [7]; note that $\frac{dl}{dt}$ is the same thing as the speed v in Hubble's law, equation [3]; then substitute for v from Hubble's law, and for l_0 from equation [7].)

As modern cosmologists reconstruct it, the cosmic expansion began with the entire universe crunched into an extremely small volume. At that time, the scaling factor R was equal to zero or close to it. Something like a big bang initiated the expansion from the original crunch, and as we see it today, distant galaxies are flying away from one another according to Hubble's law. That scenario suggests using Hubble's law to calculate how long it has taken for the universe to reach its present state of expansion. If time began with a big bang, this calculation gives us the age of the universe.

Consider any two galaxies separated by a distance l and receding from each other at the speed v. If the speed is constant, the time to achieve that separation is simply $\frac{l}{v}$. (To take a homely example, suppose two cars are driving away from each other at the constant relative speed $v = 50$ miles per hour. If they start at the same point, they are separated by $l = 100$ miles in $\frac{l}{v} = \frac{100}{50} = 2$ hours.) We see from Hubble's law, $v = Hl$, that in the cosmic realm the $\frac{l}{v}$ ratio is equal to the reciprocal of Hubble's constant,

$$\frac{l}{v} = \frac{1}{H}.$$

The time thus calculated, called the "Hubble time," and represented t_H,

$$t_H = \frac{1}{H}, \tag{9}$$

is an *estimate* of the age of the universe. It is not an accurate calculation because neither the speed of recession nor Hubble's constant is actually constant, as we have assumed. If the expansion accelerates, that is, if the speed increases with time, t_H underestimates the age of the universe; if the expansion decelerates, the speed decreasing with time, t_H overestimates the age of the universe.

For more than sixty years, astronomers have struggled with the task of measuring a reliable value of Hubble's constant. So far, they have met with only partial success because of the many difficulties inherent in measuring cosmic distances. Hubble's first measurement of his constant was in error by almost an order of magnitude because he did not realize that there is more than one kind of Cepheid variable star. In recent times, independent measurements of H have differed from each other by about 15 percent.

In the peculiar mixture of units favored by astronomers, a reliable average value of H seems to be about 65. That number requires two different units for distance: the v factor in Hubble's law is expressed in kilometers (per second) and the l factor in megaparsecs, another special astronomical distance unit (equal to 3.26 million light-years). When that violation of the rules learned in freshman physics for specifying units is repaired, the value of H calculates

$$t_H = \frac{1}{H} = 15 \text{ billion years},$$

as an approximate (apparently slightly high) age of the universe.

The Telescope

Hubble's final honor was posthumous: a great telescope was named after him. It is a space telescope that makes its observations with unprecedented clarity and wavelength range, because, from its vantage point in space, it avoids the distorting effects of Earth's atmosphere. The Hubble Space Telescope was launched in 1990. It contains a 94.5-inch primary mirror, cost about two billion dollars, and demands two hundred million dollars annually for maintenance costs.

For its first three years, the Hubble telescope threatened to be a spectacular failure. Due to an error in the manufacture of the mirror, unaccountably missed before the telescope was launched, the first images obtained were fuzzier than expected. A special shuttle mission in 1993 gave the telescope corrective optics and the vision originally desired. Since then, the Hubble telescope, like its namesake some fifty years earlier, has performed brilliantly.

A catalogue of the achievements of the Hubble telescope includes: close observation of a comet colliding with the planet Jupiter; observations suggesting that giant black holes are often located at the cores of galaxies; "deep field" views into space, penetrating more than twelve billion light-years and counting by extrapolation 120 billion galaxies in the observable universe, each containing many billions of stars; observations of collisions between galaxies; views of galaxies born less than a billion years after the big-bang event; calibration of standard distances used in calculating a value for the Hubble constant; and observations of distant supernovae (exploding stars), suggesting that expansion of the universe is accelerating.

No scientist has had a finer monument.

28

Ideal Scholar

Subrahmanyan Chandrasekhar

Birth and Death

Subrahmanyan Chandrasekhar, or "Chandra," as he was known to colleagues, friends, and relatives, often asked his wife, Lalitha, to sing a song in which the composer laments the cycles of births and deaths that follow him through life. Beginning with an odd incident provoked by Arthur Eddington, Britain's preeminent astrophysicist in the 1910s and 1920s, cycles of intellectual births and deaths became the pattern of Chandra's creative life in science. Lalitha describes this unique approach in a remembrance: "Each field, or cycle . . . took from ten to fifteen years, for the selection of the subject for investigation, study of the available scientific literature on the subject, his own research that followed, the scientific papers he wrote on the subject, and, finally, the way he gathered all the material that lay in front of him into a coherent whole that was the book on the subject."

When the book was completed, it was truly a death for Chandra: he had no more to say on the subject, refusing to spend time on the residue of minor issues remaining. "It was not in his spirit to pick up the crumbs," writes Lalitha. A "fallow period" would follow while he searched for a new field. It could be a frustrating and depressing time for him, and he would say to Lalitha, "Your friend [the composer], sing his song."

The exigencies of his life and career forced Chandra into several drastic cultural changes, each of which must have also seemed to him like a death and difficult birth. He was a native of southern India, and he said throughout his life that he felt at home only in India. Nevertheless, he left India at age nineteen and never returned except for visits. India could not offer him the graduate training he needed in astrophysics, and later afforded no suitable career opportunities for him. He went to Trinity College, Cambridge, for his graduate degree and a subsequent appointment as a Trinity fellow. The cold English climate, the bland English food, and the occasional eccentricities of the English were major and minor obstacles, but Chandra made many lasting friendships in England and would have stayed if there had been suitable job prospects. He was advised that there were none, so he moved on to a third culture, in America. First it was Harvard, then the Yerkes Observatory, and finally Chicago, where he and Lalitha remained. In America, Chandra built his unparalleled reputation as a theorist, mathematician, teacher, research adviser, editor, science historian, and storyteller.

Chandra saw himself as a man on a ladder, and kept a reminder of the metaphor on the wall of his office in a photograph taken by the artist Piero Borello. The picture shows a man who has half ascended a ladder leaning against a bright, almost featureless, but beautifully contoured wall. The man and the ladder, echoed by their shadows, seem strangely unequal to the task of ascending the wall. Chandra earned a copy of the photograph by explaining to the artist what the picture meant to him: "What impressed me about your picture was the extremely striking manner in which you visually portray one's inner feeling toward one's efforts at accomplishments; one is half-way up the ladder, but the few glimmerings of structure which one sees and to which one aspires are totally inaccessible, even if one were to climb to the top of the ladder. The realization of the absolute impossibility of achieving one's goals is only enhanced by the shadow giving one an even lowlier feeling of one's position."

These bleak words, quoted by Chandra's biographer Kameshwar Wali, provide no clue to Chandra's motivation. He simply was, in the most profound sense of the word, a scholar, "an ideal scholar of physics," Victor Weisskopf told Wali, "nothing of pushiness, nothing of job seeking, publicity seeking, or even recognition seeking. . . . His deep education, his humanistic approach . . . , his knowledge of world literature, and in particular English literature, are outstanding. I mean you'd hardly find [another] physicist or astronomer who is so deeply civilized." As a scholar, Chandra was "forever learning," remarked one of his students. "Chandra couldn't care one bit about the establishment. Everything he did was out of being curious in a productive way."

The scholar's challenge, as Chandra put it in his Nobel lecture, "is a quest after perspectives" in each chosen field. He meant by that simply "a view of my own." His urge was "to present my point of view *ab initio*, in a coherent account with order, form, and structure." He did so, repeatedly, over a period of about six decades.

Astrophysics involves the very big, the very small (in attempts to trace the history of the universe to its atomlike origins), and the very complex. Chandra had a grasp of the complexities of astrophysics that was unequaled by any of his contemporaries. In one of his cycles of study on a topic, he could assimilate the fundamentals in a field, assess their importance, build his own perspective, and express it in a comprehensive monograph. None of Chandra's colleagues in astrophysics, or in the broader physics community, could do so much.

From Madras

"Was your father a dominating influence in your life?" the journalist Vatsala Vedantam once asked Chandra. "All Indian fathers are dominating," he answered with a laugh. Wali gives us this sketch of the father, C. S. Ayyar: "A highly cultivated individual, widely read and traveled, still accepted certain customs and practices. And when it came to family matters, he was traditional and authoritarian, demanding unquestioned obedience from everyone, very much a father of his generation. Reserved and undemonstrative, he remained aloof from his children."

Ayyar was an accountant who worked in the British government service, ultimately achieving the title of chief auditor. His duties took him to the offices of most of the important British railroad companies in India. At the time of Chandra's birth in 1910, the family was living in Lahore, where Ayyar served as assistant auditor general for the Northwest Railways. Lahore was distant geographically and culturally from Ayyar's Tamil background in southeastern India, and when the opportunity arose, he established his growing family in Madras, on the southeast coast, while he traveled to his various postings. The Ayyars' home, called "Chandra Vilas," was built while Chandra was in high school; it was spacious, comfortable, and situated in an upper-middle-class suburb of Madras.

Chandra owed as much to his mother as to his father. Sitalakshmi was a strong-willed woman who bore ten children, and held her own in her husband's large family. Chandra was the oldest son. He had two older sisters, three younger brothers, and four younger sisters. Sitalakshmi's formal education was limited, but she managed nevertheless to learn English and translate Ibsen's play *A Doll's House* (a curious selection: the main character in the play, Nora Helmer, leaves her husband) into Tamil. She supported her daughters' desires for advanced education, and opposed her husband in the matter of Chandra's career decision. "You should do what you like," she told Chandra. "Don't listen to him, don't be intimidated."

Chandra's original choice was mathematics. He was fascinated by the career of Srinivasa Ramanujan. Beginning in near poverty, and with little advanced training in mathematics, Ramanujan published some papers on number theory that impressed G. H. Hardy, a leading Oxford mathematician. Hardy and his Cambridge colleague J. E. Littlewood brought Ramanujan to England through a fellowship at Trinity College, Cambridge. For three years, Hardy and Ramanujan collaborated on an important series of papers. But, as Chandra would learn later, the English climate is not friendly to transplanted Indians. Ramanujan fell ill, possibly with tuberculosis, returned to Madras, and died there at age thirty-three. Chandra remembered that when he was ten years old his mother told him about Ramanujan's brief career and tragic death.

The Ramanujan allure was not enough to convince C. S. Ayyar that mathematics was a suitable career for his son: he insisted on physics. Probably he had in mind the spectacular success of C. V. Raman, his brother and Chandra's uncle, who had in 1928 discovered the physical effect now known to physicists and physical chemists as the Raman effect. It is the scattering of monochromatic light (of a definite wavelength) by a transparent substance. Scattering data, complemented by data obtained from transmitted light, are often revealing about the shapes of the molecules interacting with the light. Raman was knighted for his work and in 1930 received a Nobel Prize.

Ayyar's plan was for Chandra to obtain the B.A. physics honors degree and then go to England to take the Indian civil-service examination. Success in the examination would guarantee a secure job in government service. Chandra agreed to the physics studies, but emphatically not to a civil-service career. "The two scientists' names I knew were Ramanujan and Raman, and to some extent they were my role models," he told Wali. Both had followed the path of pure research, and with support from his mother, that was the course Chandra took.

Chandra's talent was, no doubt, that of a prodigy. When he was eighteen and still an undergraduate at the Presidency College in Madras, he wrote a paper that got the attention of Ralph Fowler, the principal theorist at the Cavendish Laboratory in Cambridge, and Rutherford's son-in-law. The paper originated in a visit to Madras in 1928 by the German teacher and theorist Arnold Sommerfeld. Chandra went to see Sommerfeld at his hotel, hoping to make an impression with his thorough knowledge of Sommerfeld's book, *Atomic Structure and Spectral Lines.* But Sommerfeld had discouraging news: "He promptly told me that the whole of physics had been transformed after the book had been written," Chandra recalled. "[He] referred to the discovery of wave mechanics by Schrödinger, and the new developments due to Heisenberg, Dirac, Pauli, and others. I must have appeared somewhat crestfallen. So he asked me, what else did I know? I told him I had studied some statistical mechanics. He said, 'Well there have been changes in statistical mechanics too,' and he gave me galley proofs of his paper on the electron theory of metals, which had not yet been published."

Sommerfeld's paper applied the statistical method that Fermi had introduced and Dirac had generalized. Chandra quickly grasped the meaning and importance of the new statistics, and with no advice or assistance from his teachers, found an application of his own, which he developed in a paper. In one of those providential events that shape a career, he sent the paper to Fowler, whose work he had seen in *Monthly Notices of the Royal Astronomical Society.* Fowler had brought the Fermi-Dirac statistics into the field of astrophysics by developing a model of the elderly stars called "white dwarfs," which have run out of nuclear fuel and collapse to about the size of Earth. Fowler and a colleague, Nevill Mott, read Chandra's paper, recommended some changes in style, which Chandra easily made, and saw to it that the paper was published in the prestigious *Proceedings of the Royal Society.* It was, to say the least, an impressive achievement for an eighteen-year-old, unassisted undergraduate. Important people took notice, and Chandra was offered a special scholarship that would, after graduation, allow him to study and continue his research in England. But it was an unhappy time for Chandra to leave India. His mother's health was declining, and Chandra feared that if he went to England he would never see her again. Sitalakshmi herself made the painful choice. "You must go. You must pursue your own ideals to the utmost," she told him. "He is born for the world, not for me," she said to others.

To Cambridge

Chandra left India from Bombay on a sultry day in July 1930. For several days, the ship was slowed by bad weather, and Chandra was overwhelmed by seasickness. When calm weather and a settled stomach returned, his thoughts turned to physics, particularly to the strange stellar objects called white dwarfs that Fowler had studied. They have the *mass* of an ordinary star like the Sun, but their collapsed *size* is more like that of Earth. The result is that the internal stellar material has an immense density (mass per unit volume), far greater than that of any material on Earth. It occurred to Chandra that this condition placed a restriction on white-dwarf physics: the star must be relativistic—that is, its material must obey the dictates of Einstein's special relativity theory.

Chandra also thought about the physical condition that allowed white dwarfs to maintain their small size and not collapse further under the force of gravitation. According to Fowler's analysis, Pauli's exclusion principle, as elaborated by Fermi and Dirac, was crucial. It insisted that no two electrons could be squeezed close enough to each other to occupy the same state, and the result is a special electronic pressure that counters the gravitational force. The ultimate compression, whose pressure cannot be exceeded, is called a condition of "degeneracy."

Equipped with just three books among his shipboard belongings, Chandra set out to construct a relativistic version of Fowler's theory, and he came to an unanticipated conclusion: there is a limit, later called the "Chandrasekhar limit," to the mass of a star that can evolve into a white dwarf. For a star whose mass is more than about 1.4 times that of the Sun, the electronic pressure is not enough to counter the gravitational pull causing the star to collapse, and there is no mechanism for the star to pass through the white-dwarf phase before it dies. As Chandra put it later in a paper: "The life-history of a star of small mass must be essentially different from the life-history of a star with large mass. For a star of small mass the natural white-dwarf stage is an initial step towards complete extinction. A star of large mass cannot pass into the white-dwarf stage and one is left speculating on other possibilities."

The "other possibilities" did call for speculation, the wildest kind of speculation. A dying star of large mass evidently collapsed into an object even smaller, and more fantastically dense, than a white dwarf. Chandra might have entertained the weirdest possibility of all: that such a star collapsed to an ultimate condition of conceivably infinite density that allowed nothing to escape from its vicinity, not even light. If so, he was wise enough to say nothing about it. The concept of a mass limit for white-dwarf formation was enough to plunge him into a painful controversy.

When Chandra arrived in London, he was unimpressed by the sights and immediately confronted with a tangle of bureaucratic red tape. He wanted to enroll as a research student at Cambridge under Fowler, but the office of the High Commissioner for India, responsible for considering his case, was confused, uncooperative, and even insulting. Chandra wrote to his father, "I wish I had not come at all and had refused the scholarship." But, as always, he persisted, and finally his luck turned: a personal letter from Fowler gained him admission to Trinity College. He then marveled at his good fortune in another letter to his father: "I have got admission purely due to the accident that I happened to know Fowler for the last two years. Why I should have written then to Fowler, God alone knows. I suppose that was because Fowler was to help me two years later!"

Chandra found the Cambridge experience both inspiring and depressing. He recalled that "[it] was a shattering experience . . . suddenly finding myself with people like Dirac, Fowler, and Eddington, and living in a society altogether disconnected from me." Always a strict vegetarian, he resolved to "tell bold-facedly and *honestly* that it is not only possible to be a vegetarian in England for a stretch of three years, but that I have actually been one." His diet consisted mainly of potatoes spiced with chutney powders sent from home, bread and butter, and cornflakes.

Chandra's intellectual diet was richer and more stimulating. His menu of classes included Dirac on quantum mechanics, Fowler on statistical mechanics, Littlewood on function theory, and Eddington on relativity theory. At first, Chandra was not impressed by Dirac, "a lean, meek, shy young 'fellow' (FRS) who goes slyly along the streets. He walks quite close to the walls (as if like a thief!) and is not at all healthy. (A contrast to Mr. Fowler—a strong, 'big,' healthy, middle-aged man, quite happy, full of joy of life)." But Fowler was a hard man to approach, while Dirac became Chandra's mentor and a good friend. "He was very human, extremely cordial to me in a personal way," Chandra told Wali. "Even though he was not much interested in what I was doing, he used to have me for tea in his rooms at St. Johns about once a month. He also came to my rooms for tea, and some Sundays, used to drive me out to fields outside Cambridge where we used to go for long [mostly silent] walks on the Roman road." It was a meeting of minds between two gifted, reticent men.

For a time, Chandra considered switching from astrophysics to pure theoretical physics. To test the waters of contemporary theoretical work, he spent the summer of 1931 at Max Born's institute in Göttingen and the winter of 1932–33 at Bohr's in Copenhagen. Both Born and Bohr were extremely busy and Chandra saw little of them. But he made many friends among the young, freewheeling theorists who were building the great edifice of quantum mechanics. Later he would draw on those friendships. In Copenhagen, he worked on a problem Dirac had given him, and he optimistically thought he had found a solution that was "not altogether trivial." He wrote a paper and asked Bohr and Dirac for comments. Bohr approved and communicated the paper to the *Proceedings of the Royal Society*. But Chandra's cheerful mood was crushed when Dirac sent a note pointing out a fundamental error. Chandra withdrew the paper and reluctantly (but fortunately) returned to astrophysics.

Back in Cambridge, Chandra faced his doctoral oral examination. It was an informal affair. The examiners were Fowler (who had not bothered to read the thesis) and Eddington. After questions from Fowler, and objections from Eddington, then questions from Eddington, and objections from Fowler, the examination ended abruptly when Fowler looked at his watch, exclaimed, "Good heavens, I am late," and dashed out. Eddington then said, "That is all," without telling Chandra whether he had passed or failed. (He passed.)

With Ph.D. in hand, and his scholarship money running out, Chandra contemplated the future. With little hope for success, he took the examinations for a fellowship at Trinity College. A Trinity fellowship would give him four more years in England, free rooms in the college, dining privileges, and an annual allowance of three hundred pounds. Fowler thought his chances were slim at best; the only other Indian to become a Trinity fellow was Ramanujan, and his case was a special one. More realistically, Chandra planned a short stay in Oxford to work with Edward Milne, a young astrophysicist who had become a close friend and mentor. Chandra rented a room in Oxford, packed his belongings, and in a taxi on the way to the train station, decided to stop at the college and look at the list of candidates who had been elected fellows. To his complete amazement, his name was on the list. "This is it," he said to himself. "This changes my life."

Stellar Buffoonery

Chandra's life was about to change in other ways. Throughout his stay in Cambridge he had been thinking about the evolution of stars and his strange conclusion that stars of large mass were not permitted to end their lives in the way everyone at the time believed was standard, as white dwarfs. Chandra prepared a short paper on his theory and it was published in the *Astrophysical Journal* (a journal Chandra would later edit). Milne objected to some key approximations in the paper, and that prompted Chandra to develop an exact theory of white dwarfs. This work was started in 1934, after Chandra had settled into his fellowship. Eddington was curious about the work. "He took a great deal of interest in the day-to-day progress of my work," Chandra remembered. "He even got me the only calculator . . . that was around. . . . During the three months from October through December, Eddington came to my rooms quite often, at least once, sometimes twice or three times, a week."

By the end of 1934, Chandra had completed the exact theory, and arranged to present a summary of it at a meeting of the Royal Astronomical Society in London. When he looked at a program for the meeting, he noticed that immediately following his own paper Eddington was scheduled to give a lecture with the title "Relativistic Degeneracy." (Relativistic degeneracy is the technical term for the condition that in Chandra's theory leads to the white-dwarf mass limit.) "I was really very annoyed," Chandra recalled much later, "because here Eddington was coming to see me practically every day and he never told me he was giving a paper." At the tea before the meeting, Chandra was conversing with a friend, William McCrea, when Eddington joined them. "Well, Professor Eddington, what are we to understand by 'Relativistic Degeneracy'?" McCrea asked. Eddington looked at Chandra, said "That's a surprise for you," and walked away.

Chandra gave his talk and Milne added a brief comment. Then Eddington was introduced, and with his usual sarcastic wit, he quickly got to the point: "I do not know whether I shall escape from this meeting alive, but the point of my paper is that there is no such thing as relativistic degeneracy!" He summarized Chandra's position by saying that "a star of mass greater than a certain limit M remains a perfect gas and can never cool down. The star has to go on radiating and radiating and contracting and contracting until, I suppose, it gets to a few [kilometers'] radius, when gravity becomes strong enough to hold in the radiation, and the star can at last find peace."

"Dr. Chandrasekhar had got this result before," Eddington continued, "but he has rubbed it in, in his last paper; and, when discussing it with him, I felt driven to the conclusion that this was almost a *reductio ad absurdum* of the relativistic degeneracy formula. Various accidents may intervene to save the star, but I want more protection than that. I think there should be a law of Nature to prevent a star from behaving in this absurd way!" When Eddington finished, the president of the meeting hastily announced that "the arguments of this paper will need to be very carefully weighed before we can discuss it." Chandra was left silent, humiliated, and completely baffled.

After he recovered from the initial shock, Chandra began to mount a counterattack. He wrote to Léon Rosenfeld, a friend from Copenhagen who was Bohr's assistant, relating the Eddington incident. Rosenfeld responded that neither he nor Bohr could make sense of Eddington's remarks. Rosenfeld advised Chandra that his argument was correct, to "cheer up," and not worry so much about the "high priests." But Chandra could not let the matter rest. He had several conversations with Eddington that revealed little except that Eddington was relying on a distinctly unconventional view of the exclusion principle.

Chandra got informal support from Bohr, Fowler, Dirac, and Pauli, who were all mystified by Eddington's arguments. Eddington continued with his attacks on Chandra's theory, while maintaining cordial personal relations with Chandra himself. It was all very odd. In one of his last pronouncements, Eddington called Chandra's version of stellar evolution "stellar buffoonery."

Although Chandra had no doubt that he was right, he never got what he really wanted, a public statement of support from an authority in the physics community such as Bohr, Dirac, or Pauli. They were willing to give Chandra their assurances in private, but not to take on Eddington in formal debate. "It is quite an astonishing fact," Chandra told Wali, "that someone like Eddington could have such an incredible authority which everyone believed in, and it is an incredible fact that in the framework of astronomy there were not people who had boldness enough and understanding enough to come out and say Eddington was wrong. I don't think in the entire astronomy literature you will find a single sentence to say Eddington was wrong. Not only that, I don't think it is an accident that no astronomical medal I have received mentions my work on white dwarfs." It was a hard lesson in the sociology of science, or as Chandra put it, "That was protocol."

Cut off in this way from his white-dwarf theory, Chandra had no choice but to drop it altogether and to turn to another field; the Chandrasekhar mass limit was not generally accepted among astronomers for another three decades. But the incident had a surprisingly beneficial effect on Chandra. Forced to turn to a new topic (stellar structure), he discovered that he was intellectually suited to periodic changes in his fields of study. Thanks to Eddington and his stubborn denial of the white-dwarf mass limit, Chandra found his unique "birth and death" approach to scientific research.

Williams Bay

Chandra liked to tell the story of his life in two sentences: "I left India and went to England in 1930. I returned to India in 1936 and married a girl who had been waiting for six years, came to Chicago, and lived happily thereafter." Our narrative comes now to the "girl," Lalitha, and begins the long American chapter of Chandra's story.

Lalitha and Chandra were classmates in the physics department at the Presidency College in Madras. "He was one year senior to me," Lalitha recalls. "Some of the classes were common for both of us. I used to sit in the front row. Immediately behind me was Chandra. I knew his presence and he knew mine. In this way a friendship arose." While Chandra was in England, Lalitha completed her master's degree in physics and became headmistress of a school in Karaikkudi. By the fall of 1934, Chandra and Lalitha had in their correspondence reached a "mutual understanding." Chandra's father was delighted. He invited Lalitha to dinner at Chandra Vilas, and found her to be "a modest, quite reserved young lady." Ayyar fervently hoped that the marriage would bring his son back to India permanently. That plan failed, and for a while, so did the engagement. Chandra had not seen Lalitha since their college years, and he began to doubt the wisdom of asking her to accept all the uncertainties demanded by his career, probably including an extended period of living abroad. So they came to a "new mutual understanding": their commitment could wait until they had a chance to meet and discuss the future.

The only serious job opportunity for Chandra in India was an assistant professorship at the Indian Institute of Science in Bangalore, offered by Chandra's uncle, C. V. Raman. But Chandra was wary. He did not admire Raman's flamboyant style as a scientist. "While Chandra respected Raman's brilliance in physics," Wali writes, "Raman was not a role model for Chandra. Because Raman was given to sensationalism and reveling in controversies, and prone to speak in contradictory terms, he annoyed Chandra." Father and son were in agreement: "MY ADVICE KEEP OFF HIS ORBIT," Ayyar cabled Chandra.

In 1935, Chandra accepted an invitation from Harlow Shapley to go to the American Cambridge and lecture at Harvard on "cosmic physics." The lectures were a success, and Shapley offered him an attractive fellowship. At about the same time, Otto Struve, director of the University of Chicago's Yerkes Observatory, offered him a research associateship. Both Eddington and Milne gave Chandra the advice he had already given himself: to accept the position at Yerkes, one of the world's leading observatories.

With that much of his future settled, Chandra decided it was time to return to India to see his family and the patient Lalitha. In July 1936, exactly six years after his departure from India, Chandra sailed for Bombay. Lalitha met him in Madras, and they quickly found that their love for each other was stronger than ever. "Chandra's earlier decision to postpone his marriage indefinitely wilted away rather suddenly when he saw Lalitha again after six years," Wali tells us. "She was more than a dream, she was quite real. There was not even the slightest uncertainty regarding their mutual feelings. If they were ever to marry, it would be to each other and to nobody else. Lalitha shared Chandra's dedication to science. He became convinced that she would be a help rather than a hindrance to his single-minded pursuit."

Chandra and Lalitha were married in September 1936. It was a "love marriage," not arranged by the families, then and now a rarity in India. The couple sailed from Bombay in October, destined for a brief stay in England, and then to Williams Bay, Wisconsin. Chandra and Lalitha made Williams Bay and the Yerkes Observatory their home for twenty-seven years.

Chandra's presence at Yerkes was unique. He was primarily a theorist in astrophysics, while the Yerkes staff consisted mainly of astronomers, whose work was observational. His main task, in addition to research, was to develop a graduate program in astronomy and astrophysics. Chandra and Gerard Kuiper, another recent addition to the Yerkes staff, put together a scheme of eighteen courses, covering stellar atmospheres and interiors, stellar dynamics, solar and stellar spectroscopy, solar systems, and atomic physics. Of these courses, Chandra taught twelve or thirteen, one or two each quarter. According to Martin Schwarzschild, another astrophysicist and a Yerkes visitor, "Yerkes became a leading institution in every respect, including the development of one of the most outstanding, if not *the* outstanding graduate school in astronomy and astrophysics in the country. . . . Chandra was by far the most active member of the group. He just loved to give lectures and was very demanding of his students, many of whom felt enormous loyalty to him."

Chandra's energy and commitments seemed boundless. In addition to the teaching, he conducted weekly colloquia, attracted research students from all over the world, and periodically published his trademark authoritative monographs. In his first year at Yerkes, he wrote six research papers and the manuscript of his first book, *An Introduction to the Study of Stellar Structure.*

From stellar structure he turned to stellar dynamics, and then to the subject called radiative transfer, which for astrophysicists means the transport of energy by photons in star interiors. This was Chandra's favorite subject. "My research on radiative transfer gave me the most satisfaction," he told Wali. "I worked on it for five years, and the subject, I felt, developed on its own initiative and momentum. Problems arose one by one, each more complex and difficult than the previous one, and they were solved. The whole subject attained an elegance and beauty which I do not find to the same degree in any of my other work."

In the late 1930s, war broke out in Europe, "dispersing all values," as Chandra wrote to his father. Indians debated whether or not to support the British war effort, an issue that became more urgent when Japan entered the conflict with an attack on Pearl Harbor in December 1941. The Indian National Congress Party demanded a price for its support: a guarantee that India would have full independence after the war. No such agreement could be negotiated, however, and the Congress Party passed a resolution calling upon the British to "quit India" immediately, threatening civil disobedience for noncompliance. The British response was to arrest and imprison the Congress Party leaders, including Mohandas Gandhi and Jawaharlal Nehru, and that triggered uprisings all over India. Chandra thought that India should take the British side simply because the alternative was far worse, but he deplored the treatment of Gandhi, "the greatest man of our times."

The Pearl Harbor attack came while Chandra and Lalitha were visiting the Institute for Advanced Study in Princeton. Some of the Princeton scientists were contributing to the war effort, and Chandra followed suit by joining a group working on the theory of ballistics at the army's Aberdeen Proving Ground in Maryland. The work was interesting; the dense, hot gases in the explosion chamber of a gun are physically similar to those in the interior of a star. But Aberdeen was rural and racist, more so than rural Wisconsin, and Chandra did not want to ask Lalitha to cope with southern-style segregationist attitudes. From early 1943 to the end of the war in 1945, Chandra was a commuter—three weeks in Aberdeen and then three weeks at Yerkes. "It was pretty strenuous," Chandra recalled. "But the entire scientific community was behind the war effort. No two opinions as in the case of Vietnam. I didn't mind the strain."

During the 1940s, Chandra climbed the academic rungs of his metaphorical ladder. He was made an associate professor in 1942 and a full professor in 1943. He still held Indian citizenship, and in 1944 he joined the scientific elite when he was elected a fellow of the Royal Society. Despite all the earlier events, Chandra's friendship with Eddington had not been seriously damaged, and Milne reported that Eddington had supported Chandra's election to the Royal Society, "largely because of the way you have encouraged and stimulated theoretical astrophysics in America."

Chandra's fame was spreading beyond the Yerkes and Chicago academic communities. The preeminent American astronomer Henry Norris Russell retired from Princeton, and Chandra was offered a research professorship as Russell's successor. Chandra accepted, but changed his mind when the president of the University of Chicago, Robert Hutchins, who was a persuasive man and always one of Chandra's champions at Chicago, asked Chandra if Chicago was failing him in the building of his career. "If there is nothing lacking," he said, "then you should stay." If he went to Princeton, Hutchins said, the honor of succeeding Russell might be disappointing: "It is far more honorable to leave a professorship to which it is honorable to succeed than to succeed to an honorable position." Hutchins wondered if Chandra remembered who succeeded Kelvin after more than fifty years at the University of Glasgow.

An unfortunate series of events in 1952 separated Chandra from the Yerkes Observatory, first intellectually and then geographically. The trouble started when Chandra criticized the administrative abilities of Bengt Strömgren, the Yerkes director. Strömgren had been a friend for many years, and Chandra thought his advice would be taken without affront. It was not, and soon thereafter a committee appointed by Strömgren changed the graduate curriculum from the direction Chandra had been building for the last fifteen years. At a faculty meeting, Chandra told the members of the department that they had a right to make curriculum alterations, but he wanted it understood that "to the extent that I have had no role in revising the curriculum nor been consulted, I retain for myself the right to find a place in the university outside the astronomy department if I choose."

Chandra turned to Chicago's physics department, and once again the ideal scholar found a silver lining in an enforced change in the direction of his career. "I think, on the whole, this experience in the early 1950s did as much good for my science if not more than my earlier episode with Eddington," he said to Wali, "because it made me associate with people like Fermi and Gregor Wentzel, whom I could not have close contact with if I had stayed at Yerkes. I set up an experimental laboratory in hydromagnetism with Sam Allison. I taught all the standard courses in physics, quantum mechanics, electrodynamics, etc. I was the first one to teach relativity at the University of Chicago, which of course led me to research in relativity."

Chandra still had commitments to research students at Yerkes, so he and Lalitha continued to live in Williams Bay until 1964, when they moved permanently to Chicago. Before that, on the days his physics classes met, Chandra commuted to the city. Relations with his Yerkes colleagues were strained, and Chandra became vulnerable to bitter thoughts of preferential treatment given to others at Yerkes with less professional prestige. "The incredible fact is that in earlier years I was not even aware that something impolite, something improper had been done to me," he told Wali. "But I am afraid, up to a point, I was largely responsible, because people began to take me for granted, to treat me any way they liked, and I let them."

Editor

Chandra took on many burdens while he was at Yerkes, but the heaviest and most prolonged was the position of managing editor of the *Astrophysical Journal*, published by the University of Chicago. He unexpectedly got the job after a dispute with William Morgan, the previous managing editor. Morgan resigned, and Chandra, who had been associate editor for eight years, was the only one who could take over the management responsibilities. He did not want the job, but he took it and kept it for nineteen years, from 1952 to 1971.

Chandra's administration of the journal was autocratic but scrupulously fair. "He imposed upon himself an isolation from the astronomical community in order to be fair and without prejudice for or against particular individuals," Wali tells us. "He thus rejected invitations to conferences and symposia—opportunities to travel and socialize." Fermi asked him, "Why? Why do you do this?" He did not have a good answer. Later he acknowledged that "it was a mistake, a distortion of my personal life. I had no idea I would keep it for so long when I took it. I had no choice then."

Chandra's competence and objectivity were not always appreciated. He sometimes antagonized referees by disregarding negative but unsupported reviews. And authors had some strong opinions about referees. Here are a few of them:

> I consider that all the referees' comments are unimportant or sniping.
>
> You have selected a referee who is evidently not a disinterested person.
>
> The referees have not only demonstrated an incredible ignorance of the literature basic to the development of the field, but have also attempted to pad out an incompetent review with well-known material developed by the authors themselves, with irrelevant comments, and fatuous personal attacks.

Chandra took it all with remarkable equanimity, and the journal thrived under his leadership. The managing editor of the University of Chicago's journal publications expressed his gratitude in a letter to Chandra:

> You are the splendid steward of intellectual assets and your responsible exercise of these duties is demonstrated in every way, greater income, greater circulation, greater volume in pages, and all with increasing surplus.
>
> You must run a school for other editors, when you retire!

Miracles Not Welcome

Chandra was a *mathematical* physicist, perhaps more so than any of the other physicists in these chapters, with the exception of Newton. All physicists *use* mathematics, but few qualify as both mathematicians and physicists, as Chandra (and Newton) did. Even Einstein, whose creative use of the mathematics of differential geometry in arriving at his gravitational field equations was supreme, lacked the mathematical skill to find some of the important solutions to the equations. Chandra was sure that Newton would have done better.

For Chandra, mathematics was nature's language. "He talked to these equations personally and intimately till they gave up their secrets to him," as one colleague puts it. Unlike many other great physicists, Chandra had more faith in the mathematical message than in his physical intuition. And if at all possible, the mathematical account had to be clear, complete, and exact. No approximations (except as a last resort). No magic. "Miracles were not welcome, only clarity and perfection," his friend Rafael Sorkin says in a reminiscence. As another colleague put it, "Rather than being interested in new laws of Nature, Chandra strove to produce exact (and in general analytical) solutions to specific problems."

When Chandra started on a research problem, he could not let it go until he found the best solution. Sometimes, when progress was blocked, he needed inspiration from a sympathetic colleague. His favorite muse was the Oxford theorist Roger Penrose. "Whenever I meet a stumbling block," he told a friend, "I go and meet Roger Penrose." Chandra was in Chicago and Penrose in Oxford, so the meetings required transatlantic plane trips on Chandra's part, and the visits were short. "We spend an hour together in the morning when I present him with my problem. We have four or five hours of discussion after lunch," Chandra said. "Then at dinner we talk of other things—and I fly back." These "lightning visits," as Penrose called them, were always beneficial for Chandra: "In no case has he [Penrose] not cleared up my doubts in physics or mathematics. An amazing man."

Chandra did not like to leave loose ends in his work before he moved on to a new field, but there was one he could not avoid. Thanks to Eddington's opposition and Chandra's failure to muster the kind of support he needed in the physics community, Chandra had to turn his back on an intriguing question raised by his theory of white dwarfs: if a star is too massive to end its days as a white dwarf, what *is* its fate?

One answer was provided in 1939 by Robert Oppenheimer and his student George Volkoff, with some assistance from the Caltech theorist Richard Tolman. They composed a mathematical theory of "neutron stars," stellar objects resembling white dwarfs except that gravitational collapse is balanced by a neutron pressure instead of an electron pressure. Whereas white dwarfs are roughly the size of Earth, neutron stars are even smaller and denser, with diameters of less than a few hundred kilometers. Oppenheimer and Volkoff patterned their calculation after Chandra's, with the difference that they were forced to use Einstein's theory of gravitation rather than Newton's, as Chandra had been able to do. Like Chandra's conclusion for white dwarfs, they found that there is a mass limit beyond which a dying star cannot form either a white dwarf or a neutron star.

What then? Another Oppenheimer paper in 1939, this one written with his student Hartland Snyder, implied, although it did not state explicitly except in the mathematics, the possibility that a massive star could collapse gravitationally all the way to an object of incredible density that swallows everything in its vicinity, including light. At first these voracious stellar objects were too bizarre for most astrophysicists to contemplate, but by the 1950s two intrepid theorists, John Wheeler and Yakov Zel'dovich, were picking up the research trail left by Chandra and Oppenheimer. One of Wheeler's contributions was an intriguing name, "black holes," for regions of spacetime in this state of extreme gravitational collapse.

During the 1960s, Chandra's cycle of study, research, and writing concerned general relativity and relativistic astrophysics. In the 1970s he turned to black-hole research. When he did this work, it was late in his career; he was in his sixties. No doubt it gave him great satisfaction to come full circle, back to the theme that began his career as an astrophysicist. In 1983, Oxford University Press published his monumental book *The Mathematical Theory of Black Holes*. In that same year, the Swedish Academy finally caught up with history and awarded Chandra a Nobel Prize, at least partly for his white-dwarf research, done fifty years earlier. This must have been a record for the time elapsed between the work done and the awarding of the prize.

For his final study, Chandra chose a remarkable subject—Isaac Newton. Chandra was a student of science history and biography, and he had a wide acquaintance among his contemporaries in physics and astrophysics. But for him one scientist stood above all those of the past and present, and that was Newton. He decided to pay homage to Newton, and to try to fathom his genius, by translating "for the common reader" the parts of Newton's *Principia* that led to the formulation of the gravitation law.

Newton relied on geometrical arguments that are all but incomprehensible to a modern audience. To make them more accessible, Chandra restated Newton's proofs in the now conventional mathematical languages of algebra and calculus. His method was to construct first his own proof for a proposition and then to compare it with Newton's version. "The experience was a sobering one," he writes. "Each time, I was left in sheer wonder at the elegance, the careful arrangement, the imperial style, the incredible originality, and above all the astonishing lightness of Newton's proofs, and each time I felt like a schoolboy admonished by his master."

Chandra's complex personality had a dark side. I have mentioned his pessimistic fascination with the picture of the man on the ladder. He called himself a "lonely wanderer in the byways of science." This dim outlook was the result of several influences: living apart from his native culture, his intense working habits (he regularly worked thirteen hours a day), and late in his life, the ordeal of a heart attack followed by bypass surgery. But the "lonely wanderer" found rewards. He continued on his solitary path because he knew there would be breathtaking vistas. In an essay titled "Pursuit of Science" he wrote:

> The pursuit of science has often been compared to the scaling of mountains, high and not so high. But who amongst us can hope, even in imagination, to scale the Everest and reach its summit when the sky is blue and the air is still, and in the stillness of the air survey the entire Himalayan range in the dazzling white of the snow stretching to infinity? None of us can hope for a comparable vision of nature and the universe around us, but there is nothing mean or lowly in standing in the valley below and waiting for the sun to rise over Kanchenjunga.

29

Affliction, Fame, and Fortune
Stephen Hawking

1942 –

Toy Trains and Cosmology

Stephen Hawking, described accurately as "the most remarkable scientist of our time," and inaccurately as a second Einstein ("perhaps an equal of Einstein," according to *Time* magazine in 1978), was born in Oxford on January 8, 1942. On January 8, 1642, three hundred years earlier, Galileo Galilei died, and in December of the year 1642 Isaac Newton was born.

It was wartime when Stephen, the Hawkings' first child, came into the world, and his mother, Isobel, had chosen an Oxford hospital for the delivery because the university town was safe from German bombing. (The German *Luftwaffe* agreed to spare Oxford and Cambridge if the Royal Air Force would do the same for Heidelberg and Göttingen.) Oxford was not a permanent haven, however. Isobel and her husband Frank lived in Highgate, a northern London suburb, where there was a real bomb threat; a near hit by a German V-2 rocket damaged the Hawking house but none of its inhabitants.

Frank and Isobel Hawking both came from the north, Frank from Yorkshire and Isobel from Glasgow. Both had been students in Oxford, but they did not meet there. Frank studied medicine and became a researcher in tropical medicine. "The vivacious and friendly Isobel," as Hawking's biographers Michael White and John Gribbin describe her, met her future husband at the medical research institute where he was later employed. She had taken a secretarial job there, "for which she was ridiculously overqualified."

When Stephen was eight, the family moved twenty miles north of Highgate to the cathedral city of St. Albans. The Hawkings bought a large Victorian house there, "of some elegance and character," as Hawking recalls. He continues: "My parents were not very well off when they bought it and they had to have quite a lot of work done on it before we could move in. Thereafter my father, like the Yorkshireman he was, refused to pay for any further repairs. Instead, he did his best to keep it going and keep it painted, but it was a big house and he was not very skilled in such matters. The house was solidly built, however, so it withstood this neglect."

By St. Albans standards, the Hawkings were an eccentric family. Frank "cared nothing for appearances if this allowed him to save money," Stephen writes. Isobel had been a member of the Young Communist League before the war. During one of Frank's extended research trips to Africa, Isobel took her three young children to the Mediterranean island of Majorca to join her friend Beryl Pritchard, who was the wife of the expatriate English poet and novelist Robert Graves. For many years, the Hawkings drove a retired London taxi, which had cost them fifty pounds. Finally they bought a new Ford, and the entire family, except for Stephen, who could not interrupt his schooling, embarked on a yearlong car trip to India and back.

In 1952, when he was ten, Stephen began his secondary education at the St. Albans School, connected with the cathedral and academically of high quality. Unlike many of the great physicists, Hawking did not turn in an outstanding classroom performance. He writes that he "was never more than about halfway up the class," and reports that he "tended to do much better on tests and examinations than . . . on coursework." His creative energy was spent on constructing working models of trains, boats, and airplanes, and on inventing immensely elaborate games. (One of his war games was played on a board with four thousand squares.) Hawking believes that the games and the model building foreshadowed his development as a scientist. "I think these games, as well as the trains, boats, and airplanes, came from an urge to know how things worked and to control them," he wrote later in an autobiographical note. "Since I began my Ph.D., this need has been met by my research into cosmology. If you understand how the universe operates, you control it in a way."

Hawking's father, Frank, was also an important influence in his life. "I modeled myself on him," Stephen remarked in an interview. "Because he was a scientific researcher, I felt that scientific research was the natural thing to do when I grew up." Stephen's preference was for mathematics and physics, but Frank disapproved of the mathematics, which he claimed was preparation only for teaching. Chemistry took the place of mathematics, and his limited mathematical training was a handicap in Hawking's subsequent research, based on the formidable mathematics of general relativity. But when he was later facing the adversities of disease, and increasingly unable to write in the formal language of mathematics (that is, with equations), he had to start all over again and find what was for him a better route to the physical message. "I don't care much for equations myself," he says now. "This is partly because it is difficult for me to write them down but mainly because I don't have an intuitive feeling for equations. Instead, I think in pictorial terms."

Falling

In 1959, at age seventeen, Hawking went to Oxford on a scholarship to University College, his father's college. The physics course at Oxford was easy—too easy. "The prevailing attitude at Oxford at that time was very antiwork," he writes. "You were supposed to be brilliant without effort, or to accept your limitations and get a fourth-class degree. To work hard to get a better class of degree was regarded as the mark of a gray man—the worst epithet in the Oxford vocabulary." The only examinations required were the final ones. Hawking estimates that he averaged about one hour of work a day. The predictable result for Hawking and many of his fellow students was boredom and a "feeling that nothing was worth making an effort for."

One relief from the boredom was rowing, a sport with a long and serious tradition at Oxford. Hawking did not have the burly physique required to handle an oar, but with his loud voice and fascination with being in control of events, he was suited for the position of coxswain, the member of the team who sits in the stern of the boat, shouts instructions, and steers. Hawking's coach thought he was competent as a "cox," but reckless and not so devoted to winning as he might have been.

With his one-hour-a-day effort, Hawking found himself at the end of his three years at Oxford on the borderline between a first- and a second-class degree. In an interview with the examiners who would make the final decision, Hawking said he wanted to do research. He would go to Cambridge, he said, if they gave him a first, and stay at Oxford if they gave him a second. He got a first.

At Cambridge, Hawking began his career as a theoretical astrophysicist and cosmologist. His intention was to obtain his Ph.D. under Fred Hoyle, then Britain's best-known cosmologist. Instead, he was assigned to Dennis Sciama, of whom he had never heard. At first, Hawking was annoyed not to be studying under the famous Hoyle, but then he began to appreciate the friendly and stimulating environment Sciama created for his students. Kip Thorne, a Caltech astrophysicist and contemporary of Hawking's, describes Sciama's selfless relationship with his research students: "Sciama was driven by a desperate desire to know how the universe is made. He himself described this drive as a sort of metaphysical angst. The universe seemed so crazy, bizarre, and fantastic that the only way to deal with it was to try to understand it, and the best way to understand it was through his students. By having his students solve the most challenging problems, he could move more quickly from issue to issue than if he paused to try to solve them himself."

Soon after Hawking had joined Sciama and his talented band of students, he was devastated by the news that he had the incurable disorder known as amyotrophic lateral sclerosis (ALS), or (in the United States) as Lou Gehrig's disease, or (in Britain) as motor neuron disease. It attacks the nerve cells that control voluntary muscular activity. Thought and memory processes are unaffected, but muscles throughout the body atrophy, leading finally to general paralysis. The doctor who made the diagnosis gave him a grim prognosis—two years to live—and "washed his hands of me," as Hawking puts it. "In effect, my father became my doctor, and it was to him I turned for advice."

Hawking's first reaction to his disease was the most natural one: deep depression. Fortunately, he did not lose himself in drugs or alcohol. His escape was in isolation and the thundering operatic music of Wagner. He could see no sense in continuing with the Ph.D. program if he would not have the time to complete it. But he would not give in to self-pity. While he was in the hospital for tests, he saw a boy die of leukemia. "It [was] not a pretty sight," he recalls. "Clearly there were people who were worse off than me. . . . Whenever I feel inclined to be sorry for myself, I remember that boy."

Rising

Hawking lifted himself out of depression partly by the strength of his will and determination, and partly with the help of others. The help came mainly from

Jane Wilde, an extraordinary young woman who became Hawking's fiancée. She too lived in St. Albans, and the couple met at a party in 1963, soon after Hawking's ALS symptoms began to appear. She was put off by his sometimes arrogant manner, but "there was something lost, he knew something was happening to him of which he wasn't in control." Their friendship grew and they became engaged. The partnership was based on love, and because of Stephen's condition, a serious sense of purpose. "I wanted to find some purpose to my existence," Jane has said, "and I suppose I found it in the idea of looking after him. But we were in love."

For his part, Hawking recognizes that without Jane in his life the disease would have soon destroyed him. He told an interviewer: "I certainly wouldn't have managed it without her. Being engaged to her lifted me out of the slough of despond I was in. And if we were to get married, I had to get a job and I had to finish my Ph.D. I began to work hard and found I enjoyed it. Jane looked after me single-handedly as my condition got worse. At that stage, no one was offering to help us."

By the summer of 1965, Hawking had completed his Ph.D. thesis and won a research fellowship in theoretical physics at Gonville and Caius College, Cambridge, always shortened to Caius (and, for some reason, pronounced "keys"). Jane and Stephen were married in July 1965. White and Gribbin describe the wedding photograph: "Hawking looks at the camera with a proud expression, a stare of deep-rooted determination and ambition—a stance that says, 'This is just the beginning.' Jane smiles happily at the lens, equally sure, in her own gentler way, that they will make out and overcome all adversity."

Hawking had an office at the Cambridge Department of Applied Mathematics and Theoretical Physics, and the couple needed to find nearby living accommodations, so Hawking, who was becoming increasingly disabled, could commute on his own. That proved to be a challenge, particularly when Hawking offended the college bursar (an administrative officer) by asking how much his fellowship paid. Finally, with the help of a woman who had noticed their plight, they found a small, ancient, but ideally located house on a picturesque street called Little St. Mary's Lane. One of Hawking's colleagues, Brandon Carter, describes the home as a lively place with friends on hand helping with the cooking and cleaning. Mahler and Wagner provided the musical accompaniment. And so it was in this remarkably normal way that the Hawkings began their married life. Their first child, Robert, was born in 1967.

The Most Perfect Objects

Hawking's first research project centered on black holes, those astonishing stellar objects Chandrasekhar called "the most perfect macroscopic objects there are in the universe." Regardless of their size, "the only elements in their construction are our concepts of space and time." A typical black hole might have a mass of ten solar masses and a radius of only ten to fifty kilometers. Astrophysicists now surmise that there are millions of such black holes in our galaxy. At the core of our galaxy and others there are evidently gargantuan black holes, some of them having the diameter of our solar system with a mass equivalent to several *billions* of solar masses. Theorists also speculate that vast numbers of miniature black holes populate the cosmos, each with the size of an atom and the mass of a mountain.

In spite of this diversity, black holes are among the simplest objects in the universe. A black hole can be as big as the solar system, or as small as an atom, or anything between; its behavior depends *only* on its mass and rate of spin (and on its electric charge, but that is generally comparatively small). Even though they are usually macroscopic in size, they are as standardized physically as elementary particles, which are also characterized by mass, spin, and charge. Black holes are not made out of rocks, like planets, or hot gases, like stars. They are, as Martin Rees, a contemporary of Hawking's and another one of Sciama's former students, writes, "made from the fabric of space itself." It was this fundamental simplicity that fascinated Chandrasekhar.

Up to a point, black-hole theory follows from Einstein's theory of general relativity, which describes the gravitational extremity that exists within the hole. The theory reveals that the gravitational field in the hole is so powerful that anything, including light, coming closer than a certain critical radius called the "event horizon" falls into the hole and is lost forever. With care, a spaceship could safely orbit just outside the event horizon, but black-hole interiors are not for exploration. A reckless astronaut passing beneath the event horizon could never escape, and could not even communicate his or her observations to the outside, because light and all other kinds of signals are confined within the hole.

General relativity tells us everything we need to know about black holes except for the physical situation at the center of the hole. There, relativity theory prescribes a point called a "singularity," where the density and spacetime curvature are infinite. But infinities are unpopular with theoretical physicists because they are not valid numbers and are likely to indicate a flaw in the workings of the theory.

Hawking and Roger Penrose (Chandrasekhar's muse), sometimes working in collaboration, defined the problem of black-hole singularities during the period from 1965 to 1970. Hawking and Penrose worked well as team. Hawking has a penetrating physical intuition, while Penrose has the mastery of the mathematics of general relativity that Hawking lacks. As one solution to the problem, Penrose proposed a principle of "cosmic censorship": a black-hole singularity is "censored" because it is "decently hidden," as Hawking puts it, from outside observers by the event horizon. "Naked," uncensored singularities are prohibited.

The theory of black holes was well established in the 1960s by Hawking, Penrose, and others, before any observations were reported that they actually existed. Then in the early 1970s a case was made that an x-ray-emitting object called Cygnus X-1, located in the constellation Cygnus, was a black hole paired with a massive star. It was assumed that the black hole was drawing gas from the star and heating it to the point where it emitted x rays. (As the gas fell into the black hole's intense gravitational field, it lost gravitational energy and at the same time got hotter as it gained thermal energy.)

In 1974, Hawking and other astrophysicists were about 80 percent certain that Cygnus X-1 actually involved a black hole. As an "insurance policy," Hawking made a bet with his Caltech colleague Kip Thorne that Cygnus X-1 did not harbor a black hole. Hawking's "insurance" if he lost the bet was a four-year subscription to the British magazine *Private Eye*. Thorne would receive a year's subscription to *Penthouse* magazine if he won. By 1990, confidence in the Cygnus X-1 black hole had risen to about 95 percent, and Hawking cheerfully paid off the bet.

Hawking's best-known contribution to astrophysics is a theory that slightly contradicts the blackness of black holes: "Black holes ain't so black," as Hawking puts it. The mechanism by which black holes shed their blackness relies on the concept, which originated with Dirac, that electrons have antielectron counterparts called positrons. When an electron meets a positron, they annihilate each other, and gamma-ray photons are produced. The inverse of this process, in which a gamma ray photon obtained from some suitable energy source produces an electron-positron pair, is also possible.

Quantum theory permits another version of the latter process, which is, as physicists like to say, "counterintuitive," meaning weird. The energy for electron-positron pair production can be "borrowed" from the empty space of a vacuum if an electron-positron annihilation follows that repays the energy "loan." The sequence for an electron e^- and a positron e^+ is, first, pair production,

$$\text{energy} \rightarrow e^- + e^+,$$

quickly followed by pair annihilation,

$$e^- + e^+ \rightarrow \text{energy}.$$

Heisenberg's uncertainty principle shows in detail how this can happen, and allows calculation of how long the electron and positron exist before they are lost in an annihilation. A similar story can be told for any kind of particle-antiparticle pair. Particles and antiparticles involved in this coupling of pair production and pair annihilation are called "virtual" because they cannot be observed directly by a particle detector.

Hawking's idea was that the members of a virtual pair could become real and one of them observable if they were produced in the vicinity of a black hole. One might be captured by the hole and become a real particle or antiparticle, while the other, also real, might escape and be seen as emitted radiation. To the extent that these emissions occur, the hole is not literally black. Energy is required to create the particle-antiparticle pairs, and that energy comes from the black hole's gravitational field. As the energy of the field is diminished, the hole shrinks in size and eventually disappears, possibly in an immense explosion with the strength of millions of hydrogen bombs.

But black holes are, after all, *almost* black. Emission of black-hole radiation, called "Hawking radiation," is a very inefficient, slow process. The time required for a black hole with the mass of the Sun to evaporate away all its mass is predicted by Hawking's theory to be 10^{65} years; the age of the universe as we observe it is vastly less than that—roughly 10^{10} years.

Beginning and Ending

How did the world begin, if indeed it had a beginning? How will it end, if there is an ending? These questions have been asked by theologians, philosophers, and other thinkers for millennia. But not until the twentieth century did a respectable scientific research field emerge whose practitioners built theories of cosmic history. They are called cosmologists, their field is cosmology, and their tools are general relativity, quantum theory, and the observational data contributed by astronomers.

One of the first and best-known cosmologists of our time is Fred Hoyle, who succeeded Eddington at Cambridge. Hoyle, in company with Hermann Bondi and

Thomas Gold, two Austrians living in England, advocated in the late 1940s a "steady-state" universe with no beginning and no ending. In Hoyle's version, an eternal "creation field" spontaneously generated matter, usually hydrogen, which balanced the universe's expansion, and maintained a constant density. But the continuous-creation process put special demands on the theory, demands that steady-state theorists have never satisfactorily met.

The steady-state cosmology has now been superseded by its principal rival, called the big-bang theory. For big-bang theorists the universe had a beginning in an exceedingly small and dense initial state. The universe expanded with a bang from that microscopic beginning, and the further history is told in terms of the physical events accompanying the expansion. (The term "big bang" was first used—derisively—by Hoyle in an attack on his opponents.)

Some of the essentials of the big-bang theory were introduced long before Hoyle's work, in 1922, by a brilliant young Russian theorist, Alexander Friedmann, who developed dynamic models of the universe by applying Einstein's gravitational-field equations to a universe assumed to be, on the average, uniform. In one of his models, "the creation of the world," as he put it, took place at a point, and subsequent expansion brought it to its present age and size.

Friedmann's models were mathematical, and his expansion scenario was just one of several. The Belgian physicist, astronomer, and priest Georges Lemaître was unequivocally committed to the expansion model. His "fireworks theory" was proposed in the 1930s. "At the origin," he wrote, "all the mass of the universe would exist in the form of a unique atom, the radius of the universe, although not strictly zero, being relatively small. The whole universe would be produced by the disintegration of this primeval atom [into] atomic stars," and the stars into ordinary matter and cosmic radiation. What we see today are the "ashes and smoke of bright but very rapid fireworks."

Another cosmologist took the stage later, at about the same time Hoyle was developing his steady-state theory. He was George Gamow, a Russian émigré (for a brief time, he was a student of Friedmann's), who eventually went to the United States, after stops in Göttingen, Copenhagen, and Cambridge. One of Gamow's specialties, among many others, was nuclear physics, and he constructed his cosmology by adding nuclear processes to the models already developed by Friedmann and Lemaître. He believed that the big bang originated in a primordial state he called "ylem," consisting of neutrons, protons, electrons, and a sea of high-energy radiation. Gamow and his coworker, Ralph Alpher, argued in a famous letter to *Physical Review* that as the universe expanded these nuclear ingredients built atoms of ordinary matter. (Gamow could not resist adding the name of Hans Bethe, who was an innocent bystander, to the *Physical Review* paper, so the author's names were Alpher, Bethe, and Gamow. Gamow tried unsuccessfully to persuade another one of his collaborators, Robert Herman, to change his name to Delter.)

At an early stage in the chronology of Gamow's model, matter in the universe ceased to interact with radiation and thereafter the latter remained as a cosmic background radiation field. Gamow predicted that this field would have the characteristics of blackbody or thermal radiation equivalent to the very low temperature of about 5 degrees on the absolute scale (–268 on the Celsius scale). About fourteen years after Gamow made this prediction, the cosmic background radiation was observed by Arno Penzias and Robert Wilson, working for Bell Laboratories in Holmdel, New Jersey; they determined the equivalent temperature to be 3.5 degrees, remarkably close to Gamow's estimate. The Bell scientists did not attempt to develop the cosmological significance of their observation; that was done by a group at Princeton, including Robert Dicke and James Peebles, who were preparing to make the observations themselves. In more-recent work, the cosmic background radiation has been carefully observed by instruments carried by a satellite. The blackbody characteristics have been confirmed to great accuracy and an equivalent temperature of 2.735 degrees measured.

"The mid-1960s marked a watershed in cosmology," writes Helge Kragh, a chronicler of modern cosmology, "not only because of the new observational results, but also because of theoretical innovations within the theory of general relativity." The theoretical developments centered on the singularity problem. In 1965, Roger Penrose used new mathematical methods to prove that according to the principles of general relativity the gravitational collapse of a massive star ends inevitably in the singular spacetime point of a black hole. During the next five years, work by Penrose, Hawking, and others resulted in a grand cosmological theorem, which asserted that a universe controlled by general relativity begins where a black hole ends, in a spacetime singularity.

That conclusion left cosmologists with a formidable further problem. As before in black-hole theory, an exposed singularity could not be tolerated. So the story of the universe as told by general relativity was incomplete; it could not give an acceptable account of the beginning events, wrapped as they apparently were around a singularity. Somehow, theorists had to modify their picture of the microscopic world in which the universe was born. The scale was so small in that world, much smaller even than that of an atom, that it was clearly necessary to invoke the methods of quantum theory and to combine them with the gravitation theory already provided by general relativity. In short, a unified theory of "quantum gravity" was needed. Physicists have been attempting for decades now to construct that unified theory, so far without complete success.

Hawking has been, and still is, one of the leaders in the search for a quantum-gravity unification. He advocates using the version of quantum mechanics invented by Richard Feynman, in which the actual path for an event is calculated by summing all possible paths for the event, each being characterized by a different phase. He also includes a special treatment of the time dimension by giving it an abstract mathematical identity technically called "imaginary." The prize is still elusive, but Hawking, who describes himself as a "born optimist," believes that we will see a successful unified theory "by the end of the twenty-first century, and probably much sooner." He is willing to offer "fifty-fifty odds that it will be within twenty years starting now [1998]."

A Popular Book

In 1982, with medical expenses and children's school fees looming, Hawking decided to write a short "book about the universe." He would write the book for a popular audience, and hope that it would also be popular in the other sense. It certainly was; sales of the book soared into a realm no science book had ever reached.

Hawking first proposed the book to Simon Mitton at Cambridge University Press, and left no doubt that he expected a large advance against the royalties.

Mitton, who had worked with Hawking before, and had suggested that Hawking write a popular book on cosmology, was generous: he offered a ten-thousand-pound advance, more than the publisher had negotiated with any other author.

That, however, was not what Hawking had in mind. When Dennis Sciama asked him if he intended to do the book with Cambridge University Press, he answered, "Oh no. I want to make some money with this one." He found the money by way of a New York literary agent, Al Zuckerman, who saw the potential of Hawking's subject, cosmology, and just as promising, the human-interest story of Hawking's twenty-year battle with ALS. Hawking prepared a proposal for the book, and Zuckerman sent it out to interested publishers for competing bids. The competition narrowed to two publishing houses, Bantam Books and W. W. Norton. (Norton was about to publish Richard Feynman's *Surely You're Joking, Mr. Feynman.*) Bantam won the bid with an unprecedented offer, including a $250,000 advance and favorable terms on the royalties.

The editor at Bantam who worked with Hawking on the book was Peter Guzzardi. Both Guzzardi and Hawking were determined that no ghostwriting would be involved. But Hawking had some lessons to learn about how to communicate with uninformed readers; Guzzardi became the teacher. As the manuscript took shape, the editor had to say again and again in his correspondence that he did not understand what he read: could Hawking expand and clarify? It was a trying time for Hawking. Zuckerman estimates that for every page Hawking wrote he got back two or three pages of editorial comments. In the book's acknowledgments, Hawking mentions "the pages and pages of comments and queries about points [Guzzardi] felt that I had not explained properly." "I must admit," he continues, "that I was rather irritated when I received his great list of things to be changed, but he was quite right. I'm sure that it is a better book as a result of his keeping my nose to the grindstone."

Hawking's fragile life, and the book project, almost came to an end in the summer of 1985. Hawking was visiting the European center for nuclear research (CERN) in Geneva to conduct research and complete his writing task, while Jane traveled in Germany. Suddenly one night Hawking's nurse found him suffocating from a blockage of the windpipe brought on by an attack of pneumonia. Quick action by a Geneva doctor, who happened to be familiar with Hawking's condition through a television program, saved the physicist's life. Jane was hastily summoned, and she agreed with the doctors that Hawking's only hope for long-term survival was a radical procedure called a tracheotomy, involving cutting into the windpipe and implanting a breathing device. The tracheotomy restored Hawking's breathing, but also deprived him of what little use he still had of his vocal cords.

Several weeks after the tracheotomy, Hawking was at home again in Cambridge. The medical bills were now overwhelming, and Jane was forced to appeal to foundations and charitable organizations for help. She was efficient, relentless, and finally successful in raising the necessary funds. At about the same time, Hawking's voice problem was solved by a California computer programmer who supplied a program that allows Hawking to choose words and make sentences on a computer monitor with slight movements of his hand. Once a sentence is constructed, it is pronounced (in a curious accent) by a voice synthesizer.

With his financial and medical problems again under control, Hawking returned to his research and to the book, which was nearing completion. It now had a title, *A Brief History of Time*, and an explanatory subtitle, *From the Big*

Bang to Black Holes. As promised, it delivered an account of modern cosmology, background being provided where necessary in quantum theory, relativity theory, and particle physics. The book has gotten an undeserved reputation for being unreadable. Not surprisingly, many people have bought the book and read no more than a few pages; the subject is not one for casual reading. Nevertheless, it is accessible to the reader with patience and the intellectual curiosity to wonder about the events of deep space and time. A dozen or so other eminent physicists have written popular books on cosmology. Hawking's is among the best.

Whether or not *A Brief History of Time* was read, it *sold* far beyond the most optimistic expectations. It quickly appeared on the best-seller list of the *New York Times*, and stayed there for a year. Sales in Britain put it on the *London Times* best-seller list for almost four years. This astonishing performance mystified the experts. They called it a "cult book," and accused the publisher of exploiting Hawking's disability. One columnist offered a prize of £14.99 (the price of the book) "to any reader who can provide an explanation [for the book's financial success] that is at all convincing." Hawking's mother, Isobel, responded and should have earned the prize:

> The book is well-written, which makes it pleasurable to read. The ideas are difficult, not the language. It is totally non-pompous; at no time does he talk down to his readers. He believes that his ideas are accessible to any interested person. It is controversial; plenty of people oppose his conclusions on one level or another, but it stirs thought.
>
> Certainly his fight against illness has contributed to the book's popularity, but Stephen had come a long way before the book was even thought of. He did not collect his academic and other distinctions because of motor neuron disease.

In another letter to the columnist, a parent expressed the opinion that readers who could not penetrate the book (including the columnist) needed to repair some elementary deficiencies in their education: "You are mistaken in thinking that few of the purchasers of *A Brief History of Time* are able to understand the work. It is only those who . . . have had a limited education who have this problem. My 17-year-old son, a physics A Level student, found the book very easy to understand and wished that Stephen Hawking had written in greater depth."

Normal in Spirit

Hawking's biographers are not exaggerating when they say that he has attained "science superstardom." He is probably the most famous living scientist. The list of his honors and awards fills pages. He is Lucasian Professor at Cambridge, the professorship once held by Newton. He has been knighted twice (Commander of the British Empire in 1981 and the higher award, Companion of Honor, in 1989). His portrait hangs in the National Portrait Gallery in London, and he has been the subject of television documentaries. His lectures draw overflow audiences; at Caltech, his reception was likened to Einstein's in the 1930s. Recently he visited the White House and chatted with President Clinton. In short, he has become, as one rather skeptical observer put it, a "happening."

But in spite of all the fame and fortune, or perhaps because of it, the Hawking enterprise has failed in one important respect. The husband and wife senior partners in that enterprise have broken up after twenty-five years of marriage. In 1990, Stephen left Jane to live with one of his nurses, Elaine Mason. She left her husband, David, who, as it happened, had designed the computer hardware mounted on Stephen's wheelchair. The Hawkings have three children, the Masons two.

The marriage had shown signs of strain before the separation. In the late 1980s, Jane gave a Cambridge journalist a vivid glimpse of the wild ride she had taken with Stephen: "I don't think I am ever going to reconcile in my mind the swings of the pendulum that we have experienced in this house—really from the depth of a black hole to all the glittering prizes."

Jane managed to earn a Ph.D. in medieval languages, specializing in Spanish and Portuguese poetry, and then find a teaching job in Cambridge. But it was a frustrating experience: "When I was working I thought I should be playing with the children," she recalls, "and when I was playing with the children I thought I should be working." She remembers how it was to be both a mother and a father: "I have been the one who has to teach my two boys to play cricket—and I can get them out!" Jane was essential to the Hawking enterprise, but at times she wondered about her status. "I'm not an appendage," she said in a television documentary, "though Stephen knows I very much feel I am when we go to some of these official gatherings. Sometimes I'm not even introduced to people. I come along behind and I don't really know who I'm speaking to."

Religion was also a contentious issue in the marriage. Jane is deeply religious, while Stephen, like Einstein, is an atheist in the sense that he has no place for a personal God in his universe. As he put it in a television documentary, "We are such insignificant creatures on a minor planet of a very average star in the outer suburbs of one of a thousand million galaxies. So it is difficult to believe in a God that would care about us or even notice our existence."

The wonder for Hawking, as it was for Einstein, is the comprehensibility of the universe; his faith is in a "complete theory." He concludes *A Brief History of Time* by assuring us that if such a theory is discovered, "it should in time be understandable in broad principle by everyone, not just a few scientists. Then we shall all, philosophers, scientists, and just ordinary people, be able to take part in the discussion of why it is that we and the universe exist. If we find the answer to that, it would be the ultimate triumph of human reason—for then we would know the mind of God."

For Jane, this was not a path to religious enlightenment. While the marriage was still intact, she said to an interviewer, "I pronounce my view that there are different ways of approaching [religion], and the mathematical way is only one way, and he just smiles."

Hawking is now in his late fifties, and his life is as full as ever. He continues with his research activities; he teaches, travels extensively, and lectures to large audiences. Such activity would be impressive in an able-bodied man. For Hawking, under the increasingly severe constraints of his disease, it is miraculous. How does he do it? Strength of mind is certainly part of it. "If you are disabled physically," Hawking says, "you cannot afford to be disabled psychologically." His daughter Lucy puts it a bit more darkly: "[He] will do what he wants to do at any cost to anybody else." Another foundation of his character is an indestructible optimism. Despite all the grim evidence to the contrary, he sees his life as "normal." In 1992, he said to an interviewer: "I don't regard myself as cut off from normal life, and I don't think people around me would say I was. I don't feel a disabled person—just someone with certain malfunctions of my motor neurons, rather as if I were color blind. I suppose my life can hardly be described as usual, but I feel it is normal in spirit."

Chronology of the Main Events

1564	Galileo Galilei is born in Pisa, Italy.
1591	Galileo's legendary demonstration on the Tower of Pisa.
1616	Robert Cardinal Bellarmine's injunction to Galileo.
1622	Galileo publishes *The Assayer.*
1632	Galileo publishes *Dialogue concerning the Two Chief World Systems.* The Inquisition orders Galileo's publisher to cease publication of the *Dialogue.*
1633	Galileo appears before the Inquisition. Galileo in Arcetri.
1638	Galileo publishes *Discourses on Two New Sciences.*
1642	Galileo dies in Arcetri. Isaac Newton is born in Woolsthorpe, England.
1661	Newton enters Trinity College, Cambridge University.
1665	Plague in England. Newton, in Woolsthorpe, begins to think about calculus, gravity, and optics.
1668	Newton is appointed Lucasian Professor of Mathematics at Cambridge.
1671	Newton's reflecting telescope is demonstrated to the Royal Society.
1684	Newton publishes *De Motu corporum in gyrum.*
1687	Newton publishes the *Principia.*
1696	Newton moves from Cambridge to London.
1704	Newton publishes the *Opticks.*
1727	Newton dies in London.
1791	Michael Faraday is born in Newington, Surrey, now part of London.
1796	Sadi Carnot is born in Paris.
1801	Thomas Young discovers his principle of interference based on a wave model of light.
1814	Robert Mayer is born in Heilbronn, Germany.
1818	James Joule is born in Manchester, England.
1820	Hans Christian Oersted's experiment demonstrating a magnetic effect produced by an electric effect.
1821	Hermann Helmholtz is born in Potsdam, Germany. Faraday's experiment demonstrating electromagnetic rotation. Augustin Fresnel characterizes light as waves that vibrate perpendicularly to their direction of motion.
1822	Rudolf Clausius is born in Köslin, Prussia.
1824	Carnot publishes *Reflections on the Motive Power of Fire.* William Thomson is born in Belfast, Northern Ireland.
1831	Faraday discovers electromagnetic induction. James Clerk Maxwell is born in Edinburgh, Scotland.

1832	Carnot dies in Paris.
	Faraday formulates the laws of electrochemistry.
1834	Émile Clapeyron publishes a mathematical version of Carnot's theory.
1837	Faraday studies electrostatic induction.
1839	Willard Gibbs is born in New Haven, Connecticut.
1842	Mayer publishes his first paper.
1843	Joule publishes his first determinations of the mechanical equivalent of heat.
1844	Ludwig Boltzmann is born in Vienna.
1845	Mayer publishes his second paper, including a calculation of the mechanical equivalent of heat.
	Thomson develops a mathematical theory of electrostatic lines of force.
	Faraday observes the effect of a magnetic field on polarized light.
1847	Joule publishes results of his paddle-wheel experiments for determination of the mechanical equivalent of heat.
	Helmholtz publishes *On the Conservation of Force.*
1848	Thomson publishes his thermometry principle.
1850	Clausius publishes his first paper on heat theory, in which he introduces the function U and derives the equation $dQ = dU + PdV$.
1851	Thomson publishes *On the Dynamical Theory of Heat.*
1852	Faraday defends the reality of lines of force.
1854	Thomson defines absolute temperature in terms of Carnot's function.
	Clausius publishes his second paper on heat theory and derives a state function that would later represent entropy.
	Maxwell publishes his first paper on electromagnetism, *On Faraday's Lines of Force.*
1855	Thomson joins the Atlantic Telegraph Company.
1857	Clausius publishes his first paper on the molecular theory of gases.
1858	Max Planck is born in Kiel, Germany.
1860	Maxwell publishes his first paper on the molecular theory of gases.
1861	Maxwell publishes his second paper on electromagnetism, *On Physical Lines of Force.*
1864	Walther Nernst is born in Briesen, West Prussia.
1865	Clausius publishes his last paper on heat theory, in which he completes his theories of energy and entropy, and states the two laws of thermodynamics.
	Maxwell publishes his third paper on electromagnetism, *A Dynamical Theory of the Electromagnetic Field.*
1867	Faraday dies at Hampton Court, Middlesex, England.
	Maria Sklodowska is born in Warsaw, Poland.
1871	Maxwell is appointed to the Chair of Experimental Physics at Cambridge.
	Helmholtz goes to Berlin.
	Ernest Rutherford is born near Nelson, New Zealand.
1873	Maxwell publishes *A Treatise on Electromagnetism.*
	Gibbs publishes a geometrical interpretation of thermodynamics, with emphasis on the energy and entropy concepts.
1875–78	Gibbs publishes *On the Equilibrium of Heterogeneous Substances.*
1878	Mayer dies in Heilbronn, Germany.
	Lise Meitner is born in Vienna.

1879	Maxwell dies in Cambridge, England.
	Albert Einstein is born in Ulm, Germany.
1885	Niels Bohr is born in Copenhagen, Denmark.
1887	Erwin Schrödinger is born in Vienna.
1888	Clausius dies in Bonn, Germany.
1889	Joule dies in Sale, England.
	Edwin Hubble is born in Marshfield, Missouri.
1892	Louis de Broglie is born in Dieppe, France.
1893	Nernst publishes his textbook, *Theoretische Chemie.*
1894	Helmholtz dies in Berlin.
1896	Boltzmann publishes the first volume of *Lectures on Gas Theory.*
	Henri Becquerel discovers the radioactivity of uranium.
1898	Boltzmann publishes the second volume of *Lectures on Gas Theory.*
	Marie and Pierre Curie announce their discoveries of polonium and radium.
1900	Planck publishes his paper on blackbody radiation, which, in a limited way, introduces the concept of energy quantization.
	Wolfgang Pauli is born in Vienna.
1901	Gibbs publishes *Elementary Principles in Statistical Mechanics.*
	Werner Heisenberg is born in Würzburg, Germany.
	Enrico Fermi is born in Rome.
1902	Rutherford and Frederick Soddy publish a series of papers in which their transmutation theory of radioactivity is developed.
	Einstein is appointed technical expert third class in the Bern, Switzerland, Patent Office.
	Paul Dirac is born in Bristol, England.
1903	Gibbs dies in New Haven, Connecticut.
1905	Nernst goes to Berlin.
	Einstein publishes his papers on relativity, the photoelectric effect, and colloidal particles as molecules.
1906	Nernst publishes his heat theorem.
	Rutherford discovers α-particle scattering.
	Boltzmann dies in Duino, a village near Trieste, Italy.
	Pierre Curie dies in Paris.
1907	Thomson dies near Largs, Scotland.
	Rutherford goes to Manchester.
1909	Hans Geiger and Ernest Marsden publish their paper on α-particle scattering by metallic foils.
1910	Subrahmanyan Chandrasekhar is born in Lahore, then in India, now in Pakistan.
1911	Rutherford proposes the nuclear model of the atom.
1913	Einstein moves to Berlin.
	Bohr publishes his first paper on the structure of atoms and molecules.
1913–14	Henry Moseley publishes his papers on the x-ray spectra of the elements.
1915	Einstein publishes his paper on general relativity.
1918	Richard Feynman is born in Far Rockaway, New York.
1919	Rutherford becomes director of the Cavendish Laboratory in Cambridge.
1921	The Bohr Institute is inaugurated in Copenhagen.

1923	De Broglie presents his theory of wave-particle duality for matter.
1924	Hubble reports cosmic distance measurements beyond our galaxy.
1925	Heisenberg publishes his first paper on matrix mechanics.
	Max Born, Heisenberg, and Pascual Jordan publish their comprehensive paper on matrix mechanics.
	Pauli introduces his exclusion principle.
1926	Schrödinger publishes his first paper on wave mechanics.
	Born publishes his first paper on the probability interpretation of quantum mechanics.
	Fermi publishes his first paper on quantum statistics.
1927	Heisenberg proposes his uncertainty principle.
1928	Dirac introduces his relativistic electron equation.
1929	Murray Gell-Mann is born in New York.
	Hubble publishes his first paper on the linear relation between recession speeds of galaxies and their distances from Earth.
	Dirac introduces his hole theory, identifying a hole as a proton.
1931	Dirac proposes the existence of the antielectron, later called the positron.
	John Cockcroft and Ernest Walton study nuclear reactions with proton beams generated in a linear accelerator.
1933	Einstein moves to Princeton, New Jersey.
	Fermi publishes his paper on the theory of β decay.
1934	Marie Curie dies in Sancellemoz, France.
	Chandrasekhar publishes his first white-dwarf paper.
1937	The particle later identified as the μ lepton is discovered.
	Rutherford dies in Cambridge, England.
1938	Meitner and Otto Frisch propose their theory of fission.
1939	Bohr and John Wheeler publish their paper on the mechanism of fission.
	Robert Oppenheimer, George Volkoff, and Richard Tolman propose a theory of neutron stars.
	Oppenheimer and Hartland Snyder show that an idealized imploding star forms a black hole.
1941	Nernst dies at his country estate near Bad Muskau, Germany.
1942	Fermi and associates achieve the first sustained nuclear chain reaction.
	Stephen Hawking is born in Oxford, England.
1943	Los Alamos National Laboratory begins operation near Santa Fe, New Mexico.
1945	Trinity test of a plutonium bomb near Alamogordo, New Mexico.
1946	The first two "V-particles" are discovered.
	George Gamow proposes a preliminary big-bang theory.
1947	Planck dies in Göttingen, Germany.
	The Shelter Island Conference meets.
1948	The Pocono Conference meets.
	Ralph Alpher, Hans Bethe, and Gamow extend the big-bang theory.
1949	The Oldstone Conference meets.
1952	Chandrasekhar becomes managing editor of the *Astrophysical Journal*.
1953	Hubble dies in San Marino, California.
	Gell-Mann proposes the strangeness scheme.

1954	Fermi dies in Chicago.
1955	Einstein dies in Princeton.
1956	Conservation of parity in weak interactions is questioned by Tsung-Dao Lee and Chen Nin Yang.
	The electron neutrino is detected.
1958	Pauli dies in Zürich, Switzerland.
1961	Schrödinger dies in Alpbach, Austria.
	Gell-Mann proposes SU(3) symmetry for hadronic structure: the eight-fold way.
1962	Bohr dies in Copenhagen.
	The μ neutrino is detected.
1964	The Ω^- particle is discovered.
	Gell-Mann proposes the quark model, with three flavors of quarks.
	A fourth quark flavor, called "charm," is introduced.
	Roger Penrose proves that black holes must contain singularities.
1965	The "color" concept is introduced in particle physics.
1967	Wheeler coins the term "black hole."
1968	Meitner dies in Cambridge, England.
1969	Hawking and Penrose prove that the universe began in a singularity.
1972	Feynman proposes his parton model.
1973	The theory of asymptotic freedom and confinement of quarks is proposed.
1974	Discovery of the J/ψ particle.
	Experimental evidence for the charmed quark is reported.
	Hawking shows that black holes are not quite black.
1975	The τ lepton is detected.
1976	Heisenberg dies in Munich, Germany.
1977	Experimental evidence for the bottom quark is reported.
1979	Experimental evidence for gluons is reported.
1984	Dirac dies in Miami, Florida.
1987	De Broglie dies in Paris.
1988	Feynman dies in Los Angeles, California.
1989	Experimental evidence for the existence of only three generations of quarks and leptons is reported.
1995	Experimental evidence for the top quark is reported.
	Chandrasekhar dies in Chicago.
2000	The τ neutrino is detected.

Glossary

absolute temperature: Temperature reckoned on a scale that places zero at about −273 degrees on the Celsius scale.
acceleration: The rate of change of velocity with time; measured in meters per second per second, feet per second per second, etc.
acceleration of gravity: The rate of change of velocity with time due to gravitational attraction; on Earth equal to about 32.2 feet per second per second. Represented by the symbol *g*.
adiabatic system: A system insulated thermally from its surroundings.
algebra: A branch of mathematics that generalizes arithmetic by representing numbers with symbols.
alpha particles (or rays): Helium ions originating in radioactive decay.
amplitude: In quantum mechanics, a quantity calculated for an event and squared to obtain the probability for occurrence of the event.
angstrom: A very small distance unit, equal to 10^{-8} centimeter.
anion: A negatively charged ion.
anode: In electrochemistry, the positive electrode of an electrolysis cell, toward which negative ions (anions) are attracted.
antielectron: A positive electron or positron.
antiparticle: A particle that is like its corresponding particle except that it has a charge and certain other properties opposite to those of the particle. When a particle and its corresponding antiparticle meet, they annihilate each other, leaving only energy. All particles of matter have their anti counterparts.
astronomy: The study of stars, galaxies, and other celestial objects, through observations with telescopes and associated instruments.
astrophysics: The theoretical study of the physical nature of stars, galaxies, and other celestial objects.
atomic number: A number assigned to each chemical element that determines the element's place in the periodic table; also equal to the charge on the element's atomic nucleus in units of the proton charge.
atomic weight: The mass of an atom relative to the mass of a hydrogen atom taken to be about 1 (actually, 1.008).
Avogadro's number: The number of molecules of hydrogen in about 2 grams (actually, 2.016 grams) of hydrogen.
baryon: A heavy hadron composed of three quarks; examples are protons and neutrons.
base: Of a logarithm, the number that is raised to a power equal to the logarithm.
beta particles (or rays): Electrons originating in radioactive decay.
blackbody: An object that emits its own radiation when heated, but does not reflect incident radiation.

blueshift: A change in the observed color of a star or galaxy due to motion of the star or galaxy toward Earth.

Boltzmann's constant: A very small number represented by k that appears in most of the equations of statistical mechanics.

boson: An elementary particle whose spin quantum number is equal to an integer. Bosons are carriers of forces existing in fields. The best-known boson is the photon, which carries the electromagnetic force. Bosons are not constrained by the exclusion principle.

British thermal unit: A measure of energy equivalent to the heat required to raise one pound of water through one degree on the Fahrenheit scale; abbreviated Btu.

calculus: A branch of mathematics that expresses continuous change. The tools of calculus are differentiation, integration, and equations containing derivatives and integrals.

caloric theory: A now-defunct theory that considered heat to be an indestructible, noncreatable fluid called caloric.

calorie: A unit of energy equivalent to the heat required to raise one gram of water one degree on the Celsius scale.

calorimeter: An instrument for measuring heat.

capacitor: A device that stores electric charge between two metallic plates separated by an insulating material.

cathode: In electrochemistry, the negative electrode, toward which positive ions (cations) are attracted.

cathode ray: A beam of energetic electrons.

cation: A positively charged ion.

centripetal force: Newton's term for the gravitational force that holds a planet in its orbit.

cepheid variable: A star that varies regularly in brightness, with the period between minimum and maximum brightness directly related to the star's average intrinsic brightness.

Chandrasekhar limit: The principle that a massive star cannot pass through the white-dwarf phase as it dies.

chemical affinity: An early term for the force that drives chemical reactions.

chemical potential: A relative energy that specifies for a chemical component its affinity, or tendency to participate with other components in a chemical reaction. At fixed temperature and pressure, chemical reactions proceed from higher to lower chemical potentials.

classical physics: Physics before the advent of quantum physics, as it was in the nineteenth century.

cloud chamber: A device for detecting charged energetic particles by following their tracks in an atmosphere of saturated water vapor.

commutator: In algebra, the difference $xy - yx$ for two variables x and y. In ordinary algebra, commutators vanish, while in matrix algebra they may be nonvanishing.

conservation law: A law that states that a certain quantity does not change in a physical or chemical process. Examples are conservation of momentum and conservation of energy.

cosmology: The study of the structure, origin, and history of the universe.

degeneracy: In astrophysics, a star's ultimate state of gravitational collapse against a countering electron or neutron pressure.

density: Mass per unit volume; measured in kilograms per cubic meter, pounds per gallon, etc.
derivative: The mathematical entity that determines the rate of change of one quantity with respect to another. For example, velocity is the derivative of distance with respect to time, and acceleration the derivative of velocity with respect to time.
dielectric: An electrically insulating material.
differential: A very small change in a quantity.
differentiation: The mathematical procedure for determining a derivative.
diffraction: The spreading of waves (e.g., of light) after passing through a narrow opening.
diffusion: The spontaneous flow of a substance from a region of high concentration to a region of low concentration.
disorder: In general, the extent to which a system is mixed up; calculated by Boltzmann using combinatorial methods.
dynamics: The science of motion with consideration of forces and energy included.
electric current: Conveniently pictured as a flow of electrons in a conducting material such as copper; measured in amperes.
electric potential: The force driving an electric current; measured in volts.
electrochemistry: The study of chemical reactions induced by, or producing, electricity.
electrode: Any terminal through which an electric current passes in or out of an electrically conducting material.
electrodynamics: The study of moving electric charges and their fields.
electrolysis: The production of chemical changes by passing an electric current through a solution or molten material.
electrolysis cell: A device that consumes an electrical input and induces a chemical reaction.
electrolyte: A solution or molten substance that conducts electricity by the passage of ions from one electrode to another.
electromagnetism: The science of electricity and magnetism.
electrometer: A sensitive instrument for measuring electric potentials.
electron: An elementary particle; carries a negative charge equal in magnitude to that of the proton's positive charge.
electron pressure: A pressure arising from the requirement of the exclusion principle that two electrons otherwise in the same state cannot occupy the same point in spacetime.
electron-volt: A small unit of energy used to measure energies of molecules, atoms, and subatomic particles; equal to the energy acquired by an electron when it is accelerated through one volt. Abbreviated eV.
electrostatics: The study of stationary electric charges and their fields.
elementary particle: A particle that has no structure, and is viewed mathematically as a point. Examples are electrons, quarks, and neutrinos.
endothermic process: A process that proceeds with the absorption of thermal energy; an example is the melting of ice.
energy: The capacity to do work; measured in joules, calories, etc.
ensemble: In statistical mechanics, a conceptual collection of many replicas of a system of interest.
entropy: A measure of disorder. Small changes dS in entropy accompanying pas-

sage of heat dQ in or out of a system at the temperature T are calculated with $dS = \frac{dQ}{T}$.

equipartition theorem: A theorem that establishes (not always correctly) that energy added to a system is equally divided among the system's modes of motion.

equivalent weight: Of a chemical element, a weight about equal to the mass of the element that combines with one gram of hydrogen.

exclusion (or Pauli) principle: A principle that two fermions (e.g., electrons, neutrons, protons, and quarks) cannot occupy the same state.

exothermic process: A process that proceeds with the release of thermal energy; an example is a combustion reaction.

expansion coefficient: A coefficient that measures the fractional change in the volume of an object resulting from a change in temperature.

exponential function: Any function involving a variable in an exponent; examples are e^x and 10^y.

factorial: Of a positive integer n is $n! = 1 \times 2 \times 3 \ldots (n - 1) \times n$; an example is $6! = 1{\times}2{\times}3{\times}4{\times}5{\times}6 = 720$.

fermion: A particle whose spin quantum number is a half-integer. Fermions are the constituents of matter, and their behavior is restricted by the exclusion principle. The best-known fermions are electrons, protons, neutrons, and neutrinos.

field: A physical entity that exists throughout space and time. Fields cause electric, magnetic, and gravitational effects. Compare with the particle concept, which concerns a physical entity that is localized in space and time.

fission: The splitting of a heavy, unstable atomic nucleus into lighter fragments, with the release of a large amount of energy.

fluxion: Newton's term for rate of change of any quantity with time.

foot-pound: A measure of work; equivalent to the work required to lift one pound through one foot. Abbreviated ft-lb.

force: Any influence that causes an object to change its motion from a state of rest or from uniform motion in a straight line.

free energy: Energy that can be converted to work; also called Gibbs energy.

friction: The force that resists the relative motion of two objects in contact.

function: A mathematical term for a quantity that depends on another quantity or quantities; for example, the area A of a circle is a function of the circle's radius r, according to $A = \pi r^2$.

galaxy: A more or less independent system of stars.

galvanometer: An instrument for measuring small electric currents.

gamma rays: High-energy electromagnetic radiation originating in radioactive decay.

gas constant: The constant R in the ideal gas law $PV = nRT$, which states that the volume V of an ideal gas is directly proportional to the molar amount n and the absolute temperature T, and inversely proportional to the pressure P.

Geiger counter: An instrument for measuring radioactivity, developed by Hans Geiger.

Gibbs energy: Energy that can be converted to work; also called free energy.

gram: A unit of mass.

gravitational constant: A number represented by G that appears in most of the equations of gravity theory.

hadron: Any subatomic particle held together by strong interactions. Hadronic structural units are quarks.
heat: In general, thermal energy. In thermodynamics, heat is thermal energy passing in or out through the boundary of a system; measured in calories or joules.
heat capacity: The heat required to raise the temperature of a certain amount of a substance one degree in temperature.
heat engine: Any device for producing work from heat.
***H* theorem:** A theorem developed by Boltzmann, which proves that a property H of a macroscopic system does not increase. Related to the second law of thermodynamics.
Hubble's constant: The constant of proportionality in Hubble's law; represented by H. The reciprocal of H estimates the age of the universe.
Hubble's law: A law that expresses a linear relationship between a galaxy's speed of recession from Earth and the galaxy's distance from Earth.
ideal gas: A gas whose volume is directly proportional to absolute temperature and molar amount, and inversely proportional to pressure, hence following the gas law $PV = nRT$, where P, V, T, and n are the pressure, volume, absolute temperature, and molar amount of the gas, and R is a constant.
induction: An electrical or magnetic effect produced by a field.
inertia: The tendency of an object to remain at rest or in uniform motion unless influenced by a force.
integral: The mathematical entity that sums very small changes in a quantity; represented by the symbol $\int$.
integration: The mathematical procedure for determining an integral.
internal energy: The energy of an object possessed by its constituent molecules.
ion: An atom or molecule that is electrically charged.
irreversible process: In thermodynamics, a nonideal process whose direction cannot be reversed without changes in a system's surroundings. All real processes are to some degree irreversible.
isolated system: A system completely disconnected from its surroundings.
isothermal system: A system held at constant temperature.
isotope: An atom that has a different mass but the same nuclear charge or atomic number as another atom.
joule: A unit of energy; equivalent to about 0.239 calorie.
kilocalorie: One thousand calories.
kilogram: One thousand grams.
kilogram-meter: A measure of work; equal to the work required to lift one kilogram one meter.
kinematics: The science of motion without consideration of forces or energy.
kinetic energy: The energy of an object due to its motion; equal to $\frac{mv^2}{2}$, where m is the mass of the object and v its speed.
lepton: An elementary particle that does not participate in strong interactions; examples are electrons, neutrinos, and muons.
light-year: An astronomical distance unit; the distance traveled by a light ray in a vacuum in one year, equal to 5.88 trillion miles.
line element: A measure of the distance in spacetime between two nearby events.
lines of force: Faraday's representation of the forces inherent in an electric or magnetic field.

logarithm: Of a number, the exponent of a base—usually 10—that calculates the number. For example, because $10^3 = 1{,}000$, the logarithm of 1,000 is 3.
mass: The property of an object that measures its resistance to a change in its motion; also the property that results in gravitational attraction. Measured in grams, kilograms, etc.
matrix mechanics: The version of quantum mechanics originated by Heisenberg, Born, and Jordan.
mechanical equivalent of heat: The mechanical effect (dropping of weights in Joule's experiments) equivalent to a unit of heat. Represented by *J;* measured by Joule in foot-pounds per British thermal unit.
mechanics: The science of motion.
megaparsec: An astronomical distance unit; equal to a million parsecs or 3.26 million light-years.
meson: A hadron of intermediate mass composed of a quark and an antiquark; an example is a pion.
metaphysics: The study of nature beyond physics.
Mev: An energy unit favored by particle physicists; equal to a million electron-volts.
mole: In chemistry, the quantity of a chemical component containing Avogadro's number of molecules.
molecular weight: The mass of a molecule relative to the mass of a hydrogen atom taken to be about 1 (actually, 1.008).
momentum: The mass of an object multiplied by its velocity.
multiplet: In particle physics, a group of particles whose members all have the same energy or nearly the same energy.
natural logarithm: A logarithm whose base is the number *e*.
nebula: A large cloud of gas and dust in space; also, in Hubble's terminology, a galaxy.
neutrino: An elementary particle that carries no charge and hardly any mass.
neutron: An uncharged particle whose mass is approximately equal to that of the proton; one of the fundamental constituents of all nuclei.
neutron star: An elderly star that has consumed its nuclear fuel, and for a star of the mass of the Sun, collapsed to a diameter of 50 to 1,000 kilometers. The gravitational force in the star is countered by a neutron force.
nuclear chain reaction: The nuclear process in which fission events produce as many neutrons as, or more neutrons than, they consume, and these neutrons induce further fissions.
nuclear reactor: A device for sustaining a controlled nuclear chain reaction.
parsec: An astronomical distance unit equivalent to 3.26 light-years.
particle: A physical entity that is localized in space and time. Compare with the field concept, which concerns a physical entity that exists throughout space and time.
partition function: A summation of exponential terms that is fundamental in statistical mechanics.
Pauli principle: See exclusion principle.
perfect gas: Another name for an ideal gas.
period: Applied to periodic motion, the time required for completion of one cycle of motion.
periodic table: The arrangement of the chemical elements in a table whose columns contain elements with similar chemical properties. Usually (but not al-

ways), the rows of the table list the elements in order of increasing atomic weight.

perpetual motion: The concept that a machine can be designed that continues to provide useful output forever, even though it requires no energy input. Such a machine is prohibited by the laws of thermodynamics.

phase: In wave theory, a certain stage in wave motion. Two waves reinforce each other if they are in phase, and cancel each other if they are out of phase.

photon: An elementary particle that carries the electromagnetic force in radiation fields; endowed with wave as well as particle properties.

pile: Fermi's term for a graphite-moderated nuclear reactor using natural uranium.

pion: A meson. Three different kinds of pions are observed, π^-, π^0, and π^+, with the charges -1, 0, $+1$.

Planck's constant: A small number represented by h that appears in most of the equations of quantum theory.

polarized light: Light whose waves vibrate in a certain plane.

potential: A measure of the energy available from a field at a certain point, measured per unit of the physical property affected by the field (e.g., mass or electric charge).

potential energy: The energy of an object due to its position. For example, the potential energy of an object of mass m held a distance z above ground level has the gravitational potential energy mgz, with g representing the acceleration of gravity.

proportionality: In mathematics, a relationship between two quantities such that if one quantity changes the other changes proportionately. If x and y are proportional to each other ($x \propto y$), doubling x doubles y and vice versa, tripling x triples y, and so forth.

proportionality constant: In mathematics, a constant that converts a proportionality into an equation. Thus the proportionality constant k converts the proportionality $y \propto x$ into the equation $y = kx$.

proton: A hydrogen nucleus; one of the fundamental constituents of all nuclei. Carries a positive charge equal in magnitude to that of the electron's negative charge.

quantization: As applied to a physical property such as energy, a change in a property, which change occurs in discrete steps rather than continuously.

quantum electrodynamics (QED): The study of electrons, photons, and their interactions.

quantum mechanics: The generic term for matrix mechanics, wave mechanics, and the synthesis defined by Dirac.

quantum number: An integer or half-integer number that specifies a state determined by quantum theory.

radioactive decay: The nuclear event in which a radioactive element spontaneously emits an energetic particle, usually an alpha particle, or a beta particle, or a gamma ray, or some combination of these.

radioactivity: The process of radioactive decay. May be accompanied by the spontaneous transmutation of an atom of one element into an atom of another element, by emission of an alpha or beta particle.

radiochemistry: The branch of chemistry that deals with chemical techniques that separate radioelements.

radioelement: A radioactive element.
redshift: A change in the observed color of a star or galaxy due to motion of the star or galaxy away from Earth.
reflecting telescope: A telescope that collects light and brings it to focus with a concave mirror.
reflection: The deflection of waves (e.g., of light) when they meet a surface.
refracting telescope: A telescope that collects light and brings it to focus with a convex lens.
refraction: The bending of a wave (e.g., of light) when it passes from one medium to another.
relativity: The study of the mechanics of objects in relative motion to each other.
resonators: Planck's term for the vibrating molecules in the walls of a blackbody oven.
reversible process: In thermodynamics, an idealized process whose direction can be reversed with no net changes in the system of interest or in its surroundings.
scalar: A quantity that has a magnitude but no directional aspect.
singularity: In general relativity, a point in spacetime where physical quantities such as density become infinite.
slow neutron: A neutron with low energy.
specific heat: The heat required to raise the temperature of a unit mass of a substance one degree.
spectral line: A particular wavelength, frequency, or energy in a spectrum.
spectroscope: An instrument that displays a spectrum.
spectroscopy: The study of the spectra of atoms, molecules, atoms, or particles.
spectrum: The separation of radiation or particles into component wavelengths, frequencies, or energies.
speed: The magnitude of velocity; measured in meters per second, miles per hour, etc.
state function: A function that expresses a property of a system strictly in terms of the state of the system, as determined, for example, by pressure and temperature.
statistical mechanics: The study of macroscopic systems from the point of view of the average behavior of the system's constituent molecules.
strong interaction: In particle physics, the force that binds together quarks in hadrons, and protons and neutrons in nuclei.
thermochemistry: The study of exothermic and endothermic reactions.
thermodynamics: At first, the science of heat, but finally broadened to include such things as the calculation of chemical driving forces.
trigonometry: A branch of mathematics that solves problems relating to triangles.
vector: Any quantity that has both direction and magnitude.
velocity: The rate of change of distance with time, including both magnitude and direction.
viscosity: The property of a substance that measures its resistance to flow.
voltaic cell: A chemical device whose output is an electric current.
wave mechanics: The version of quantum mechanics originated by Schrödinger and de Broglie.
weak interaction: In nuclear physics, the interaction involved in β decay.
weight: The gravitational force exerted on an object.

white dwarf: An elderly star that has consumed its nuclear fuel, and for a star with the mass of the Sun, collapsed to a diameter about equal to that of Earth. The star's gravitational force is balanced by electron pressure.

work: In physics, what is accomplished when a force is applied to an object to move it over a distance, as in lifting, pushing, or pulling the object; measured in joules, calories, etc.

x rays: High-energy electromagnetic radiation.

Invitation to More Reading

As biographies, the chapters in this book are necessarily brief. The suggestions that follow are intended to afford the reader an opportunity to become better acquainted with the main characters in this story. Full-length biographies and related material are given for the subject of each chapter. The list is far from comprehensive; the books selected are those that were preferred as sources in the writing of the book. The abbreviation *DSB* stands for *Dictionary of Scientific Biography* (New York: Scribner, 1971–90), an invaluable source of short but authoritative biographies of most of the subjects herein.

Chapter 1

The Galileo literature is enormous. A few selections are: Stillman Drake, *Galileo at Work: His Scientific Biography* (New York: Dover, 1995), *Galileo* (Oxford: Oxford University Press, 1980), and *Galileo: Pioneer Scientist* (Toronto: University of Toronto Press, 1990); James Reston, Jr., *Galileo* (New York: HarperCollins, 1994); and Dava Sobel, *Galileo's Daughter* (New York: Walker, 1999).

Chapter 2

Like Galileo, Newton has been popular with scholars. A recent biography is Richard Westfall, *The Life of Isaac Newton* (Cambridge: Cambridge University Press, 1994), which is a shortened version of Westfall's earlier *Never at Rest: A Biography of Isaac Newton* (Cambridge: Cambridge University Press, 1980). To get a taste of Newton's *Principia,* see Subrahmanyan Chandrasekhar, *Newton's* Principia *for the Common Reader* (Oxford: Oxford University Press, 1995). François De Gandt also analyzes the *Principia* in *Force and Geometry in Newton's* Principia (Princeton: Princeton University Press, 1995).

Chapter 3

Little is known about Sadi Carnot's personal life. Biographical commentary mainly concerns his scientific work. See J. F. Challey's article on Carnot in *DSB*, and D. S. L. Cardwell, *From Watt to Clausius* (Ithaca, N.Y.: Cornell University Press, 1971). Carnot's *Reflections on the Motive Power of Fire* has been translated several times, most recently by R. Fox (Manchester: Manchester University Press, 1986). R. H. Thurston's translation (London: Macmillan, 1890) includes portions of Hippolyte Carnot's biography of his brother. Clifford Truesdell gives a critical account of the history of thermodynamics (including the work of Carnot and his successors) in *The Tragicomical History of Thermodynamics* (New York: Springer-Verlag, 1980).

Chapter 4

Biographical material on Mayer is scarce. Try R. Bruce Lindsay, *Men of Physics: Julius Robert Mayer* (New York: Pergamon Press, 1973), and R. Steven Turner's *DSB* article on Mayer. Biographies of the "three Ts" who stirred up the great Joule-Mayer controversy are Silvanus Thompson, *The Life of William Thomson* (London: Macmillan, 1910), Arthur S. Eve, *Life and Work of John Tyndall* (London: Macmillan, 1945), and C. G. Knott, *Life and Scientific Work of Peter Guthrie Tait* (Cambridge: Cambridge University Press, 1911).

Chapter 5

Principal biographies of Joule are D. S. L. Cardwell, *James Joule: A Biography* (Manchester: Manchester University Press, 1989), and Osborne Reynold, *Memoir of James Prescott Joule* (Manchester: Manchester University Press, 1892). J. G. Crowther's book of short biographies, *Men of Science* (New York: Norton, 1936), contains a readable chapter on Joule.

Chapter 6

Leo Königsberger, *Hermann Helmholtz* (New York: Dover, 1965), is the principal Helmholtz biography. Also see R. Steven Turner's *DSB* article on Helmholtz. Some of Helmholtz's writings are collected in Russell Kahl, *Selected Writings of Hermann Helmholtz* (Middletown, Conn.: Wesleyan University Press, 1971).

Chapter 7

The most complete Thomson (Kelvin) biography is Crosbie Smith and M. Norton Wise, *Energy and Empire: A Biographical Study of Lord Kelvin* (Cambridge: Cambridge University Press, 1989). Silvanus P. Thompson, *The Life of William Thomson, Baron Kelvin of Largs* (London: Macmillan, 1910), is valuable for its many quotations from correspondence and diaries.

Chapter 8

Clausius is another physicist (like Carnot) who is on the biographically endangered list. The *DSB* article by Edward Daub is recommended. D. S. L. Cardwell, *From Watt to Clausius* (Ithaca, N.Y.: Cornell University Press, 1971), places Clausius's work in its historical context.

Chapter 9

Gibbs has two biographies: Muriel Rukeyser, *Willard Gibbs* (Woodbridge, Conn.: Ox Bow Press, 1988), and Lynde Phelps Wheeler, *Josiah Willard Gibbs: The History of a Great Mind* (Woodbridge, Conn.: Ox Bow Press, 1998). Neither biography does justice to Gibbs's work as a scientist. For that, see Martin Klein's *DSB* article on Gibbs. J. G. Crowther has written briefly about Gibbs's life in *American Men of Science* (New York: Norton, 1937).

Chapter 10

For an entertaining account of Nernst's life and times, see Kurt Mendelssohn, *The World of Walther Nernst: The Rise and Fall of German Science, 1864–1941* (Pittsburgh: University of Pittsburgh Press, 1973). The comments of Franz Simon quoted in the chapter can be found in the *Yearbook of the Physical Society* (1956), 2.

Chapter 11

Two nineteenth-century Faraday biographies are Henry Bence Jones, *The Life and Letters of Faraday* (London: Longmans, Green, 1870), and John Tyndall, *Faraday as Discoverer* (New York: Appleton, 1868). The principal twentieth-century Faraday biography is Pearce Williams, *Michael Faraday: A Biography* (New York: Basic Books, 1964). For a more condensed version of Faraday's work, see Williams's *DSB* article on Faraday. Geoffrey Cantor discusses the religious dimension of Faraday's life in *Michael Faraday: Sandemanian and Scientist* (New York: St. Martin's Press, 1991). J. G. Crowther tells about the ups and downs of the Davy-Faraday relationship in *Men of Science* (New York: Norton, 1936). Faraday's most famous lecture at the Royal Institution was published as *The Chemical History of a Candle* (Atlanta: Cherokee, 1993).

Chapter 12

The main Maxwell biography, written by his friend Lewis Campbell, is *The Life of James Clerk Maxwell* (London: Macmillan, 1882). Two more-recent biographies are C. W. Everitt, *James Clerk Maxwell: Physicist and Philosopher* (New York: Scribner, 1975), and Martin Goldman, *The Demon in the Aether: The Story of James Clerk Maxwell* (Bristol, England: A. Hilger, 1983). Maxwell's papers have been collected in *The Scientific Papers of James Clerk Maxwell*, ed. W. D. Niven (New York: Dover, 1952). For the story of Hertz's brief but remarkable life, see Charles Susskind, *Heinrich Hertz: A Short Life* (San Francisco: San Francisco Press, 1995).

Chapter 13

Boltzmann has only two short biographies in English, Englebert Broda, *Ludwig Boltzmann*, trans. Engelbert Broda and Larry Gay (Woodbridge, Conn.: Ox Bow Press, 1983), and Carlo Cercignani, *Ludwig Boltzmann: The Man Who Trusted Atoms* (Oxford: Oxford University Press, 1998). The story of Boltzmann's confrontation with the antiatomists is told in David Lindley's recent *Boltzmann's Atoms: The Great Debate That Launched a Revolution in Physics* (New York: Free Press, 2000). Boltzmann's writings on gas theory are translated by Stephen Brush in *Lectures on Gas Theory* (New York: Dover, 1995).

Chapter 14

The Einstein literature is overwhelming. The best of the many biographies is Abraham Pais, *Subtle Is the Lord: The Science and Life of Albert Einstein* (Oxford: Oxford University Press, 1982). Ronald Clark, *Einstein: The Life and Times* (New

York: World, 1971), tells about Einstein's public life. Einstein told his own story (briefly) as "Autobiographical Notes" in *Albert Einstein: Philosopher-Scientist*, ed. P. A. Schilpp (New York: Harper and Row, 1951). A recent Einstein biography is Albrecht Fölsing, *Albert Einstein*, trans. Ewald Osers (London: Penguin, 1997). In collaboration with Leopold Infeld, Einstein wrote an excellent introduction to modern physics, *The Evolution of Physics* (New York: Simon and Schuster, 1938).

Chapter 15

There is no full-length biography of Planck in English. John Heilbron writes about Planck's preeminent role in the German scientific community in *The Dilemmas of an Upright Man: Max Planck as Spokesman for German Science* (Berkeley and Los Angeles: University of California Press, 1986). Planck's own remarks in his *Scientific Autobiography and Other Papers*, trans. F. Gaynor (New York: Philosophical Library, 1949), are revealing.

Chapter 16

The best biography of Bohr in English is Abraham Pais, *Niels Bohr's Times, in Physics, Philosophy, and Polity* (Oxford: Oxford University Press, 1991). Also see the collection of reminiscences about Bohr edited by Stefan Rozental, *Niels Bohr: His Life and Work as Seen by His Friends and Colleagues* (Amsterdam: North-Holland, 1967); Léon Rosenfeld's *DSB* article on Bohr; and Ruth Moore, *Niels Bohr* (New York: Knopf, 1966). The story of Bohr's efforts on behalf of an open nuclear policy is told in Alice Kimball Smith, *A Peril and a Hope* (Chicago: Chicago University Press, 1965).

Chapter 17

There is no biography of Pauli in English. Glimpses of the great critic are seen in Rudolf Peierls, *Bird of Passage* (Princeton: Princeton University Press, 1985). The *DSB* article by Markus Fierz outlines Pauli's scientific work.

Chapter 18

The enigmatic Heisenberg has a full-length biography, David Cassidy's aptly titled *Uncertainty: The Life and Science of Werner Heisenberg* (New York: Freeman, 1991). Heisenberg tells part of his own story in *Physics and Beyond: Encounters and Conversations* (New York: Harper and Row, 1971). Elisabeth Heisenberg, in *Inner Exile* (Boston: Birkhäuser, 1984), emphasizes her husband's precarious status during the war years. Max Born, *Physics in My Generation* (New York: Springer-Verlag, 1969), is an account of the revolution that started with Einstein's relativity theory and continued with the matrix mechanics created by Born, Heisenberg, and others.

Chapter 19

De Broglie has no full-length biography in English, not even a *DSB* entry. Schrödinger, on the other hand, has Walter Moore's revealing *Schrödinger: Life and Thought* (Cambridge: Cambridge University Press, 1989). An earlier biography is

William T. Scott, *Erwin Schrödinger: An Introduction to His Writings* (Amherst: University of Massachusetts Press, 1967). Schrödinger was a prolific writer and lecturer. Samples of his work can be found in *What Is Life?* (Cambridge: Cambridge University Press, 1967), *My View of the World* (Cambridge: Cambridge University Press, 1964), and *Science and Humanism* (Cambridge: Cambridge University Press, 1961).

Chapter 20

Marie Curie's remarkable life is told in Susan Quinn, *Marie Curie: A Life* (Reading, Mass.: Perseus, 1995). Also see Eve Curie, *Madame Curie*, trans. Vincent Sheehan (New York: Doubleday, 1937); and, in one volume, Marie Curie's *Autobiographical Notes* and her loving biography of her husband, *Pierre Curie* (New York: Dover, 1963).

Chapter 21

The main Rutherford biography is Arthur S. Eve, *Rutherford: Being the Life and Letters of the Rt. Hon. Lord Rutherford, O.M.* (Cambridge: Cambridge University Press, 1939). A collection of reminiscences edited by J. B. Birks, *Rutherford at Manchester* (New York: Benjamin, 1963), shows Rutherford in his middle period. Mark Oliphant, *Rutherford: Recollection of the Cambridge Days* (Amsterdam: Elsevier, 1972), tells about Rutherford at the Cavendish Laboratory. John Campbell's new biography, *Rutherford* (Christchurch, New Zealand: AAS Publications, 1999), emphasizes Rutherford's years in New Zealand.

Chapter 22

Ruth Lewin Sime sets the record straight on the discovery of the nuclear fission concept in *Lise Meitner: A Life in Physics* (Berkeley and Los Angeles: University of California Press, 1996). Otto Frisch's account of his inspired conversation with Meitner on a Swedish ski trail is told in *What Little I Remember* (Cambridge: Cambridge University Press, 1979). Otto Hahn has written several autobiographies. One of them is *Otto Hahn: My Life, The Autobiography of a Scientist*, trans. Ernst Kaiser and Eithne Wilkins (New York: Herder and Herder, 1970).

Chapter 23

Laura Fermi, in *Atoms in the Family* (Chicago: University of Chicago Press, 1954), tells about life with her husband and his physics. Emilio Segrè writes about Fermi as a colleague in *Enrico Fermi: Physicist* (Chicago: University of Chicago Press, 1970). The story of the Manhattan Project has been told many times. One of the most recent accounts, and probably the best, is Richard Rhodes, *The Making of the Atomic Bomb* (New York: Simon and Schuster, 1986). The moral and ethical legacy of nuclear weaponry is explored by Mary Palevsky in a series of recent interviews with physicists who participated in the Manhattan Project, *Atomic Fragments: A Daughter's Questions* (Berkeley and Los Angeles: University of California Press, 2000).

Chapter 24

For Dirac's story, see Helge Kragh, *Dirac: A Scientific Biography* (Cambridge: Cambridge University Press, 1990), and the collection of appreciations edited by Behram Kursunoglu and Eugene Wigner, *Reminiscences about a Great Physicist: Paul Adrien Maurice Dirac* (Cambridge: Cambridge University Press, 1987). Kragh has also written a general history of quantum theory, *Quantum Generations: A History of Physics in the Twentieth Century* (Princeton: Princeton University Press, 1999).

Chapter 25

Remarkably for a scientist, Feynman has two outstanding biographies: James Gleick, *Genius: The Life and Science of Richard Feynman* (New York: Pantheon, 1992), and Jagdish Mehra, *The Beat of a Different Drum: The Life and Science of Richard Feynman* (Oxford: Oxford University Press, 1994). Feynman's monologue books, autobiographies of a kind, are *Surely You're Joking, Mr. Feynman: Adventures of a Curious Character*, as told to Ralph Leighton, ed. Edward Hutchings (New York: Norton, 1985), and *What Do* You *Care What Other People Think? Further Adventures of a Curious Character*, as told to Ralph Leighton (New York: Norton, 1988). Feynman was one of the best science teachers of his time. Many of his lectures have been collected into books. Most remarkable are Richard Feynman, *The Character of Physical Law* (New York: Modern Library, 1994), *QED: The Strange Theory of Light and Matter* (Princeton, Princeton University Press, 1985), and, with Robert Leighton and Matthew Sands, *The Feynman Lectures on Physics* (Reading, Mass.: Addison-Wesley, 1963). The story of QED, including short biographies of Feynman, Schwinger, Tomonaga, and Dyson, is told by Silvan S. Schweber in *QED and the Men Who Made It* (Princeton: Princeton University Press, 1994). John Wheeler writes about his unconventional career, with assistance from Kenneth Ford, in *Geons, Black Holes, and Quantum Foam: A Life in Physics* (New York: Norton, 1998).

Chapter 26

The principal Gell-Mann biography is George Johnson's recent *Strange Beauty: Murray Gell-Mann and the Revolution in Twentieth-Century Physics* (New York: Knopf, 1999). Gell-Mann's book is *The Quark and the Jaguar: Adventures in the Simple and the Complex* (New York: Freeman, 1994). For an account of the people and history of modern particle physics, see Robert Crease and Charles Mann, *The Second Creation: Makers of the Revolution in Twentieth-Century Physics* (New York: Macmillan, 1986). Abraham Pais's more detailed *Inward Bound: Of Matter and Forces in the Physical World* (Oxford: Oxford University Press, 1986) is also recommended.

Chapter 27

The main Hubble biography is Gale Christianson, *Edwin Hubble: Mariner of the Nebulae* (New York: Farrar, Straus, Giroux, 1995). Helge Kragh chronicles the dispute (still in progress) between the proponents of the big-bang and steady-state cosmologies in *Cosmology and Controversy: The Historical Development of Two Theories of the Universe* (Princeton: Princeton University Press, 1996).

Chapter 28

Chandrasekhar has Kameshwar Wali's biography, *Chandra: A Biography of S. Chandrasekhar* (Chicago: University of Chicago Press, 1984), and the collection of reminiscences edited by Wali, *S. Chandrasekhar: The Man behind the Legend* (London: Imperial College Press, 1997). See Oystein Ore's *DSB* article on Chandrasekhar's boyhood role model, Srinivasa Ramanujan.

Chapter 29

There are several Hawking biographies. Try Michael White and John Gribbin, *Stephen Hawking: A Life in Science* (New York: Dutton, 1992). Hawking's famously popular book on cosmology is *A Brief History of Time: From the Big Bang to Black Holes* (New York: Bantam, 1988).

Index

*** Note—Page numbers in **bold** refer to the main entry for each subject.
*** Note—Page numbers in *italics* refer to diagrams and illustrations.